Pollution Science

Pollution Science

Editors

Ian L. Pepper
Charles P. Gerba
Mark L. Brusseau

Department of Soil, Water and Environmental Science
The University of Arizona
Tucson, Arizona

Technical Editor & Illustrator

Jeffrey W. Brendecke

Anagennesis Custom Graphic Solutions
Tucson, Arizona

Academic Press

San Diego New York Boston
London Sydney Tokyo Toronto

The chapter entitled "Surface Water Pollution" by D. J. Baumgartner is a U.S. government work in the public domain.

Photograph courtesy of Jeffrey W. Brendecke

Find Us on the Web! http://www.apnet.com

This book is printed on acid-free paper. ♾

Academic Press
A Division of Harcourt Brace & Company
525 B Street, Suite 1900, San Diego, California 92101-4495

United Kingdom Edition published by
Academic Press Limited
24-28 Oval Road, London NW1 7DX

Library of Congress Cataloging-in-Publication Data

Pollution science / edited by Ian L. Pepper, Charles P. Gerba, Mark L. Brusseau.
p. cm.
Includes index.
ISBN 0-12-550660-0 (alk. paper)
1. Pollution. I. Pepper, Ian L. II. Gerba, Charles P., date. III. Brusseau, Mark L.
TD174.P68 1996
628.5--dc20 96-12826

PRINTED IN CANADA
98 99 00 01 FR 9 8 7 6 5 4 3 2

"This work is dedicated to my father, Leonard Pepper, who struggled all of his adult life with muscular dystrophy, but was still cheerful in the face of adversity. He never gave up, no matter how daunting the problem. Len worked in a coal mine as a youth, and consequently never had a chance to be a scientist—but he gave me mine. Thanks, Len."

— *Ian L. Pepper*

"I dedicate this work to my son, Stephen, who is an endless source of joy, and to my grandfather, Louis Ketchum, whom I wish to thank for passing to me his 'love of the land.'"

— *Mark L. Brusseau*

Table of Contents

Chapter 3 Physical Processes Affecting Atmospheric Pollution

A.D. Matthias and H.L. Bohn

Chapter 4 Biotic Activity in Soil and Water

I.L. Pepper and K.L. Josephson

Chapter 5 Physical Processes Affecting Contaminant Fate and Transport in Soil and Water

P.J. Wierenga

Chapter 6 Chemical Processes Affecting Contaminant Fate and Transport in Soil and Water

M.L. Brusseau and H.L. Bohn

Chapter 7 Biological Processes Affecting Contaminant Fate and Transport

R.M. Miller

Part 2 · Monitoring and Remediation of Environmental Pollution

Chapter 8 Statistics in Pollution Science

A.W. Warrick, S.A. Musil, and J.F. Artiola

Chapter 9 Environmental Monitoring in Pollution Science

J.F. Artiola, M.L. Brusseau, A.D. Matthias, and S.A. Musil

Chapter 10 Waste Disposal

J.F. Artiola

Chapter 11 Remediation

M.L. Brusseau and R.M. Miller

Part 3 · Sources, Extent, and Characteristics of Pollution

Chapter 12 Atmospheric Pollution

A.D. Matthias

Chapter 13 Surface Water Pollution

D.J. Baumgartner

Chapter 14 Agricultural Fertilizers as a Source of Pollution

T.L. Thompson

Chapter 15 Sediments (Soil Erosion) as a Source of Pollution

D.F. Post

Chapter 16 Animal Wastes

R.J. Freitas and M.D. Burr

Chapter 17 Pesticides as a Source of Pollution

J.E. Watson

Chapter 18 Industrial Sources of Pollution

J.F. Artiola

Chapter 19 Pathogens in the Environment

C.P. Gerba

Chapter 20 Municipal Waste and Drinking Water Treatment

C.P. Gerba

Part 4 • Risk Assessment and Risk Management

Chapter 21 Principles of Toxicology

C.P. Gerba

Chapter 22 Risk Assessment

C.P. Gerba

Chapter 23 Federal Environmental Regulations and Laws in the United States

C.P. Gerba

Chapter 24 Pollution in the Twenty-First Century

I.L. Pepper, C.P. Gerba, and M.L. Brusseau

Preface

To The Student

Pollution is an inevitable product of all industrial economies—it is, in effect, one of the prices we pay for our way of life. As such, it is something that you and your elected decision-makers will have to manage now and in the future. The concepts and guidelines in this book will provide you with a scientific rationale for decisions that directly affect the environment around you. Such decisions may be as simple as the option to recycle the plastic and glass components of your trash, or could involve the complex issues that are debated in Congress.

As far as possible, we have avoided the intrusion of any political ideology that would confuse the scientific assessment of environmental quality. We have adhered to scientific principles, models, and observations. In the text, we have attempted to challenge you with problems and questions relevant to each topic. Real-life situations are introduced by including case studies in several different areas of pollution science. The text does not attempt to define answers or approaches to specific pollution problems. Rather we have laid out a frame work of principles that can be used to understand any pollution problem. We also include sections on environmental monitoring, sampling design, and the use of appropriate statistics.

Notice that this preface does not describe the different sections of the book. If you want to know what is in the book—read it! With any luck you will view this text as a valuable resource long after the course is over. We wish you well and hope that some of you will pursue a career aimed at ensuring environmental health for all of us.

To the Instructor

Environmental issues are attracting ever-increasing attention at local, national, and international levels. Part of the dilemma facing society and government alike is choosing strategies that maintain a safe environment without excessive regulation and cost. The realization that there are a limited number of individuals with sufficient background to make such decisions has resulted in the development of new "Environmental Science" majors at many universities and colleges. These programs are, for the most part, rigorous science-based curriculum that require a broad background in earth science, chemistry, physics, biology, and mathematics.

When we initiated our Environmental Science major at the University of Arizona in the fall of 1993, we realized that a capstone course on pollution that would integrate knowledge from several scientific disciplines was needed. We also felt that a textbook with this scientific base would be critical to the success of the course. However, it was quickly discovered that no such text existed that covered the appropriate material in sufficient breadth and depth. The prospect of teaching this class without a text galvanized us into action, and the concept of creating a multi-authored textbook emerged, with prospective authors being chosen from our own Department of Soil, Water and Environmental Science. We feel that contributions from several different scientific disciplines is essential for tracing critical aspects related to the fate and mitigation of pollutants across traditional disciplinary boundaries. As such, this textbook is a pioneering effort

This textbook has been designed to be understood by a wide range of students with diverse backgrounds. Conceptually, it is accessible to almost any student with a good general science background. In addition, however, there is sufficient rigor for the text to be a relevant reference book for advanced students. The text focuses on pollution of the atmosphere, of surface water, and of soil and groundwater. The chapters have been written by several contributors, many of whom are nationally-known, with each author writing on his or her own specific area of expertise. Normally, multi-authored texts are cumbersome due to inadequate consistency and style. However, drawing all contributors from one institution facilitated the effort to coordinate, cross-reference, and integrate different parts of the text. While the level of science presented is challenging, we have attempted to make this book stimulating to students. Numerous color illustrations and photographs help to clarify the key concepts of the text.

Therefore, we hope that this text will inspire and encourage you to teach the wide range of subjects that make up the discipline of Pollution Science.

Ian Pepper
Chuck Gerba
Mark Brusseau

January 1996

The Editors

All three editors are professors in the Department of Soil, Water and Environmental Science at the University of Arizona.

Ian L. Pepper Ph.D., The Ohio State University, 1975. Currently Professor of Environmental Microbiology. Dr. Pepper's diverse research interests are reflected in the fact that he is Fellow of the Soil Science Society of America, The American Academy of Microbiology, and the American Society of Agronomy. He is a pioneer in the use of molecular analyses for the detection of microbial diversity and activity in environmental samples. He has also been heavily involved in the evaluation of the benefits and hazards of waste utilization.

"I was born in Tonypandy, Wales, where the world consisted of coal mines, cricket, soccer, and tea. Now I live in the Sonoran Desert, a far cry from those green valleys of my youth. Neither ecosystem has been immune to the ravages of pollution including the effects of mining and the influence of locally high population densities. We must however, meet the formidable challenges and problems of pollution through a combination of science and common sense."

Charles P. Gerba Ph.D., University of Miami, 1973. Currently Professor of Microbiology. Dr. Gerba is a Fellow of the American Academy of Microbiology and a Member of the EPA Science Advisory Board. As such he has been influencing national research areas of study within the environmental science arena. He has an international reputation for his methodologies for pathogen detection in water and food, pathogen occurrence in households, and risk assessment.

"My interest in microbiology was sparked by Paul DeKruf's inspiring tales of the scientific achievements of early microbiologists in the book *The Microbe Hunters* and my mother's error in giving me a microscope for Christmas instead of the chemistry set I wanted. In my first summer job out of college I was introduced to environmental microbiology by studying sewage disposal. Later I examined the fate of viruses in sewage discharged into the ocean. These beginnings led me to an exciting and adventurous career in environmental microbiology where every day brings a new problem to be addressed."

Mark L. Brusseau Ph.D., University of Florida, 1989. Currently Associate Professor of Subsurface Hydrology and Environmental Chemistry. Dr. Brusseau's research is focused on developing a fundamental understanding of the factors and processes influencing the transport and fate of chemicals in the subsurface. He is especially well known for integrating physical, chemical, and microbiological processes into subsurface hydrology using both experimental and model-based approaches. He is also involved in the development and evaluation of subsurface remediation systems. Dr. Brusseau has received the National Academy of Sciences Young Investigator Award and the U.S. Department of Energy Distinguished Young Faculty Award.

"I have been interested in earth science since high school. However, it took many years and a tortuous path before I finally discovered the "name" for my interests—Environmental or Pollution Science. I hope this text will help others to discover their interest in this field."

Editor Photographs: R.F. Walker.

Technical Editor

Jeffrey W. Brendecke Technical Editor and Scientific Graphic Artist. President of Anagennesis Custom Graphic Solutions. M.S., The University of Arizona, 1992. Mr. Brendecke incorporates his computer graphics expertise with an extensive background in chemistry, biology, agriculture, and environmental science to design and illustrate conventional and multimedia materials for education and commerce. In addition to this text, some of his other published works are found in the journal *Microcontamination*, *The Guidebook: Nitrogen Best Management Practices for Concentrated Animal Feeding Operations in Arizona*, and *Environmental Microbiology: A Laboratory Manual*, which he also co-authored.

Technical Editor Photograph: R.F. Walker.

Acknowledgments

The editors and authors extend their sincere thanks to many individuals for their assistance with this textbook. We thank Emily Thompson of E.T. Editorial Services, Ft. Worth, Texas, for her invaluable copyediting. Don Armstrong, Carlos Enriquez, Dave Garret, Eileen Maura Jutras, Kelly Reynolds, and Ken Weber all helped find information, at times on an emergency basis. We gratefully acknowledge the tireless word-processing skills of Elenor Loya, who took the various drafts and diligently formatted them. Peter Gerba, Deborah Newby, Barbara Pepper, Timberley Roane, and James Tanguay all assisted in the proofing of the text, and Kelly Tate and Beth Petruccio helped in the development of the subject index. We appreciate the constructive suggestions on the fine tuning of the design and layout supplied by Jean Brendecke, Linda Shapiro, Emily Thompson, and James Tanguay. Mridula Gupta, M.T. Macari, Sheri A. Musil, Joe Vinson, and Y. Zhang assisted with the initial drafts of several of the illustrations. Olivia Montaño and Jean Brendecke helped prepare the digital information for pre-press.

Contributing Authors

Janick F. Artiola Associate Research Scientist
Ph.D., The University of Arizona, 1980

Don J. Baumgartner Professor
Ph.D., Oregon State University, 1967

Hinrich L. Bohn Professor
Ph.D., Cornell University, 1963

Mark D. Burr Research Specialist
Ph.D., The University of Arizona, 1996

Robert J. Freitas Assistant in Extension
M.S., The University of Arizona, 1991

Karen L. Josephson Senior Research Specialist
B.S., The University of Arizona, 1980

Allan D. Matthias Associate Professor
Ph.D., Iowa State University, 1979

Raina M. Miller Associate Professor
Ph.D., Rutgers University, 1988

Sheri A. Musil Senior Research Specialist
B.S., The University of Arizona, 1979

Donald F. Post Professor
Ph.D., Purdue University, 1967

Thomas L. Thompson Assistant Professor
Ph.D., Iowa State University, 1991

Arthur W. Warrick Professor
Ph.D., Iowa State University, 1967

Jack E. Watson Associate Extension Specialist
Ph.D., The University of Arizona, 1982

Peter J. Wierenga Professor and Department Head
Ph.D., University of California at Davis, 1968

Author Photographs: R.F. Walker.

The Department of Soil, Water and Environmental Science

The roots of our department were in Soil Science, but over the past decade we have developed a broader focus on Environmental Science. In addition to the traditional undergraduate degree in Soil and Water Science the University of Arizona now offers a Bachelor of Science degree in Environmental Science.

Two years after the inception of this new degree program, we have some 150 undergraduates and 100 graduate students, the majority of whom are involved in Environmental Science research projects. Examples of pollution-related research projects in our department include:

- development of innovative methods for cleanup of waste sites contaminated by toxic organic compounds,
- molecular-based detection of bacterial and viral pathogens in Mamala Bay, Hawaii,
- development of technologies to clean up heavy-metal contaminated soils and groundwater,
- basic research to understand the transport and fate of contaminants in the subsurface,
- development of strategies for the safe storage of radionuclides, and
- development of strategies to minimize nitrate concentrations in groundwater.

The twenty-seven faculty members in the department form an interdisciplinary team of microbiologists, chemists, physicists, hydrologists, and earth scientists. It is this integrated approach, together with the dedication of all of our faculty members, that led to this textbook. We hope this book will allow a variety of scientists to teach Pollution Science in all of its breadth and complexity.

Part 1

Processes Affecting Fate and Transport of Contaminants

On Reverse *Situated in an area characterized by an intense mixture of industry, residences, and agriculture, the Saar River in Germany can continue to be used both as a safe drinking water source and for such recreational activities as boating and fishing through proper management possible only through an understanding of the fate and transport of pollutants.*

Pollution science gives us the knowledge base necessary to manage pollution despite ever increasing population pressure. Photograph: J.W. Brendecke.

1.1 Underlying Causes

To understand pollution we must first define what pollution is and what causes it. In general, pollution is the accumulation and adverse interaction of contaminants with the environment. These contaminating substances normally result from two sources: animate activities and inanimate processes. The first source is the buildup of waste materials that result from the activity of living organisms. The living organisms most commonly associated with pollution are humans, who can cause pollution either directly or indirectly through their activities with other organisms. For example, we can pollute directly by applying excess fertilizers or pesticides, or we can pollute indirectly by mismanaging cattle feedlots. The second major source of pollutant materials is the result of natural processes such as metal accumulation from rock dissolution or the intrusion of seawater into fresh water owing to drought conditions. The most extreme examples of pollution, however, are usually associated with or caused by human activities. Although pollution has been present throughout the history of humanity, in the last half of the twentieth century, human population increased in magnitude so that its population now poses a severe threat to the world environment.

To understand the relationship between population and pollution, let us examine a typical curve for the growth of a pure culture of bacteria in a liquid medium (Figure 1-1 on page 4). Early on, the bacteria growing in the medium do not increase significantly in number, due to low population densities, which results in organisms operating as separate entities. This initial low-growth phase is known as the **lag period**. Next, the number of organisms increases exponentially for a finite period of time. This phase of growth is known as the **exponential phase** or **log phase**. After this exponential phase of growth, a **stationary phase** occurs, during which the total number of organisms remains constant as new organisms are constantly being produced while other organisms are dying. Finally, we observe the **death phase**, in which the total number of organisms decreases. We know that bacteria reproduce by binary fission, so it is easy to see how a doubling of bacteria occurs

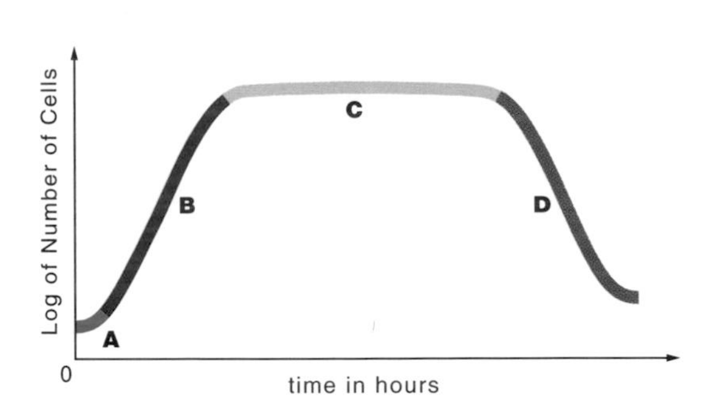

Figure 1-1 Typical growth curve for a pure culture of bacteria. A = lag period, B = exponential phase, C = stationary phase, D = death phase.

during exponential growth. But what causes the stationary and death phases of growth?

Two mechanisms prevent the number of organisms from increasing *ad infinitum*: first, the organisms begin to run out of nutrients; second, waste products build up within the growth medium and become toxic to the organisms. An analogous situation exists for humans. Initially, in prehistoric times, population densities were low and population numbers did not increase significantly or rapidly (see Figure 1-2). During this time resources were plentiful; thus the environment could easily accommodate the amount of wastes produced. Later, populations began to increase very rapidly. Although not exponential, this phase of growth was comparable to the microbial log phase of growth. During this period then, large amounts of resources were utilized, and wastes were produced in ever greater quantities. This period of growth is still under way. However, we seem to be approaching a period in which lack of resources or buildup of wastes (*i.e.*, pollution) will limit continued growth—hence the renewed interest in recycling materials as well as in controlling, managing, and cleaning up waste materials. To do this, we must arrive at an understanding of the predominant biotic and abiotic characteristics of the environment.

1.2 The Environment as a Continuum

The environment plays a key role in the ultimate fate of pollutants. The environment consists of soil, surface water, and the atmosphere; all sources of pollution are initially released or dumped into one of these phases of the ecosystem. As pollutants interact with the environment, they undergo physical and chemical changes and are ultimately incorporated into the environment. The environment thus acts as a continuum into which all waste materials are placed. The pollutants, in turn, obey the second law of thermodynamics: matter cannot be destroyed; it is merely converted from one form to another. Thus, taken together, the way in which substances are added to the environment, the rate at which these wastes are added, and the subsequent changes that occur determine the impact of the waste on the environment. It is important to recognize the concept of the environment as a continuum because many physical and chemical processes occur not within one of these phases, such as the air alone, but rather at the interface between two phases such as the soil/water interface.

The concept of the continuum relies on the premise that resources are utilized at a rate at which they can be replaced or renewed, and that wastes are added to the environment at a rate at which they can be incorporated without disturbing the environment. Historically, natural wastes were generated that could easily be broken down or transformed into beneficial, or at least benign, compounds. However, post-industrial contamination has resulted in the formation of **xenobiotic waste**—compounds that are foreign to natural ecosystems—which are less subject to degradation. In some cases, natural processes can actually enhance the toxicity of the pollutants. For example, organic compounds that are not themselves carcinogenic can be microbially converted into carcinogenic substances. Other compounds, even those not normally considered pollutants, can cause pollution if they are added to the environment at rates that result in high concentrations of these substances. An excellent example here is nitrate fertilizer, often added to soil at a high rate. Such nitrates can end up in drinking water supplies and cause methemoglobinemia (blue baby disease) in newborn infants.

Some pollutants, such as microbial pathogens are entirely natural and may be present in the environment at very low concentrations. Even so, they are still capable of causing pathogenic diseases in humans or animals. Such natural microorganisms are also classified as pollutants, and their occurrence within the environment needs to be carefully controlled.

1.3 Pollution and Population Pressures

If the world's population were limited to a few million individuals, pollution would not be nearly the problem that it is, even with today's variety of human activities. But, given our current population (in excess of five billion and increasing rapidly), pollution is a problem of increasing proportion. Not only are we producing substances that are hazardous in their own right through our industrial and agricultural activities, but we are also producing pollutants that have become a hazard because of the sheer quantities involved, as is the case with municipal wastes.

Two trends associated with population are currently leading to increased levels of pollution. First, the overall world population has increased dramatically (Figure 1-2). Between 1930 to 1995 alone, the world population increased from 2 to 5.5 billion. At the same time, population concentrations within different sections of the world have also changed. Developing countries—those characterized as having a rural population and in which the majority of the world's inhabitants live—have increased in overall population compared to industrialized or urban (developed) countries (Figure 1-3a on page 7). Now,

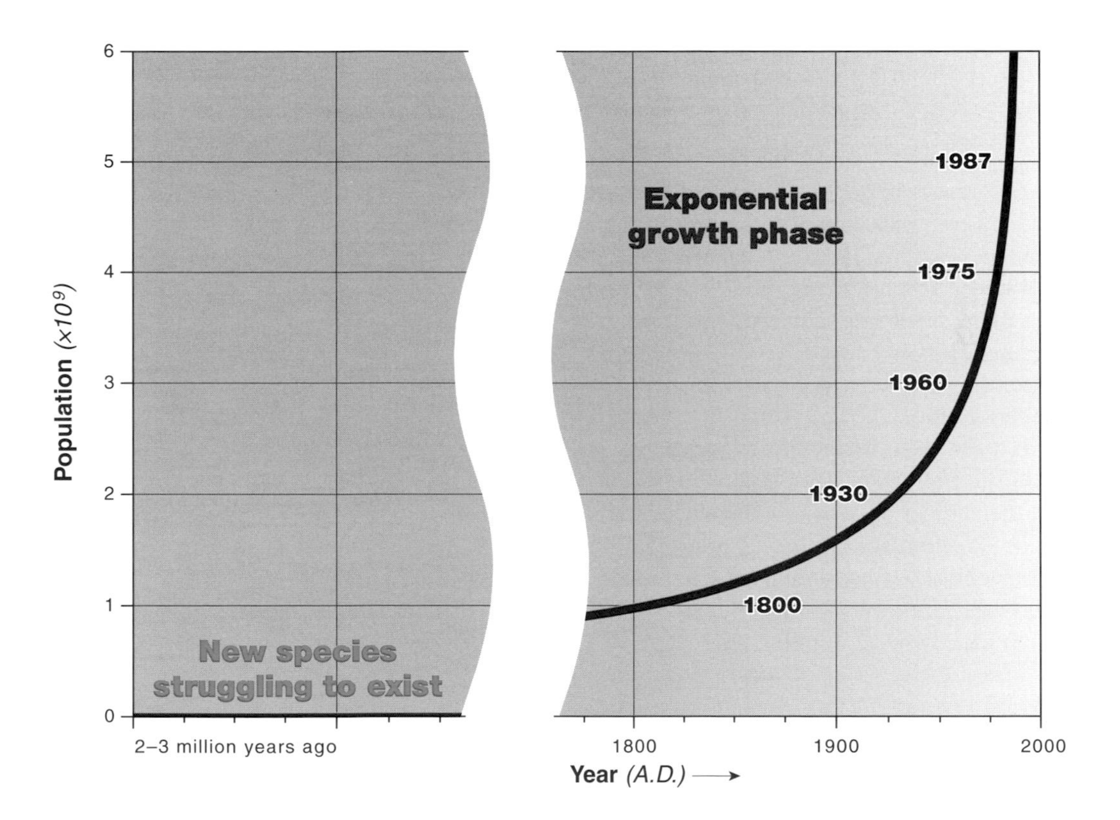

Figure 1-2 World population increases from the inception of the human species to the 1993. Source: Population Reference Bureau, Inc.,1990.

however, a new trend is emerging. As the world population increases, and the lures of material well-being become increasingly hard to resist, people are relocating from sparsely populated rural areas to more congested urban centers (Figure 1-3b). Typically, urbanized areas consume more natural resources and produce more waste products per capita than rural areas. In an urban setting, where production of food is not the primary function, people produce more waste than in rural areas, and this waste is also more hazardous. Increased urbanization also puts pressure on the agrarian sector to produce more food on less land, leading to increased agricultural pollution. In developed countries, where the level of affluence has been relatively high, the level of urbanization increased by only 7% from 1970 to 1990 (Figure 1-3b), while the rate of population increase declined from 1.5% per year in 1970 to 0.8% per year in 1990 (Figure 1-3c). In developing countries, however, the level of urbanization increased by 27% from 1970 to 1990 (Figure 1-3b), with the rate of population increase staying roughly the same at about 3.7% annually (Figure 1-3c).

The main differences between pollution in industrialized and rural countries, however, lie not only in the sheer numbers of people involved, but also in the population's stance toward maintaining a healthy environment. Some developed countries, such as the United States and parts of Europe, which have been industrialized for a century or more, felt the adverse effects of pollution early on. Such countries have begun to modify their polluting activities and also to clean up existing pollution. This change has occurred while the number of people in the developed countries has been relatively low. Developing countries, on the other hand, are escalating their polluting activities now—at a time when their numbers are relatively high and their economic resources low. They can ill afford the costs of pollution reduction and cleanup. It is important to note that pollution problems are generally exacerbated by economic problems or the lack of financial resources.

1.4 Science and Pollution

Given the rate of communication in the present age, it is all too easy to acquire a fashionable sense of impending doom with respect to pollution. The news media are very adept at transforming any environmental problem into one more step toward an uninhabitable planet. Crisis reporting tends to polarize the population, galvanizing it into creating cycles of action and reaction that more often hinder than help in finding a rational resolution of the current pollution problem. Science, however, offers us a useful tool for understanding the world about us, as well as a vehicle for reasoned discourse about pollution.

- **Pollution science** is the study of the physicochemical and biological processes fundamental to the fate, transport, and mitigation of contaminants that arise from human activities as well as natural processes.

By studying pollution through science, and balancing the answers it gives us with prudent economic and legislative action founded on scientifically sound conclusions, we can come closer to maintaining an environment that is healthful not only for us, but for all organisms on earth.

To decide when and how to act in any given pollution situation, we need to pose some well guided questions about the pollution.

What is the pollutant? What is the pollutant's chemical identity? What are its physical and chemical properties? How is the pollutant entering the environment, and what happens to it when it is in the environment? Is the pollutant being emitted at a point source or in low levels over a large area (nonpoint source)? By answering these questions, we can formulate procedures either to stop production of the pollutant or to reduce its concentrations in the environment by finding other methods of manufacturing or waste handling. We can also use this information to select and develop cleanup strategies if necessary: is a chemical approach, physical treatment, or biological transformation feasible in a given situation?

Is the pollutant harmful? Is the substance itself harmful, or is it transformed into something harmful either in the environment or in living organisms? How harmful is it? The answers to such questions are invaluable for setting guidelines for allowable levels of the pollutant in the environment and regulating the conditions under which the pollutant is allowed to reach those levels. For example, certain

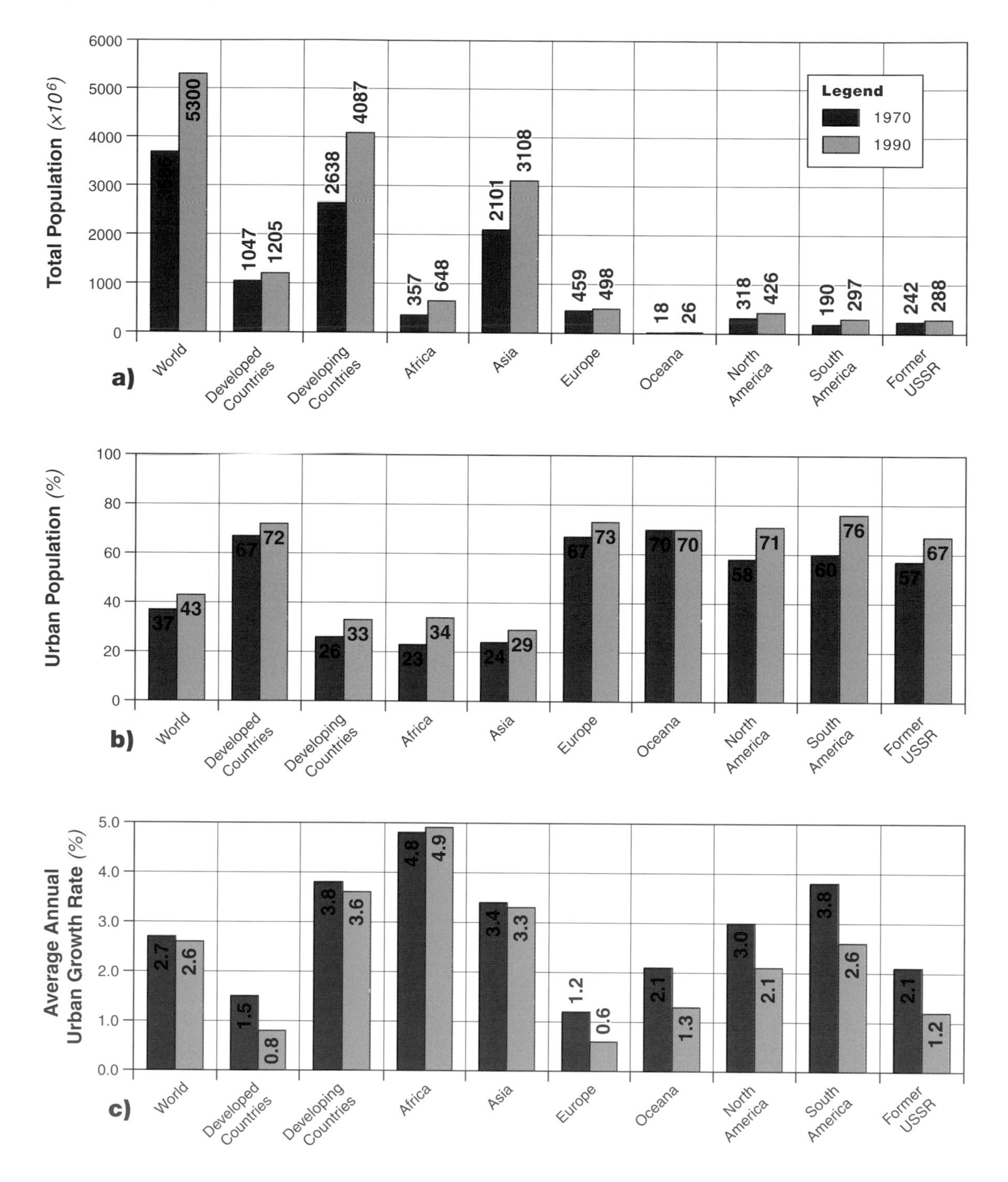

Figure 1-3 World population statistics by region showing (a) total population, (b) percent urbanization, and (c) average annual urban growth rate for 1970 and 1990. Source: UNEP, 1991.

types of metals are toxic under some conditions but not under others. Moreover, some adverse effects of pollutants, such as cancer induction, ozone layer deterioration, or global warming, may not be definitively ascertained until after many years of research and study; thus decisions based on preliminary data may have to be modified.

What are the costs associated with the pollutant? Both social and monetary costs apply here. Is the pollutant adversely affecting human or environmental health? What would be the costs of switching to alternative production or waste-handling methods? Are there alternative, cost-effective technologies to reduce concentrations of the pollutant? Could legislation be enacted or international agreements reached to encourage or compel compliance with regulatory guidelines without economically devastating the segment of the population involved? For example, with the advent of low-cost unleaded gasoline and the catalytic converter in the early 1970s, legislation that enforced a clean air standard became economically feasible. Since then, concentrations of lead in air have dropped by as much as 89% at certain sites, as measured between 1982 and 1991. Use of low-sulfur alternative fuels, such as natural gas in place of coal in energy production, has led to significant decreases in sulfur dioxide emissions, a pollutant that can contribute to respiratory ailments, acid rain, and the erosion of buildings and monuments. But, what were the costs to the coal producers? To the mine workers? Could the polluted area be cleaned up? What would be the most cost-effective method to remediate the site? Can polluted sites, such as designated Superfund sites in the United States, be cleaned up to a safe level at a reasonable cost, or would containment of the pollutant be better?

1.5 Pollution Science and the Global Future

All of the above questions can be answered only with the help of a solid foundation in the science of pollution. Therefore, it should not be surprising that contributions to the knowledge base that pollution science provides are essential if we are to keep pollution levels down to safe, manageable levels. Governing bodies and advisory councils of cities, states, countries, continents, and the world need help in managing pollution. Thus, pollution science can help an informed public make educated and environmentally sound decisions. It is these decisions that can add up to a global pollution problem, or to a clean and healthy environment.

References and Recommended Reading

Population Reference Bureau, Inc. (1993) *World Population Data Sheet of the Population Reference Bureau, Inc.* Population Reference Bureau, Inc., Washington, D.C.

UNEP (1991) *United Nations Environmental Programme Environmental Data Report*, 3rd Edition. Basil Blackwell, Oxford.

Problems and Questions

1. What are two major factors that ultimately limit population growth?
2. Describe the major phases associated with a typical growth curve for a pure culture of bacteria.
3. Where are the largest populations in the world in terms of developed, undeveloped, urban, and rural areas?

Soils are diverse in nature and exist even in extreme environments. This soil developed within the ecosystem of the Sonoran Desert in the southwestern USA. Photograph: I.L. Pepper.

2.1 The Soil

The human environment is located at the earth's surface and is heavily dependent on the soil/water/atmosphere continuum. Conceptually, this continuum controls all of our activities, and the physical, chemical, and biological properties of each phase are interactive. Although it may be intuitively obvious that soil and water processes are intimately linked, it is not so obvious that these, in turn, are linked to the atmosphere. Soil itself is composed of solids, liquids, and gases, each of which is in equilibrium with the atmosphere, as well as with rivers, lakes, and the oceans. Therefore, the fate and transport of pollutants are influenced by each of these components.

Soil is an intricate, yet durable entity that directly and indirectly influences our quality of life. Colloquially known as dirt, soil is taken for granted by most people, and yet it is essential to our daily existence. It is responsible for plant growth, for the cycling of all nutrients through microbial transformations, and for maintaining the oxygen/carbon dioxide balance of the atmosphere. It is also the ultimate site of disposal for most waste products. Soil is a complex mixture of weathered rock particles, organic residues, water, and billions of living organisms. It can be as thin as 15 centimeters, or it may be a hundred or more meters thick. Because soils are derived from unique sources of parent material under specific environmental conditions, no two soils are exactly alike. Hence there are literally thousands of different kinds of soils just within the United States. These soils have different properties that influence the way soils are utilized optimally.

Because many different types of pollutants are added to soil, and soil is in direct contact with water and the atmosphere, the ultimate fate of a pollutant and its impact on the environment are directly dependent on the particular soil type. Generally, we refer to the uppermost portion of land as soil, which is the medium for plant growth. The region beneath this is known as the vadose zone.

2.2 Solid Phase

2.2.1 Definition of Soil and the Vadose Zone

Soil is the weathered end product of the action of climate and living organisms on soil parent material with a particular topography, over time. We refer to the above listed factors as the five soil-forming factors. The biotic component consists of both microorganisms and plants. The **vadose zone** is the water-unsaturated and generally unweathered material between groundwater and the land surface.

There are several parameters of soil that vitally affect the fate and transport of environmental pollutants. We will now discuss these parameters while providing an overview of soil as a natural body as it affects pollution. Many texts are available for those who wish more detailed information (see *References and Recommended Reading* on page 18).

2.2.2 Soil Profiles

The process of soil formation generates different horizontal layers, or **soil horizons**, that are characteristic of that particular soil. It is the number, nature, and extent of these horizons that give a particular soil its unique character. A typical soil profile is illustrated in Figure 2-1. Generally, soils contain a dark, organic-rich layer, designated as the O horizon, then a lighter colored layer, designated as the A horizon, where some humified organic matter accumulates. The layer that underlies the A horizon is called the E horizon because it is characterized by eluviation, which is the process of removal or transport of nutrients and inorganics out of the A horizon. Beneath the E horizon is the B horizon, which is characterized by illuviation. Illuviation is the deposition of the substances from the E horizon into the B horizon. Beneath the B horizon is the C horizon, which contains the generally unweathered parent material from which the soil was derived. Below the C horizon is the R horizon, a designation applied to bedrock. Although certain horizons are common to most soils, not all soils contain each of these horizons.

2.2.3 Primary Particles and Soil Texture

Soil normally consists of about 95% inorganic and 1–5% organic material by weight. The primary inorganic material is in turn composed of three primary particles—**sand**, **silt**, and **clay**, which are delineated on a size basis (Table 2-1).

Table 2-1 Size classifications of the three primary particles of soil (USDA definition).

Primary Particle	Size Range (diameter)
Sand	2 mm–0.05 mm
Silt	0.05 mm–0.002 mm
Clay	<0.002 mm (2 μm)

The percentage of sand, silt, and clay in a particular soil determines its **soil texture**, which affects many of the physical and chemical properties of the soil. Various mixtures of the three primary components result in different textural classes (Figure 2-2 on page 12). Of the three primary particles, clay is by far the dominant factor in determining a soil's properties. This is because there are more particles of clay per unit weight than of sand or silt, due to the smaller size of the clay particles. In addition, the clay particles are the only soil particles that have an associated electrical charge (see Section 6.3.1, beginning on page 67). The predominance of clay particles explains why any soil with greater than 35% clay by weight has the term *clay* in its textural class.

2.2.4 Soil Structure

The three primary particles do not normally remain as individual entities. Rather, they aggregate to form secondary structures, which occur because microbial gums, polysaccharides, and other microbial metabolites bind the primary particles together. In addition, particles can be held together physically by fungal hyphae and plant roots. These secondary aggregates, which are known as **peds**, can be of different sizes and shapes, depending on the particular soil. Soils with even modest amounts of clay usually have well defined peds, and hence a well defined soil structure. These aggregates of primary particles usually remain intact as long as the soil is not disturbed, for example, by plowing. In contrast, sandy soils with low amounts of clay generally have less well defined soil structure.

The phenomenon of soil structure has a profound influence on the physical properties of the soil. In Figure 2-3 on page 13, we see an illustration of the

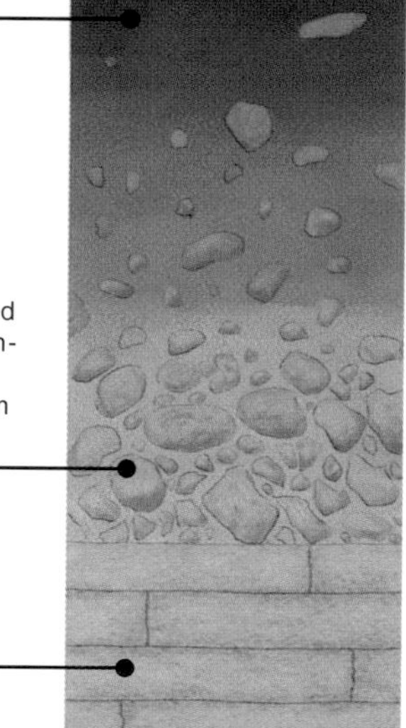

Location: High-altitude plateau in Arizona.

Vegetation: Pine forest.

Uses: Timber.

Horizon Notes

O Pine needles in various stages of decomposition.

A Shallow horizon enriched with humic materials.

E Leached horizon with less organic matter and clay than the horizons above and below it.

B Horizon marked by accumulated clays: some limestone parent material present in the lower part.

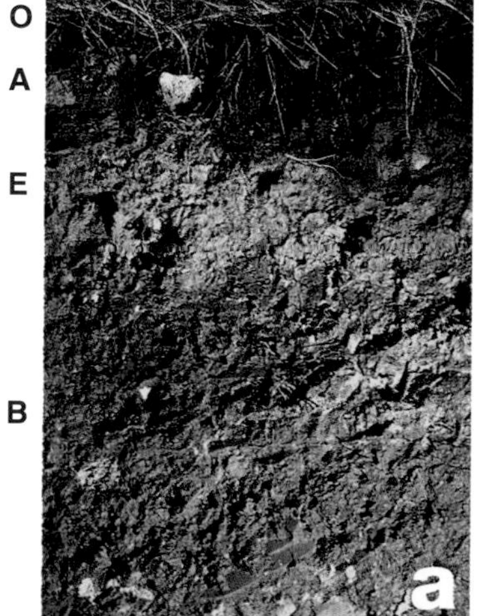

Location: Montana.

Vegetation: Grassland.

Uses: Wheat farming.

Horizon Notes

O Native grass residues.

A Moderately deep zone of built-up humic materials.

B Horizon of heavy clay accumulation.

C Calcareous glacial till parent material.

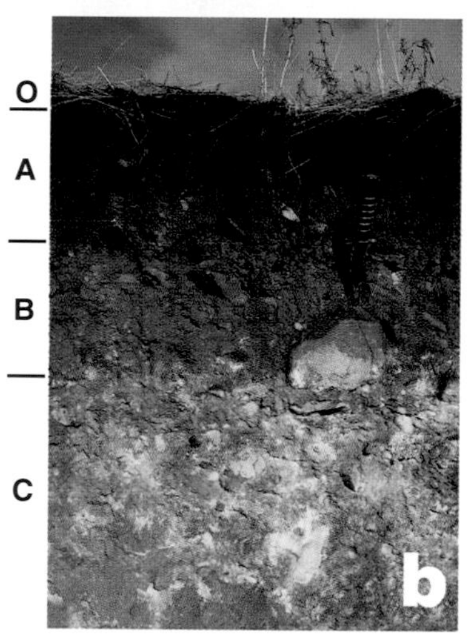

Location: South-eastern desert of Arizona.

Vegetation: Creosote.

Uses: Limited grazing.

Horizon Notes

A Shallow A horizon with a small amount of organic material.

C Alluvial deposits. The numbered horizons, C1–C5, here denote successive deposition events that vary significantly in mineral composition and texture.

Figure 2-1 Typical soil profiles illustrating different soil horizons. These horizons develop under the influence of the five soil-forming factors and result in unique soils.

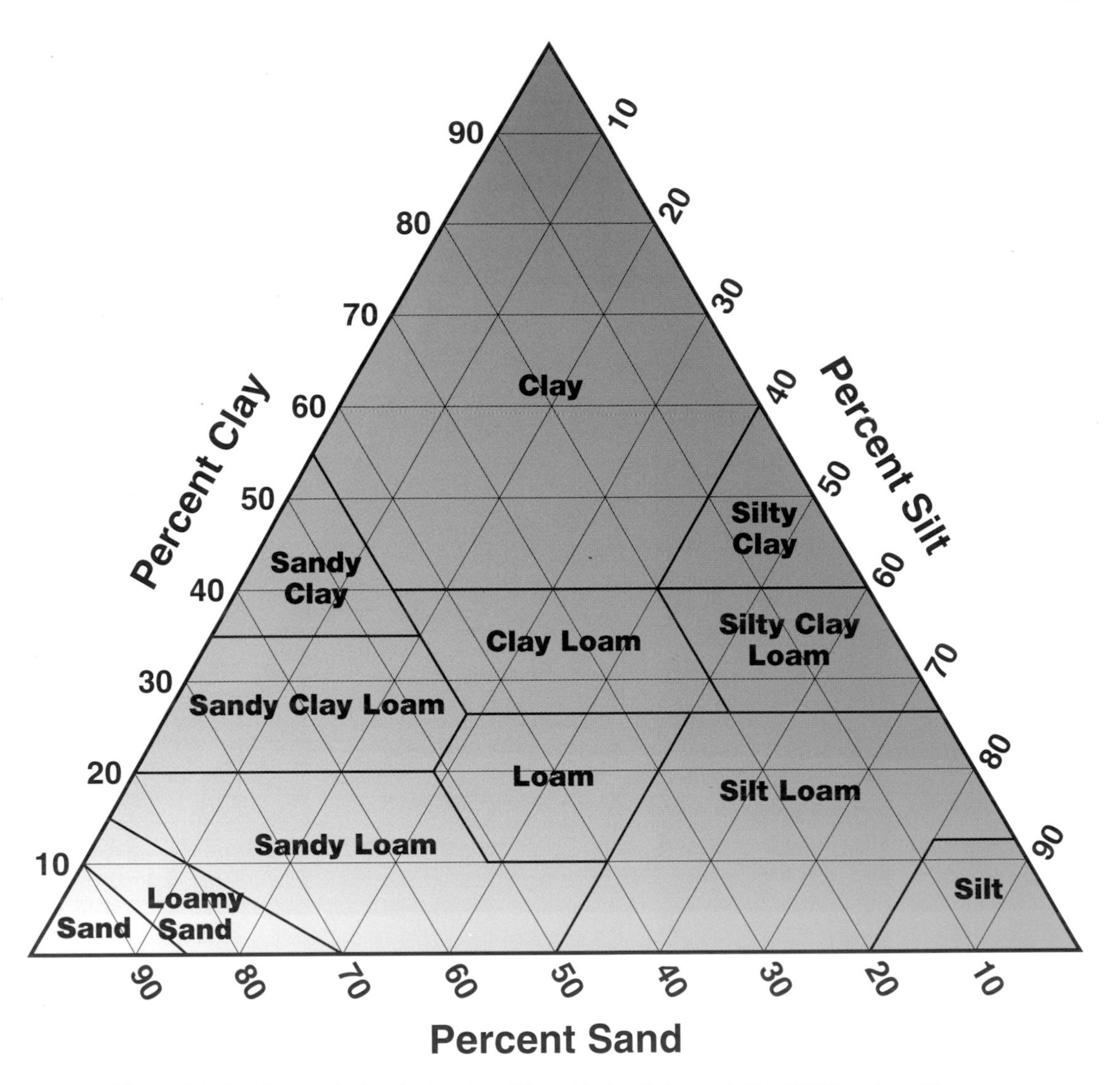

Figure 2-2 A soil textural triangle showing different textural classes in the USDA system. These textural classes characterize soils with respect to many of their physical properties. For a description of the size classes of the particles, see Table 2-1 on page 10.

concept of soil as a discontinuous environment, consisting of a matrix of organics and inorganics of different sizes, bound together and populated by soil organisms. Because its particles are arranged in secondary aggregates, a certain volume of the soil includes voids that are filled with either soil air or soil water. Soils in which the structure has many voids within and between the peds offer favorable environments for soil organisms and plant roots, both of which require oxygen and water. Soils with no structure, that is, those consisting of individual primary particles, are characterized as **massive**. Massive soils have very few (and very small) void spaces and therefore little room for air or water.

We call the void the **pore space**. Individual pores may be open or closed to the exterior of the ped. Closed pores that contain oxygen-using microorganisms generally have lower oxygen concentrations than open pores since atmospheric air can not freely

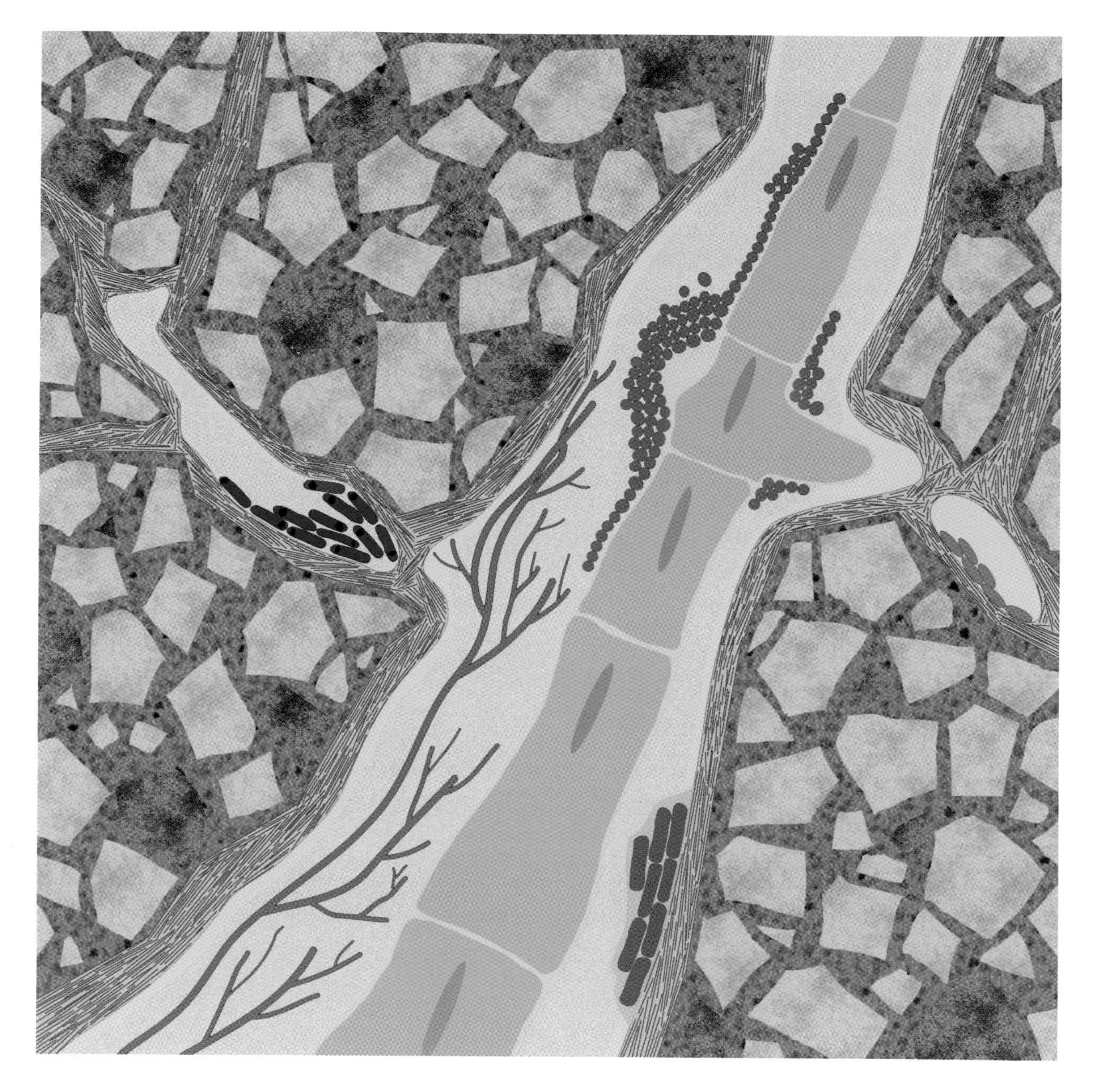

Figure 2-3 Individual primary particles bind together to form secondary aggregates with resultant soil pores. Here, the pores, lined with a thin layer of clay (clay skin, shown in gray), are occupied by fungi, actinomycetes, and bacteria. The main pore (*center*) is open to the atmosphere and contains a thin water film. Within the pore are numerous organisms including a large fungal filament (orange) and a branched actinomycete (purple). One colony of bacteria adheres to the wall of the pore with its own secretion (*lower right*). Other bacteria (spherical in shape) adhere directly to the fungus. Two closed pores are visible as well. One closed pore (*left*) contains some air, is lined with a film of water, and is inhabited by bacteria. A second closed pore (*right*) is likewise inhabited by bacteria but is completely filled with water. For further information on soil microorganisms, see Chapter 4, beginning on page 31.

diffuse into closed pores. Pore space may be increased by plant roots, worms, insects, and small mammals, whose root channels, holes, and burrows create macro openings. These larger openings can result in significant aeration of surface and subsurface soils and sediments, as well as preferential flow of water through the soil.

2.2.5 Cation Exchange Capacity

The parameter known as **cation exchange capacity (CEC)** arises because of the charge associated with clay particles. Normally, this is a negative charge that occurs for one of two reasons:

1. **Isomorphic substitution:** Clay particles exist as inorganic lattices composed of silicon and aluminum oxides. Substitution of a divalent magnesium cation (Mg^{2+}) for a trivalent aluminum cation (Al^{3+}) can result in the loss of one positive charge, which is equivalent to a gain of one negative charge. Other substitutions can also lead to increases in negative charge.

2. **Ionization:** Hydroxyl groups ($-OH$) at the edge of the lattice can ionize, resulting in the formation of negative charge:

$$\text{Al}-\text{OH} \rightleftharpoons \text{Al}-\text{O}^- + \text{H}^+ \quad (2\text{-}1)$$

These are also known as **broken-edge bonds**. Ionizations such as these usually increase as the pH increases, and are therefore known as **pH-dependent charge**. The functional groups of organic matter, such as carboxyl groups, are also subject to ionization and can contribute to the total pH-dependent charge. The total amount of negative charge is usually measured in terms of millimoles of positive charge per kilogram of soil (mmol (+) kg^{-1}) and is a measure of the potential CEC of the soil.[1] A CEC of 150–200 mmol (+) kg^{-1} of soil is considered to be average, whereas a CEC > 300 is considered high.

[1] Formerly, CEC was measured in milliequivalents per 100 g of soil. A milliequivalent (meq) is one-thousandth of an equivalent weight. Equivalents of chemicals are related to hydrogen, which has a defined equivalent weight of 1. Here, the equivalent weight of a chemical is the atomic weight divided by its valence. For example, the equivalent weight of the calcium ion is 40/2 = 20 g. 1 meq 100 g^{-1} = 10 mmol (+) kg^{-1}.

Table 2-2 Soil pH regimes.

Soil	pH Regime
Acidic	<5.5
Neutral	6–8
Alkaline	>8.5

Note that it is the clays and organic particles that are negatively charged. Due to their small particle size, they are collectively called **soil colloids**. The existence of CEC allows the phenomenon of cation exchange to occur (see Section 6.3.1, beginning on page 67).

2.2.6 Soil pH

We define pH as the negative logarithm of the hydrogen ion concentration:

$$\text{pH} = -\log[\text{H}^+] \quad (2\text{-}2)$$

Usually, water ionizes to H^+ and OH^-:

$$\text{HOH} \rightleftharpoons \text{H}^+ + \text{OH}^- \quad (2\text{-}3)$$

The **dissociation constant** (K_{eq}) is defined as

$$K_{eq} = \frac{[\text{H}^+][\text{OH}^-]}{[\text{HOH}]} \quad (2\text{-}4)$$

Since the concentration of HOH is large relative to that of H^+ or OH^-, it is normally given the value of 1; therefore

$$[\text{H}^+][\text{OH}^-] = 10^{-14}\,M \quad (2\text{-}5)$$

For a neutral solution

$$[\text{H}^+] = [\text{OH}^-] = 1 \times 10^{-7}\,M \quad (2\text{-}6)$$

and

$$\text{pH} = -\log[\text{H}^+] = -(-7) = 7 \quad (2\text{-}7)$$

A pH value of less than 7 indicates acidity whereas a pH value greater than 7 indicates alkalinity (or basicity). Soils are normally characterized by pH as described in Table 2-2.

Table 2-3 Major constituents of plant residues.

Constituent	% Dry Weight
Cellulose	15–60
Hemicellulose	10–30
Lignin	5–30
Protein and nucleic acids	2–15
Soluble substances, *e.g.*, sugars	10

In areas with high rainfall, basic cations tend to leach out of the soil profile; moreover, soils developed in these areas have higher concentrations of organic matter, which contain acidic components and residues. Thus, such soils tend to have decreased pH values and are acidic in nature. Soils in arid areas do not undergo such basic leaching, and the concentrations of organic matter are lower. In addition, water tends to evaporate in such areas, allowing salts to accumulate. These soils are therefore alkaline, with higher pH values.

Soil pH affects the solubility of chemicals in soils by influencing the degree of ionization of compounds and their subsequent overall charge. The extent of ionization is a function of the pH of the environment and the dissociation constant (K_{eq}) of the compound. Thus, soil pH may be critical in affecting transport of potential pollutants through the soil and vadose zone.

2.2.7 Organic Matter

Organic compounds are incorporated into soil at the surface via plant residues such as leaves or grassy material. These organic residues are degraded microbially by soil microorganisms, which utilize the organics as food or microbial substrate. The main plant constituents, shown in Table 2-3, vary in degree of complexity and ease of breakdown by microbes. In general, soluble constituents are easily metabolized and break down rapidly, whereas lignin, for example, is very resistant to microbial decomposition. The net result of microbial decomposition is the release of nutrients for microbial or plant metabolism, as well as the partial breakdown of complex plant residues. These microbially modified complex residues are ultimately incorporated into large macromolecules that form the stable basis of soil organic matter. This stable organic matrix is slowly metabolized by indigenous soil organisms, a process that results in about 2% breakdown of the complex materials annually. Owing to the slow but constant decomposition of the organic matrix and annual fresh additions of plant residues, an equilibrium is achieved in which the overall amount of soil organic matter remains constant. In humid areas with high rainfall, soil organic matter contents can be as high as 5% on a dry-weight basis. In arid areas with high rates of decomposition and low inputs of plant residues, val-

Table 2-4 Terms used to define soil organic matter

Term	Definition
Organic residues	Undecayed plant and microbial biomass and their partial decomposition products.
Soil biomass	Live microbial biomass.
Soil organic matter or humus	All soil organic matter, except organic residues and soil biomass.
Humic substances	High-molecular-weight complex stable macromolecules with no distinct physical or chemical properties. These substances are never exactly the same in any two soils because of variable inputs and environments. This is the stable backbone of soil organic matter, and is degraded only slowly (2% per year).
Nonhumic substances	Known chemical materials such as amino acids, organic acids, carbohydrates, or fats. They include all known biochemical compounds, and have distinct physical and chemical properties. They are normally easily degraded by microbes.

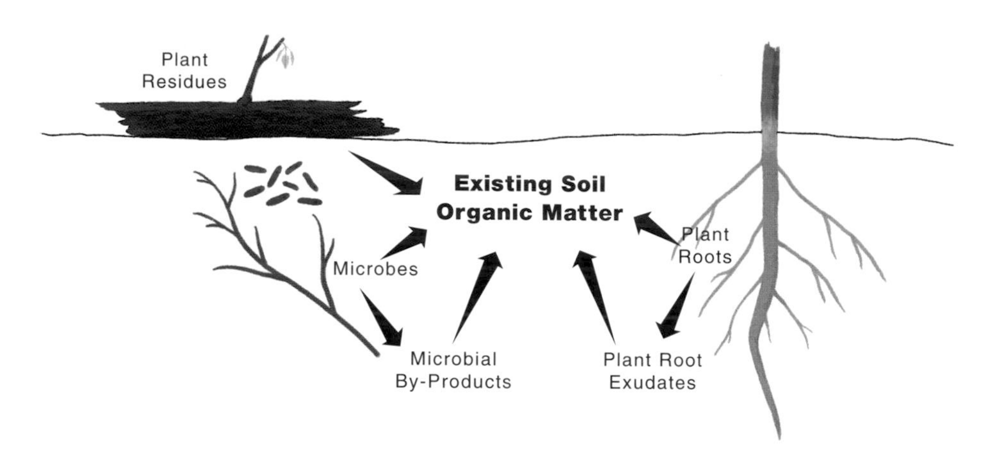

Figure 2-4 Schematic representation of the formation of soil organic matter.

ues are usually less than 1%. Terms used to define soil organic matter are shown in Table 2-4 on page 15, and the formation of soil organic matter is illustrated in Figure 2-4.

The release of nutrients that occurs as plant residues degrade has several effects on soil. The enhanced microbial activity causes an increase in soil structure, which affects most of the physical properties of soil, such as aeration and infiltration. The stable humic substances contain many functional groups that contribute to the pH-dependent CEC of the soil. In addition, many of the humic and nonhumic substances can complex or chelate heavy metals, affecting their availability to plants and soil microbes as well as their potential for transport into the subsurface.

2.3 Gaseous Phase

2.3.1 Constituents of Soil Atmosphere

Soil and the atmosphere are in direct contact; therefore, most of the gases found in the atmosphere are also found in the soil atmosphere, but at different concentrations. The main gaseous constituents are oxygen, carbon dioxide, nitrogen, and other volatile compounds such as hydrogen sulfide or ethylene. The concentrations of oxygen and carbon dioxide in the soil atmosphere are normally 19–20% and 1% by volume, respectively, as compared to 21% and 0.035% in the atmosphere. This variation reflects the utilization of oxygen by aerobic soil organisms and subsequent release of carbon dioxide. The gaseous concentrations in soil are normally regulated by diffusion of oxygen into soil and of carbon dioxide from soil.

2.3.2 Availability of Oxygen and Soil Respiration

The oxygen content of soil is vital for aerobic microorganisms, which utilize oxygen as a **terminal electron acceptor** during degradation of organic compounds. Facultative anaerobes can utilize oxygen or combined forms of oxygen (such as nitrate) as a terminal electron acceptor. Anaerobes cannot utilize oxygen as an acceptor. Strict anaerobes are lethally affected by oxygen because they do not contain enzymes that can degrade toxic peroxide radicals. Since microbial degradation of many organic compounds in soil, including xenobiotics, is carried out by aerobic organisms, the presence of oxygen in soil is necessary for such decomposition. Physically, oxygen is found either dissolved in the soil solution or in the soil pores, but soil oxygen concentrations in solution are much lower than in soil pores.

The total amount of pore space depends on soil texture and soil structure. Soils high in clays have more total pore space, but smaller pore sizes. In contrast, sandy soils have larger pore sizes, allowing more rapid water and air movement. In any soil, as the amount of soil structure increases, the total pore space of the soil increases. Aerobic soil microbes require both water and oxygen, which are both found within pore space. Therefore, the soil moisture content controls the amount of available oxygen in a soil. In soils saturated with water, all pores are full of water and the soil oxygen content is very low. In dry soils, all pores are essentially full of oxygen, so the soil moisture content is very low. In soils at **field capacity**, that is, soils having moderate soil moisture, both oxygen and moisture are readily available to soil microbes. In such situations, soil respiration via aerobic microbial metabolism is normally at a maximum. It is important to note, however, that closed pores allow anaerobic microsites to exist even in aerobic soils, thereby allowing facultative anaerobes and strict anaerobes to metabolize. This is an excellent example of how soil can function as a discontinuous environment of great diversity.

2.4 Liquid Phase

2.4.1 Properties of Water

Water is, of course, essential for all biological forms of life, in part because of the unique nature of its structure. The fact that the oxygen moiety of the water molecule is slightly more electronegative than the hydrogen counterparts results in a polar molecule. This polarity, in turn, allows water to hydrogen-bond both to other water molecules and to other polar molecules. This capacity to bond with almost anything has a profound influence on biological systems, and it explains why water is a near-universal solvent. It also explains the hydration of cations and the adsorption of water to soil colloids.

2.4.2 Soil Water Potential

In soil, the amount of water present is characterized by the soil water potential, which is actually a measure of the free energy of the soil water. All water in soil has an associated energy comprising either kinetic or potential types. In unsaturated soils, water does not move freely; thus the associated kinetic energy is negligible, even though the potential energy may be large.

- **Soil water potential** (ψ) is the amount of potential energy of soil water relative to the potential energy of pure water under standard temperature and pressure.

We designate the potential difference between two locations in a soil as $\Delta\psi$. In general, water always moves from higher potential to lower potential. In soil, water normally exists as moisture films around soil colloids, which hold onto the water molecules and thus restrict their movement. As such, water in soil is less mobile than free water in a pond. Thus energy must be provided to remove the water from colloids, enabling it to move. Because water is under tension in soil, the soil water potential is negative relative to that of free water. As soil moisture contents decrease, resulting in thinner moisture films around colloids, the remaining moisture is held more tightly by the colloids and its ψ value becomes more negative with less potential energy to do work. The units of ψ are normally megapascals or bars:

$$1 \text{ bar} \cong 0.987 \text{ atmospheres}$$
$$\cong 10^5 \text{ pascals (Pa)} \cong 0.1 \text{ megapascal (MPa)}$$

Of these, the megapascal is the unit of choice. If two locations in soil have ψ values of -0.1 MPa and -2 MPa respectively, water moves from the former point (more moist) to the latter (less moist), that is, from higher to lower soil moisture potentials. A discussion of soil water potential is given in Section 5.1, beginning on page 45.

We can calculate the amount of water within a soil sample by weighing the moist soil, then reweighing the sample after heating the soil for 24 hours to evaporate all water.

$$\%\text{ moisture} = \frac{w - d}{d} \times 100 \qquad \text{(2-8)}$$

where w is the wet (moist) weight of the soil and d is the weight of the soil after drying. This results in the moisture content of the soil on a dry weight basis. Expressing the moisture content of a soil in this fashion makes the value independent of the moisture content of any particular samples:

$$\% \text{ moisture} = \frac{\text{amount of } H_2O}{\text{dry weight of soil}} \quad (2\text{-}9)$$

For example, if $d = 66.6$ g and $w = 100$ g,

$$\% \text{ moisture} = \frac{100 \text{ g} - 66.6 \text{ g}}{66.6 \text{ g}} \times 100 \quad (2\text{-}10)$$

$$= 50\ \%$$

References and Recommended Reading

Brady N.C. (1990) *The Nature and Property of Soils*, 10th Edition. MacMillan Publishing Company, London.

Hassett J.J. and Banwart W.L. (1992) *Soils and Their Environment*. Prentice Hall, Englewood Cliffs, New Jersey.

Problems and Questions

1. The hydrogen ion concentration of the soil solution from a particular soil is 3.0×10^{-6} *M*. What is the pH of the soil solution?
2. What is the textural class of a soil with 20% sand, 60% silt, and 20% clay?
3. A 100 g sample of a moist soil initially has a moisture content of 15% on a dry weight basis. What is the new moisture content if 10.0 g of water are uniformly mixed into the soil?
4. Which of the factors listed in Sections 2.2.3–2.2.7 affect the cation exchange capacity (CEC) of a soil? Explain why.
5. Which of the factors listed in Sections 2.2.3–2.2.7 can potentially affect the transport of contaminants through soil and the vadose zone? Explain why.

Wildland fires, such as this one outside of Tucson, Arizona, can contribute particulate pollutants to the atmosphere. Photograph: I.L. Pepper.

3.1 Chemical Composition

By mass and by volume, more than 99% of the atmosphere is made up of nitrogen (N_2), oxygen (O_2), and argon (Ar) gases (Table 3-1 on page 20). The concentrations of these atmospheric gases, together with neon (Ne), helium (He), and krypton (Kr), have probably been constant for many millions of years and are unlikely to change markedly, either by natural or anthropogenic means.

The trace gas concentrations, on the other hand, are variable. These gases, which are the subject of considerable concern, are listed in Table 3-2 on page 20. [*Note:* Although water vapor is listed among the variable components, it will not be discussed here.] We know that all of these gases are affected by human activities as well as by reactions with the soil, biosphere, and oceans. But how much change is due to human activity and how much to lesser-known natural causes is still unclear. These changes have generated much controversy—and many fears; however, some of these fears may be groundless as they are based on inadequate and short-term investigations. It is quite conceivable that short-term fluctuations will prove to have little significance over the long run.

The debate is further complicated because soils, for example, serve as both source and sink for virtually all of the gases in Table 3-2. Whether soils function as source or sink can vary between day and night, with the season, with water content, with cultivation, with fertilization, and with the gas being considered, as can the strength of that function. The oceans also fluctuate in their source/sink behavior.

The concentration[1] of carbon dioxide [CO_2] is about 355 $\mu L\ L^{-1}$ as of the mid-1990s, which repre-

---

[1] Various units are used to express gas concentrations in the atmosphere. Here, $\mu L\ L^{-1}$ and percentages by mass and volume will be used. The unit $\mu L\ L^{-1}$ corresponds to the commonly used unit of ppmv (parts per million by volume). We can also interpret both $\mu L\ L^{-1}$ and ppmv as the number of gas molecules or atoms of a component gas in one million air molecules or atoms (as we can show through derivation). For air pollutants and regulatory purposes, a mass per volume unit, mg m^{-3}, is also used.

Table 3-1 Constant atmospheric components.

Gas	Concentration ($\mu L\ L^{-1}$)
Nitrogen (N_2)	780,840
Oxygen (O_2)	209,460
Argon (Ar)	9,340
Neon + helium + krypton (Ne + He + Kr)	24

sents an increase from about 300 $\mu L\ L^{-1}$ since 1900. The CO_2 concentration decreases by a few parts per million each summer because of photosynthesis by terrestrial plants. From fall through spring, [CO_2] rises because of microbial decomposition of organic matter and plant respiration. The heights of the annual peaks and valleys are buffered by the less seasonal photosynthesis-degradation cycle in the oceans. In the southern hemisphere, the [CO_2] peaks and valleys are six months out of phase with those of the northern hemisphere. The amplitude is also much smaller because the southern land areas are much smaller and a larger percentage is arid.

We usually attribute the [CO_2] increase to combustion of fossil fuels. Fossil-fuel carbon burning results in only about 10% of the CO_2 annually released by the decomposition of organic matter in soils and oceans. Other, less quantifiable CO_2 sources are (1) cultivation of native soils, which converts some of the soil organic matter into CO_2; (2) agricultural plants, which accumulate less biomass carbon because they are annuals, are less dense, and have a shorter growing season than do native plants; and (3) clearing of forests, which decreases biomass carbon. Because the increase in atmospheric CO_2 amounts to only about half of the total amount of CO_2 produced by fuel consumption alone, we can infer that the capacity of the biosphere, soils, and oceans to buffer changes in atmospheric [CO_2] is quite large. The sinks include increased photosynthesis and biomass, increased CO_2 absorption by the oceans, and probably increased soil organic matter.

Table 3-2 Variable gas concentrations in the atmosphere.

Gas	Concentration ($\mu L\ L^{-1}$)
Water vapor (H_2O)	Saturation–10,000
Carbon dioxide (CO_2)	355
Methane (CH_4)	1.5
Hydrogen (H_2)	0.50
Nitrous oxide (N_2O)	0.27
Ozone (O_3)	0.02
Carbon monoxide (CO)	<0.05
Ammonia (NH_3)	0.004
Nitrogen dioxide (NO_2)	0.001
Sulfur dioxide (SO_2)	0.001
Nitric oxide (NO)	0.0005
Hydrogen sulfide (H_2S)	0.00005

Another carbon-based gas is methane (CH_4), whose major sources are swamps, natural gas seepage, and termite activity. Methane concentrations in the atmosphere have been increasing over the past several decades. Hydrogen is also liberated in small amounts from swamps.

Nitrous oxide (N_2O) and ammonia (NH_3) are released from and absorbed by soils naturally, and releases of these gases are higher after fertilization. Nitrous oxide, whose concentration has also been increasing slowly, is a rather unreactive gas that has a long residence time (>100 years) in the atmosphere. Atmospheric ammonia has also been observed in higher concentrations, especially in industrial regions, where it often takes the form of $(NH_4)_2SO_4$.

Nitrogen dioxide (NO_2), its dimer N_2O_4, and nitric oxide (NO) (which are often combined and written as NO_x) are produced by combustion of coal and petroleum, as is sulfur dioxide (SO_2). While sulfur is a constituent of coal and oil, the nitrogen oxides are by-products of high-temperature furnaces and internal-combustion engines. In addition, NO_x is produced by lightning. These gases are highly reactive in air: they rapidly oxidize to nitric and sulfuric acid, which quickly dissolve in water and wash out as **acid rain** (see also Section 12.2, beginning on page 172). These gases are also absorbed directly from the air by plants and calcareous soils.

Concerns about acid rain have begun to diminish. Recent studies have shown that the European trees are dying from old age rather than acid rain (see the photograph at the top of page 171). In fact, young forests are flourishing. Similarly, lakes in Scandinavia and eastern North America are being acidified by reversion of agricultural lands to natural forest and lack of soil liming rather than by acid rain. The extent of lake acidification previously observed has proved to be partly an artifact of the methodology used in measuring lake acidity. However, acid rain docs cxist and to some extent remains a pressing environmental hazard.

On the global scale, carbon monoxide (CO) is not considered an air pollutant because soil microorganisms adsorb it relatively rapidly and oxidize it to CO_2. In urban areas, however, carbon monoxide can accumulate during rush hour traffic.

To decrease air pollution, regulatory agencies worldwide have put increasing strictures on SO_2, NO_x, and organic chemical emissions. London, Pittsburgh, San Francisco, Los Angeles, and many other North American and European cities have already shown obvious improvements (see Figure 23-3 on page 369). Unfortunately, in other cities, such as Mexico City (see Figure 12-10 on page 183), the situation will probably worsen before it improves.

Ozone (O_3) is considered an air pollutant in the lower atmosphere because it is harmful to plants and humans. Ozone is produced by the action of ultraviolet (UV) sunlight on polluted air that contains organic gases such as industrial solvents, fuels, and partially oxidized hydrocarbons. The ozone concentration has therefore been adopted by the U.S. Environmental Protection Agency (EPA) as an index of air pollution.

In the upper atmosphere, a distinctive pollution problem is the **ozone hole**, a seasonal decrease in stratospheric ozone measured over the South Pole. The size of the ozone hole has been increasing in recent years. Stratospheric ozone absorbs much of the UV fraction of sunlight; increased UV light tends to increase human and animal skin cancer and stunt plant growth.

At higher altitudes ozone is attacked by chlorine (Cl) and to a lesser extent by NO_x. In the more intense UV of that region, chlorofluorocarbons (CFCs, such as CCl_2F_2) decompose to Cl_2, which degrades O_3 to oxygen. NO_x degrades O_3 as well, but to a lesser extent; N_2O at higher altitudes is converted by UV light to NO_x (see Section 12.5.2, beginning on page 184).

3.2 Physical Properties and Structure

The ability of the atmosphere to accept, disperse, and remove pollutants is strongly related to its various physical and dynamic properties. Atmospheric winds, for example, determine the pathways and speeds at which pollutants are transported away from sources such as cars and smoke stacks. Another physical process, the condensation of water vapor into rain and fog droplets, scavenges water-soluble pollutants from the atmosphere, ultimately determining the rate of their removal. In addition, the vertical variation of temperature greatly influences atmospheric stability and hence the turbulent mixing of polluted air with clean air. Temperature also affects reaction rates between chemical species, such as those involved in ozone formation in polluted urban environments.

The following sections provide a brief introduction to the physical and dynamic properties of the atmosphere that are most relevant to our understanding of air pollution processes. The purpose here is to gain an overall understanding of air density, pressure, wind, water vapor, precipitation, radiation transfer, and temperature. We will not cover several important topics concerning the atmosphere, such as large-scale weather disturbances (*e.g.*, hurricanes) or forecasting weather conditions for air pollution advisories. Interested readers should consult more comprehensive textbooks in atmospheric science for detailed information. Suggested references are listed at the end of this chapter.

3.2.1 Density, Pressure, and Wind

Air is a multicomponent mixture of gaseous molecules and atoms, which are constantly moving about and undergoing frequent collisions. The mass and kinetic energy of each of these moving molecules imparts a force upon collision, which gives rise to **atmospheric pressure**. The horizontal variations of pressure, which result in air flow (winds) across the earth's surface, are an important factor in air pollution dispersal.

Pressure is the force per unit area exerted by air molecules. Usually expressed in units of newtons per square meter (N m^{-2}) or pascals (Pa)[2], pressure is exerted equally in all directions because molecular-scale motion is uniformly distributed in all directions. Thus, at any height in the atmosphere, pressure is the cumulative force (weight) per unit area exerted by all molecules above that height. Under static equilibrium conditions, the weight of the atmosphere pushing down on an air parcel at any height is exactly balanced by a pressure gradient force pushing upward.

The weight of the atmosphere compresses air molecules near the earth's surface. In fact, nearly two-thirds of all atmospheric molecules are contained within a one scale-height distance of about 9 km above the surface. Air density (measured in kg m^{-3}) and pressure are both highest at sea level about 1.2 kg m^{-3} and 101.3 kPa (1013 mbar) on average, respectively. From sea level upward, both decrease exponentially with height, as illustrated in Figure 3-1. At heights greater than about 100 km, so few molecules (and atoms and ions) are present that both density and pressure become almost negligible.

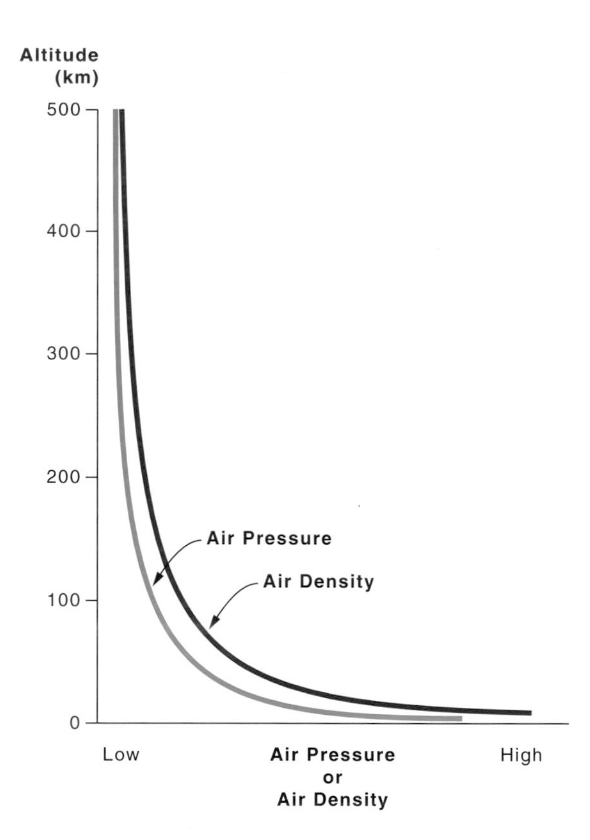

Figure 3-1 Density and pressure variations with altitude in the earth's atmosphere. Reprinted by permission from page 7 of *Essentials of Meteorology: An Invitation to the Atmosphere* by C. Donald Ahrens; copyright © 1993 by West Publishing Company. All rights reserved.

Sea-level pressure varies both temporally and spatially across the earth's surface. For example, at any time of day, the pressure at Seattle, Washington may be several millibars higher or lower than the pressure at Miami, Florida, even though both cities are at sea level. Surface pressure differences (gradients) can result from several factors, such as variations in the heating of air molecules by the sun, fluctuations in atmospheric water vapor and cloud cover, and rotation of the earth.

Belts of semipermanent high (H) and low (L) surface pressure circle the earth at various latitudes. These belts are a result of the general circulation of the atmosphere, as illustrated simplistically in Figure 3-2. Large-scale circulation is composed, on average, of six main convective cells known as **Hadley cells**. The energy to drive the motion within these cells comes from the uneven latitudinal distribution of surface heating by sunlight. More sunlight is absorbed per unit surface area at the equator than at the poles; thus the cells transport energy from the warm equator toward the cold polar regions. In addition, the cells facilitate long-range transport of particulate matter and gaseous air pollutants over the earth's surface.

Low surface pressure results when warm, moist, buoyant air, such as that over the equator, ascends from the earth's surface. Moist air rises above the equator because the temperature is relatively high and because moist air, containing perhaps 2–4% water vapor, is less dense (and lighter) than dry air. The relatively low molecular weight of water vapor (18) relative to dry air (29) lowers the average molecular weight of moist air. The lighter air moves upward and exerts relatively less pressure at the surface.

[2] For a comparison of commonly used units of pressure see Section 2.4.2 on page 17.

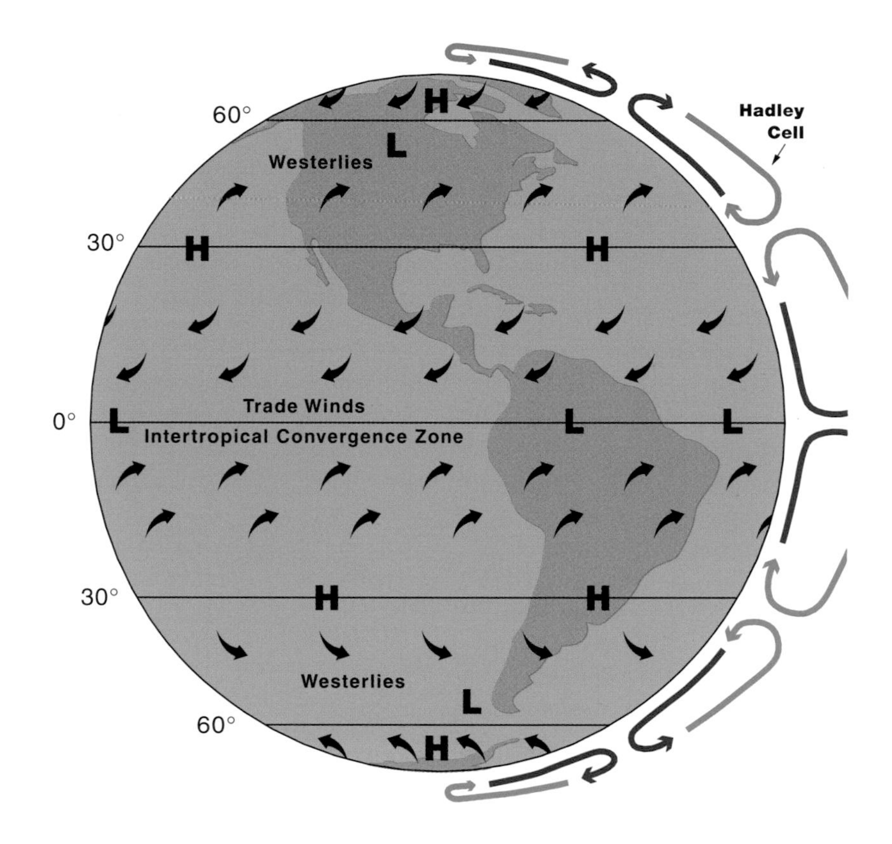

Figure 3-2 General circulation of the earth's atmosphere showing large-scale convective (Hadley) cells, predominant surface wind directions, and long-term surface pressure features. H = high pressure, L = low pressure. Black arrows indicate the direction of surface winds.

Surface flows of moist air associated with the trade winds converge at the **intertropical convergence zone (ITCZ)** near the equator (see Figure 3-2). As the flows converge from the northern and southern hemispheres, the air is heated by the equatorial sun and thus rises. As the air flow rises, it cools and loses its moisture by condensation and precipitation. At high altitudes the rising air current diverges northward and southward. At subtropical latitudes (about 30° north and south), the dry upper-level air flow subsides (sinks) toward the (mostly) ocean surfaces where it again becomes moist and flows back to the ITCZ. The subsiding air compresses (and thus warms) the atmosphere and increases pressure. Subsidence may occur at a rate of about 1 km per day at subtropical latitudes. High pressure is therefore associated with relatively warm, dry, subsiding air. This process has important implications for dispersal of air pollutants.

Surface wind patterns associated with atmospheric circulation are more complex than the simple idealized flow patterns shown in Figure 3-2. In general, surface winds are influenced by several forces acting on air masses, including pressure-gradient (flow from high to low pressure), coriolis (deflection of air to the right in the northern hemisphere due to the rotation of the earth), frictional, and centrifugal forces. In the northern hemisphere, these forces generally combine to cause air flow counterclockwise around low pressure and clockwise around high pressure.

Flow around low pressure is called **cyclonic flow**. Flow around high pressure is termed **anti-cyclonic flow**. At low latitudes, prevailing surface **trade winds** are generally from northeast to southwest. At midlatitudes, prevailing surface winds are generally from southwest to northeast—the **westerlies** shown in Figure 3-2 on page 23. At high latitudes, air flow is generally northeast to southwest. Because of convergence at the ITCZ, winds tend to be light over the equator.

The general circulation and resultant spatial pattern of high and low pressure influence long-range pollutant transport. In regions with semipermanent high pressure features (such as those within the subtropical high-pressure belt at about 30° latitude), calm, stagnant conditions often persist for long periods, thereby amplifying air pollutant concentrations near the surface. For example, air subsidence associated with high pressure over the eastern Pacific Ocean (see Figure 3-2) markedly influences air quality in the coastal cities of southern California (see Figure 12-4 on page 176). Similarly, high pressure over the southwestern United States also adversely affects air quality in the region, particularly over large urban areas such as Phoenix, Arizona. Similar high pressures can cause air quality problems over Cairo, Egypt.

3.2.2 Temperature

Temperature is a measure of the kinetic energy (heat content) of molecules and atoms. Air temperature affects nearly all physical, chemical, and biological processes within the earth-atmosphere system. A good example is the influence it has on the atmosphere of polluted urban environments, where high temperature greatly increases the rate of smog (smoke plus fog) formation. Furthermore, once smog is formed, it may be dispersed upward and downward by buoyancy-generated atmospheric turbulence resulting from temperature (and hence density) differences between individual air parcels and their surrounding environment. We can see the effects of buoyancy on air motion by watching the erratic motion of a helium-filled balloon once it is released into the atmosphere.

Air temperature near the earth's surface varies markedly over different time scales, ranging from seconds to years. By midsummer, for example, air temperature at 2 m above the Sonoran Desert floor in Arizona may vary diurnally from about 45°C maximum (at midafternoon) to 20°C minimum (at dawn). In midwinter the daily variation in the desert may range about 10–30°C. Temporal variations of air temperature are caused mainly by varying solar energy input to the surface.

Air temperature at a given height, say 2 m, also varies markedly across the earth's surface owing to spatial variations in energy input. Obviously, the lowest temperatures occur in the polar regions where energy input is low. The highest temperatures occur in low-latitude deserts, such as the Sahara in Africa, where energy input is very large and little water is available for evaporative cooling of the ground.

The question of how air temperature changes with increasing height above the ground is important when considering how air pollution is dispersed near the ground. To answer this question, we must recognize that energy exchange takes place almost continuously between the surface and the atmosphere. Some heat exchange occurs by **conduction** through a very thin layer a few millimeters thick of air over the surface; however, most heat exchange occurs by means of **convection**, which is the turbulent exchange caused by buoyancy and shear stress. Convection becomes increasingly more efficient with increasing height. This increase is due to the decreased effect of surface frictional drag at greater heights. At midday the change in temperature with height above the surface is often very large. Temperature gradients above a hot desert soil surface may be as high as –1°C per millimeter. Note that a negative temperature gradient means that air temperature decreases with increasing height from the ground surface. Because convection quickly becomes very efficient in mixing air with increased height, the temperature gradients within turbulent air rapidly decrease in magnitude with increasing height. It is important to remember that air in contact with the earth's surface during daylight hours is generally warmer than air aloft because of strong surface heating. Thus, air temperature generally decreases with increased height within the lower part of the atmosphere during the daytime.

Atmospheric scientists use the term **lapse rate** to describe the observed change (generally a decrease) of air temperature with height ($\Delta T/\Delta z$). The lapse rate at a given height (z, in meters) and location may vary greatly throughout the day in response to changes in heat flow between the surface and the atmosphere. At a height of 2 m at midday above a hot desert soil surface, for example, the lapse rate may range from about –0.01 to –0.2°C per meter. Within the first few kilometers of the lower atmosphere, however, the lapse rate is, on average, about –0.0065°C per meter.

During the night, however, the situation is generally the reverse of daytime conditions. The ground surface may quickly lose energy to space by infrared radiation emission (see Section 3.2.4, beginning on page 26) and become relatively cool. This cooling process also cools the air in direct contact with the surface. Thus at night, air temperature often increases with increasing height, typically on the order of 0.1 to 1°C per meter. An **air temperature inversion** occurs when temperature increases with height up to a level (called the **inversion height**) of maximum air temperature. Above the inversion height, the temperature decreases with height. Radiation inversions are particularly common in the dry desert environment of the southwestern United States and northeastern Africa, where nocturnal loss of radiant energy from the ground to space causes cooling of the air in contact with the ground. Inversions can also occur as a result of subsidence associated with anti-cyclonic flow, which is also common over the southwestern United States. (These and other causes of inversions are discussed further in Section 12.4.1, beginning on page 178.) As discussed in the following paragraphs, the stable atmospheric conditions associated with inversions tend to trap pollutants near their source. The stability of air defines its ability to mix and disperse pollutants. Air can be unstable, stable, or neutral.

Unstable air results in turbulent motion associated with free convection due to buoyancy within the atmosphere. Buoyant motion enhances upward penetration of air parcels into the atmosphere, thus helping to disperse pollutants. Under unstable conditions an air parcel that is displaced adiabatically (without heat exchange with its surroundings) upward or downward a short distance is accelerated away from its initial position by buoyancy. Air is unstable because the net buoyancy force acting on the parcel accelerates it either upward or downward, depending upon the temperature (density) difference between the parcel and its surrounding environment. Unstable conditions are prevalent during daytime when convection carries heat upward from the soil surface.

In **stable air**, turbulence is suppressed or even absent. In stable conditions, buoyancy tends to restore an adiabatically displaced parcel to its original height. In other words, the buoyancy force acts in the direction opposite to the motion of the displaced parcel. Stable conditions occur most often at night, when convective heat flow is downward from the atmosphere to the soil surface.

Neutral stability means that the buoyancy force is zero and that a balance exists between gravity (acting downward) and the pressure gradient force (acting upward) on the parcel. The pressure gradient force is the difference between the pressures at the top and bottom of the parcel divided by the distance between top and bottom. Thus, under neutral conditions an air parcel displaced upward or downward from its initial height remains at its new height unless acted upon by an external force. Neutral conditions often occur briefly after sunrise, and before sunset, when convective heat flow is zero. Cloudy, windy days are also favorable for neutral stability.

We base the assessment of the pollutant-dispersal ability of the atmosphere on quantification of the stability of the atmosphere. Stability is largely determined by the value of the measured lapse rate $\Delta T/\Delta z$ relative to the adiabatic lapse rate (Γ). The constant Γ is defined as the change in the temperature of the air parcel when the parcel is displaced upward or downward adiabatically from a base height z_b (see Figure 3-3 on page 27). This change in temperature results from a change in pressure, as described by the ideal gas law. When the atmosphere is relatively dry, Γ is equal to $-g/c_p$ where g is the acceleration due to gravity and c_p is the specific heat of air at constant pressure. Γ thus has a value of about –0.01°C per meter. This means that the temperature (T_b) of an air parcel adiabatically lifted from height z_b will decrease by 0.01°C per meter of displacement. Likewise, if the parcel is lowered from z_b it will warm by 0.01°C per meter of displacement.

- When $\Delta T/\Delta z < \Gamma$ (*e.g.*, −0.05°C per meter is less than −0.01°C per meter), the atmosphere is unstable, as shown in Figure 3-3 (*center*). It is unstable because a parcel adiabatically displaced upward from its initial position at z_b is always warmer than its surroundings. When it is adiabatically displaced downward, it is cooler than its surroundings. Thus, it may be accelerated up or down from z_b by the buoyancy force, causing turbulence.
- When $\Delta T/\Delta z > \Gamma$, conditions are said to be stable (Figure 3-3, *bottom*). During stable conditions, a parcel adiabatically displaced upward from z_b becomes cooler than its surroundings. When it is displaced downward, it becomes warmer than its surroundings. Thus, buoyancy restores the parcel to z_b, suppressing turbulent motion.
- When $\Delta T/\Delta z = \Gamma$, neutral conditions are present (Figure 3-3, *top*), and the buoyancy force acting on the parcel is zero at any height.

The criteria for characterizing stability are summarized as follows:

unstable conditions: $\Delta T/\Delta z < \Gamma$

stable conditions: $\Delta T/\Delta z > \Gamma$

neutral conditions: $\Delta T/\Delta z = \Gamma$.

3.2.3 Water Vapor and Precipitation

Water vapor is a highly variable part of the atmosphere. In warm, humid, tropical rain forests, high rates of evaporation of water from the earth's surface keep the lower atmosphere almost continuously saturated. On the other hand, in dry, hot deserts little evaporation occurs and the amount of water vapor in the atmosphere is almost negligible.

Atmospheric water vapor is characterized by various parameters, including vapor pressure, relative humidity, dew point temperature, water vapor density, and specific humidity. **Relative humidity** is probably the most familiar. It is defined as the ratio of the actual vapor pressure to the saturation vapor pressure of the air, which is solely a function of air temperature.

Condensation of water vapor into cloud and fog droplets occurs when air is cooled to saturation at the **dew point temperature**. Cooling occurs by various processes, such as the radiational cooling of the surface at night, the upward convective movement, advective motion of the atmosphere (in which a cold air front displaces warm moist air upward), and orographic lifting (in which air rises over mountain ranges).

Precipitation in the form of rain or snow rids the air of many types of particulate matter and gaseous pollutants. This removal by scavenging is known as **wet deposition**. Pollutants may be solubilized directly in the water droplets, or they may be adsorbed on the droplets. Soluble pollutants include, among other compounds, various oxides of sulfur and nitrogen. Removal of these pollutants increases the acidity of precipitation, producing acid rain.

3.2.4 Radiative Transfer

The radiation environment of the earth is largely a function of three main radiative transfer processes:

1. the flux of radiant energy reaching the earth's surface from the sun,
2. the redistribution of the radiant energy within the atmosphere,
3. the loss of radiant energy to space.

The overall radiation balance of the earth-atmosphere system involves absorption, scattering, transmission, and emission processes, which are described briefly in this section. Two wavelength intervals of the electromagnetic spectrum are of primary importance to the overall radiation balance: shortwave (solar) and longwave radiation. Shortwave consists of the wavelength interval from about 0.15–3.0 μm. Within this portion are the components of **ultraviolet** (0.15 – 0.36 μm), **visible** (0.36–0.75 μm), and **near infrared** (0.75–3 μm) radiation. Shortwave energy is emitted by the sun, which is an almost perfect blackbody radiator with a temperature of about 6000 K. Ultraviolet and visible

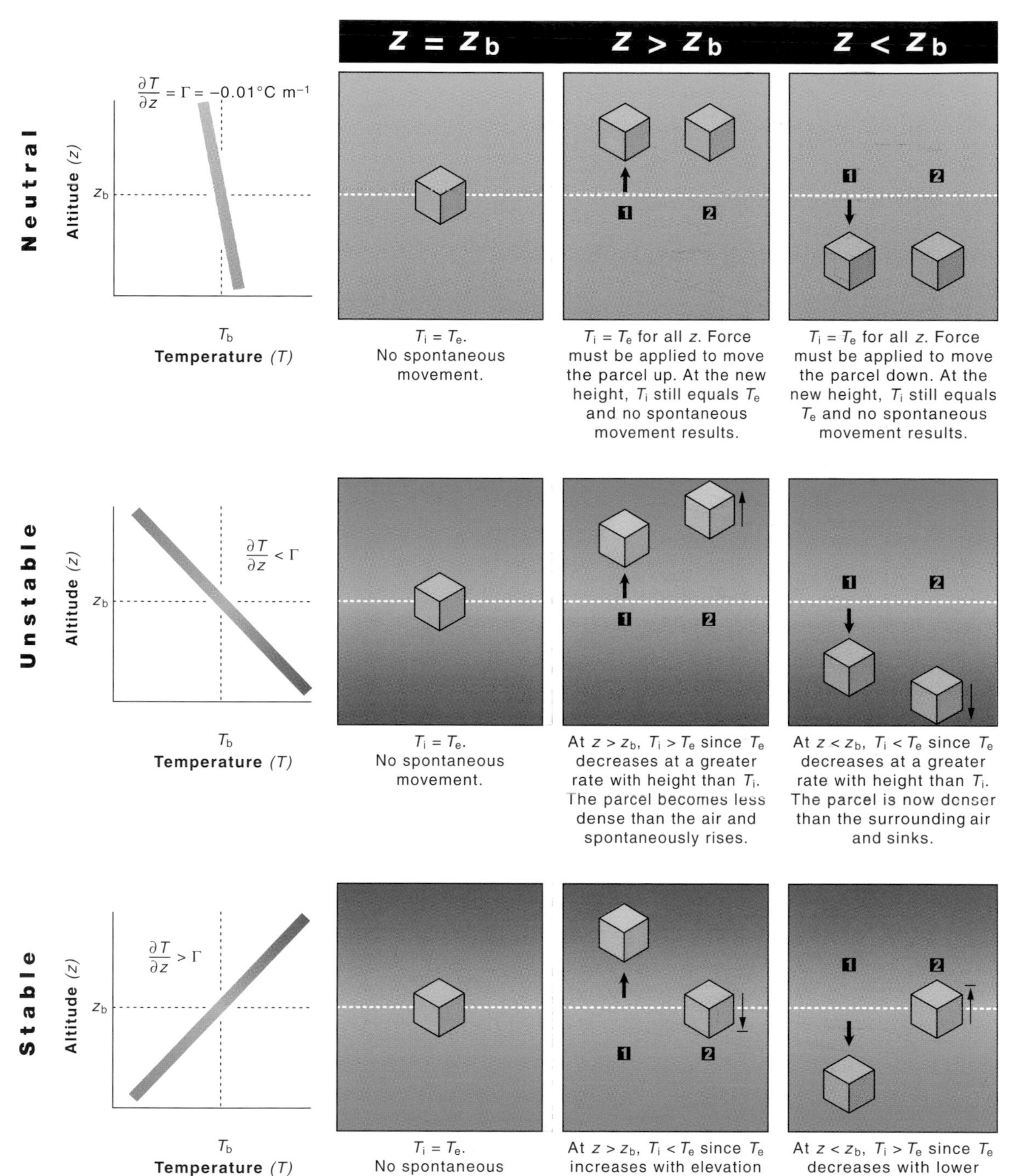

Figure 3-3 Air-temperature variations with height during neutral, unstable, and stable conditions.

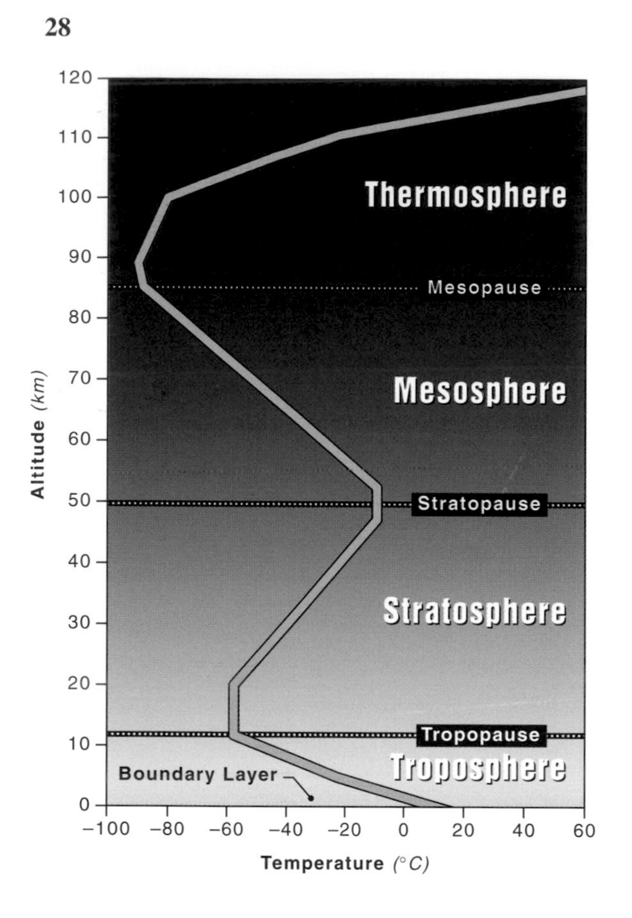

Figure 3-4 The structure of the atmosphere as defined by average variation of temperature with altitude. Adapted by permission from page 9 of *Essentials of Meteorology: An Invitation to the Atmosphere* by C. Donald Ahrens; copyright © 1993 by West Publishing Company. All rights reserved.

light are the shortwave components most significant to the global environment.

On a per-unit-wavelength basis, ultraviolet (UV) is very high-energy radiation. Fortunately, most of the high-energy UV wavelengths are selectively absorbed by ozone and other constituents in the earth's upper atmosphere. Some UV, however, reaches the earth's surface, where it can be harmful to life and contributes to the production of photochemical smog in the lower atmosphere.

Most shortwave radiation is within the visible portion of the spectrum. In fact, the wavelength of maximum energy flux from the sun is at 0.48 μm, which is visible to the human eye as green light. Most visible light from the sun passes through clear air without significant loss. Scattering of visible light by atmospheric molecules, clouds, and aerosols does occur, however. Visible light reaching the land surface is either absorbed (about 75%) or reflected (about 25%) by surface matter (*e.g.*, plants, water, and soil). The absorbed radiant energy heats the soil and air and evaporates water. Some of the absorbed energy is re-emitted back to the atmosphere in the form of longwave radiation.

Longwave radiation encompasses the spectrum from about 3.0–100 μm and is emitted by matter within the earth-atmosphere system. Since terrestrial temperatures are about 290 K, the wavelength of maximum longwave emission is about 10 μm. Longwave radiation is commonly referred to as terrestrial or **infrared** radiation.

The earth's atmosphere is largely opaque to most of the longwave spectrum. A window exists, however, between about 8 and 11 μm that permits escape of a portion of longwave energy to space. Absorption and re-emission of longwave energy within the atmosphere occur within the vibrational energy mode of various molecular species, including water vapor, carbon dioxide, nitrous oxide, and methane. These and a few other species are the well known "greenhouse" gases, and are responsible for the earth's **greenhouse effect**.

Without the warming by the natural greenhouse effect, the earth would be about 33°C colder than its present mean temperature of about 15°C. There is, however, much public and scientific concern that an additional 1–3°C global warming will occur by the end of the twenty-first century owing to gradual accumulation of atmospheric greenhouse gases from anthropogenic sources.

3.2.5 Lower Atmosphere

Temperature variation with height defines the various layers of the atmosphere. The major atmospheric layers are shown in Figure 3-4. The **troposphere** is the lowest major layer. Within the troposphere, vertical variation of temperature is characterized by lapse-rate conditions; thus the troposphere is generally unstable and well mixed. The troposphere extends upward from the surface to a height of about 10–15 km, depending upon latitude and season of the year. The troposphere is certainly familiar to us, since it is tropospheric air we breathe. Also, most of our weather occurs in the troposphere, including cloud formation, rain, winds, and other meteorological processes. The tropopause is the upper limit of the troposphere, which separates the troposphere from the stratosphere above.

The **atmospheric boundary layer**, which is an important sublayer at the bottom of the troposphere, forms the atmospheric interface between the troposphere and the ground surface. In this region of the atmosphere, airflow patterns are strongly affected by buoyancy (free convection) and surface shear forces (forced convection). Within the first few meters above the ground surface, vertical gradients of air temperature, wind speed, humidity, and other scalar quantities are often large and variable with time. These gradients are due mainly to the temporal variability of energy and mass exchanges (*e.g.*, evaporation of water) between the surface and the atmosphere. The depth of the boundary layer varies over the course of the day. By midafternoon, rising air from the heated ground may extend the boundary layer up to the 1-km height. This height is often referred to as the **mixing depth** or **mixed layer**. The turbulent air parcels, often referred to as **eddies**, undergo eddying motion. By night, however, the atmosphere cools, and the boundary may shrink to a thickness of only about 0.1 km. Most of the important atmospheric pollutant transport and transformations occur within the boundary layer. However, some chemically stable gases are dispersed upward throughout much of the troposphere. Some, such as N_2O and CFCs, eventually diffuse upward into the stratosphere.

At the very bottom of the boundary layer, directly above the earth's surface, is a sublayer known as the **surface layer**. This sublayer generally extends upward to about one-tenth of the boundary-layer depth. The properties of the surface layer are most directly affected by surface roughness and surface heat exchange. Energy and mass fluxes are nearly constant with height in the surface layer; thus this sublayer is sometimes called the **constant flux layer**.

3.2.6 Upper Atmosphere

The **stratosphere** is the stable (stratified) layer of atmosphere extending from the tropopause upward to a height of about 50 km (Figure 3-4). The stratosphere is highly stable because the air temperature increases with height up to the stratopause, which is the height of the temperature inversion. The increased temperature in this layer is due mainly to UV absorption by various chemical species, including ozone and molecular oxygen present in the stratosphere. Maximum heating takes place in the upper part of the stratosphere. Because of the stable air, pollutant mixing is suppressed within this layer. Thus, natural (*e.g.*, N_2O) and synthetic (*e.g.*, CFC) chemicals that reach the stratosphere from the troposphere tend to diffuse upward very slowly within the stratosphere.

Ozone is formed naturally and photochemically within the stratosphere. Ozone is considered a pollutant in the troposphere, but in the stratosphere it is essential to life on earth because it absorbs biologically harmful UV radiation.

The **mesosphere** and the **thermosphere** are two additional atmospheric layers above the stratosphere. These layers are largely decoupled from the stratosphere and troposphere below; therefore, they exert little influence on our weather and on pollutant transport processes. Likewise, pollution has little or no effect upon these two upper layers.

References and Recommended Reading

Ahrens C.D. (1993) *Essentials of Meteorology: An Invitation to the Atmosphere*. West Publishing Company, Minneapolis/St. Paul, Minnesota.

Mitchell J.F.B. (1989) The greenhouse effect and climate change. *Reviews of Geophysics*. **27**, 115–139.

Problems and Questions

1. What are the most important atmospheric greenhouse gases? How do these gases increase the mean temperature of the earth?
2. What causes wind? How does wind affect the movement of heat, water vapor, and pollution in the atmosphere?
3. How can surface pressures differ between Venice and New York even though both cities are at sea level?
4. Describe how and why air temperature varies with increasing height in the troposphere and stratosphere.
5. Why is air generally well-mixed in the troposphere but not in the stratosphere?

Chapter 4

Biotic Activity in Soil and Water

I.L. Pepper and K.L. Josephson

Bacteria, such as these rods, are the cause of many of the biochemical transformations in soil and water. Photograph: I.L. Pepper.

4.1 Major Groups of Organisms

The abiotic characteristics of soil and water strongly influence chemical reactions in the environment. Biochemical reactions in soil and water are controlled by biotic activity: specifically, all environments contain living organisms that mediate biochemical transformations. Soil and the vadose zone are no exceptions. Soil itself usually contains billions of organisms, with the vadose zone containing smaller, but still significant, populations. The major groups of organisms in soil include viruses, bacteria, fungi, algae, and macro fauna such as protozoa and arthropods. All of these organisms have specific ecological niches and functions, and each contributes to the overall biotic activity of the environment. The bacteria and fungi are particularly important with respect to biochemical transformations (see also Chapter 7, beginning on page 77). Because of their critical role in influencing the fate and mitigation of many pollutants, we will concentrate on these two groups. In addition, a large subdivision of bacteria includes the actinomycetes, which are often treated as a separate group of organisms because of their unique characteristics. In the following discussion we will give a broad overview of bacteria, actinomycetes, and fungi and examine their significance with respect to pollution.

The importance of the soil microflora is illustrated by their numbers and biomass, as shown in Table 4-1 on page 33. We can see that very large populations can be sustained in soil. In addition, the groups are very diverse, so that large groups of different organisms can mediate an almost infinite number of biochemical transformations. The different sizes and shapes of these organisms are illustrated in Figure 4-1 on page 32.

4.1.1 Bacteria

Bacteria, which are the most numerous organisms in soil or anywhere else on earth, are prokaryotic organisms lacking a nuclear membrane. They are characterized by a complex cell envelope, which contains cytoplasm but no cell organelles. Bacteria are capable of rapid growth and reproduction, both of which occur by binary fission. Genetic exchange

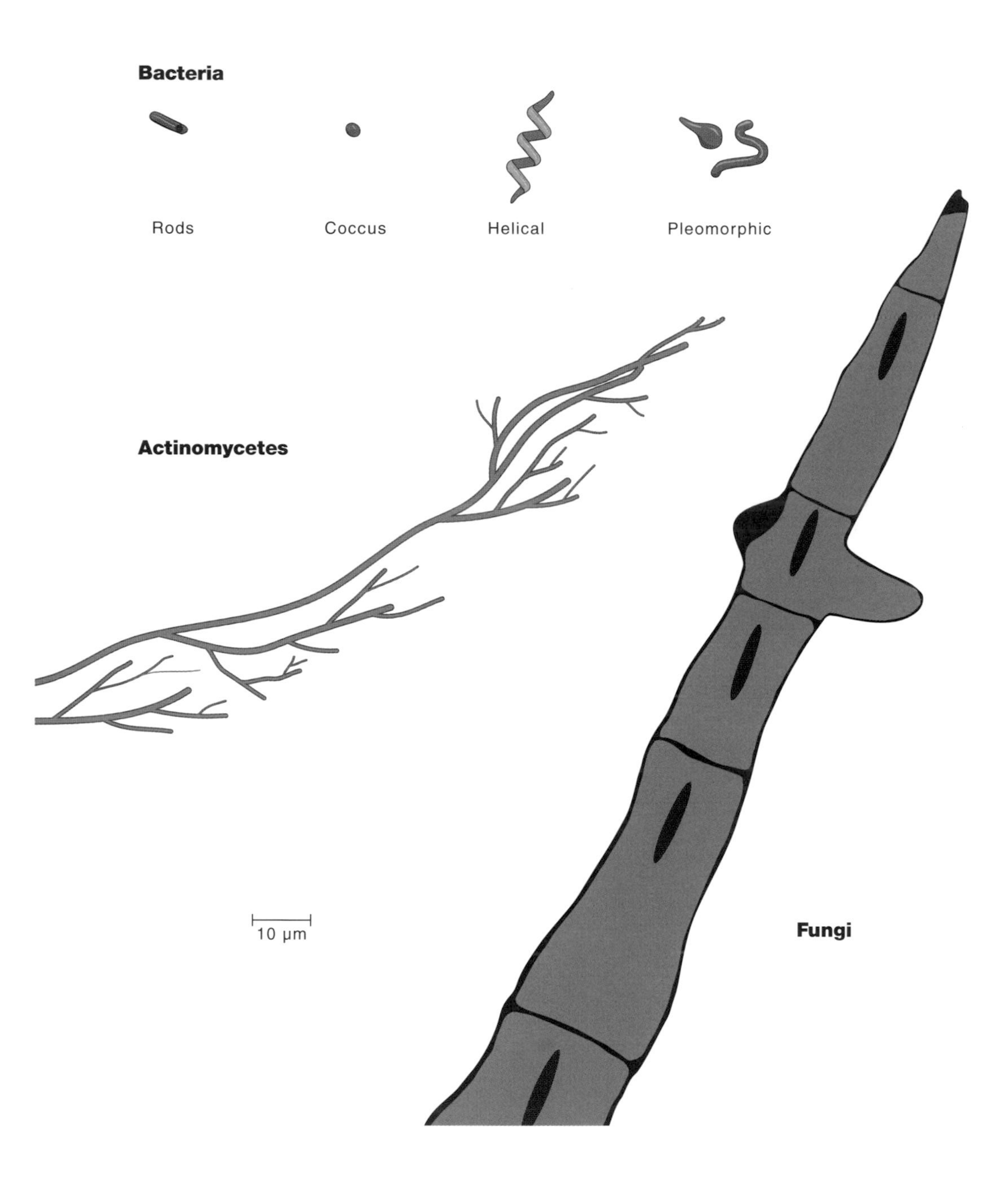

Figure 4-1 Relative size and shape of representative bacteria, actinomycetes, and fungi.

Table 4-1 Relative estimates of the abundance of soil microbes in the environment.

Microbe	Number (per gram of soil)	Biomass within Root Zone (kg ha^{-1})
Bacteria	10^8	500
Actinomycetes	10^7	500
Fungi	10^6	1500

occurs predominantly by **conjugation** (cell-to-cell contact) or **transduction** (exchange via viruses), although **transformation** (transfer of naked DNA) also occurs. The size of bacteria generally ranges from 0.1 to 2 μm. Soil bacteria can be rod-shaped, coccoidal, helical, or pleomorphic (see Figures 4-1 and 4-2).

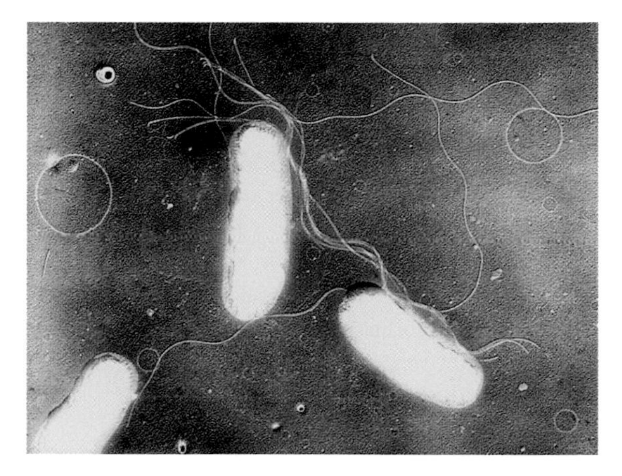

Figure 4-2 Bacteria are a vital component of the soil microflora. The above scanning electron micrograph depicts *Rhizobium meliloti*, a bacterium with multiple flagella. The circles are detached flagella that have spontaneously assumed the shape of a circle. Photograph courtesy of I.L. Pepper.

4.1.1.1 Mode of Nutrition

We can classify bacteria according to their mode of nutrition:

Autotrophic mode: Strict soil autotrophs obtain energy from inorganic sources and carbon from carbon dioxide. These kinds of organisms generally have few growth-factor requirements. Chemoautotrophs obtain energy from the oxidation of inorganic substances (Eq. T4-9 through Eq. T4-13 on page 39 and Eq. T4-15 on page 40), whereas photoautotrophs obtain energy from photosynthesis (Eq. T4-7 on page 38).

Heterotrophic mode: Heterotrophs obtain energy and carbon from organic substances. Chemoheterotrophs obtain energy from oxidations (Eq. T4-8 on page 39 and Eq. T4-14 and Eq. T4-16 on page 40); however, photoheterotrophs obtain energy from photosynthesis with an organic electron donor requirement. Figure 4-3 on page 34 depicts the colonies of heterotrophic bacteria cultured on a growth medium.

In soil, chemoheterotrophs and chemoautotrophs predominate; phototrophs of either variety are not as numerous because soil is not permeable to sunlight.

4.1.1.2 Type of Electron Acceptor

Aerobic organisms utilize oxygen as a terminal electron acceptor and possess superoxide dismutase or catalase enzymes that are capable of degrading peroxide radicals (Eq. T4-8 through Eq. T4-13). **Anaerobic organisms** do not utilize oxygen as a terminal electron acceptor (Eq. T4-16 on page 40). Strict anaerobes do not possess superoxide dismutase or catalase enzymes and are thus poisoned by the presence of oxygen. Although other kinds of anaerobes do possess these enzymes, they utilize terminal electron acceptors other than oxygen, such as nitrate or sulfate. **Facultative anaerobes** can use oxygen or combined forms of oxygen as terminal electron acceptors (as seen in Eq. T4-14 and Eq. T4-15 on page 40).

4.1.1.3 Ecological Classification

Indigenous soil bacteria can be **autochthonous** or **zymogenous**. The former metabolize slowly in soil, utilizing soil organic matter as a substrate. The latter are adapted to intervals of dormant and rapid growth, depending on substrate availability, following the addition of fresh substrate or amendment to the soil. **Allochthonous** organisms, which are organisms introduced into the soil, usually survive only for short periods of time. This type of ecological classification was introduced by Sergei Winogradsky (1856–1953), the "father of soil microbiology." However, a more recent theory of classification is founded on the concept of **r and K selection**. Organisms adapted to living under conditions in which substrate is plentiful are designated as K-selected. **Rhizosphere** organisms living off root exudates are examples of K-selected organisms. Organisms that are r-selected live in environments in which substrate is the limiting factor except for occasional flushes of substrate. r-Selected organisms rely on rapid growth rates when substrate is available, and generally occur in uncrowded environments. In contrast, K-selected organisms exist in crowded environments and are highly competitive.

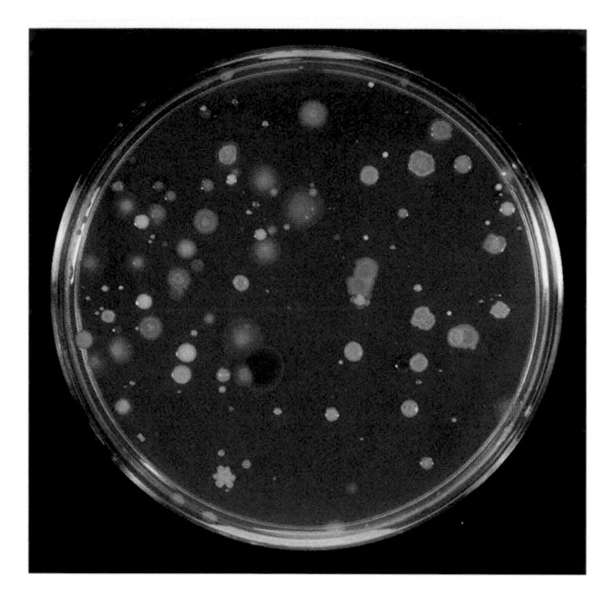

Figure 4-3 Here, a variety of bacteria with some actinomycetes are isolated from a field soil on a petri dish on a general heterotrophic medium. Photograph: E.M. Jutras.

4.1.1.4 Dominant Soil Bacteria

Arthrobacter: The most numerous bacteria in soil, as determined by plating procedures, arthrobacters represent as much as 40% of the culturable soil bacteria. These autochthonous organisms are pleomorphic and Gram-variable. Young cells are Gram-negative rods, which later become Gram-positive cocci.

Streptomyces: These organisms are actually actinomycetes (which are discussed later). They are Gram-positive, chemoheterotrophic organisms that can comprise 5–20% of the bacterial count in soil. These organisms produce antibiotics, including streptomycin (discovered by Selman Waksman, who was awarded the Nobel Prize in medicine in 1942).

Pseudomonas: These Gram-negative organisms, which are also known as pseudomonads, are ubiquitous and diverse in nature. They are generally heterotrophic and aerobic, but some are facultative autotrophs. As a group, they possess many different enzyme systems and are capable of degrading a wide variety of organic compounds, including recalcitrant compounds. These organisms can comprise 10 to 20% of the bacterial population.

Bacillus: About as prevalent as the pseudomonads, bacilli are characterized as Gram-positive aerobic organisms that produce endospores. This genus is heterotrophic and diverse. Bacilli often comprise 10% of the bacterial population.

4.1.1.5 Distribution and Function

Bacteria are the most abundant soil organisms, with a biomass of about 500 kg ha^{-1} to the depth of the root zone. Generally, aerobes are more prevalent than anaerobes, particularly in the A horizon (see Section 2.2.2 on page 10). As we decrease in depth, the number of anaerobes increases relative to aerobes.

Bacteria are critically involved in almost all soil biochemical transformations, including the metabolism of both organics and inorganics. The importance of soil bacteria in the fate and mitigation of pollutants cannot be overestimated. Because of their prevalence and diversity, as well as fast growth rates and adaptability, they have an almost unlimited ability to degrade most natural products and many xenobiotics. We will examine, in detail, the influence of bacteria on waste disposal and pollution mitigation in the succeeding chapters of this text.

4.1.2 Actinomycetes

These are organisms that technically are classified as bacteria, but are unique enough to be discussed as an individual group. They have some characteristics in common with bacteria, but are also similar in some respects to fungi. For the most part they are aerobic, chemoheterotrophic organisms consisting of elongated single cells. They display a tendency to branch into filaments, or hyphae, that resemble fungal mycelia; these hyphae are morphologically similar to those of fungi, but are smaller in diameter (about 0.5–2 μm). The total number of actinomycetes in soils is often about 10^7 per gram of soil. Generally, the population of actinomycetes is 1 to 2 orders of magnitude less than that of other bacteria in soil. They are not known to reproduce sexually, but all produce asexual spores called **conidia**. The genus *Streptomyces* dominates the actinomycete population, and these Gram-positive organisms may represent 90% of the total actinomycete population.

4.1.2.1 Similarities to Bacteria and Fungi

Like all bacteria, actinomycetes are prokaryotic organisms. In addition, the adenine-thymine and guanine-cytosine contents of bacteria and actinomycetes are similar, as are the cell wall constituents of both types of organisms. Actinomycete filaments are also about the same size as those of bacteria.

Like fungi, however, actinomycetes display extensive mycelial branching, and both types of organisms form aerial mycelia and conidia. Moreover, growth of actinomycetes in liquid culture tends to produce fungus-like clumps or pellets rather than the turbidity produced by bacteria. Finally, growth rates in fungi and actinomycetes are not exponential as they are in bacteria; rather, they are cubic.

4.1.2.2 Distribution and Function

Actinomycetes can metabolize a wide variety of organic substrates, including organic compounds that are not usually metabolized, such as phenols and steroids. Eq. (T4-13) on page 39 is an example of the degradation of cyanide by *Streptomyces*. They are also important in the metabolism of heterocyclic compounds such as complex nitrogen compounds and pyrimidines. The breakdown products of their metabolites are frequently aromatic, and these metabolites are important in the formation of humic substances and soil humus. The earthy odor associated with most soils is due to **geosmin**, a compound produced by actinomycetes.

Actinomycetes often comprise about 10% of the total bacterial population, with a biomass of about 500 kg ha^{-1} to the depth of the root zone. More tolerant of alkaline soils ($pH > 7.5$) but less tolerant of acidic soils ($pH < 5.5$), actinomycetes are also more tolerant of low moisture contents than other bacteria. Because of this and their tolerance of alkaline soils, actinomycete populations tend to be higher in desert soils, such as the southwestern USA.

Many actinomycetes interact with plants in symbiotic and pathogenic associations. For example, the genus *Frankia* initiates root nodules with nonleguminous, nitrogen-fixing plants, whereas the species *Streptomyces scabies* is the causative agent of potato scab. On the other hand, many streptomycetes produce antibiotics, including streptomycin, chloramphenicol, tetracycline, and cycloheximide. These compounds can be important in the control of other soil organisms.

4.1.3 Fungi

The fungi are the third major group of soil organisms. However, they differ from bacteria and actinomycetes in that they are eukaryotic. We refer to fungal organisms colloquially as molds, mildews, rusts, yeasts, or mushrooms. They are all heterotrophic, and most are aerobic—the exception is yeasts, which are fermenting organisms. Although fungi are eukaryotic, they contain no chlorophyll; therefore, they do not photosynthesize. Among the fungi in soil, the filamentous molds are most critically involved in the degradation of organic substrates and hence in the fate and mitigation of pollution. These fungi are characterized by extensive branching and mycelial growth, as well as by the production of sexual and asexual spores. Ascomycetes and Basidiomycetes are known to produce sexual spores in specialized fruiting structures. The Deuteromycetes are also complex fungi, but they are not known to reproduce sexually. The Deuteromycetes are also known as the *Fungi imperfecti* because they lack sexual reproduction; however, a sexual stage is quite often discovered later, in which case the organism must be reclassified into other known classes. Some of the most common genera of soil fungi are *Penicillium, Aspergillus, Fusarium, Rhizoctonia, Alternaria*, and *Rhizopus*.

Table 4-2 Characteristics of bacteria, actinomycetes, and fungi.

Parameter	Bacteria	Actinomycetes	Fungi
Population	Most numerous	Intermediate	Least numerous
Biomass	Bacteria and actinomycetes have similar biomass		Largest biomass
Degree of branching	Slight	Filamentous, but some fragment to individual cell	Extensive filamentous forms
Aerial mycelium	Absent	Present	Present
Growth in liquid culture	Yes—turbidity	Yes—pellets	Yes—pellets
Growth rate	Exponential	Cubic	Cubic
Cell wall	Murein, teichoic acid, and lipopolysaccharide	Murein, teichoic acid, and lipopolysaccharide	Chitin or cellulose
Complex fruiting bodies	Absent	Simple	Complex
Competitiveness for simple organics	Most competitive	Least competitive	Intermediate
Fix N	Yes	Yes	No
Aerobic	Aerobic, anaerobic	Mostly aerobic	Aerobic except yeast
Moisture stress	Least tolerant	Intermediate	Most tolerant
Optimal pH	6–8	6–8	6–8
Competitive pH	6–8	>8	<5
Competitiveness in soil	All soils	Dominate dry, high-pH soils	Dominate low-pH soils

4.1.3.1 Distribution and Function

As measured by plate counts, the populations of fungi are of the order of 10^6 per gram of soil, although such estimates are biased toward sporulating species. The diameter of fungal hyphae can be 10–50 μm. This size, which allows us to distinguish them morphologically from the smaller actinomycetes, results in a total biomass of about 1500 kg ha^{-1}. Thus their biomass is greater than that of the bacteria and actinomycetes, even though they are numerically less prevalent in most soils.

Fungi are heavily involved in the degradation of soil organics. As a group, they contain extremely diverse enzyme systems and are therefore competitive for simple sugars, organic acids, and complex compounds such as cellulose or lignin. Fungi, then, are very important in controlling the ultimate fate of organic pollutants added to soil. They are not involved directly in the fate of inorganic pollutants.

Because they are more tolerant of acidic soils ($pH < 5.5$) than are bacteria or actinomycetes, the fungi would be expected to be more actively involved in the degradation of organics in acidic soils. (Note that the optimal pH for most fungi is actually between 6 and 8 in pure culture, as shown in Table 4-2.) However, in soils, they are more competitive than other organisms in low-pH soils. Like the filamentous actinomycetes, fungi are also tolerant of low soil moisture.

Fungi form pathogenic associations with plants, as we see in the *Fusarium spp.*, for example, and also form beneficial associations with almost all plants through the mycorrhizal fungi. Fungal organisms can produce beneficial antibiotics for humans, such

Applied Theory: Biological Generation of Energy

Biological activity requires energy, and all microorganisms generate energy. This energy is subsequently stored as **adenosine triphosphate** (ATP), which can then be utilized for growth and metabolism as needed, subject to the second law of thermodynamics.

The Second Law of Thermodynamics:

> *In a chemical reaction, only part of the energy is used to do work. The rest of the energy is lost as entropy.*

Gibbs free energy ΔG is the amount of energy available for work for any chemical reaction. For the reaction $A + B \rightleftharpoons C + D$, the thermodynamic equilibrium constant is defined as:

$$K_{eq} = \frac{[C][D]}{[A][B]} \quad \text{(T4-1)}$$

Case 1 If product formation is favored:

That is, if

$$[C][D] > [A][B] \quad \text{(T4-2)}$$

then $K_{eq} > 1$ and $\ln K_{eq}$ is positive. For example, if $K_{eq} = 2.0$, then

$$\ln K_{eq} = 0.69 \quad \text{(T4-3)}$$

Case 2 If product formation is *not* favored:

$$[C][D] < [A][B] \quad \text{(T4-4)}$$

then $K_{eq} < 1$ and $\ln K_{eq}$ is negative. For example, if $K_{eq} = 0.20$, then

$$\ln K_{eq} = -1.61 \quad \text{(T4-5)}$$

The relationship between the equilibrium coefficient K_{eq} and the free energy ΔG is given by

$$\Delta G = -RT \ln K_{eq} \tag{T4-6}$$

where R is the universal gas constant and T is the absolute temperature (K).

The following table illustrates the effect of the Gibbs Free Energy on the spontaneity of a chemical reaction and the resulting release of energy.

ΔG	K_{eq}	$\ln K_{eq}$	Energy Status in Reaction
negative	$K_{eq} > 1$	positive	energy is released, reaction proceeds spontaneously
positive	$0 < K_{eq} < 1$	negative	energy must be added to promote the reaction

Thus, we can use ΔG values for any biochemical reaction mediated by microbes to determine whether energy is liberated for work, and how much energy is liberated.

Soil organisms can generate energy via several mechanisms, which can be divided into two main categories:

1. Photosynthesis

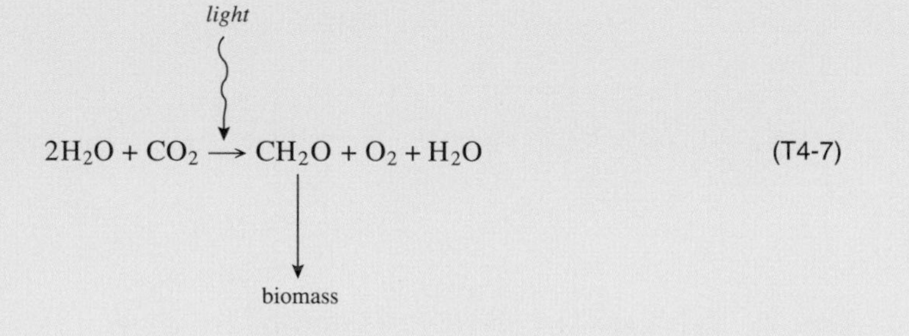

$$2H_2O + CO_2 \xrightarrow{light} CH_2O + O_2 + H_2O \tag{T4-7}$$

$$\Delta G \approx +481 \text{ kJ mol}^{-1}$$

For this reaction, energy supplied by sunlight is necessary. The fixed organic carbon is then used to generate energy via respiration. Examples of soil organisms that undergo photosynthesis are *Rhodospirillum*, *Chromatium*, and *Chlorobium*.

2. Respiration

(a) Aerobic, heterotrophic respiration: Many soil organisms undergo aerobic, heterotrophic respiration, for example, *Pseudomonas* and *Bacillus*.

$$C_6H_{12}O_6 + 6O_2 \longrightarrow 6CO_2 + 6H_2O \quad \text{(T4-8)}$$

$$\Delta G = -2870 \text{ kJ mol}^{-1}$$

(b) Aerobic, autotrophic respiration: The reactions carried out by *Nitrosomonas* (Eq. T4-9) and *Nitrobacter* (Eq. T4-10) are known as **nitrification**:

$$NH_3 + \tfrac{3}{2}O_2 \longrightarrow HNO_2 + H_2O \quad \text{(T4-9)}$$

$$\Delta G = -280 \text{ kJ mol}^{-1}$$

$$KNO_2 + \tfrac{1}{2}O_2 \longrightarrow KNO_3 \quad \text{(T4-10)}$$

$$\Delta G = -73.2 \text{ kJ mol}^{-1}$$

The following two reactions carried out by *Beggiatoa* (Eq. T4-11) and *Thiobacillus thiooxidans* (Eq. T4-12) are examples of **sulfur oxidation**:

$$2H_2S + O_2 \longrightarrow 2H_2O + 2S \quad \text{(T4-11)}$$

$$\Delta G = -350 \text{ kJ mol}^{-1}$$

$$2S + 3O_2 + 2H_2O \longrightarrow 2H_2SO_4 \quad \text{(T4-12)}$$

$$\Delta G = -992 \text{ kJ mol}^{-1}$$

The next reaction involves the degradation of cyanide by *Streptomyces*:

$$2KCN + 4H_2O + O_2 \longrightarrow 2KOH + 2NH_3 + 2CO_2 \quad \text{(T4-13)}$$

$$\Delta G = -230 \text{ kJ mol}^{-1}$$

All of the above reactions illustrate how soil organisms mediate reactions that can cause or negate pollution. For example, nitrification and sulfur oxidation can result in the production of specific pollutants, *i.e.*, nitrate and sulfuric acid, whereas the destruction of cyanide is obviously beneficial with respect to the mitigation of pollution.

(c) Facultative anaerobic, heterotrophic respiration: The bacterium *Pseudomonas denitrificans* is capable of this kind of metabolism utilizing nitrate as a terminal electron acceptor rather than oxygen. Note that this bacterium can use oxygen as a terminal electron acceptor if it is available, and that aerobic respiration is more efficient than anaerobic respiration.

$$5C_6H_{12}O_6 + 24KNO_3 \longrightarrow 30CO_2 + 18H_2O + 24KOH + 12N_2 \quad \text{(T4-14)}$$

$$\Delta G = -150 \text{ kJ mol}^{-1}$$

(d) Facultative anaerobic autotrophic respiration: *Thiobacillus denitrificans*

$$S + 2KNO_3 \longrightarrow K_2SO_4 + N_2 + O_2 \quad \text{(T4-15)}$$

$$\Delta G = -280 \text{ kJ mol}^{-1}$$

The reactions in Eqs. T4-14 and T4-15 are known as **denitrification**.

(e) Anaerobic heterotrophic respiration: *Desulfovibrio*

$$\underset{\textit{Lactic acid}}{CH_3CHOHCOOH} + SO_4^{2-} \longrightarrow \underset{\textit{Acetic acid}}{2CH_3COOH} + HS^- + H_2CO_3 + HCO_3^- \quad \text{(T4-16)}$$

$$\Delta G = -170 \text{ kJ mol}^{-1}$$

Soil organisms can also engage in fermentation, which is also an anaerobic process. But fermentation is not widespread in soil. Overall, there are many ways in which soil organisms can and do generate energy. The above-listed mechanisms illustrate the diversity of soil organisms and explain the ability of the soil community to break down or transform almost any natural substance. In addition, enzyme systems have evolved to metabolize complex molecules, be they organic or inorganic. These enzymes can also be used to degrade xenobiotics with similar chemical structures. Xenobiotics, which do not degrade easily in soil, are normally chemically different from any known natural substance. Hence soil organisms have not evolved enzyme systems capable of metabolizing such compounds. There are many other factors that influence the breakdown or degradation of chemical compounds. These will be discussed in Chapter 7, beginning on page 77.

as penicillin, which was discovered by Sir Alexander Fleming (1881–1955). However, they are also capable of causing disease, and the diseases they cause can be difficult to treat. (As both fungi and humans are eukaryotic, fungal metabolism is similar to that of humans, and hence, it is difficult to eradicate fungi in a human host.) For example, the genus *Coccidioides* causes the chronic pulmonary disease of valley fever in the southwestern deserts of the United States.

We can view soil organisms in total as a vast biological entity whose parts live in unison, with diverse capabilities for the degradation of all natural organics and many xenobiotics. In Table 4-2 on page 36, where the major characteristics of bacteria, actinomycetes, and fungi are compared, we can begin to see how the parts of this entity fit together.

4.2 Soil Factors Affecting the Growth and Activity of Soil Microbes

In Chapter 2, beginning on page 9, we discussed the abiotic characteristics of soil. Using this framework, we can now look at this environment from a soil microbial perspective in order to understand the factors that limit microbial activity.

4.2.1 Biotic Stress

Since indigenous soil microbes are in competition with one another, the presence of large numbers of organisms results in biotic stress factors. Competition can be for substrate, water, or growth factors. In addition, microbes can secrete inhibitory or toxic substances, including antibiotics, that harm neighboring organisms. Finally, many organisms are predatory or parasitic on neighboring microbes. For example, phages infect both bacteria and fungi. Because of biotic stress, nonindigenous organisms that are introduced into a soil environment often survive for fairly short periods of time (days to several weeks). This effect has important consequences for pathogens (Chapter 19, beginning on page 279) and for other organisms introduced to aid biodegradation (Section 11.4.1.5, beginning on page 164). This process of adding organisms to assist biodegradation is called **bioaugmentation**.

4.2.2 Abiotic Stress

4.2.2.1 *Light*

Soil is impermeable to light, that is, no sunlight penetrates beyond the top few centimeters of the soil surface. Phototrophic organisms are therefore limited to the top few centimeters of soil. At the surface of the soil, however, such physical parameters as temperature and moisture fluctuate significantly throughout the day and also seasonally. Hence most soils tend to provide a harsh environment for photosynthesizing organisms. A few phototrophic organisms, including algae, have the ability to switch to a heterotrophic respiratory mode of nutrition in the absence of light. Such "switch-hitters" can be found at significant depths within soils. Normally, these organisms are not competitive with other indigenous heterotrophic organisms for organic substrates.

4.2.2.2 *Soil Moisture*

Typically, soil moisture content varies considerably in any soil, and soil organisms must adapt to a wide range of soil moisture contents. (See Section 2.4.2, beginning on page 17, for more information on soil moisture content and how it relates to water availability.) Soil aeration is dependent on soil moisture: saturated soils tend to be anaerobic, whereas dry soils are usually aerobic. But soil is a heterogeneous environment; even saturated soils contain pockets of aerobic regimes, and dry soils harbor anaerobic microsites that exist within the centers of secondary aggregates. Although the bacteria are the least tolerant of low soil moisture, as a group they are the most flexible with respect to soil aeration. They include aerobes, anaerobes, and facultative anaerobes, whereas the actinomycetes and fungi are predominantly aerobic.

4.2.2.3 *Soil Temperature*

Soil temperatures vary widely, particularly near the soil surface. Most soil populations are resistant to wide fluctuations in soil temperature although soil populations can be psychrophilic, mesophilic, or thermophilic, depending on the geographic location of the soil. Most soil organisms are mesophilic because of the buffering effect of soil on temperature, particularly lower in the soil profile.

Table 4-3 Redox potential at which soil substrates are reduced. See also Figure 7-13 on page 87.

Redox Potential (mV)	Reaction	Type of Organism
+800	$O_2 \longrightarrow H_2O$	Aerobes
+740	$NO_3 \longrightarrow N_2, N_2O$	Facultative anaerobes
–220	$SO_4 \longrightarrow S^{2-}$	Anaerobes
–300	$CO_2 \longrightarrow CH_4$	Anaerobes

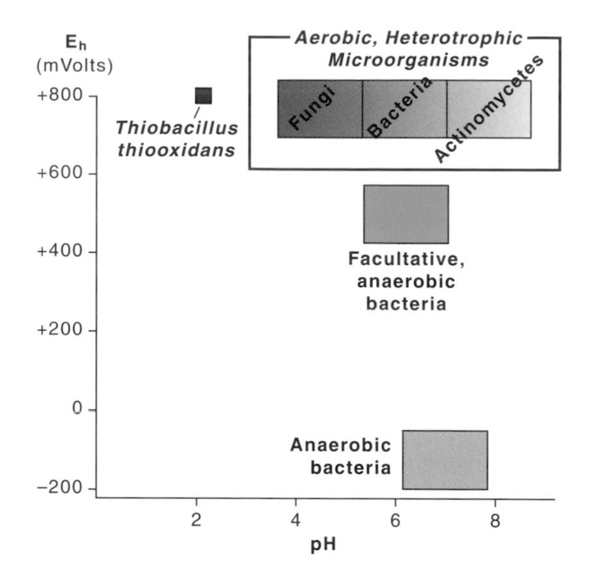

Figure 4-4 Optimal E_h/pH relationships for various soil organisms.

4.2.2.4 Soil pH

Undisturbed soils usually have fairly stable soil pH values within the range of 6–8, and most soil organisms have pH optima within this range (see Section 2.2.6 and Table 2-2 on page 14). There are, of course, exceptions to this rule, as exemplified by *Thiobacillus thiooxidans*, an organism that oxidizes sulfur to sulfuric acid and has a pH optimum of 2–3. Microsite variations of soil pH can also occur due to local decomposition of an organic residue to organic acids. Here again, we see that soil behaves as a heterogeneous or discontinuous environment, allowing organisms with differing pH optima to coexist in close proximity. Normally, soil organisms are not adversely affected by soil pH unless drastic changes occur. Drastic change can happen, for example, when lime is added to soil to increase the pH or when sulfur is added to decrease the pH.

4.2.2.5 Soil Texture

Almost all soils contain populations of soil organisms regardless of the soil texture (see Section 2.2.3 on page 10). Even soils whose textures are extreme, such as pure sands or clays, usually contain populations of microbes, albeit in lower numbers than in soils with less extreme textures. Most nutrients are associated with clay or silt particles, which also retain soil moisture efficiently. Thus soils with at least some silt or clay particles offer a more favorable habitat for organisms than do soils without these materials.

4.2.2.6 Soil Carbon and Nitrogen

Carbon and nitrogen are both nutrients that are found in soils. Since both of these nutrients are present in low concentrations, the growth and activity of soil organisms is limited. In fact, many organisms exist in soil under limited starvation conditions and hence are dormant. Without added substrate or amendment, soil organisms generally metabolize at low rates. The major exception to this is the rhizosphere, whose root exudates maintain a high population level. In most cases, all available soil nutrients are immediately utilized. Soil humus represents a source of organic nutrients that is mineralized slowly by autochthonous organisms. Similarly, specific microbial populations can utilize xenobiotics as a substrate, even though the rate of degradation is sometimes quite low. This is the basic mechanism of biodegradation (see Section 7.2, beginning on page 77).

4.2.2.7 Soil Redox Potential

Redox potential (E_h) is the measurement of the tendency of an environment to oxidize or reduce substrate. In a sense, we can think of it as the availability of different terminal electron acceptors that are necessary for specific organisms. Such electron acceptors exist only at specific redox potentials, which are measured in millivolts (mV). An aerobic soil, which is an oxidizing environment, has a redox potential or E_h of +800 mV; an anaerobic soil, which is a reducing environment, has an E_h of about 0 to –300 mV. Oxygen is found in soils at a redox potential of about +800 mV. When soil is placed in a closed container, oxygen is used by aerobic organisms as a terminal electron acceptor until all of it is depleted. As this process occurs, the redox potential of the soil decreases, and other compounds can be used as terminal electron acceptors. Table 4-3 illustrates the redox potential at which various substrates are reduced, and the activity of different types of organisms in a soil. (See also Figure 7-13 and Section 7.4.2, beginning on page 87.)

The fact that different terminal electron acceptors are available for various organisms having diverse pH requirements means that some soil environments are more suitable than others for various groups of organisms. Figure 4-4, which illustrates optimal E_h / pH relationships for various groups of organisms, shows that redox potential affects the activity of all organisms.

4.3 Overall State and Activity of Microbes in Soil

Is the soil environment a favorable one for soil microbes? There are actually two possible answers to this question. On the one hand, soil is a very harsh environment; on the other hand, soil contains very large populations of microbes. How can this be? It appears that soil organisms are well adapted to this harsh environment, so they must have mechanisms to exploit the resources available. These mechanisms also allow organisms to survive long periods of time when resources are not available. Viewing soil as a community having large numbers of organisms and great microbial diversity, we can infer that each species has a habitat and a niche, rather like a home and a job. Because soil is a heterogeneous environment, many different kinds of organisms can coexist (see Figure 2-3 on page 13). Diversity also ensures that all available nutrients are utilized in soil.

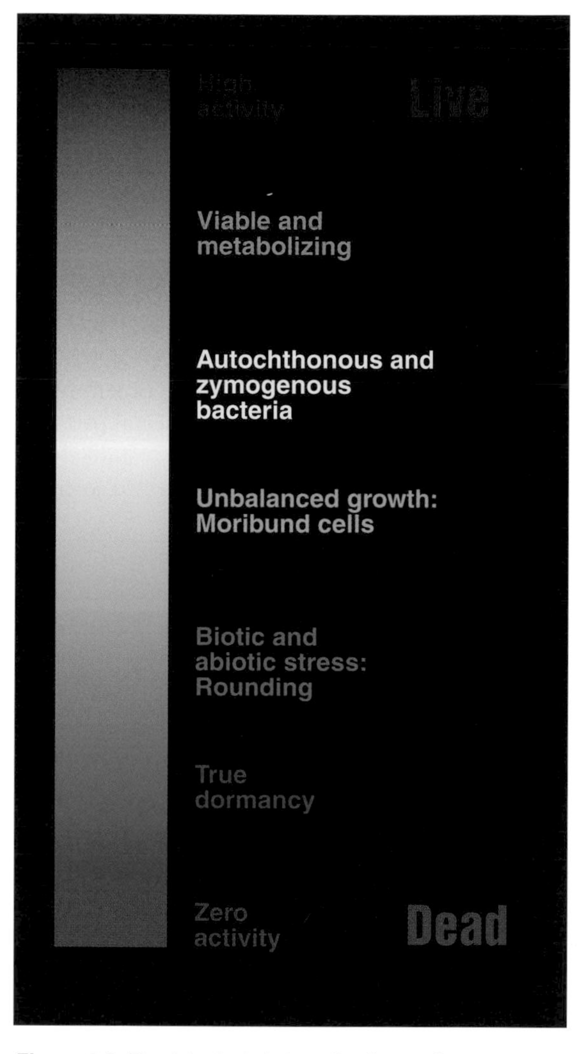

Figure 4-5 Physiological states of soil organisms.

Owing to the harsh physical soil environment and the fact that nutrients are usually limiting, soil organisms do not actively metabolize most of the time. In fact, they exist under stress, so they may exist injured or even be killed. Figure 4-5 shows the various potential states of soil organisms, the two extremes being live microbes and dead ones. Between these two extremes, other physiological states are possible, including metabolically active and dormant states. Because many organisms are in fact injured in soil and have diverse, specific nutritional needs, many soil organisms cannot be cultured by conventional methods. These are the so-called **viable but**

nonculturable organisms. [*Note*: In practice, perhaps 99% of all soil organisms may be nonculturable (Roszak and Colwell, 1987). Thus, any methodology that relies on obtaining soil organisms via a culturable procedure may in fact be sampling a very small subsection of the soil population.]

Based on our discussion of the soil as an environment for microbes, we may reasonably infer indigenous organisms within a particular soil are selected by the specific environment in that soil. Indigenous organisms are often capable of surviving wide variations in particular environmental parameters. Introduced organisms, however, are unlikely to be as well adapted and cannot be expected to compete with indigenous organisms unless a specific niche is available.

References and Recommended Reading

Atlas R.M. and Bartha, R. (1993) *Microbial Ecology*, 3rd Edition. The Benjamin/Cummings Publishing Company, New York.

Paul E.A. and Clark F.E. (1988) *Soil Microbiology and Biochemistry*. Academic Press, San Diego, California.

Roszak D.B. and Colwell R.R. (1987) Survival strategies of bacteria in the natural environment. *Microbiological Reviews*. **51**, 365–379.

Problems and Questions

1. For the chemical rection

$$A + B \rightleftharpoons C + D$$

the thermodynamic equilibrium constant $K_{eq} = 0.38$. Deduce whether ΔG for the reaction is negative or positive. Is energy liberated from the reaction, or must energy be added to promote the reaction?

2. For the following organisms, identify the substrate that can be oxidized as well as the terminal electron acceptor utilized by the organism:

(a) *Nitrosomonas spp.*

(b) *Thiobacillus thiooxidans*

(c) *Thiobacillus denitrificans*

(d) *Desulfovibrio spp.*

3. Briefly discuss the statement: "Soil is a favorable environment as a habitat for microorganisms."

4. Briefly discuss the statement: "Soil is a favorable environment for metabolizing autochthonous and zymogenous organisms."

5. What are the major factors that limit microbial activity in most soils?

6. Which of the following organisms are most likely to be involved in the degradation of organic compounds: *Pseudomonas, Nitrosomonas, Thiobacillus*, filamentous fungi, actinomycetes, *Bacillus*? Give a rationale for each of your answers.

Chapter 5

Physical Processes Affecting Contaminant Fate and Transport in Soil and Water

P.J. Wierenga

An understanding of the physical properties of a soil is indispensable when developing mathematical models that accurately describe the rate of leaching of a pollutant. Photograph: J.W. Brendecke.

5.1 Water in Soil and Groundwater

5.1.1 Water in Soil

Water is omnipresent. It is in the atmosphere, soil, and groundwater, as well as in oceans, lakes, and rivers. As the sun warms the oceans and lakes, water evaporates to the atmosphere, falls as precipitation (rain or snow), infiltrates the upper crust of the earth (soil), and leaches downward to groundwater. From there, it seeps back into rivers and lakes or is pumped up to be used for municipal water supplies or crop irrigation. Finally, river water and excess effluent water from urban and rural areas may again reach oceans, completing the cycle. This process, which is illustrated in Figure 5-1 on page 46, is called the hydrologic cycle. In this chapter we concentrate on water in soil and groundwater.

We can readily observe water in soil by drilling a vertical borehole or excavating a pit (Figure 5-2 on page 47). In the upper part of the borehole or pit, soil may be moist, that is, the soil obviously contains water, but not so much that it runs freely out of the wall of the pit. Near the soil surface, only the finer pores are filled with water, and the soil is said to be **unsaturated**. Deeper in the borehole, water may accumulate and we may observe standing water. This water is called **groundwater**, and the elevation or level of the water in the borehole is called the **groundwater table**. Below the groundwater table the pore space in the soil is completely filled with water, and the soil is **saturated**. The region between the groundwater table and ground surface is called the **unsaturated zone**. We generally use the term vadose zone for the region between the groundwater table and the lower end of the rooting zone.

The groundwater depth varies from site to site, depending, to a large extent, on the ratio of precipitation to evaporation. In areas where rainfall is high and evaporation low or moderate, the depth of the groundwater table may vary from less than 0.5 m to 3 m or more. In arid areas, the groundwater depth may be 100 m or more. However, even in arid areas there may be many specific sites, such as river basins, that have shallow water tables. Thus for each specific location, groundwater depth is different. Groundwater depth is an important factor in pollu-

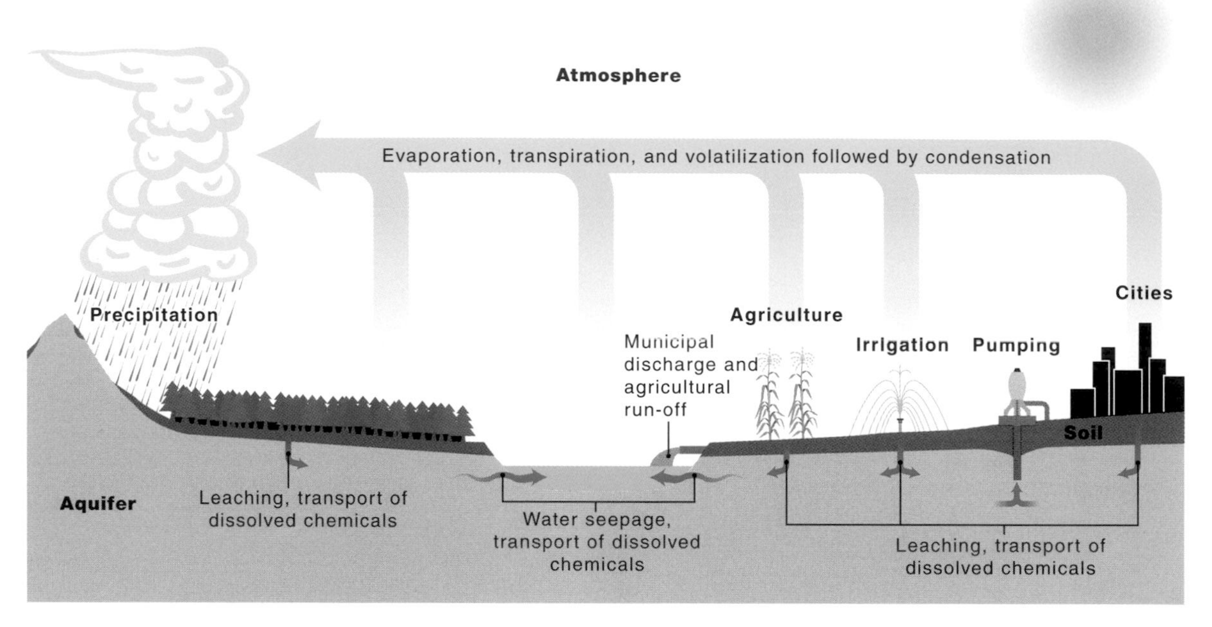

Figure 5-1 The hydrologic cycle showing the interactions between air, soil, groundwater, and surface water. Adapted from Maidment, D.R., Editor, *Handbook of Hydrology*; copyright © 1993 by the McGraw-Hill Company. Material adapted by permission of the McGraw-Hill Company.

tion. In areas with high rainfall and shallow water tables, for example, the short distance between the soil surface and groundwater means that groundwater can become polluted rapidly. In arid areas with low rainfall and deep water tables, groundwater pollution occurs less quickly.

Nearly all land areas are underlain by groundwater in **aquifers**, which are permeable geologic units that can store and transmit significant quantities of water. Aquifers can be relatively small, from a few square kilometers, or very large, up to several thousand square kilometers. An example of a very large aquifer is the Ogalalla aquifer, which underlies large areas of North Texas, Kansas, and Nebraska. Water is pumped from this aquifer for agricultural, municipal, and industrial purposes.

When water percolates from the surface through the unsaturated zone to the groundwater, aquifers become replenished or recharged. As a result of the gravitational pull, the direction of water movement through the unsaturated zone is mostly vertical. However, water movement in aquifers (see Figure 5-1) is mostly horizontal and results from differences in pressure or elevation. For example, if large amounts of water are pumped from one end of an aquifer, the water table there will decrease; this decrease causes water to slowly flow away from the other end of the aquifer where the water table is higher.

5.1.2 Soil Water Potential

When we measure the water held in the soil through the sides of a pit (see Figure 5-2), we see that, under static conditions, the amount of water in the soil increases from the surface down to the groundwater. This "bottoms-up" increase in water content is primarily the result of capillary action. Near the water table, the larger pore spaces and the large-diameter capillary spaces are filled with water (see Section 2.3.2, beginning on page 16). Farther above the water table, the larger pores are empty, and water is held only in the finer pores. Capillary action is demonstrated in Figure 5-3 on page 48, which shows that water rises higher in narrow pores than in wider pores. The capillary rise of water in pores of different diameter is explained by the capillary rise equation:

$$h = \frac{2s \cos \alpha}{Dgr} \quad (5\text{-}1)$$

where h is the capillary rise (m) of water in a glass tube, r is the tube diameter or pore radius (m), D is the density of water (Mg m^{-3}), g is the acceleration due to gravity, α is the contact angle (degrees) between the glass and the water, and s is surface tension of the water (N m^{-1}). For soil-air interfaces, $\alpha = 0$, $\cos \alpha = 1$, the density of water is 0.998 Mg m^{-3}, $s = 0.0728$ N m^{-1} at 20°C, and $g = 9.81$ J kg^{-1} m^{-1}. With these values, Eq. (5-1) can be reduced to

$$h = \frac{3 \times 10^{-5}}{d} \quad (5\text{-}2)$$

where d is pore diameter in meters. Although Eqs. (5-1) and (5-2) do not strictly apply to soils—in soils we never see straight, clean capillary pores—the analogy to pores is useful. For example, Eq. (5-2) explains why capillary rise decreases as pore diameter increases, and vice versa. This phenomenon has been observed in field soils, where coarse, sandy soils (with large pores) exhibit a much shorter capillary rise than do fine-textured loam soils (with fine pores).

The capillary action of soils is one of the main reasons why water is held in unsaturated soils and does not come gushing out of the pit wall above the water table (Figure 5-2). In addition, water is often held tightly in unsaturated soil because it is adsorbed on soil particle surfaces. Water molecules are dipolar; that is, they have both a negative and a positive charge. Thus they can attach strongly to soil surfaces, which are also electrically charged. In fact, there are usually several molecular layers of water around soil particles. The joint action of capillarity and adsorption also explain why water is held in a moist sponge: a slight pressure on the sponge forces water out of the larger pores while a stronger squeeze causes the finer pores to lose water. Similarly, we have to apply energy, by squeezing or by suction, to get water out of unsaturated soil. In the case of the soil above the groundwater table of Figure 5-2, we would have to apply suction to the exposed soil surface to obtain a water sample. It does not come out freely on its own.

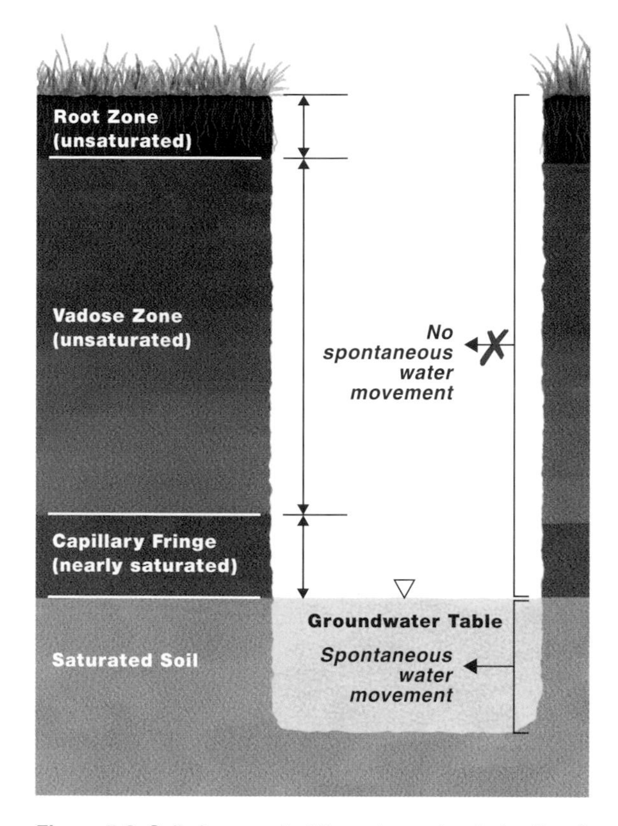

Figure 5-2 Soil pit excavated through unsaturated soil and into saturated soil. Note the groundwater in the bottom of the pit and the level of the groundwater.

The energy status of water in soil—for example, how much effort it takes to move water in soil—is expressed through the **total soil-water potential energy**, often abbreviated to **soil-water potential**. The soil-water potential is formally defined as the mechanical work required to transfer water from soil to a standard reference state (*i.e.*, a container with water at the same elevation), where the total soil-water potential energy is zero by definition.

The soil-water potential is made up of three component potentials: the gravitational potential energy, the pressure potential energy, and the solute poten-

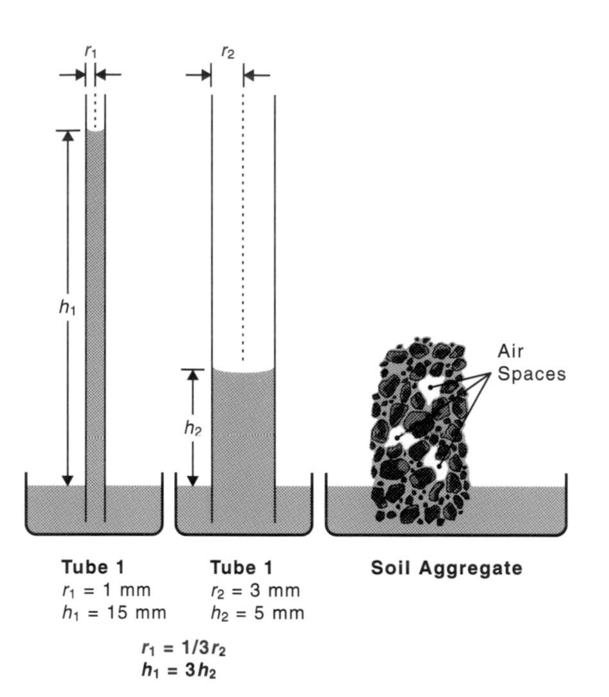

Figure 5-3 Capillary rise of water in capillaries of different size and in a soil aggregate. Adapted from Brady N.C. and Weil R.R., *The Nature and Properties of Soil*, 1996, page 147; reprinted by permission of Prentice Hall, Upper Saddle River, New Jersey.

tial energy. There are additional energy terms that contribute to the total energy status of water in soil, but these suffice for the present discussion:

- The **gravitational potential energy**, or **gravity potential**, results from the gravitational force field. Water that is situated at higher elevations in a soil profile or riverbed has greater potential energy than water lower in the soil profile or riverbed. Water flows from high to low potential energy; thus, water moves downward in a soil profile, and downstream in a riverbed, as a result of this difference in potential energy.

- The **pressure potential energy**, or **pressure potential**, results from capillary, adsorptive, and other forces in unsaturated soils. It indicates how tightly water is held in a soil. Its value is always negative in unsaturated soils (remember, it takes energy to remove water from unsaturated soil). In saturated soils, capillary forces are zero, but water below the water table surface is subject to pressure from the overlying water. As a result, the pressure potential has a positive value below the water table.

- The **solute potential energy**, or **solute potential**, results from the difference in potential energy between soil solution and pure free water. As soil solution contains a variety of dissolved minerals and nutrients, its energy status is decreased relative to pure water. This decrease in potential energy from dissolved minerals can be quite substantial, and accounts for the large amount of energy required to convert seawater to drinking water. Differences in solute potential between water in soil and water in plant roots cause water to flow into roots. Within roots the solute potential is much lower due to dissolved minerals and organic substances.

In discussions on water movement through soils and groundwater, we often ignore the driving forces for uptake by plant roots (solute potential differences). The remaining components of water movement are gravity and pressure potentials. These are generally expressed in pressure units (bars or megapascals) or length units (meters of a water column): 1 bar is approximately 0.1 MPa, which is approximately 10.2 m H_2O. Expressing these two component potentials in terms of length units, we obtain:

$$H = h + z \tag{5-3}$$

where H is the soil-water potential (*i.e.*, total soil-water potential energy in m H_2O), h is the pressure potential (m), and z is the gravity potential (m). If the component potentials are expressed in length units, the term **head** (which refers to distance) is often used. Equation (5-3) then reads as follows: Total head H equals pressure head h plus gravity head z. The difference in soil-water potential H between two locations in a soil indicates in which direction water may be moving.

5.1.3 Measurement of Soil Water

To determine the status of water in a soil, we can determine either the soil-water potential or the soil-water content.

The soil-water potential can be measured with a variety of instruments. One of the simplest instruments is a **tensiometer**, which measures the pressure

Figure 5-4 Tensiometer with handheld digital readout device, or Tensimeter™. After inserting the needle of the Tensimeter through the septum stopper, a transducer senses the negative pressure in the tensiometer and hence in the soil. Photograph: J.W. Brendecke.

potential (h). It consists of a plastic tube with a porous ceramic cup on one end. The tube is filled with water, closed with a rubber septum stopper, and installed horizontally or vertically in the soil (Figure 5-4).

The tensiometer shown in Figure 5-4 is installed above the water table where the soil is unsaturated. The pressure potential there is negative, so water will move from the tensiometer through the porous cup out into the soil. A slight vacuum (negative pressure), called tension, will develop in the air space at the top of the tensiometer. Water flow out of the tensiometer will continue until the pressure in the soil and that in the tensiometer are equal. At that time, we can determine the negative pressure or tension by using a pressure gauge attached to the tensiometer or an electronic pressure transducer. By adding the length of the water column in the plastic tensiometer tube to the gauge or transducer reading (in cm H_2O), we obtain a measure of the pressure potential of the soil water at the tensiometer cup.

The standard method for determining the water content of soil is to take soil samples, dry the samples in an oven, and calculate water contents by the difference between wet and dry soil mass. The **volumetric water content (θ)** is the volume of water per unit volume of soil ($m^3\ m^{-3}$). The **mass water content**, also called the gravimetric water content, is the mass of water per unit mass of dry soil ($kg\ kg^{-1}$). The volumetric water content is calculated from the mass water content by multiplying the latter by the soil bulk density ϱ (ϱ = mass of dry soil per volume of soil, $Mg\ m^{-3}$). Determining water content by sampling is both destructive and time-consuming. Moreover, this kind of sampling is sometimes impossible. For example, the soil below a landfill is not readily accessible.

Where we need to take repetitive measurements of water content, we may advantageously use indirect methods. The most common indirect technique for determining water contents in field soils is the **neutron probe** method. This method involves installing an access tube in the soil and lowering a radioactive source emitting fast neutrons through it. The count of neutrons slowed by hydrogen is a measure of the soil-water content. In newer methods for measuring soil water in the field, such as the **time domain reflectometry (TDR)** method, we employ the dielectric constant of water, which is much larger (*i.e.*, 60–80) than the dielectric constant of dry soil (*i.e.*, 2–5). The neutron probe and TDR are good methods for repeated measurement of soil moisture at the same location; they are particularly suitable for detecting changes in soil moisture below landfills or other waste-disposal facilities.

5.1.4 Relationships Between Soil-Water Potential and Soil-Water Content

The relationship between the amount of water held in soil and its energy status is known as the soil-water characteristic curve or soil-water retention curve.

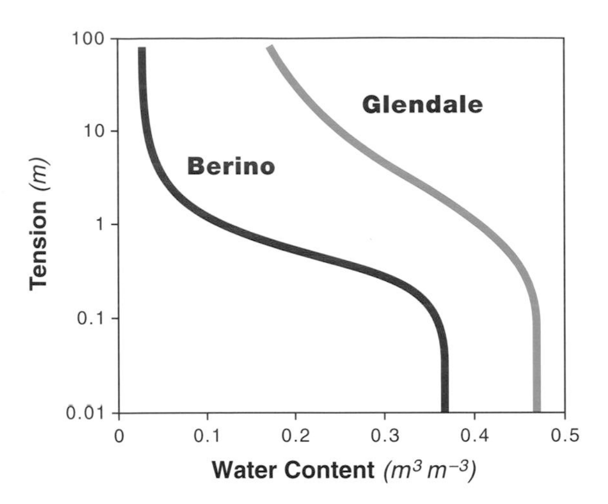

Figure 5-5 Soil-water retention curves for Berino loamy fine sand and Glendale clay loam. The Glendale clay loam contains more clay and silt, has finer pores, and holds more water at all tensions. Adapted from Wierenga, P.J., *Water and Solute Transport Storage*, 41–60, in the *Handbook of Vadose Zone Characterization & Monitoring*, Wilson, L.G., Everett, L.G., Cullen, S.J., Eds., Lewis Publishers, an imprint of CRC Press, Boca Raton, Florida, 1995. With permission.

This relationship is characteristic for each soil. Examples of water-retention curves for a loamy fine sand and for a clay loam are shown in Figure 5-5. In this figure we see that, as soil dries, its tension increases (pressure potential becomes more negative). The differences between the two curves, as shown in Figure 5-5, are significant. The clay loam has a much higher water content at saturation as well as at intermediate tensions. The water retention curve for the fine sand is very steep at tensions above 10 m.

Water-retention data are used extensively in modeling water movement through soil. For example, in order to make a computer model predict how fast water moves from the surface through the unsaturated zone to the groundwater, we need to know the soil-water retention properties of the soils through which the water moves. Such modeling exercises are often required for determining the impacts of major projects, such as regional landfills, on water and contaminant flow to groundwater.

When modeling water flow in unsaturated soils, it is most convenient to express the soil-water retention data mathematically. One of the most common

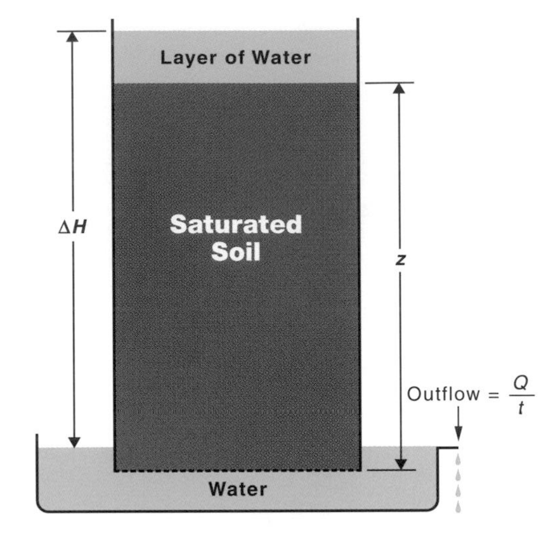

Figure 5-6 Saturated soil column with cross-sectional area *A*. By measuring the outflow, as well as ΔH and z, the saturated hydraulic conductivity can be calculated.

mathematical expressions for this purpose is the **van Genuchten equation**:

$$S_e(h) = \frac{\theta - \theta_r}{\theta_s - \theta_r} = [1 + (\alpha h)^n]^{-m} \tag{5-4}$$

where S_e is the effective saturation (dimensionless because it is equal to the ratio of θ / θ_s for very small θ_r); θ is the volumetric water content ($m^3\,m^{-3}$); θ_r is the residual water content ($m^3\,m^{-3}$), which is the water content at extreme dryness; θ_s is the saturated water content ($m^3\,m^{-3}$), which is the water content at full saturation; and α (m^{-1}), n (dimensionless), and m (dimensionless) are empirical parameters. This function fits observed data well for a wide range of soils.

5.2 Movement of Water in Soil and Groundwater

5.2.1 Saturated Flow

Early laboratory studies on water movement were conducted by packing soil in cylinders and letting water flow through them as illustrated in Figure 5-6. From measurements of inflow/outflow and of water pressure at the inlets/outlets of these containers, it was found that the volume of water Q moving through a soil column is proportional to the hydraulic head difference ΔH ($\Delta H = \Delta h + \Delta z$) between the inlet and outlet, the cross-sectional area A of the column, and the time t; and is inversely proportional to the length of the column z. Thus

$$Q = K \frac{\Delta H A t}{z} \tag{5-5}$$

where the proportionality constant K is called **hydraulic conductivity**. Equation (5-5) is known as **Darcy's law**, first formulated by the French engineer Henri Darcy in 1856. Practical units are ΔH in meters, A in m^2, t in days, z in meters, Q in m^3, and K in m day^{-1}. The experimental setup shown in Figure 5-6 is used to determine saturated hydraulic conductivity of subsurface materials. For example, if $z = 0.50$ m, $\Delta H = 0.60$ m, $A = 0.01$ m^2, $t = 1$ day, and $Q = 0.012$ m^3, then

$$0.012 = K\left(\frac{0.60 \cdot 0.01 \cdot 1}{0.50}\right)$$

and

$$K = 1 \text{ m day}^{-1}$$

Exploratory bore holes may be drilled to determine the suitability of an aquifer for a water well or to take a soil or water sample. Often, soil removed during such drilling is packed in a cylinder and its K value determined. If K is found to be high (*i.e.*, >10 m day^{-1}) the site may be suitable for a water well. But if K is low (*i.e.*, <0.10 m day^{-1}), the site may not be suitable for a well.

5.2.2 Unsaturated Flow

Darcy's law is also used to describe water flow in unsaturated soils. The only difference is that in unsaturated soils the hydraulic conductivity is not constant, but decreases rapidly with decreasing water content. This is because as soil loses water, pores empty (drain) and the cross-sectional area for flow decreases, with large pores draining more easily than small pores. Thus sandy soils, which drain more rapidly, show a greater decrease in hydraulic conductivity with decreasing water content than clay soils. Because clay soils have more pores of intermediate size and drain more slowly, the hydraulic conductivity of

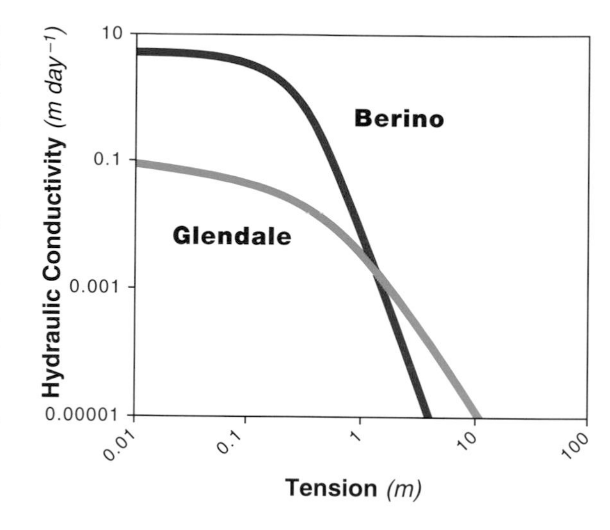

Figure 5-7 Hydraulic conductivity versus tension for Berino loamy fine sand and Glendale clay loam. Note that at tensions above approximately 1.5 m, the Glendale clay loam has a higher conductivity than the sand. Adapted from Wierenga, P.J., *Water and Solute Transport Storage*, 41–60, in the *Handbook of Vadose Zone Characterization & Monitoring*, Wilson, L.G., Everett, L.G., Cullen, S.J., Eds., Lewis Publishers, an imprint of CRC Press, Boca Raton, Florida, 1995. With permission.

clay soils decreases less rapidly with decreasing water content. An example of changes in hydraulic conductivity with tension is shown in Figure 5-7.

The following equation describes the dependence of the unsaturated hydraulic conductivity $K(h)$ on pressure potential (h):

$$K(h) = K e^{\alpha h} \tag{5-6}$$

where α (m^{-1}) is a soil constant. Values for the saturated hydraulic conductivity K and α must be determined by fitting Eq. (5-6) to measured values of unsaturated conductivity and pressure potential.

Equation (5-6) shows a strong dependence of $K(h)$ on tension, and hence on water content, because water content and tension are related. As a result of this dependency, a relatively small decrease in water content or tension may cause a large decrease in hydraulic conductivity. For example, if the Berino soil shown in Figure 5-7 were to dry so that its tension increased from 0.6 (quite moist) to 5 m (quite dry), its hydraulic conductivity would decrease from 0.1 to

Figure 5-8 Lysimeters are indispensable in the controlled *in situ* study of solute movement through an actual column of soil. In (a) and (b) a lysimeter is lowered into the study site with the upper surface level with the ground. Panel (c) shows the lysimeter in position in the ground. Tensiometers and solution collection probes dot the side of the lysimeter now filled with soil. The entire lysimeter rests on a scale enabling accurate measurements of water content changes in the soil caused in particular by water lost through surface evaporation and root water uptake. Photographs: M. Young (a) and (b), M.R. Stoklos (c).

10^{-5} m day^{-1}, a 10,000-fold decrease. Such an increase in tension occurs in soils when plant roots remove water for transpiration. Typically, plants can cause soil water tensions to increase up to 100 m, but in desert soils it is not uncommon to see tensions of up to 300 m. It is clear that at such high tensions the hydraulic conductivity is very low (see Figure 5-7 on page 51) and, as a result, liquid flow is nearly zero. This is why desert soils are often considered excellent sites for waste disposal facilities. The high tensions found in these soils cause the hydraulic conductivity to be extremely low, thus greatly reducing the chances for significant water flow to the groundwater table.

5.2.3 Transient Flow

Darcy's equation was originally developed for steady flow, where Q is constant over time. If Q is constant, the water content and pressure potential are generally also constant. However, in many practical situations, water content, pressure potential, and flow all change with time. For such transient flow conditions, Eq. (5-5) on page 51 needs to be written in differential form:

$$q = -K \frac{\partial H}{\partial z} \tag{5-7}$$

where $q = Q/At$ is the **Darcy flux** (m day^{-1}), and $\partial H/\partial z$ is the hydraulic gradient (m m^{-1}). The parameter q is defined as the volume of water moving through 1 m^2 face area per unit time. Its units are m^3 m^{-2} day^{-1}, which is equivalent to m day^{-1}. For comparison, rainfall is also expressed in m day^{-1} (or mm day^{-1}), which is the volume of water falling over 1 m^2 in 1 day. Note that the actual velocity of water moving through soil is considerably higher than q. Water moves only through pore space and not through the solid materials. Therefore, the pore water velocity is considerably higher than the Darcy flux q. Although pore water velocity depends on the pore size (higher in larger pores, smaller in finer pores), the average pore water velocity is generally defined as

$$v = \frac{q}{\theta} \tag{5-8}$$

where v is pore water velocity. For saturated soils, θ is equal to the total pore space or porosity. If the porosity is 50% (*i.e.*, if the pore space is 50% of the total volume), then according to Eq. (5-8) pore water velocity $v = q / 0.50$ or twice the flux. For unsaturated sands θ may be as low as 0.10 $m^3\ m^{-3}$, in which case the average velocity or the rate at which water travels through pores is 10 times the flux. More specifically, if rain falls on an unsaturated soil at a steady rate of $q = 0.025$ m day^{-1}, and the soil has a water content of 10% (*i.e.*, $\theta = 0.10\ m^3\ m^{-3}$), we can solve Eq. 5-8 for the pore water velocity in the soil v: 0.25 m day^{-1}.

It is important to understand the concept of pore water velocity because it is also the rate at which many contaminants travel through soils and aquifers.

In order to fully describe water flow through soil, Eq. (5-7) needs to be combined with the **equation of continuity**. This equation is

$$\frac{\partial \theta}{\partial t} = -\frac{\partial q}{\partial z} \tag{5-9}$$

where t is time (days or seconds) and θ is the water content. Equation (5-9) simply states that the change in water content $\partial\theta$ of a unit volume of soil over a small time increment ∂t is equal to the difference between inflow and outflow ∂q over the length of the sample ∂z. Combining Eqs. (5-7) and (5-9) yields

$$\frac{\partial \theta}{\partial t} = \frac{\partial}{\partial z}\left(K(h)\frac{\partial h}{\partial z}\right) + \frac{\partial K(h)}{\partial z} \tag{5-10}$$

Equation (5-10), which describes unsaturated vertical flow through soil, is called the **Richards' equation**. The equation states that changes in water content over time, $\partial\theta/\partial t$, result from pressure gradients, the first term on the right-hand side of Eq. (5-10), and gravity flow, the second term on the right-hand side.

Equation (5-10), or variations thereof, are used as the basis for computer models. Computer models are used extensively to predict water flow and to determine the results of various management decisions on flow of water through soil and groundwater. For example, we might need to predict how quickly rain falling on a soil during a summer rainstorm reaches the groundwater table. And we might want to know how fast this water will move to an adjacent lake or river once it has reached the groundwater table. These and similar predictions are important for preparing environmental impact statements. They can be made by using appropriate computer models based on Eq. (5-10), provided certain physical properties of the soils such as water retention (Eq. 5-4 on page 50) and hydraulic conductivity (Eq. 5-6 on page 51) are known.

For saturated soils, water content θ and hydraulic conductivity K in Eq. (5-10) are constant. This simplifies the equation somewhat, and allows for a wide range of groundwater flow predictions.

5.3 Movement of Contaminants in Soil and Groundwater

Having discussed how water moves through the soil, we can now discuss how chemicals move through soil and groundwater.

In an industrial society, chemicals and chemical waste can be found almost everywhere. For example, chemicals are applied as fertilizer to make crops grow better. They are also used for insect and weed control. They may be transported through pipelines (oil) or stored in large underground tanks (industrial chemicals). In the past, waste chemicals were dumped in landfills or disposed of in rivers, lakes, or oceans. Now such wastes are often stored in open ponds or underground tanks. Unfortunately, pipelines, tanks, ponds, and landfills leak, and chemicals enter the subsurface environment. Chemicals also enter the subsurface following atmospheric deposition (*e.g.*, tritium from nuclear bomb testing) or surface runoff and subsequent infiltration (lead from car exhaust, road salts). Thus, chemicals can enter the subsurface environment from a large number of sources. However, what happens to the chemicals once they are in the soil or groundwater? Fifty years ago, disposing of chemicals in the subsurface was considered an acceptable disposal practice. Now we know that such chemicals will eventually reach groundwater, making it unsuitable as a source of drinking water. Many chemicals can move to groundwater because they dissolve in soil water and move down to groundwater in solute form (see Figure 5-1 on page 46). Other chemicals, such as petroleum products, are less water-soluble, but the downward movement of water still enhances their flushing into groundwater. The rate at which this happens varies greatly, depending to a large extent on climatic conditions. In areas with high rainfall, water movement and chemical transport in soils is much faster than in low rainfall areas. However, many other factors, such as the properties of particular chemicals, soil type, climate, and vegetation, affect the movement of chemicals through soil and groundwater.

Figure 5-9 Trench sites, such as this one in New Mexico, are useful for the study of water and solute movement through a soil profile. Photograph courtesy of P.J. Wierenga.

5.3.1 Mechanisms of Chemical Transport

Chemicals often move through soil with water by a process called **mass transport** or **convection**. Mass or convective transport refers to the passive movement of a dissolved chemical with water. The rate of this passive movement is called **flux density (J_m)** with units of kilograms per square meter per day (kg m^{-2} day^{-1}). Flux density is the rate of chemical transport across a unit surface area per unit time:

$$J_m = qC \qquad (5\text{-}11)$$

where q is the Darcian velocity (m day^{-1}) and C is the concentration of the chemical in the pore water (mg L^{-1}). The calculation of C is explained in Section 6.2, beginning on page 63.

In addition to mass flow, chemicals may slowly redistribute within the soil pore water by molecular diffusion. **Diffusive transport (J_D)** results from the natural thermal motion of dissolved ions and molecules. This process can be described by **Fick's law of diffusion**, as follows:

$$J_D = -\theta D_m \frac{\partial C}{\partial z} \qquad (5\text{-}12)$$

where D_m (m^2 day^{-1}) is the porous medium diffusion coefficient and z the distance (m). Its value depends on many factors, including temperature, but a useful estimate for soil and groundwater is 10^{-4} m^2 day^{-1}. Equation (5-12) states that diffusive transport is equal to the gradient in concentration $\partial C/\partial z$, multiplied by a diffusion coefficient D_m. In soils and groundwater, the effect of molecular diffusion on chemical transport is minimal for short times, except over very small distances.

The porous nature of soils and vadose zone materials results from the existence of pores, fissures, wormholes, and other features, as well as from the loose stacking of the soil material itself. The different shapes, sizes, and orientation of the open spaces and pores result in fluid velocities that differ from place to place. In addition, fluid velocities even differ within an individual pore, depending on where the velocity is determined with respect to the pore wall. Velocities are lowest near the pore wall. These velocity variations cause chemicals to be transported at different rates, leading to mixing, which is macroscopically similar to molecular diffusion. This mechanical spreading arising from local velocity variations is called **mechanical dispersion**. The process can be described by an equation similar to Eq. (5-12), where J_h is **dispersive transport** (g m^{-2} day^{-1}) and D_h is the mechanical dispersion coefficient (m^2 day^{-1}):

$$J_h = -\theta D_h \frac{\partial C}{\partial z} \quad (5\text{-}13)$$

The value of the mechanical dispersion coefficient increases with increasing water velocity. Thus, as water moves more quickly through the pores, spreading of the chemical increases, and the coefficient D_h in Eq. (5-13) becomes greater. In its simplest form the relationship between pore water velocity v (m day^{-1}) and dispersion coefficient D_h is linear; thus

$$D_h = \alpha v \quad (5\text{-}14)$$

where α (m^{-1}) is the **dispersivity**.

Because D_m and D_h are macroscopically similar, these terms are usually added. Thus

$$D = D_m + D_h \quad (5\text{-}15)$$

where D is the longitudinal hydrodynamic dispersion coefficient.

Combining Eqs. (5-11), (5-12), (5-13), and (5-15) yields the **chemical flux J_s**:

$$J_s = -\theta D \frac{\partial C}{\partial z} + q C \quad (5\text{-}16)$$

Equation (5-16) states that total flux density of chemicals through soil J_s is caused by diffusion plus dispersion ($D (\partial C / \partial z)$) and convective flow ($q C$).

In order to use Eq. (5-16) for modeling purposes, we have to combine it with an equation of continuity, as was done for water flow (Eq. 5-9 on page 53). For chemical transport this equation is

$$\theta \frac{\partial C}{\partial t} = -\frac{\partial J_s}{\partial z} \quad (5\text{-}17)$$

where t is time. Equation (5-17) states that for a small soil volume (*e.g.*, a 10^{-6}-m^3 block of soil), the change in concentration in that block of soil with time, $\partial C / \partial t$, should equal the amount coming in minus the amount going out, $\partial J_s / \partial z$. In other words, no chemical is "lost." If more is coming in than going out, the amount inside the block must increase.

Combining Eq. (5-16) and Eq. (5-17) yields the general chemical transport equation in one dimension:

$$\frac{\partial C}{\partial t} = \frac{\partial}{\partial z}\left[\left(D \frac{\partial C}{\partial z}\right) - v C\right] \quad (5\text{-}18)$$

where v is the average linear pore water velocity (see Eq. 5-8 on page 52).

The leftmost term of Eq. (5-18), $\partial C / \partial t$, represents the rate at which the concentration at a fixed point changes in a flowing fluid (*e.g.*, at a point in a pore with fluid flowing through). The first term on the right-hand side of Eq. (5-18) represents the contribution of diffusion and dispersion, while the second term represents mass transport advection. Equation (5-18) was derived for a noninteracting chemical; however, in practice, many chemicals adsorb to soil while others degrade by microbial action. For this reason, Eq. (5-18) is often expanded to account for chemical interaction and transformation reactions, as we'll see in more detail in forthcoming sections.

5.3.2 Movement through One-Dimensional Columns

It is clear that chemical movement through soil is rather complex. Scientists and engineers have conducted laboratory column studies for many years to improve their understanding of the processes of chemical transport. In such studies, water containing one or more tracers or contaminants is applied to the top of a column (Figure 5-10 on page 56). Then additional water is applied so that the contaminant is carried through the soil to the lower end of the column. Here, a fraction collector is used to collect the water in small fractions, and the concentration of the contaminant is determined. Then the concentration of the contaminant is plotted versus time of sample collection or versus cumulative outflow volume. Figure 5-11 on page 56 is a plot of three such effluent concentration distributions. Note that in Figure

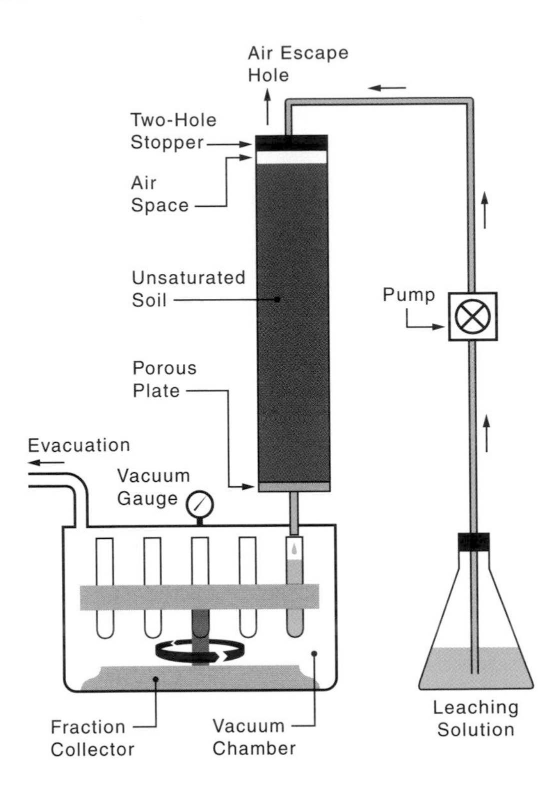

Figure 5-10 The experimental setup for column studies. Adapted with permission from *Methods of Soil Analysis, Part 1: Physical and Mineralogical Methods*, 2nd Edition. Published by American Society of Agronomy, Inc., Soil Science Society of America, Inc., 1986.

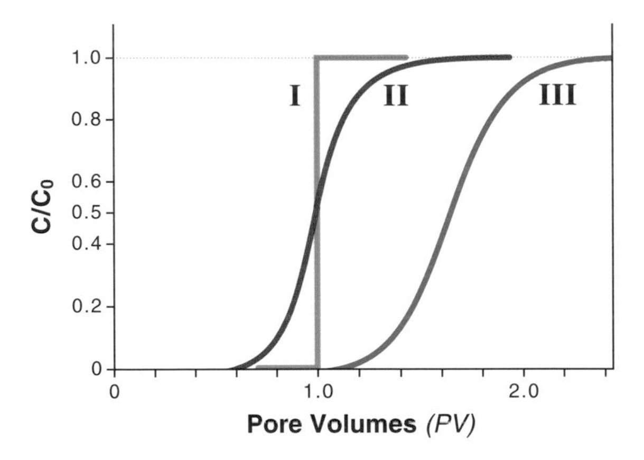

Figure 5-11 Column effluent concentration distributions for three scenarios: (I) piston flow; (II) dispersion and diffusion; (III) dispersion, diffusion, and retardation.

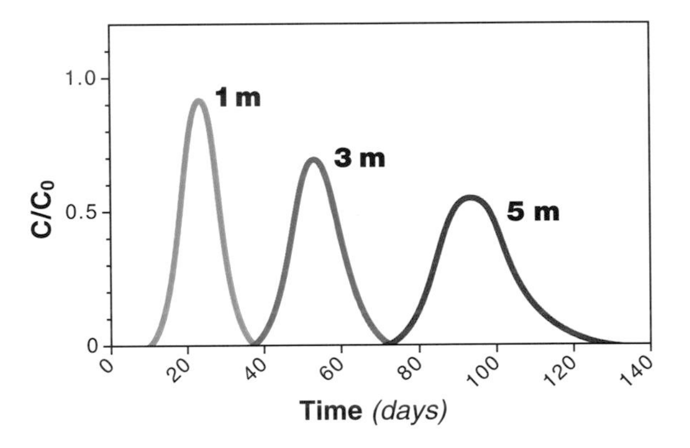

Figure 5-12 Concentration versus time for tritiated water moving at a steady rate through a deep, unsaturated soil. I. Porro, P.J. Wierenga, R.G. Hills, 1993. *Water Resources Research* 29, No 13: 1321–1330. Copyright by the American Geophysical Union.

5-11 the relative concentration C/C_0 is plotted rather than the measured effluent concentration. The relative concentration is obtained by dividing the effluent concentration C by the concentration of the input solution C_0. When relative concentrations are plotted, all the data are on the same scale, so results can be compared more easily.

For the same reasons, relative pore volumes are often used for the horizontal scale rather than time. One pore volume (PV) is equal to the total volume of fluid in the column. If, for example, the cumulative effluent is equal to one-half of the total volume of fluid in the column, then PV = 0.5. This definition of relative pore volume applies both to saturated and unsaturated soil. For convenience, the three effluent concentration distributions in Figure 5-11, also called **breakthrough curves**, are labeled I, II, and III.

Curve I in Figure 5-11 shows that the effluent concentration remains zero until one PV of fluid has moved through the entire column, for example, after all the original fluid in the column has been replaced by the incoming solution. We can visualize this if we think in terms of adding a colored substance to the top of the column in Figure 5-10. If the substance does not interact with the soil, the color of the effluent initially remains colorless. But after exactly one pore volume, for example, all the original water in the column, has been replaced by colored water, and the color in the column changes suddenly. The process that causes such a sudden change in concentra-

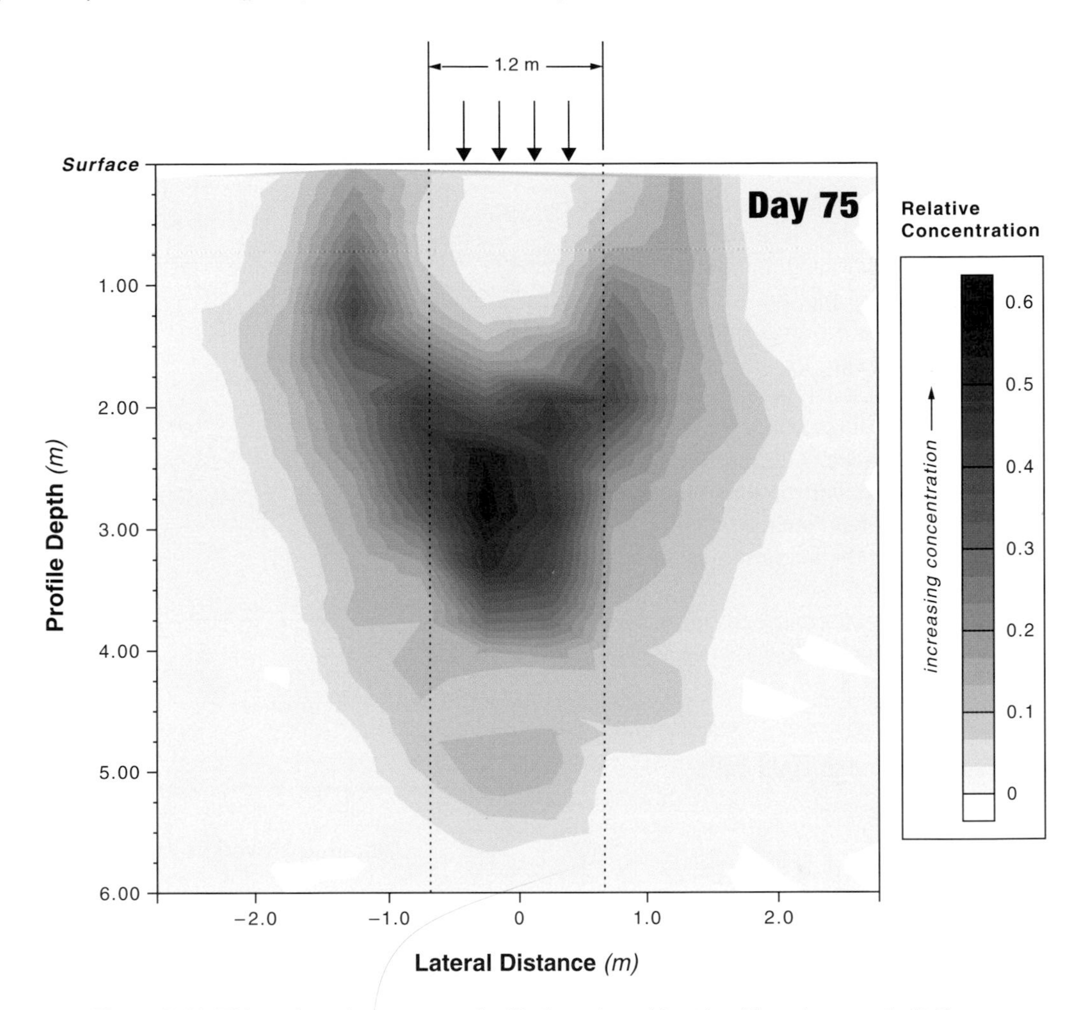

Figure 5-13 Tritium plume in unsaturated soil below a 1-m wide strip with surface-applied tritium. The darker the color, the higher the concentration. The data was collected using the trench site shown in Figure 5-9 on page 54.

tion or color is called **piston flow**. In reality, true piston flow, or total fluid displacement, seldom occurs.

Soils contain pores of different diameters; therefore, the traced water moves faster through some pores than through others. The result is a distribution as shown by breakthrough curve II. The effluent is initially without tracer (colorless), but gradually changes concentration (color) after about 0.5 PV, reaching a relative concentration of 1.0 (full color) after about 1.5 PV. Thus, the different fluid velocities in the column cause the breakthrough curve to be dispersed. The process causing the effluent concentration curve to be spread out or dispersed is called **dispersion**. Solute dispersion occurs in all soil and groundwater.

The third curve in Figure 5-11, curve III, represents a chemical that interacts with the soil in the column. The chemical is delayed, and the breakthrough curve is displaced to the right. Reasons for chemical interaction are presented in Section 6.3, beginning on page 67.

Figure 5-12 presents an example of a chemical (tritium) moving through a deep, unsaturated field soil column. Water was artificially rained on this column at a rate of 21 mm day^{-1} for several months. The rainwater contained tritium (^3H) during the first

11 days only. Used as a tracer for groundwater, tritium was measured at 1, 3, and 5 m below ground surface. The figure shows an increase followed by decrease in relative tritium concentration at each of the three depths over time. For example, at a depth of 1 m, the tritium concentration started increasing 10 days after rainfall started. It reached a maximum of about 0.97 after 22 days and then started decreasing again. In contrast, the maximum concentration at 3 m was reached after 53 days and at 5 m after 93 days. Thus, on average, the peak concentration moved down at a rate of 0.056 m day^{-1}. Note that the maximum tritium concentration decreases with distance from the soil surface. The farther down from the top of the soil, the lower the maximum concentration, and the more spread out the tritium pulse becomes. The decrease in the maximum concentration and the spreading of the pulse are caused by diffusion and dispersion, mathematically described by the dispersion coefficient D in Eq. (5-16) on page 55.

5.3.3 Movement through Field Soils

Figure 5-12, which depicts the fate of a tracer applied on the surface of a deep, unsaturated soil profile, uses tritium as an example to demonstrate the principles that apply to a chemical spill or to a contaminant leaking from a storage tank. Once a chemical contaminant has entered the soil, downward-moving water transports it through the root zone and the underlying unsaturated zone into the groundwater. Once in the groundwater, it may flow in a horizontal direction. Figure 5-13 on page 57 shows the two-dimensional distribution of tritium in the subsoil of a soil with a deep groundwater table. The tritium was applied with water over a 1-m wide strip for a 15-day period, followed by a 60-day period during which tritium-free water was applied. The figure clearly shows that tritium has mostly moved downward, the maximum tritium concentration being found at 2.5 m below the surface. However, considerable lateral spreading has also occurred beyond the 1-m application area, as indicated by the presence of tritium beyond the two dotted lines in Figure 5-13. Like downward movement, lateral spreading is a result of diffusion and dispersion and occurs in soils as well as in groundwater.

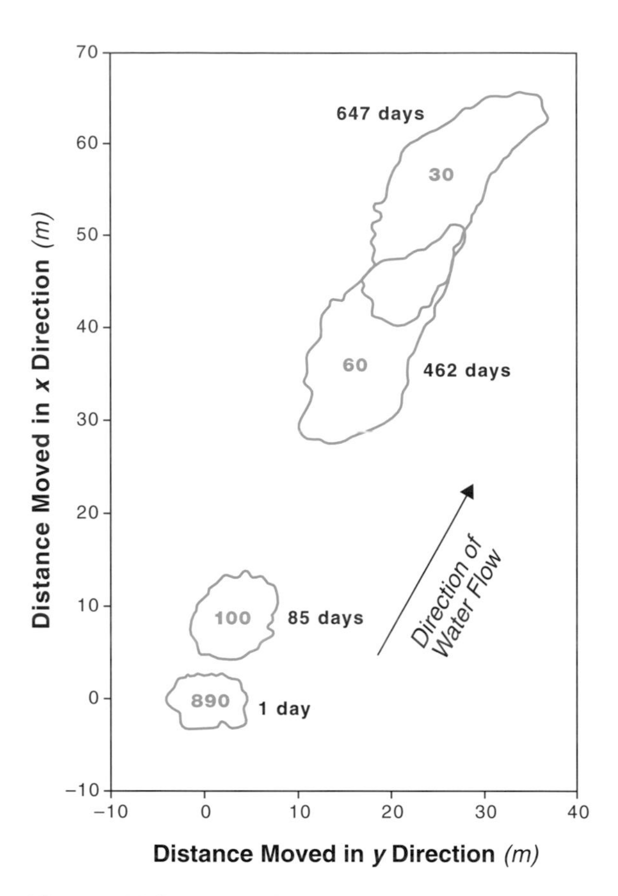

Figure 5-14 Spreading of chloride plume in a sand aquifer in Borden, Ontario, Canada. The chloride pulse injected at $(x,y) = (0,0)$, had a concentration of 892 mg L^{-1}. Note the decrease in maximum concentration as the plume moves with the groundwater in the direction of flow. Adapted from D.M. Mackay, D.L. Freyberg, P.V. Roberts, and J.A. Cherry, 1986. *Water Resources Research* 22, No 13: 2017–2029. Copyright by the American Geophysical Union.

Figure 5-14 shows how contaminants travel through groundwater. Chemical contaminants were introduced into the aquifer at $(x, y) = (0,0)$. The figure shows contaminant distributions in the groundwater after 1, 85, 462, and 647 days in the x,y-plane parallel to the groundwater table. As in Figure 5-13, considerable spreading has occurred, especially in the direction of flow, as is evidenced by the elongation of the plume. Notice that, in order to investigate groundwater for contaminants, we have to sample the water within the plume area for best results. If sample wells are wrongly placed or if sampling is done at the wrong depths, the information obtained is unlikely to be correct.

Case Study: Low-Level Radioactive Waste Disposal in Ward Valley

Low-level radioactive waste comes from radioactive material used in medicine, scientific research, in some industrial processes, and in nuclear power plants. Most of the waste, including 95% of the volume, decays to acceptable levels in a hundred years or less.

Subject to the Low-Level Radioactive Waste Policy Act of 1980, the individual states in the USA are responsible for their own radioactive waste. In 1988, California joined with three other states—Arizona, North Dakota, and South Dakota—to establish a low-level disposal site in California. After extensive studies of the climate, geology, ecology, soils, and groundwater hydrology, a site was selected. This site is Ward Valley, which is located in the Mojave Desert, west of Needles, California, on the Colorado River. The Ward Valley site has several characteristics that make it suitable for low-level nuclear waste disposal. These include low rainfall (125 mm yr^{-1}) and a thick unsaturated zone (190–213 m). In addition, it is situated within a low-earthquake zone, away from an important groundwater aquifer, and it has a long flow path to a major source of water (Colorado River, 130 km).

After site selection studies, environmental review, and state licensing were completed, the State of California made a request to the federal government that the U.S. Department of the Interior transfer the federal land to the state. At this time three California hydrologists (the Wilshire group) brought up seven issues that they claimed had not been adequately addressed in the environmental impact report. To resolve these issues the U.S. Department of the Interior asked that the National Research Council (NRC) convene a committee to evaluate the Wilshire concerns. A complete discussion of these seven issues can be found in the Ward Valley report prepared by the NRC committee. For the present example, we restrict ourselves to two of the seven issues discussed:

1. Transfer of contaminants through the unsaturated zone and
2. Potential for hydrologic connection between the Ward Valley site and the Colorado River

Issue No. 1: Transfer of contaminants through the unsaturated zone

The NRC committee concluded that "recharge or potential transfer of contaminants through the unsaturated zone to the water table is highly unlikely." The committee based its conclusion on several lines of evidence:

(a) Water contents, as determined by gravimetric sampling and neutron probe measurements, at the site were very low; for example, 94% of the water contents were less than 0.10 m^3 m^{-3}. At these low water contents, the hydraulic conductivities of the sandy loam subsurface soils are very low (see Figure 5-7 on page 51), and subsurface water flux rates are near zero.

(b) The chloride concentrations in the soil water in the upper 30 m were very high, and such high levels suggest that downward fluxes must be low; high downward fluxes would have flushed chloride out of the soil.

Rainwater contains small concentrations of chloride (from the ocean), and plants use rainfall on soils for transpiration. Plants do not, however, take up chloride; therefore, the total chloride in the soil profile increases with time. The time t needed to accumulate chloride to any depth Z is given by

$$t = \frac{\text{total chloride to depth } Z}{J_{\text{Cl}}} \tag{C5-1}$$

where the total chloride is calculated from the chloride concentration in the soil multiplied by the water content by volume, and J_{Cl} is the average chloride input, estimated at 0.164 g m^{-2} yr^{-1}. At this input rate, it would take nearly 50,000 years to accumulate the chloride currently found in the profile down to $Z = 30$ m. Water fluxes estimated from these chloride data range from 0.03 to 0.05 mm yr^{-1}, indicating very little to no recharge.

Although the majority of the Ward Valley committee supported the above conclusions, two of the seventeen panel members disagreed. They contended that the detection of tritium (originating from bomb tritium and hence no more than 40 years old) in soil at 30 m indicated a much faster downward migration of soil water than assumed for the license application. The majority of the committee attributed the tritium found at 30 m to sampling procedures.

Issue No. 2: Potential hydrologic connection to the Colorado River

The committee concluded that "the potential impacts on the river water quality would be insignificant relative to present natural levels of radionuclides." The committee based its conclusions in part on travel time calculations.

The contaminant travel time can be estimated using Darcy's equation (Eq. 5-7 on page 52). Groundwater measurements show that groundwater below the site is 360 m above the Colorado River water. The average saturated hydraulic conductivity of the aquifer is 5.8 m day^{-1}. The most likely pathway for flow to the river is 130 km long, resulting in a hydraulic gradient of $360 / 130{,}000 = 0.0028$ m m^{-1}. By applying Darcy's equation, we calculate the groundwater flux as $5.8 \times 0.0028 = 0.016$ m day^{-1}. Assuming an effective porosity of 0.1, this results in a flow velocity of $0.016 / 0.1 = 0.16$ m day^{-1} or 58 m yr^{-1}. The travel time for a noninteracting contaminant, or the time needed to reach the river, is then $130{,}000 / 58 = 2240$ years. Most radionuclides interact (sorb) with aquifer material, and thus the actual travel time may be longer. Furthermore, as shown above, the time needed for contaminant transport through the unsaturated zone could also be hundreds of years or more. Decay of short-lived radionuclides is much less than 2240 years; therefore, short-lived radionuclides should pose no direct threat to the water in the Colorado River.

The examples shown in Figures 5-12 through 5-14 show fairly well-behaved contaminant transport. However, chemicals frequently move through soil and groundwater in a more erratic, less predictable manner.

Surface soils and deeper geologic materials are heterogeneous. This is a direct result of their formation during past geologic times. The structure of soils may also differ from place to place. In structured soils and in soils containing root holes, worm holes, cracks, and other large pores, such irregularities serve as pathways permitting water and contaminants to move preferentially through them. This type of movement, called **preferential flow**, allows water and contaminants to penetrate to greater depths more quickly or to travel longer distances than they would in soils without these irregularities. Preferential flow can be an important factor contributing to the pollution of underground aquifers.

References and Recommended Reading

Brady N.C., Weil R.R. (1996) *The Nature and Properties of Soils*. Prentice Hall, New Jersey.

Guymon G.L. (1994) *Unsaturated Zone Hydrology*. Prentice Hall, Englewood Cliffs, New Jersey.

Hanks R.J. (1992) *Applied Soil Physics*, 2nd Edition. Springer Verlag, New York.

Hemond H.F. and Fechner, E.J. (1994) *Chemical Fate and Transport in the Environment*. Academic Press, New York.

Hillel, D. (1980). *Fundamentals of Soil Physics*. Academic Press, New York.

Jury, W.A., Gardner, W.R., and Gardner, W.H. (1991). *Soil Physics*, 5th Ed. John Wiley and Sons, New York.

Mackay D.M., Freyberg D.L., Roberts P.V., and Cherry J.A. (1986) A natural gradient experiment on solute transport in a sand aquifer: 1. Approach and overview of plume movement. *Water Resources Research*. **22**, 2017–2029.

Maidment D.R., Editor (1993) *Handbook of Hydrology*. McGraw-Hill, New York.

National Research Council (1995). *Ward Valley: An Examination of Seven Issues in Earth Sciences and Ecology*. National Academy Press, Washington, D.C.

Porro I., Wierenga P.J., and Hills R.G. (1993) Solute transport through large uniform and layered soil columns. *Water Resources Research*. **29**, 1321–1330.

van Genuchten, M.Th. and Wierenga P.J. (1986) Solute dispersion coefficients and retardation factors. In *Methods of Soil Analysis, Part 1: Physical and Mineralogical Methods* (A. Klute, Editor), 2nd Edition, p.p. 1025–1054. American Society of Agronomy, Inc., Soil Science Society of America, Inc., Madison Wisconsin.

Wierenga P.J. (1995) Water and solute transport and storage. In *Handbook of Vadose Zone Characterization & Monitoring* (L.G. Wilson, L.G. Everett, and S.J. Cullen, Editors), p.p. 41–60. Lewis Publishers, Boca Raton, Florida.

Problems and Questions

1. The length of the water column in the tensiometer tube must be added to the tensiometer reading (expressed in m H_2O) to obtain the pressure potential at the location of the tensiometer cup. Explain why.

2. Tensiometers are used to measure the pressure potential (h) of soil water. When two tensiometers are placed at different depths in a soil profile, the pressure potential can be calculated at each depth. Given the vertical distance to a preselected reference level, we can use this information to calculate soil-water potentials (H), and from these, the gradient in soil-water potential. The gradient determines the direction of water flow (up or down).

   (a) Tensiometer A has the ceramic cup at 1.0 m below soil surface, while tensiometer B has its ceramic cup at 1.5 m below soil surface. Tensiometer readings at A and B are –1.5 and –2.0 m H_2O, respectively.

   Calculate pressure potentials (h) at A′ and B′, and identify the depth at which the soil is driest. Explain.

   (b) Calculate soil-water potentials (H) at A′ and B′. Does the water move up, down, or not at all in this unsaturated soil? Discuss.

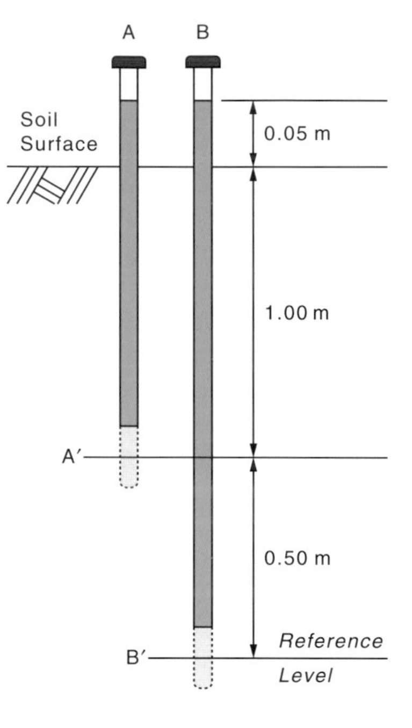

Figure 5-15 Illustration of the tensiometers required to answer Problem 2 a) and b).

   [*Hint*: Gravity plays a large role in soil-water movement. In order to solve problems on water movement, we have to establish a reference level. For convenience, the reference is often taken at the lowest tensiometer cup. The gravity potential (z) is then the vertical distance between the measurement position and the reference level.]

3. Water is added to an unsaturated soil column (as in Figure 5-10 on page 56) at a rate of 100 mL h^{-1}. The column is 0.5 m long and has a cross-sectional area of 2×10^{-3} m^2. After 1 day, a nonreactive tracer is added to the column. Assuming that the volumetric water content of the soil is 0.25 m^3 m^{-3}, calculate the following: (a) pore water velocity, (b) one pore volume, (c) the time in hours for the tracer to travel from the top of the column (assume piston flow) to the bottom of the column.

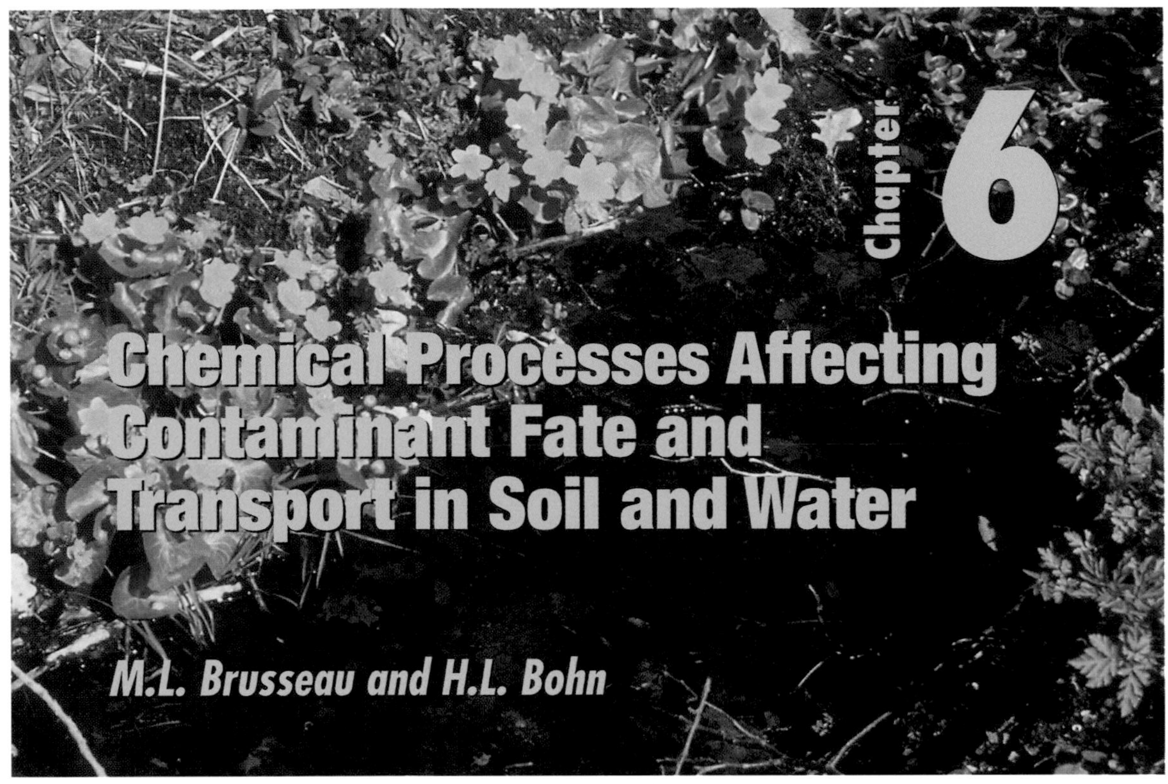

Chemical properties of soil determine whether a pollutant, such as nitrate, is retained by soil or leached to the groundwater. Above, yellow-flowered Ranunculus *grows in drainage water contaminated with nitrate. Photograph, J.W. Brendecke.*

6.1 Soil Phases

Soil consists of at least three phases: solid (soil particles), liquid (water), and gas (soil atmosphere). When soil is contaminated by a pollutant, the pollutant may associate with one or more of these phases. The extent to which a pollutant is distributed among the soil phases depends on the characteristics of both the pollutant and the soil phases. Knowing if, why, and how a pollutant will associate with soil phases is important to understanding the fate of pollutants in soil. Thus, in the first portion of this section we will focus on the distribution of pollutants between two or more phases (*i.e.*, phase-transfer processes).

We will begin by examining the following phase-transfer processes of pollutants:

- **evaporation**: transfer between pure pollutant and gas phases
- **solubilization**: transfer between pure pollutant and water phases
- **volatilization**: transfer between water and gas phases
- **sorption**: transfer between water and solid phases

These processes are illustrated in Figure 6-1 on page 64.

The phase-transfer behavior of inorganic and organic pollutants can be quite different, so we will discuss them separately. In the second portion of this section, we will examine the movement or transport of contaminants undergoing sorption. Then we will briefly discuss two types of chemical reactions that can transform the physical and chemical properties of pollutants.

6.2 Solubility and Volatility

6.2.1 Solubility of Inorganic Pollutants

The extent to which pollutant molecules or ions transfer from their pure forms into water is the **aqueous solubility**. Owing to their electric charge,

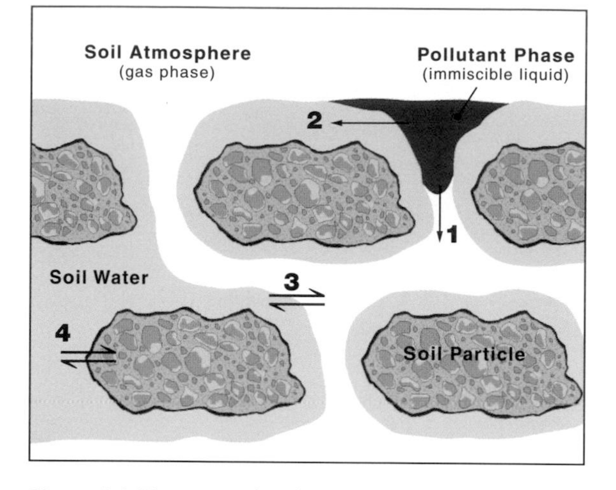

Figure 6-1 Phase transfer of pure pollutant with water and air phases: (1) evaporation, (2) solubilization, (3) volatilization, (4) sorption.

inorganic ions interact strongly with the oxygen atom of the water molecule (HOH). Thus, inorganic ions that are readily solvated by water exhibit high potential solubility. Usually, however, there are competing interactions that limit the solubility of many inorganic compounds. One major factor in solubility is the energy that is required to break the bonds between molecules of a solid. Another major factor is the presence of other solution components that can bond with the ion.

Under most natural conditions, the interactions of alkali (Li^+, Na^+, K^+) and alkaline-earth (Mg^{2+}, Ca^{2+}) ions with other solution components are relatively weak, which means these inorganic ions can remain dissolved in water. However, many multivalent ions react more strongly with the HOH molecule, so strongly that these ions can extract H^+ ions from the water molecule to form hydroxyoxides such as FeOOH, $Al(OH)_3$, and SiO_2. Because these hydroxyoxide molecules are uncharged, they tend to precipitate out from aqueous solution. Some highly charged cations (*e.g.*, S^{6+} and C^{4+}) react even more strongly with oxygen and are found only as oxyanions (SO_4^{2-} and CO_3^{2-}). Although the charge of oxyanions should make them water-soluble, some oxyanions (PO_4^{3-}, MoO_4^{2-}, CrO_4^{2-}, PbO_4^{4-}) form uncharged compounds with multivalent cations (Al^{3+}, Fe^{3+}, and Ca^{2+}) and are therefore quite insoluble.

The aqueous solubility of inorganic compounds is expressed as the **solubility product**, K_{sp}, which is determined under ideal conditions using pure compounds and pure water. In complex solids such as soils, however, the solubility product is a poor indicator of the aqueous solubility of inorganic ions. In soils an ion's solubility is usually less than the value predicted by the K_{sp}. Moreover it usually decreases with the amount of the ion in the system, generally decreasing asymptotically with time after the ion is added to the soil.

6.2.2 Solubility of Organic Pollutants

The aqueous solubility of organic pollutants depends strongly on the degree to which water and pollutant molecules interact. In most cases, we can use the well known rule of thumb that "like dissolves like" to predict solubility. Water is very polar; therefore, the aqueous solubility of organic compounds depends strongly on the degree of polarity of their molecules. Because water can interact easily with other polar compounds, the aqueous solubilities of ionic or polar organic compounds are relatively high. Conversely, it requires much more energy for water to interact with, or solvate, nonpolar organic compounds, so the solubilities of nonpolar compounds are generally much lower than those of the polar and ionic compounds. The solubilities for representative organic pollutants are presented in Table 6-1, where we see that the highly polar liquid phenol is considerably more water-soluble than the nonpolar liquid benzene.

Under "natural" conditions (temperature T = 25°C, pressure P = 1 atm[1]), organic compounds in their pure form exist as solids (*e.g.*, naphthalene, anthracene), liquids (*e.g.*, benzene, toluene, xylene), or gases (*e.g.*, methane, propane). Dissolution (solubilization) requires energy to break the bonds between molecules; therefore, the solubility of organic compounds depends, in part, on the form of the compound. For example, more energy is required to break the stronger bonds in solids than the weaker bonds in liquids. Therefore, the solubilities of solid

[1] 1 atm ≅ 0.1 MPa

Table 6-1 Aqueous solubilities and vapor pressures for selected organic compounds. Source: Verschueren, 1983.

Compound	Aqueous Solubility ($mg\ L^{-1}$)	Vapor Pressure (mm Hg[a])	Henry's Constant (*H*—unitless)	State (STP[b])
Benzene	1780	76	0.18	nonpolar liquid
Toluene ($-CH_3$)	515	22	0.23	nonpolar liquid
Naphthalene	30	~0.5	0.02	nonpolar solid
Phenol ($-OH$)	82,000	0.2	0.00005	polar liquid

[a] 1 mm Hg ≅ 133 Pa
[b] $T = 20°C$, $P = 1$ atm; 1 atm ≅ 0.1 MPa

organic compounds are usually lower than those of liquid compounds (see Table 6-1).

An important property of liquid organic compounds is their capacity to mix with water—that is, their **miscibility**, or lack thereof. A miscible organic liquid is one that can be mixed with water such that a single liquid phase results. Simple alcohols such as methanol and ethanol are examples of miscible liquids; in other words, they are considered to be infinitely soluble in water. Conversely, an immiscible liquid is one that cannot be mixed with water. For example, benzene, an aromatic hydrocarbon that is a major component of gasoline, is considered an immiscible liquid. When a volume of pure liquid benzene is mixed with water, the two liquids quickly separate after the mixing action stops. But, like most chemical processes, miscibility is not an all-or-nothing phenomenon. Thus, a relatively small fraction of the benzene molecules will transfer into the water phase; that is, this fraction dissolves. The maximum amount of benzene that can be dissolved in water is the (aqueous) solubility of benzene. Even though the maximum amount of immiscible liquid that can dissolve in water may be very small, it can be very significant. For instance, the solubility of benzene in water is $< 2\ g\ L^{-1}$, but the federal maximum contaminant level for benzene is $5\ \mu g\ L^{-1}$.

The solubility of organic pollutants in water is a function of water-pollutant and pollutant-pollutant molecular interactions, and depends primarily on the properties of the compounds involved. However, other factors may also affect solubility. One such factor is temperature. Most organic compounds become more soluble as the temperature increases, although a few behave in the opposite way. In general, however, solubility changes by a factor of < 2 in the temperature range of most natural systems (0 to 35°C). Salinity or ionic strength can also cause a small decrease in the solubility of nonpolar organic compounds through a process known as the **salting-out effect**. This effect is also moderate for typical environmental conditions, changing the solubility by a factor of < 2.

In the preceding discussion, we dealt with the solubility of single organic pollutants, which involves transfer of pollutant molecules from the pure contaminant phase (solid, liquid, or gas) to water. However, many important sources of pollutants contain multiple pollutants. The primary examples of this type of pollution are multicomponent immiscible liquids such as gasoline, diesel fuel, and coal tar. Knowledge of the partitioning behavior of multicomponent liquids is essential to the prediction of their impact on environmental quality.

The partitioning of components into water is controlled by the aqueous solubility of the component and the composition of the liquid. A simple approach to estimating the solubility of multicomponent liquids involves an assumption of ideal behavior in both water and organic phases and the application of **Raoult's law**

$$C_{w,i} = X_{o,i} \; S_{w,i} \qquad (6\text{-}1)$$

where $C_{w,i}$ is the aqueous concentration (mol L^{-1}) of component i, $X_{o,i}$ is the mole fraction of component i in the organic liquid, and $S_{w,i}$ is aqueous solubility (mol L^{-1}) of component i. The mole fraction is the concentration of component i in the immiscible liquid. Raoult's law tells us that the aqueous concentration obtained for any given component is proportional to the amount of that component in the immiscible liquid.

Assume, for example, that we have a two-component, immiscible liquid, in which the mole fraction of each component (A and B) equals 0.5. This means that there is an equal amount of each component. Let us further assume that the aqueous solubility of component A is 100 mg L^{-1} and that of component B is 10 mg L^{-1}. We now wish to calculate the concentrations of A and B in a volume of water that is in contact with the immiscible liquid. Using Raoult's law, we find that the aqueous concentration for component A is 50 mg L^{-1} and that for component B is 5 mg L^{-1}. Thus, the aqueous concentration for each component is one-half of its aqueous solubility. This is because the two components are "competing" with each other to dissolve in water. Inspection of Eq. (6-1) shows us that when the mole fraction is equal to 1 (*i.e.*, a single-component liquid), the aqueous concentration is equal to the aqueous solubility.

6.2.3 Evaporation of Organic Pollutants

Evaporation is the transfer from the pure liquid or pure solid phase to the gas phase. The **vapor pressure** of a pollutant, then, is the pressure of its gas phase in equilibrium with the solid or liquid phase, and is an index of the degree to which the compound will evaporate. In other words, we can think of the vapor pressure of a compound as its "solubility" in air. Evaporation can be an important transfer process when a pure-phase pollutant is present in the vadose zone, whereby pollutant molecules transfer into the soil atmosphere.

While solubility is governed both by pollutant-pollutant and water-pollutant molecular interactions, evaporation depends almost solely on pollutant-pollutant interactions—that is, the energy of bonding—in the solid or liquid phase. Gas-phase intermolecular interactions are negligible because the space between gas molecules is usually quite large. Thus, at a simple level, we can say that the greater the bonding energy between the molecules of a pollutant, the lower its vapor pressure will be. The vapor pressures of liquids are therefore typically much higher than those of solids (see Table 6-1 on page 65). Unsurprisingly, then, the vapor pressure of a pollutant is also a strong function of temperature, owing to the strong influence of temperature on gas-phase processes.

6.2.4 Volatilization of Organic Pollutants

Volatilization, which is the transfer of pollutant between water and gas phases, is an important component of the transport of many organic compounds in the vadose zone. Volatilization is different from evaporation, which specifies a transfer of pollutant molecules from their pure phase to the gas phase. For example, the transfer of benzene molecules from a pool of gasoline to the atmosphere is evaporation, whereas the transfer of benzene molecules from water (where they are dissolved) to the atmosphere is volatilization. The vapor pressure of a compound gives us a rough idea of the extent to which a compound will volatilize, but volatilization also depends on the solubility of the compound as well as environmental factors.

At equilibrium, the distribution of a pollutant between gas and water phases is described by **Henry's law**:

$$C_g = HC_w \qquad (6\text{-}2)$$

where C_g is concentration of pollutant in the gas phase (in mg L^{-1}), C_w is concentration of pollutant in the water phase (mg L^{-1}), and H is Henry's constant (dimensionless).

We can use Henry's law to evaluate the preference of a pollutant for water and gas phases. For example, consider the preference of three pollutants,

A, B, and C, confined in three separate, closed containers, each of which holds equal volumes of water and air. Suppose pollutant A has a Henry's constant $H = 1$. A value of 1 means that the pollutant concentration in the air is equal to the concentration in the water. Thus, pollutant A "likes" water and air equally. Pollutant B has a Henry's constant of $H = 0.1$. This value means that the concentration of the pollutant in the air is ten times less than the concentration in the water; that is, pollutant B prefers the water. The third pollutant C has a Henry's constant of $H = 10$, which means that its concentration in air is ten times greater than it is in water.

When we use Henry's law, we assume the condition of equilibrium, which implies that transfer between gas and water phases is instantaneous. Even though instantaneous transfer is not guaranteed in natural systems, the rate of pollutant transfer between water and gas phases in soils tends to be quite rapid relative to other transport processes. Thus, the assumption of instantaneous transfer is rarely a major problem.

6.3 Sorption (Retention) of Pollutants

Sorption is a major process influencing the transport of pollutants in soils. The broadest definition of **sorption** (or **retention**) is the association of pollutant molecules with the solid phase of the soil (soil particles). In the following sections we will discuss how pollutants sorb to soil and how to quantify sorption. Although our examination of sorption mechanisms will focus on each mechanism in isolation, it is important to understand that the sorption of a given pollutant often involves more than a single mechanism. [*Note*: Soils retain organic and inorganic substances by several complex mechanisms; thus, retention may be a better word than sorption. We will, however, use both terms here.]

6.3.1 Sorption Mechanisms for Inorganic Pollutants

Because soil components such as clay are predominantly negatively charged, cation exchange is a major sorption mechanism for inorganic pollutants (Figure 6-2 on page 68). Generally, positively charged ions are attracted to the negatively charged sites on the clay particles. Common soil cations include Ca^{2+}, Mg^{2+}, K^+, Na^+, and H^+, which can exist on the exchange sites or in the soil solution. If solubility products of these cations in the soil solution are exceeded, these cations can combine with negative species and precipitate out as insoluble salts. Usually, the cations in the soil solution are in equilibrium with the cations on the exchange sites, while the cations on the exchange sites are more numerous than those in solution. The number of particular cations on the exchange sites controls the number of those cations in the soil solution through the mechanism of cation exchange. Thus soil solution cations that come into the vicinity of a cation on an exchange site are able to replace that cation. For example, a monovalent Na^+ can replace or exchange a monovalent K^+, whereas a divalent Mg^{2+} cation can replace a divalent Ca^{2+} or two monovalent cations. Whether an ion is exchanged or not largely depends on the charge density of both cations involved: in general, cations of higher valence are held more tightly to exchange sites. We use the term **adsorption affinity** to describe the force with which a cation is held sorbed to an exchange site. The adsorption affinities of several common cations are given as follows, in decreasing order:

$$Al^{3+} > Ca^{2+} = Mg^{2+} > K^+ = NH_4^+ > Na^+$$

The size of the cation also affects the adsorption affinity. Hence, Na^+, a small cation which is normally highly hydrated, has a very low adsorption affinity and can be exchanged rather easily.

As ions in the soil solution are utilized by microorganisms or taken up by plants, the equilibrium between the soil-solution cation concentrations and the exchange-site cation concentrations is disturbed, resulting in exchange of the cation from the colloids to the soil solution. When exchange sites no longer contain a particular cation, soil solution concentrations of that cation can be replenished by dissolution of precipitated salts (Figure 6-3 on page 69). Regardless of the mechanism for replacing cations in the soil solution, the maximum concentration of particular cations is dependent on the solubility products of all cation/anion species in solution. In a sense,

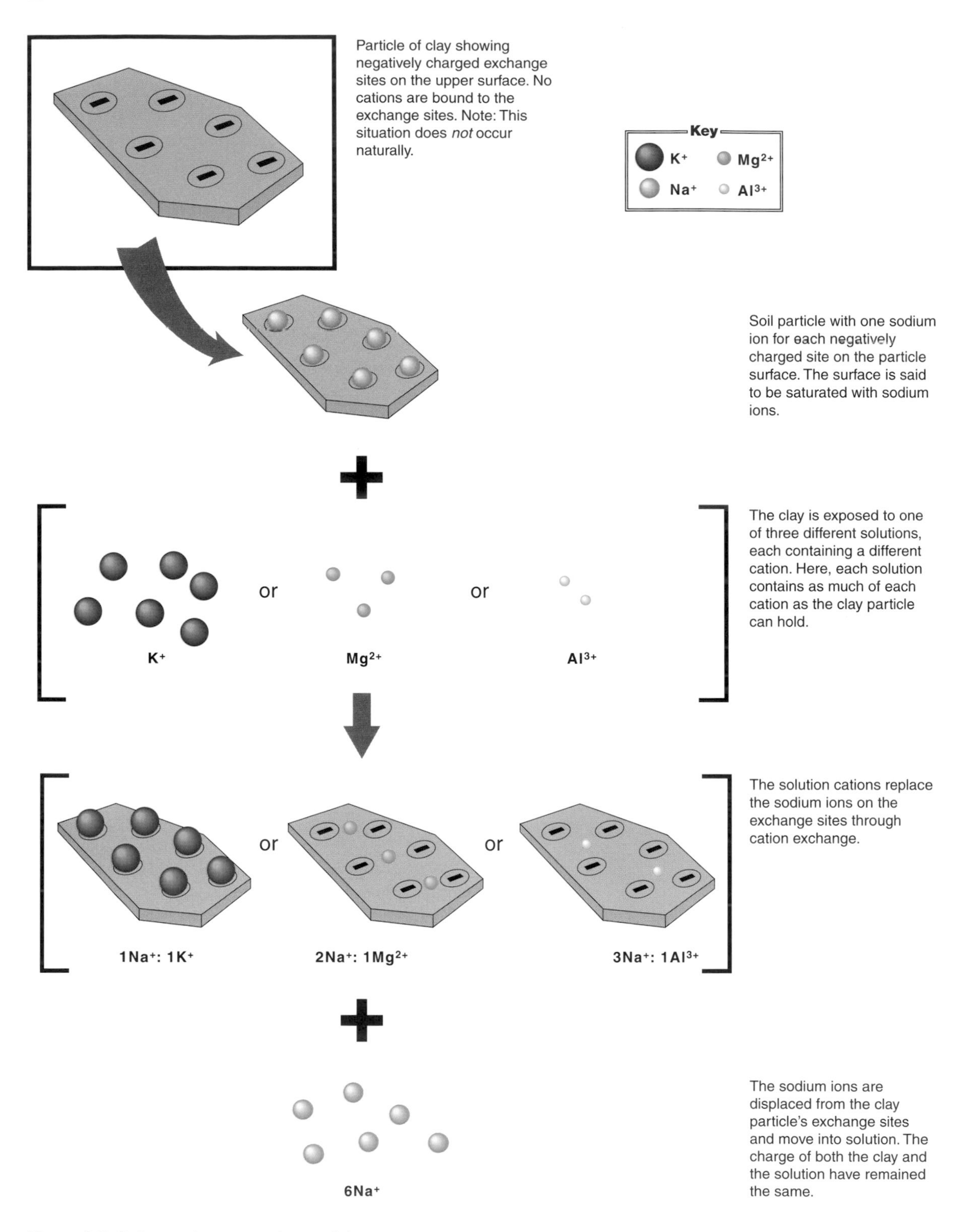

Figure 6-2 Cation exchange on clay particles.

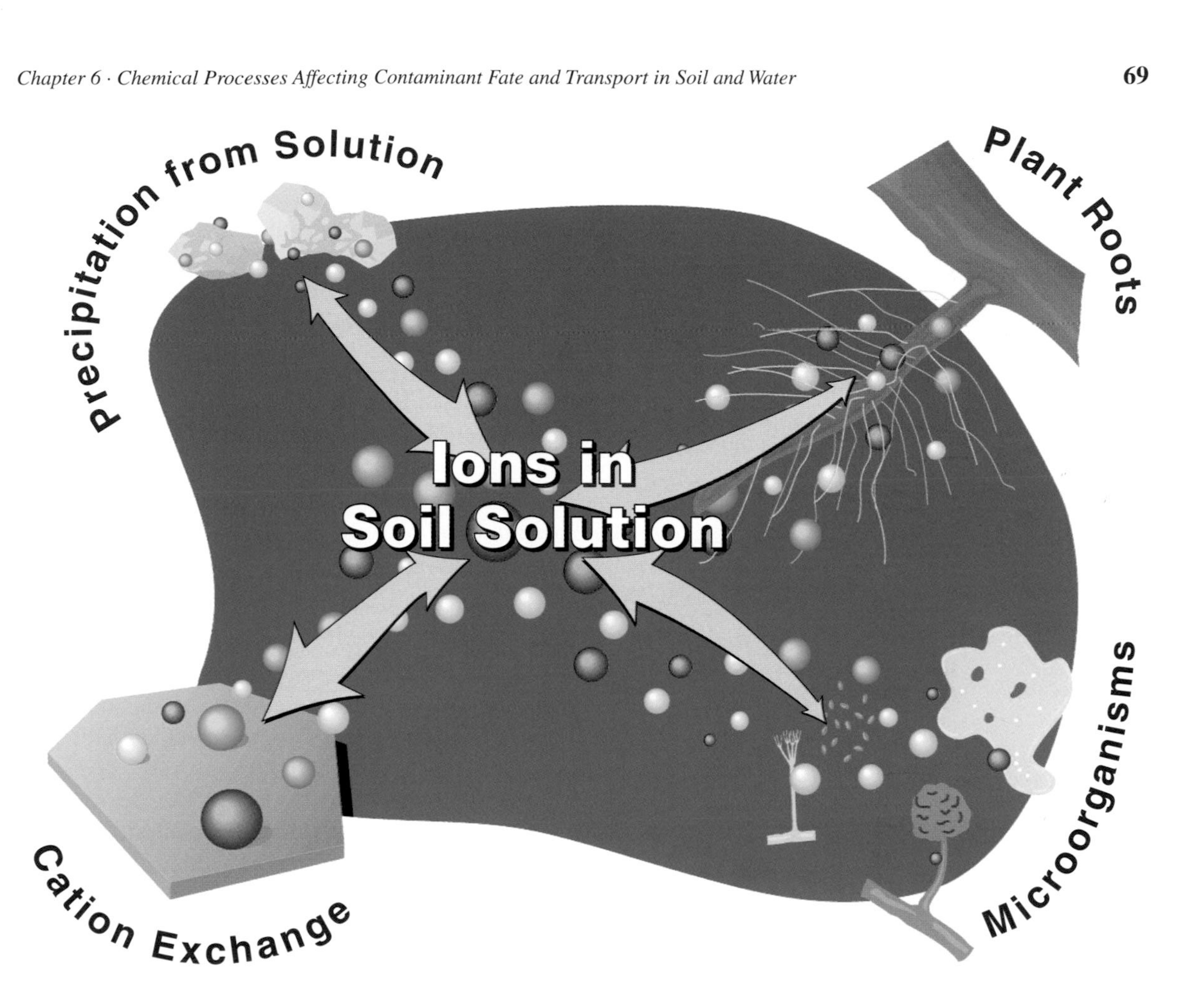

Figure 6-3 Paths of dissolution and uptake of minerals in the soil.

then, we can think of the exchange sites and precipitated salts as reserve supplies of cations to replenish the soil solution.

Cation exchange at negatively charged sites on soil surfaces is a major retention mechanism for cationic heavy metals, an important class of pollutants. These include cadmium, lead, and zinc. The fact that cations are attracted to soil also influences the mobility of those cations, which in turn affects the potential of those cations to be leached through the soil profile. Two important factors affecting the mobility of ions in soil are the solubility of the ion in question and the charge of the ion. In general, anionic (negatively charged) ions are not attracted to exchange sites on soil, which are predominantly negatively charged. Thus, anionic ions are usually very mobile. Examples of mobile ions include Cl^- and NO_3^-. Chemicals that are not as mobile tend to be cationic or positively charged and/or fairly insoluble. For example, ions such as Al^{3+} or Cd^{2+} tend to be sorbed to cation exchange sites, whereas insoluble species such as $H_2PO_4^-$ tend to precipitate out of the soil solution. The overall mobility of chemicals has profound implications on their potential for transport through soil and the vadose zone to groundwater, and thus directly affects the potential for pollution of groundwater.

6.3.2 Sorption Mechanisms for Organic Pollutants

The mechanisms by which many organic pollutants are sorbed or retained by soil are usually quite different from those involved for inorganic pollutants.

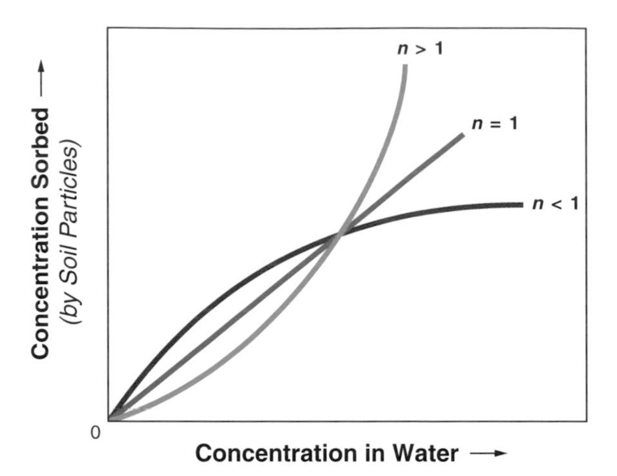

Figure 6-4 (a) Linear ($n = 1$) and (b) nonlinear isotherms ($n > 1$, $n < 1$) for sorption of pollutants.

Here we can use the "like dissolves like" rule to help explain the sorption of nonpolar compounds. The predominant sorbent for organic pollutants is the organic material associated with soils. Organic material is generally less polar than water and provides a more favorable environment for nonpolar organic pollutants. Thus, the sorption of nonpolar organics is driven primarily by the hydrophobic effect—the incompatibility between water and the organic compound. Organic pollutants, then, prefer to associate with the soil organic material because their polarities are more similar.

Not all organic compounds, however, are nonpolar. Some organic pollutants are polar or ionic, and the mechanisms governing their retention are similar to some of those governing retention of inorganic pollutants (*e.g.*, ion exchange). Many of the important ionizable organic pollutants (*e.g.*, phenols and chlorophenols) are negatively charged under environmental conditions. Because most soil particles have a net negative charge, they repel negatively charged organic pollutants. This means that negatively charged pollutants usually exhibit very little sorption.

The polarity of any pollutant, and whether it is organic or inorganic, is a key determinant of the degree to which the pollutant is sorbed or retained by the soil. However, the physical and chemical properties of the soil are also important factors. For example, the sorption of nonpolar organic pollutants is often controlled by the amount of organic matter associated with the soil. The cation exchange capacity, clay content, and metal oxide content of the soil are significant for sorption of ionizable and ionic pollutants.

6.3.3 Sorption Isotherms

A **sorption isotherm** (which is measured at one temperature) is simply a quantitative description of the relationship between the concentration of pollutant sorbed by the soil and the concentration in water. Of the many different forms of isotherms that are used to describe sorption, the simplest is the linear form (which is based on the same idea as Henry's law):

$$S = KC_{\mathrm{w}} \tag{6-3}$$

where S is the concentration of pollutant sorbed by the soil (mg kg^{-1}) and K is the sorption coefficient (L kg^{-1} or mL g^{-1}). The larger the value of K, the greater the degree to which a pollutant is sorbed by the soil. Pollutants for which the sorption to soil is described by a linear isotherm exhibit a direct proportionality between the concentration of the pollutant in water and the sorbed concentration of the pollutant. The sorption of many nonpolar organic pollutants is linear. An example of a linear isotherm is shown in Figure 6-4 (where $n = 1$). Some organic and many inorganic pollutants exhibit nonlinear sorption. One widely used nonlinear isotherm is the **Freundlich isotherm**, given by

$$S = K_{\mathrm{f}}C_{\mathrm{w}}^{n} \tag{6-4}$$

where K_{f} is the Freundlich sorption coefficient and n is a power function related to the sorption mechanism(s). Examples of nonlinear isotherms for the cases of $n > 1$ and $n < 1$ are presented in Figure 6-4.

Temperature has a small, but measurable effect on the sorption of organic pollutants. The effect of temperature depends on the type of sorption mechanism involved. As we have seen, the sorption of low-polarity organic compounds is governed by aqueous-solubility interactions. So, if a change in temperature can cause a change in solubility, it's reasonable to expect that a change in temperature may also cause a change in sorption. Relatively large organic compounds such as anthracene, for example, increase in water solubility with increasing temperature; thus, the sorption of anthracene may decrease with increasing temperature. The sorption mechanisms for inorganic pollutants, however, involve strong pollutant-sorbent interactions. An increase in temperature can increase the energetics of this interaction and hence produce an increase in sorption. The effect of salinity or ionic strength on sorption also depends on the type of sorption mechanism involved. For nonpolar, organic pollutants, the effect is usually relatively small, whereas it can be significant for inorganic pollutants.

The use of sorption isotherms is based on the assumption of equilibrium between the solid and water phases. This assumption is often valid for pollutants that sorb by cation exchange. However, it may not be valid for many organic pollutants. For example, research has shown that the rate of sorption of many nonpolar, organic pollutants is very slow, taking anywhere from several hours to several months to reach equilibrium. This slow rate of sorption is often due to the slow diffusion of pollutant molecules into spaces in the soil that have very small openings or pores. We know, for example, that the sorption of many organic pollutants is dominated by soil organic matter, which has a polymeric structure. Pollutant molecules can take a long time to move from the surface to the inside of such polymeric organic material. Also, some soils have solid particles that have small pore spaces, into which pollutant molecules diffuse very slowly.

6.4 Transport of Sorbing Pollutants

Sorption is such an important process because it reduces the transport velocity of a pollutant relative to that of water. That is, as long as a pollutant molecule is sorbed by soil, it is prevented from moving with the water. Thus, the pollutant takes longer to travel a given distance than water does. To account for the effect of sorption on pollutant transport, we have to modify the general chemical transport equation introduced in Chapter 5 (Eq. 5-18 on page 55). For simplicity assume a one-dimensional, homogeneous, porous medium and a constant water flow. Then the equation governing the transport of a water-soluble pollutant that can sorb to the soil particles is

$$\theta \frac{\partial C_w}{\partial t} + \varrho \frac{\partial S}{\partial t} = -q \frac{\partial C_w}{\partial z} + \theta D \frac{\partial^2 C_w}{\partial z^2} \tag{6-5}$$

where ϱ is bulk density of the soil (Mg m^{-3}); θ is the moisture content of the soil, which is equal to porosity when the soil is saturated (m^3 m^{-3}); z is distance (m), and the other terms are as defined previously.

In Eq. (6-5) there are two unknown concentration terms, C_w and S; thus, we need an additional equation to solve the problem—one that relates the concentration of pollutant in the sorbed phase to the aqueous concentration. This relationship is the isotherm function, which was introduced earlier. To take the simplest approach, we use the linear isotherm, replacing the S term in Eq. (6-5)

$$\frac{\partial S}{\partial t} = K \frac{\partial C_w}{\partial t} \tag{6-6}$$

and rewriting as follows (with rearrangement):

$$\left[1 + \frac{\varrho K}{\theta}\right] \frac{\partial C_w}{\partial t} = -v \frac{\partial C_w}{\partial z} + D \frac{\partial^2 C_w}{\partial z^2} \tag{6-7}$$

where v is equal to the velocity of the moving water. The term in the brackets in Eq. (6-7) is called the **retardation factor (*R*)**, which is defined by

$$R = \left[1 + \frac{\varrho K}{\theta}\right] = \frac{v_w}{v_p} = \frac{d_w}{d_p} \tag{6-8}$$

where v_w is the velocity of water (m s^{-1}), v_p is velocity of pollutant (m s^{-1}), d_w is the distance traveled by water, and d_p is the distance traveled by the pollutant. Inspection of Eq. (6-8) shows that, as sorption increases (*i.e.*, K increases), the retardation factor increases.

The retardation factor is an important component describing the transport of pollutants in soil. The effect of retardation on transport is illustrated in Figure 6-5 on page 72. When the pollutant is not sorbed by the soil (*i.e.*, $K = 0$), the retardation factor $R = 1$. This means that the pollutant moves at the same velocity as water ($v_p = v_w$). When $R = 2$, the pollutant moves at an effective velocity that is half that of water ($v_p = 0.5v_w$). In other words, the pollutant moves

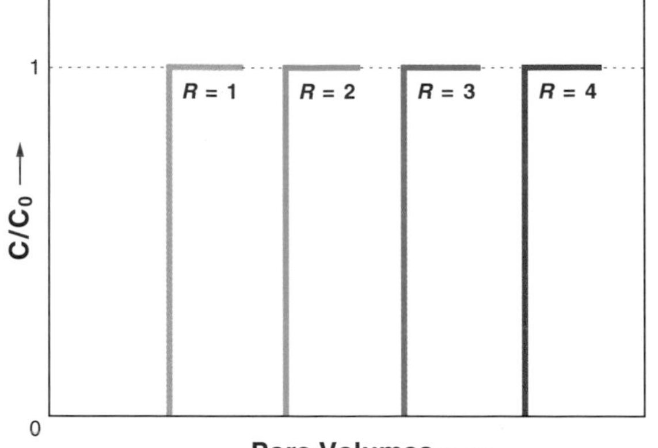

Figure 6-5 Impact of sorption on solute transport; note that the system is simplified by assuming no dispersion.

only half as fast as water. When $R = 10$, the pollutant moves at one-tenth the velocity of water ($v_p = 0.1\ v_w$).

Pollutants that have small retardation factors (<10) are considered to be relatively mobile. They can move rapidly from a spill site to contaminate a relatively large area. Conversely, pollutants that have very large retardation factors (>1000) move very slowly with respect to water and tend to create contaminated zones that are less extensive than those produced by mobile pollutants.

Some systems present a special case because the retardation factor R is less than 1. In this case the solute moves faster than the water in which it is dissolved. This behavior is sometimes seen in the transport of anions such as Cl^- or Br^-. As discussed previously, the surfaces of many soils have a net negative charge, which repels negatively charged solutes. For soil domains characterized by very small pores, this repulsion prevents the solute from entering the water residing in the pores. Thus, because the solute travels through only a portion of the soil, the result is solute transport that appears to be more rapid than the velocity of water movement. This **anion exclusion** phenomenon has been demonstrated in solute transport experiments. (See Figure 6-6 where Br^- is seen to move more quickly than tritiated water.) Another cause of $R < 1$ behavior is **size exclusion**, in which extremely large solutes are too large to pass through the smallest pores. Thus, as in the case of anion exclusion, the solute travels through only a portion of the soil, resulting in transport whose rate appears faster than the rate of water movement. Size exclusion has been observed in transport experiments for such solutes as dissolved humic acid and bacteria.

6.5 Estimating Phase Distributions of Pollutants

It is often important to know how a pollutant will distribute among the gas, liquid, and solid phases of soil. A simple way to estimate pollutant distributions is to use phase distribution coefficients. We are already familiar with two key distribution coefficients: the sorption coefficient K, which describes the distribution of a pollutant between solid and water phases, and Henry's constant H, which describes distribution between gas (air) and water phases. Now, how do we use these coefficients to obtain useful information?

The first piece of information of interest is obvious: How much pollutant is present in the soil system? We can define total mass of pollutant contained in a soil system as

$$M = V_w C_w + M_s S + V_g C_g \tag{6-9}$$

where M is total pollutant mass, V_w is the volume of the water phase, M_s is the mass of soil particles, and V_g is the volume of soil gas. Then we can define a unit-mass of pollutant (M^*) by dividing Eq. (6-9) by V_T, the total volume of the soil system. We can also define M^* in terms of one concentration by substituting the distribution equations (Eq. 6-2 on page 66, and Eq. 6-3 on page 70) into Eq. (6-9). We use C_w as the key concentration because it is usually the concentration measured in subsurface monitoring programs.

The modified equation for estimating pollutant mass is

$$M^* = (\theta + \varrho K + \theta_{gas} H) C_w \tag{6-10}$$

where θ_{gas} is the soil-gas content (volume of soil gas per volume of soil) and the other parameters have

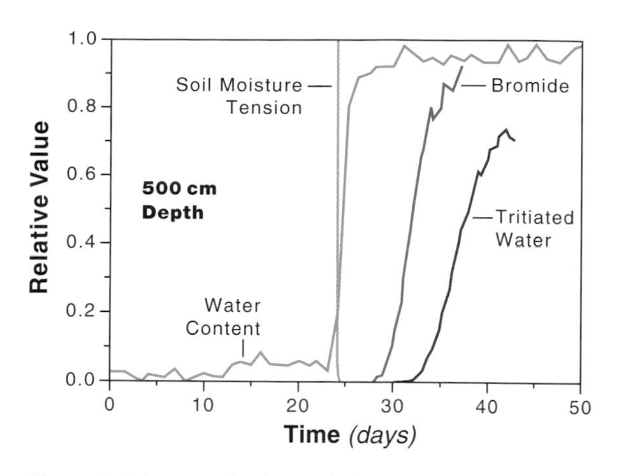

Figure 6-6 Impact of anion exclusion on solute transport. Adapted from Figure 6 on page 196 of I. Porro and P.J. Wierenga, 1993, *Ground Water,* volume 31, by permission of the Ground Water Publishing Company. All rights reserved.

already been defined. Remember that M^* is mass of pollutant per soil-component volume; thus, total pollutant mass (M) is calculated by

$$M = M^* V_T \quad (6\text{-}11)$$

The fraction of pollutant residing in the water (X_w), soil (X_s), and gas (X_{gas}) phases can be calculated by the following equations:

$$X_w = \frac{\theta C_w}{M^*} \quad (6\text{-}12)$$

$$X_s = \frac{\varrho K C_w}{M^*} \quad (6\text{-}13)$$

$$X_{gas} = \frac{\theta_{gas} H C_w}{M^*} \quad (6\text{-}14)$$

It is important to remember that this approach is based on the assumption that the distribution processes have reached equilibrium. If this is not true, the estimates obtained by using Eq. (6-12) through Eq. (6-14) can be erroneous.

Suppose we have the following values for the soil properties: $\theta = 0.25$ m^3 m^{-3}, $\theta_{gas} = 0.25$ m^3 m^{-3}, and $\varrho = 1.5$ g cm^{-3} (or, g mL^{-1}). For simplicity, let's say that $K = H = 1$ and assume that $C_w = 1$ mg L^{-1}. Substituting these values into Eq. (6-10) through Eq. (6-14), $M^* = 2$ mg L^{-1}, $X_w = 0.125$, $X_s = 0.75$, and $X_{gas} = 0.125$.

Thus, 75% of the pollutant mass is associated with the soil particles while 12.5% is associated each with water and with air.

6.6 Abiotic Transformation Reactions

Some pollutants can be transformed by **abiotic** physical and chemical processes. In this section we briefly examine two major abiotic transformation processes: hydrolysis and radioactive decay. (Biotic transformation of pollutants is discussed in Chapter 7, beginning on page 77.)

6.6.1 Hydrolysis

Because water is a ubiquitous component of soil systems, pollutants nearly always come into contact with water to some extent. It is important to know, therefore, whether and when a pollutant will react with water. Pollutants react with water by **hydrolysis**, a process that is generally written as

$$RX + H_2O \longrightarrow X^- + H^+ + ROH \quad (6\text{-}15)$$

where RX is an organic compound and X represents a functional group such as a halide (*e.g.*, Cl). By reacting with water, the original compound (RX) is transformed into another compound (ROH).

The two key factors in hydrolysis reactions are the charge properties of the pollutant molecules and the pH. Hydrolysis is essentially an interaction between **nucleophiles**, which are substances with excess electrons (*e.g.*, OH^-), and **electrophiles**, which are substances deficient in electrons (*e.g.*, H^+). Thus, the charge properties of its molecules govern the pollutant's reactivity with water. For many compounds, hydrolysis may be catalyzed or enhanced under acidic or basic conditions. This means that both the occurrence and rate of hydrolysis are often pH-dependent. For example, a hydrolysis reaction catalyzed by OH^- occurs more rapidly at higher pH values, where OH^- concentrations are higher. Hydrolysis can also be influenced by sorption interactions. For example, the pH at the surface of many soils is lower than the pH of the water surrounding the soil particles; thus, an acid-catalyzed hydrolysis reaction is enhanced when the pollutant is associated (sorbed) with the soil.

At a fixed pH, the rate of hydrolysis can be described with a first-order equation:

$$\frac{\partial C}{\partial t} = -k_h C \quad (6\text{-}16)$$

where k_h is the hydrolysis rate constant ($1/T$, where T = time). For transformation reactions it is useful to define a **half-life**, which is the time required for half of the original pollutant mass to be transformed. For reactions that follow first-order kinetics, the half-life $T_{1/2}$ is defined as

$$T_{1/2} = \frac{0.693}{k_h} \quad (6\text{-}17)$$

This equation can be used to estimate the time required for a pollutant to be transformed, and this time is related to the pollutant's persistence in the environment. For example, less than 1% of the original pollutant mass remains after a time period equal to 7 half-lives. After 10 half-lives, less than 0.1% remains. For a pollutant that has a half-life of 1 day, this means that only 0.1% would remain untransformed after 10 days.

6.6.2 Radioactive Decay

Radioactive decay is an important transformation reaction for a special class of pollutants, namely the radioactive elements. Radioactive decay is caused by an instability of the nucleus in the atom, such that either protons and neutrons or electrons are emitted in the form of radiation. This transformation is spontaneous, but the rate at which it occurs varies widely, depending on the element concerned. The rate of radioactive decay is represented by first-order kinetics:

$$\frac{\partial C}{\partial t} = -k_d C \quad (6\text{-}18)$$

where k_d is the radioactive decay constant ($1/T$). Half-lives for radioactive decay can be calculated by using Eq. (6-17), substituting k_d for k_h.

References and Recommended Reading

Brusseau M.L. and Rao P.S.C. (1989) Sorption nonideality during organic contaminant transport in porous media. *CRC Critical Reviews in Environmental Control*. **19**, 33–99.

Brusseau M.L. (1993) Complex mixtures and groundwater quality. *Environmental Research Brief*. EPA/600/S-93/004. U.S. Environmental Protection Agency, Washington, D.C.

Brusseau M.L. (1994) Transport of reactive contaminants in heterogeneous porous media. *Reviews of Geophysics*. **32**, 285–314.

Dragun J. (1988) *The Soil Chemistry of Hazardous Materials*. Hazardous Materials Control Research Institute, Silver Spring, Maryland.

Porro I. and Wierenga P.J. (1993) Transient and steady-state solute transport through a large unsaturated soil column. *Ground Water*. **31**, 193–200.

Schwarzenbach R.P., Gschwend P.M., and Imboden D.M. (1993) *Environmental Organic Chemistry*. Wiley, New York.

Verschueren K. (1983) *Handbook of Environmental Data on Organic Chemicals*, 2nd Edition. Van Nostrand Reinhold Company, New York.

Problems and Questions

1. The following chemicals are undergoing hydrolysis:

Chemical	K_h (h^{-1})
1	0.1
2	1
3	10
4	0.6
5	3

Calculate the half-life ($T_{1/2}$) of each chemical in hours.

2. You are given the following data:

$V_T = 1\ m^3$ $\quad H = 0.5$
$\theta = 0.3\ m^3\ m^{-3}$ $\quad K = 2\ cm^3\ g^{-1}$
$\theta_{gas} = 0.1$ $\quad C_w = 0.1\ mg\ L^{-1}$
$\varrho = 1.5\ g\ cm^{-3}$

(a) Calculate the total contaminant mass.

(b) Calculate the fraction of contaminant present in water, in soil atmosphere, and sorbed to the soil.

3. You are given the following data with $\varrho = 1.5\ g\ cm^{-3}$ and $\theta = 0.3\ m^3\ m^{-3}$:

Chemical	K ($cm^3\ g^{-1}$)
benzene	0.1
trichloroethane	0.2
chlorobenzene	0.4
naphthalene	0.6
PCB	10

(a) Calculate retardation factors for these chemicals.

(b) Calculate the distances traveled by each chemical, given the following information:

(i) the velocity of water moving through the soil is 1 m day^{-1}

(ii) the elapsed time is 300 days.

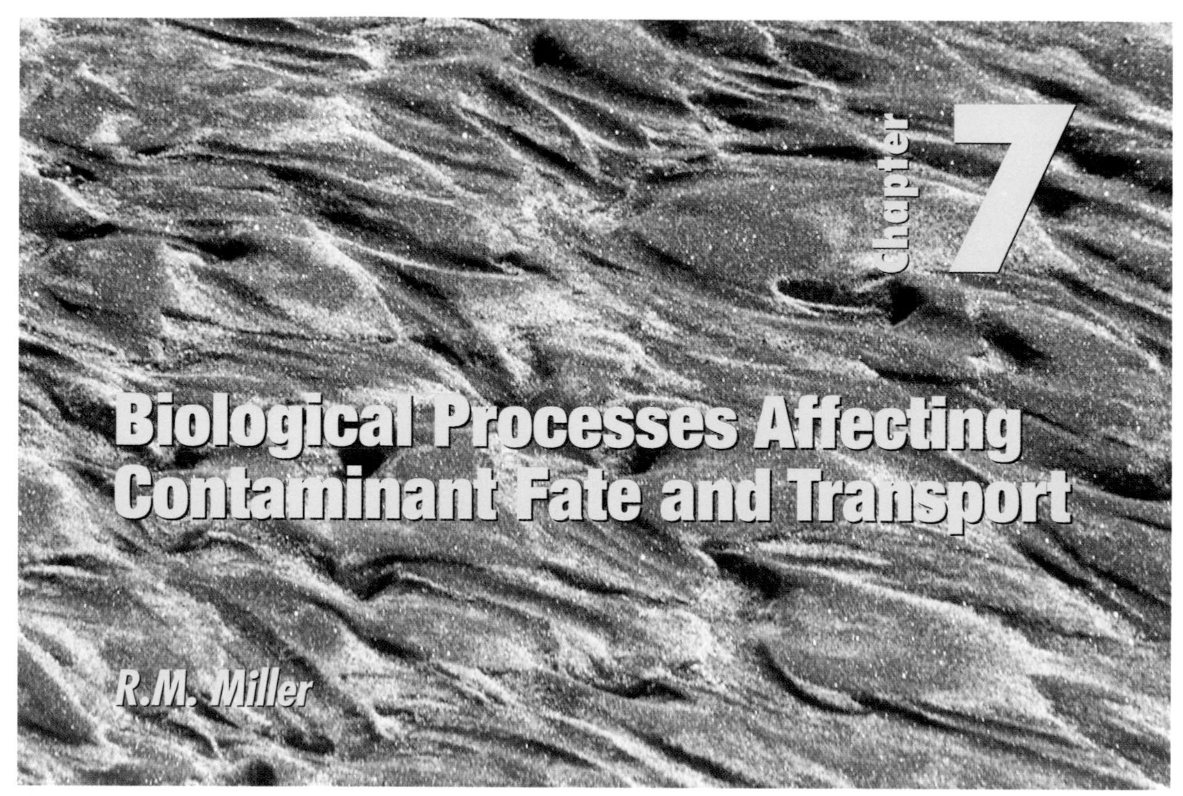

Above, sand contaminated with oil residues from a nearby oil refinery forms ripples on a southern California beach. Photograph: A.B. Brendecke.

7.1 Biological Effects on Pollutants

Although physical and chemical factors affect the fate of pollutants in soil and water, it is apparent that these are not the only factors governing the fate of pollutants. If they were, then large-scale accumulation of organic matter, including contaminants and natural organic substances, would dominate the earth, rendering our current pollution problems minuscule by comparison. Thus we must look at a third basic factor—the biological component of soil and water. This component, which is responsible for degradation of naturally occurring organic matter, also mitigates the impact of pollutants on the environment. Although less well studied by pollution scientists than the physical and chemical factors, the biological factor is receiving increasing attention owing to growing interest in the use of biological approaches to the remediation of contaminated sites. The presence of microorganisms in soil and water can affect the distribution, movement, and concentration of pollutants through a process called biodegradation. Indeed, some pollutants have very short lifetimes under normal environmental conditions because they serve as sources of food for actively growing microorganisms. For other pollutants, the effect of microorganisms may be limited for a variety of reasons. Low numbers of degrading microorganisms, microbe-resistant pollutant structure, or adverse environmental conditions can all cause extremely low rates of biodegradation.

In this chapter we will focus on understanding the interaction between microorganisms and pollutants in the environment. We will examine biodegradation of both organic and inorganic pollutants, as well as the effects that environmental parameters and pollutant structure have on these biodegradation reactions and reaction rates.

7.2 The Overall Process of Biodegradation

Biodegradation is the breakdown of organic compounds ("organics") through microbial activity. Biodegradable organic compounds serve as the food source, or **substrate** for microbes, and the availabil-

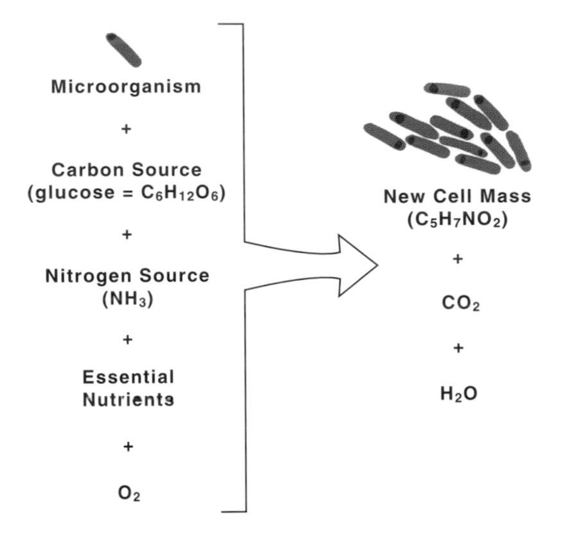

Figure 7-1 Aerobic mineralization of an organic compound.

ity of an organic to such microbes is the **bioavailability** of that organic. Bioavailability, which is one important aspect of the biodegradation of any substrate, depends largely on water. Microbial cells are 70–90% water, and the food they obtain comes from the water surrounding the cell. Thus, the bioavailability of a substrate refers to the amount of substrate in the water solution around the cell. One important factor that reduces bioavailability is sorption of substrate by soil (see Section 6.3, beginning on page 67).

We can think of biodegradation of organic compounds as "a series of biological degradation steps or a pathway which ultimately results in the oxidation of the parent compound." Often, the degradation of these compounds generates energy (as described in *Applied Theory: Biological Generation of Energy,* beginning on page 37 in Chapter 4). Complete biodegradation, or **mineralization**, involves oxidation of the parent compound to form carbon dioxide and water, a process that provides both carbon and energy for growth and reproduction of cells. Figure 7-1 illustrates the mineralization of any organic compound under aerobic conditions. The series of degradation steps comprising mineralization is similar, whether the carbon source is a simple sugar such as glucose, a plant polymer such as cellulose, or a pollutant molecule. Each degradation step in the pathway is facilitated by a specific catalyst, or **enzyme**, made by the degrading cell. Enzymes are most often found within a cell, but they are also made and released from the cell to help initiate degradation reactions. Enzymes found external to the cell are known as **exoenzymes**. Exoenzymes are important in the degradation of macromolecules such as the plant polymer cellulose because macromolecules must be broken down into smaller subunits to allow transport into the microbial cell. Both internal enzymes and exoenzymes are essential to the degradation process: degradation will stop at any step if the appropriate enzyme is not present (Figure 7-2). Lack of appropriate biodegrading enzymes is one common reason for the persistence of some pollutants, particularly those with unusual chemical structures that existing enzymes do not recognize. Thus, we can see that degradation depends on chemical structure. Pollutants that are structurally similar to natural substrates usually degrade easily while pollutants that are dissimilar to natural substrates often degrade slowly, or not at all.

Mineralization can also be described as a chemical reaction. As Figure 7-1 shows, in the presence of oxygen and a nitrogen source (such as ammonia, NH_3), glucose is converted to new cell mass (estimated by the formula $C_5H_7NO_2$), carbon dioxide, and water. Like glucose, many pollutant molecules—such as most gasoline components and many of the herbicides and pesticides used in agriculture—can be mineralized under the correct conditions.

Some organic compounds are only partially degraded. Incomplete degradation can result from the absence of the appropriate degrading enzyme, or it may result from **cometabolism**. In cometabolism, a partial oxidation of the substrate occurs, but the energy derived from the oxidation is not used to support growth of new cells. This phenomenon arises when organisms possess enzymes that coincidentally degrade a particular pollutant; that is, their enzymes are nonspecific. Cometabolism can occur not only during periods of active growth, but also during periods in which resting (nongrowing) cells interact with an organic compound. Although difficult to measure in the environment, cometabolism has been demonstrated for some environmental pollutants.

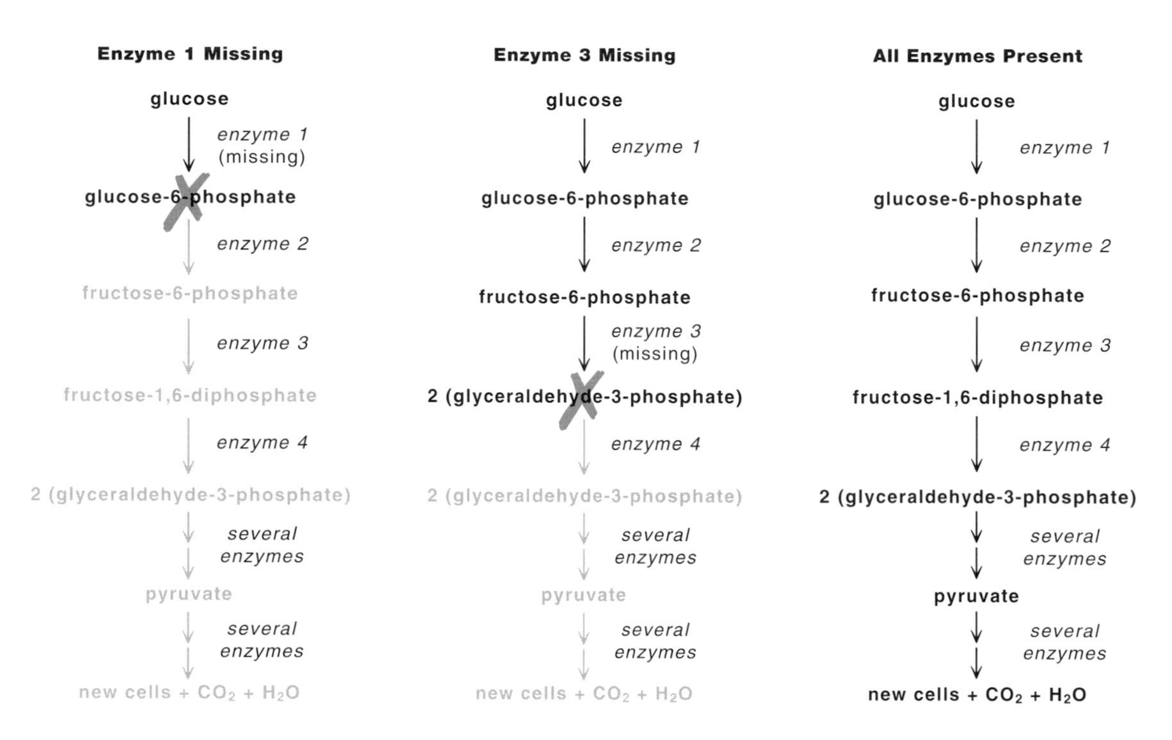

Figure 7-2 Stepwise degradation of organic compounds. A different enzyme catalyzes each step of the biodegradation pathway. If any one enzyme is missing, the product of the reaction it catalyzes is not formed (denoted with a red cross). The reaction stops at that point and no further product is made (shown in gray).

For example, the industrial solvent trichloroethene (TCE) can be oxidized cometabolically by **methanotrophic bacteria**, whose sole carbon substrate is methane. Trichloroethene is currently of great interest for several reasons. It is one of the most frequently reported contaminants at hazardous waste sites, it is a suspected carcinogen, and it is generally resistant to biodegradation.

As shown in Figure 7-3 on page 80, the first step in the oxidation of methane is catalyzed by **methane monooxygenase**, the enzyme produced by methanotrophic bacteria. This enzyme is so nonspecific that it can also cometabolically catalyze the first step in the oxidation of TCE when both methane and TCE are present. The bacteria receive no energy benefit from this cometabolic degradative step of the reaction. The subsequent degradation steps shown in Figure 7-3 may be catalyzed spontaneously, by other bacteria, or in some cases by the methanotrophs themselves. This cometabolic reaction may have great significance in remediation. Currently, research is under way to investigate the application of these methanotrophs to TCE-contaminated sites. Other cometabolizing microorganisms that grow on toluene, propane, and even ammonia are also being evaluated for use in bioremediation.

Partial or incomplete degradation can also result in **polymerization**, that is, the synthesis of compounds more complex and stable than the parent compound. This occurs when initial degradation steps, often catalyzed by exoenzymes, create highly reactive intermediate compounds, which can then combine either with each other or with other organic matter present in the environment. As illustrated in Figure 7-4 on page 80, which shows some possible polymerization reactions that occur with the herbicide propanil during biodegradation, these include formation of stable dimers or larger polymers, both

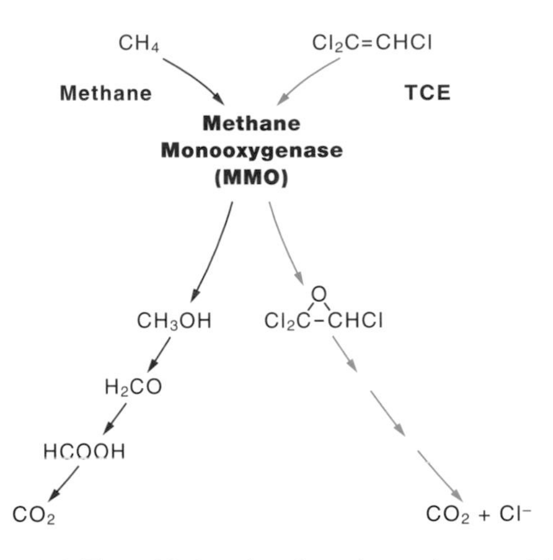

Figure 7-3 The oxidation of methane by methanotrophic bacteria is catalyzed by the enzyme methane monooxygenase. The same enzyme can act nonspecifically on TCE. Subsequent TCE degradation steps may be catalyzed spontaneously, by other bacteria, or in some cases by the methanotroph.

of which are quite stable in the environment. Such stability may be the result of low bioavailability (high sorption), or the absence of degrading enzymes.

7.3 Microbial Activity and Biodegradation

It is often difficult to predict the fate of a pollutant in the environment because the interactions between the microbial, chemical, and physical components of the environment are still not well understood. Total microbial activity depends on a variety of factors, such as microbial numbers, available nutrients, environmental conditions (including soil), and pollutant structure. In this section we will discuss the impact of some of the most important factors affecting microbial activity, with the implicit understanding that

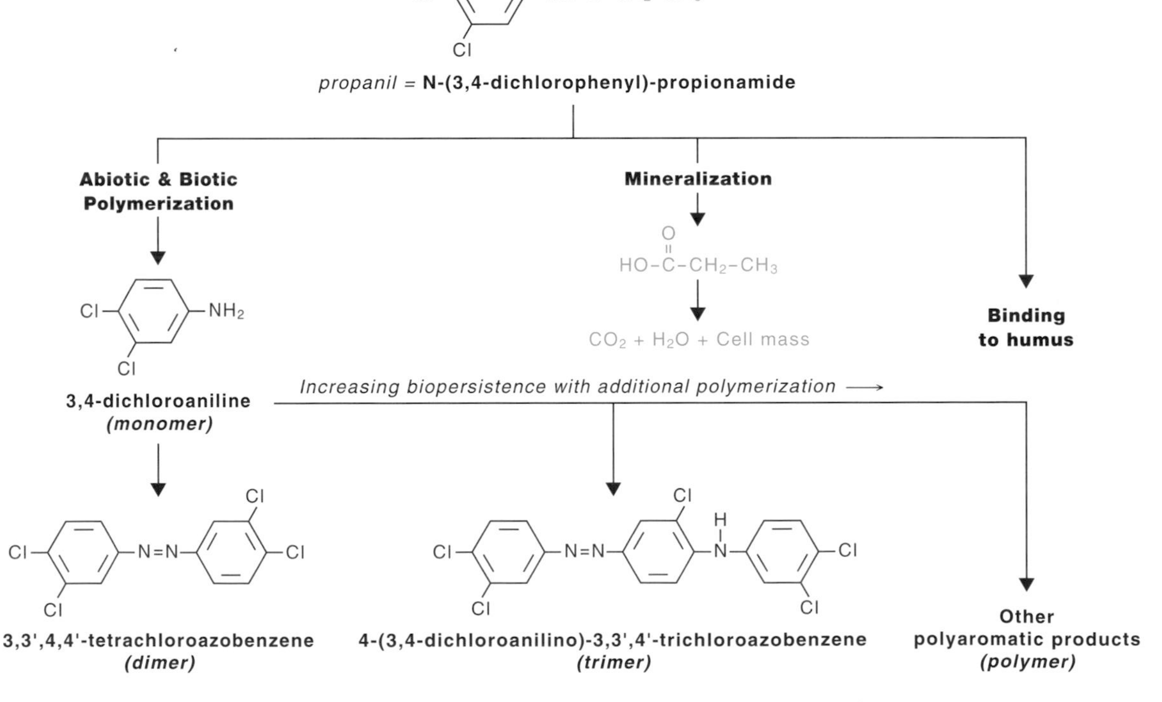

Figure 7-4 Polymerization reactions that occur with the herbicide propanil during biodegradation. Propanil is a selective post-emergence herbicide used in growing rice. It is toxic to many annual and perennial weeds. The environmental fate of propanil is of concern because it, like many other pesticides, is toxic to most noncereal crops. It is also toxic to fish. Care is used in propanil application to avoid contamination of nearby lakes and streams. From *Microbial Ecology, Fundamentals and Applications,* Third Edition, by R. Atlas and R. Bartha. Copyright (©) 1993 by The Benjamin/Cummings Publishing Company. Reprinted by permission.

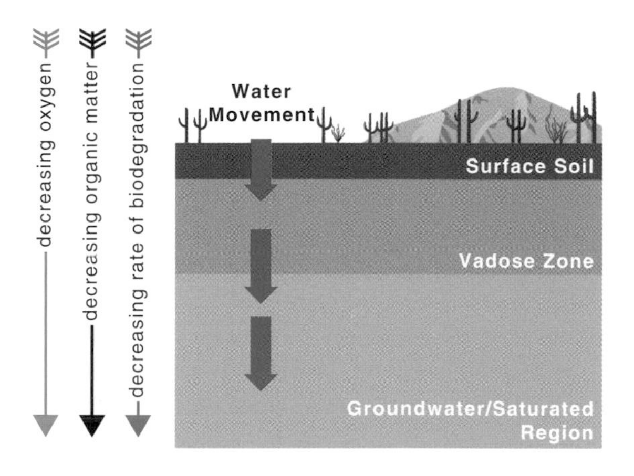

Figure 7-5 There are three major locations where contamination can occur in terrestrial ecosystems: surface soils, the vadose zone, and the saturated zone. The availability of both oxygen and organic matter varies considerably in these zones. As indicated, oxygen and organic matter both decrease with depth, resulting in a decrease in biodegradation activity with depth.

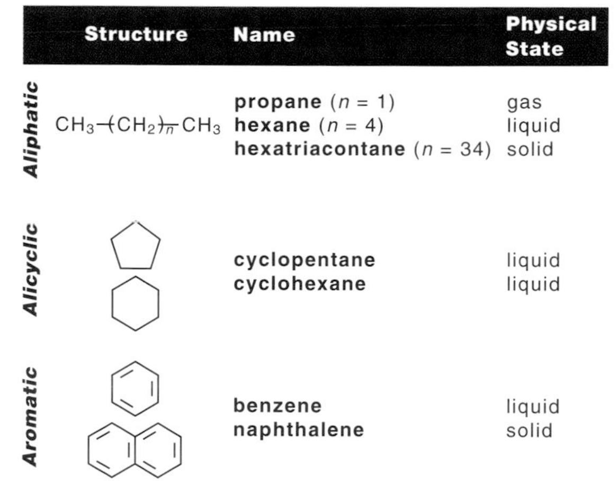

	Structure	Name	Physical State
Aliphatic	$CH_3-(CH_2)_n-CH_3$	propane (n = 1)	gas
		hexane (n = 4)	liquid
		hexatriacontane (n = 34)	solid
Alicyclic		cyclopentane	liquid
		cyclohexane	liquid
Aromatic		benzene	liquid
		naphthalene	solid

Figure 7-6 Aliphatic, alicyclic, and aromatic hydrocarbons.

microbial activity can be inhibited by any one of these factors, even if all other factors are optimal.

7.3.1 Environmental Effects on Biodegradation

The environment around a microbial community—that is, the sum of the physical, chemical, and biological parameters that affect a microorganism—determines whether a particular microorganism will survive and/or metabolize. The occurrence and abundance of microorganisms in an environment are determined by nutrient availability, as well as by various physicochemical factors such as pH, redox potential, temperature, and soil texture and moisture. (These factors are described in detail in Chapter 2, beginning on page 9, and Chapter 4, beginning on page 31.) Because a limitation imposed by any one of these factors can inhibit biodegradation, the cause of the persistence of a pollutant is sometimes difficult to pinpoint.

Oxygen availability, organic matter content, nitrogen availability, and bioavailability are particularly significant in controlling pollutant biodegradation in the environment. Interestingly, the first three of these factors can change considerably, depending on the location of the pollutant. As Figure 7-5 shows, contamination can occur in terrestrial ecosystems in three major locations: surface soils, the vadose zone, and the saturated zone. The availability of both oxygen and organic matter varies considerably in these zones. In general, oxygen and organic matter both decrease with depth, so biodegradation activity also decreases with depth. There are exceptions to this rule in shallow-groundwater regions, which can have relatively high organic matter contents because the rates of groundwater recharge are high.

7.3.1.1 Oxygen

Oxygen is very important in determining the extent and rate of biodegradation of pollutants. In general, biodegradation is much faster under **aerobic** (oxygen is present) conditions than under **anaerobic** (no oxygen is present) conditions. Also, some pollutants that degrade aerobically are not degradable anaerobically. Thus the saturated aliphatic hydrocarbons found in petroleum, some of which are shown in Figure 7-6, are readily degraded aerobically; but, unless an oxygen atom is present in the structure initially, these compounds are quite stable under anaerobic conditions (Figure 7-7 on page 82). This anaerobic stability explains why underground petroleum reservoirs, which contain no oxygen, have remained intact for thousands of years, even though microorganisms are present. In contrast, highly chlorinated organic compounds are more stable under aerobic conditions. That is, increasing chlorine content fa-

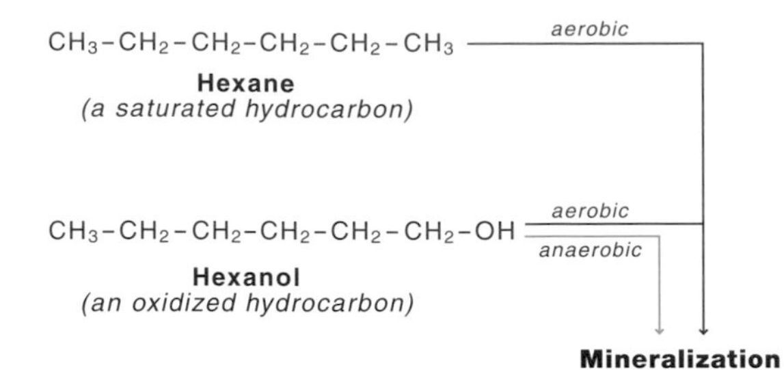

Figure 7-7 The effects of oxidation on the biodegradability of aliphatic compounds. Hydrocarbons with no oxygen, such as hexane, are only degraded aerobically, while the addition of a single oxygen atom (hexanol) enables both aerobic and anaerobic degradation to occur.

vors anaerobic dehalogenation (removal of chlorines) over aerobic dehalogenation.

In terms of oxygen availability, surface soils and the vadose zone are similar, being primarily aerobic regions. Thus, these regions tend to favor aerobic degradation of pollutants. However, these regions may contain pockets of anaerobic activity generated by localized conditions (*e.g.*, high biodegradative activity or confined zones of saturation) that reduce oxygen levels. In contrast, the oxygen concentrations in the groundwater or saturated region are low. The only oxygen that exists in this region is dissolved oxygen, and the oxygen levels are low because oxygen is not very water-soluble. Therefore, if significant microbial activity occurs, the limited supply of oxygen is rapidly used up, causing anaerobic conditions to develop (see also Section 11.4.1, beginning on page 160). Addition of air or oxygen can often improve biodegradation rates, particularly in subsurface areas with high clay content.

7.3.1.2 Microbial Populations and Organic Matter Content

Surface soils have large numbers of microorganisms. Bacterial numbers generally range from 10^6 to 10^9 organisms per gram of soil. Fungal numbers are somewhat lower, 10^4 to 10^6 per gram of soil. In contrast, microbial populations in deeper regions, such as the vadose zone and groundwater region, are often lower by two orders of magnitude or more. This large decrease in microbial numbers with depth is primarily due to differences in organic matter content. Whereas the soil surface may be rich in organic matter, both the vadose zone and the groundwater region often have low amounts of organic matter. One consequence of low total numbers of microorganisms is that the population of pollutant degraders initially present is also low. Thus, biodegradation of a particular pollutant may be slow until a sufficient biodegrading population has been built up. A second reason for slow biodegradation in the vadose zone and groundwater region is that the organisms in this region are often dormant owing to the low amount of organic matter present. If microorganisms are dormant, their response to an added carbon source is slow, especially if the carbon source is a pollutant molecule to which they have not been exposed.

Given these two factors, oxygen availability and organic matter content, we can make several generalizations about surface soils, the vadose zone, and the groundwater region (see Figure 7-5 on page 81):

1. Biodegradation in surface soils is primarily aerobic and rapid.
2. Biodegradation in the vadose zone is also primarily aerobic, but significant acclimation times may be necessary for sufficient biodegrading populations to build up.
3. Biodegradation in the groundwater region is initially slow owing to low numbers, and can rapidly become anaerobic owing to lack of available oxygen.

7.3.1.3 Nitrogen

Nitrogen is another macronutrient that often limits microbial activity because it is an essential part of many key microbial metabolites and building blocks, including proteins and amino acids. As shown by the chemical formula for a cell (Figure 7-1 on page 78), nitrogen is a large component by mass of microorganisms. It is also subject to removal from the soil/water continuum by various processes such as leaching or denitrification (see Section 14.2.2, beginning on page 212). Many xenobiotics are carbon-rich and nitrogen-poor; thus nitrogen limitations can inhibit their biodegradation while the simple addition of nitrogen-rich compounds can often improve it. For example, in the case of petroleum oil spills, where nitrogen shortages can be acute, biodegradation can be significantly accelerated by adding nitro-

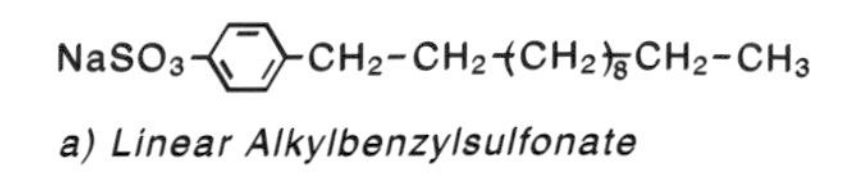

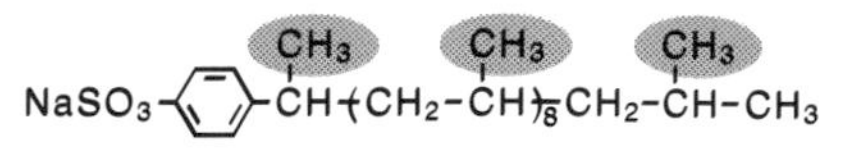

Figure 7-8 Linear and branched alkylbenzylsulfonates (ABS) are commonly used surfactants. As the linear variant (a) is readily biodegradable while the branched form (b) is not, and both work equally well as detergents, the linear ABS has entirely supplanted the branched ABS in environmentally conscious markets. From *Microbial Ecology, Fundamentals and Applications,* Third Edition, by R. Atlas and R. Bartha. Copyright (©) 1993 by The Benjamin/Cummings Publishing Company. Reprinted by permission.

gen fertilizers. In general, microbes have an average C:N ratio within their biomass of about 5:1 to 10:1, depending on the type of microorganism, so the C:N ratio of the material to be biodegraded must be 20:1 or less. The difference in the ratios is due to the fact that approximately 50% of the carbon metabolized is released as carbon dioxide, whereas almost all of the nitrogen metabolized is incorporated into the microbial biomass.

7.3.2 Pollutant Structure

The rate at which a pollutant molecule is degraded in the environment depends largely on its structure. If the molecule is not customarily found in the environment—or if its structure does not resemble that of a molecule usually found in the environment—a biodegrading organism may not be present. In this case, chances for biodegradation to occur are slim. The bioavailability of the pollutant is also extremely important in determining the rate of biodegradation. If the water solubility of the pollutant is extremely low, it will have low bioavailability (see Section 11.4.1, beginning on page 160). Many pollutant molecules that are persistent in the environment share the property of low water solubility. Examples include **d**ichloro**d**iphyenyl**t**richloroethane (DDT), a pesticide which is now banned in the United States; polychlorinated biphenyls (PCBs), which are used as heat-exchange fluids; and petroleum hydrocarbons. Both PCBs and petroleum hydrocarbons are liquids at room temperature and actually form a hydrophobic phase that is separate from the aqueous phase. Although microorganisms are not excluded from this phase, active metabolism seems to occur only in the aqueous phase or at the oil water interface. The second factor that reduces bioavailability is sorption of the pollutant by soil. Compounds that have low water solubility, such as DDT, PCBs, and petroleum constituents, are also prone to sorption by soil surfaces (see Section 6.3, beginning on page 67).

Many pollutants have extensive branching or functional groups that block or sterically hinder the pollutant carbon skeleton at the reactive site, that is, the site at which the substrate and enzyme come into contact during a transformation step. For example, we now use biodegradable detergents, namely the linear alkylbenzylsulfonates (ABSs). The only difference between these readily biodegradable detergents and the slowly biodegradable nonlinear ABSs is the absence of branching (Figure 7-8).

As a result of our increasing knowledge of the effect of pollutant structure on biodegradation in the environment, efforts are focusing on developing and utilizing "environmentally friendly" compounds. For example, slowly biodegradable pesticides are being replaced by rapidly biodegradable ones, which are used in conjunction with integrated pest-management approaches (see Section 17.5, beginning on page 263). This approach means that pesticides are not used on a yearly basis, but rather, are rotated. Thus, insects do not become fully acclimated to these easily degraded pesticides, and soil microorganisms degrade them so rapidly that they are active only during the intended time frame.

7.4 Biodegradation Pathways

The vast majority of the organic carbon available to microorganisms in the environment is material that has been photosynthetically fixed (plant material). Anthropogenic activity has resulted in the addition of many industrial and agricultural chemicals, including petroleum products, chlorinated solvents, and pesticides. Many of these chemicals are readily degraded in the environment because of their similarity to photosynthetically produced organic material. This allows degrading organisms to utilize preexisting biodegradation pathways. However, some

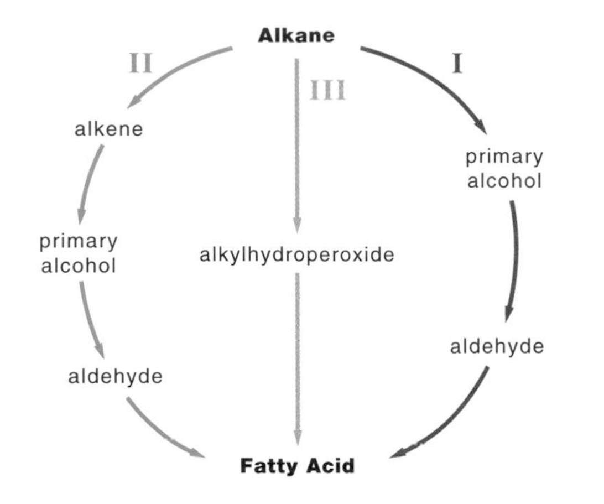

Figure 7-9 Aerobic biodegradation pathways for aliphatic compounds.

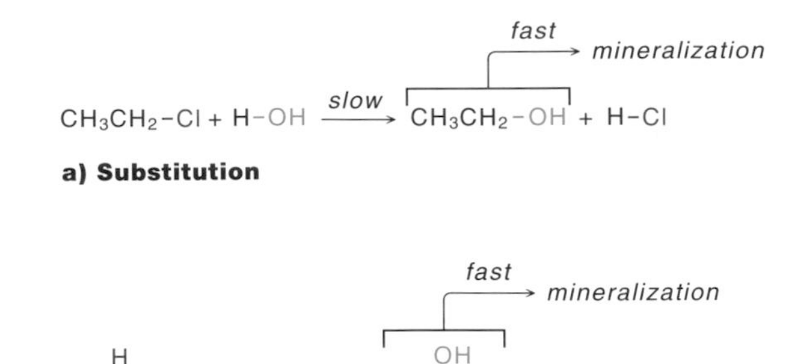

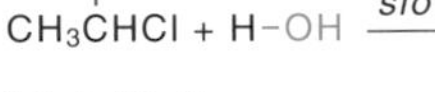

Figure 7-10 Aerobic biodegradation of chlorinated aliphatic compounds.

chemical structures are unique, or have unique components, which result in slow or little biodegradation. To help us understand and predict biodegradation of organic contaminants in the environment, we classify organic contaminants into one of three basic structural groups: the aliphatics, the alicyclics and the aromatics (see Figure 7-6 on page 81). Constituents of each of these groups can be found in all three physical states—gaseous, solid, and liquid. The general degradation pathways for each of these structural classes are delineated below. These pathways differ for aerobic and anaerobic conditions and can be affected by contaminant structural modifications.

7.4.1 Biodegradation under Aerobic Conditions

In the presence of oxygen, many heterotrophic microorganisms rapidly mineralize organic compounds (see Chapter 4, beginning on page 31). During degradation some of the carbon is completely oxidized to carbon dioxide to provide energy for growth, and some carbon is used as structural material in the formation of new cells (see Figure 7-1 on page 78). Energy used for growth is produced through a series of oxidation-reduction (redox) reactions in which oxygen is used as the terminal electron acceptor and reduced to water.

7.4.1.1 Aliphatic Hydrocarbons

Aliphatic hydrocarbons are straight-chain and branched-chain structures. Most aliphatic hydrocarbons introduced into the environment come from industrial solvent waste and the petroleum industry. Liquid aliphatics readily degrade under aerobic conditions, especially when the number of carbons is between 8 and 16 (see Figure 7-6 on page 81). Longer-chain aliphatics are usually waxy substances. Biodegradation of these longer chains is slowed due to limited water solubility, while biodegradation of shorter chains may be impeded by the toxic effects of the short-chain aliphatic on microorganisms. In addition several common structural modifications can result in severely reduced biodegradation. One of these modifications is extensive branching in the hydrocarbon chain. Commonly found in petroleum, branched hydrocarbons comprise one of the slowest degraded fractions therein. Another common modification that slows biodegradation is halogen substitution, as seen in the compound TCE, for example. Chlorinated solvents such as TCE have become a serious environmental pollution problem. Although such severely branched and highly chlorinated hydrocarbons degrade slowly, these solvents pose a more severe problem because of their toxicity. Thus, both the rate of biodegradation and toxicity must be considered in evaluating the potential hazard of pollutants in the environment.

Biodegradation of aliphatic compounds generally occurs by one of the three pathways shown in Figure 7-9. The most common is a direct enzymatic incorporation of molecular oxygen O_2 (pathway I). All three of these pathways result in the formation of a

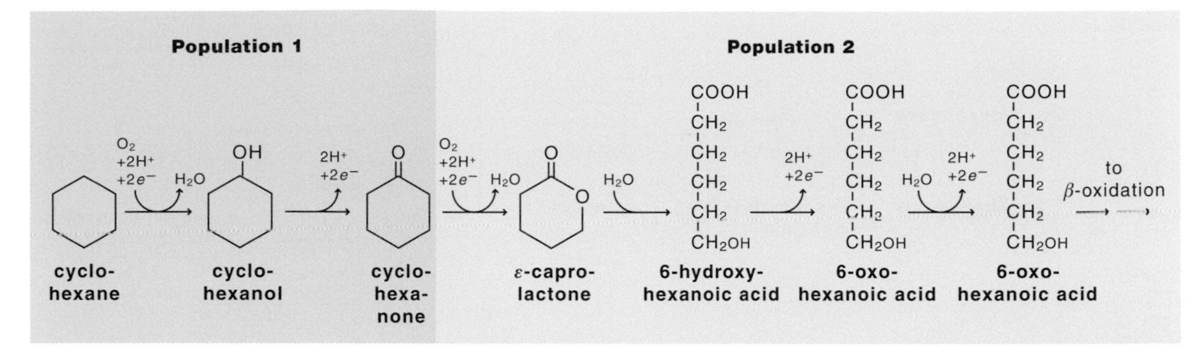

Figure 7-11 Aerobic biodegradation of cyclohexane by a microbial consortium. From *Microbial Ecology, Fundamentals and Applications,* Third Edition, by R. Atlas and R. Bartha. Copyright (©) 1993 by The Benjamin/Cummings Publishing Company. Reprinted by permission.

primary fatty acid. The fatty acid formed in degradation of an alkane is subject to normal cellular fatty acid metabolism. This includes β-oxidation, which cleaves off consecutive two-carbon fragments. Each two-carbon fragment is removed by coenzyme A (CoA) as acetyl-CoA and shunted to the tricarboxylic acid (TCA) cycle for complete degradation to CO_2 and H_2O. If the alkane has an even number of carbons, acetyl-CoA is the last residue. If the alkane has an odd number of carbons, propionyl-CoA is the last residue, which is also shunted to the TCA cycle after conversion to succinyl-CoA.

We know that both branching and halogenation slow biodegradation. In the former case, we can see that extensive branching causes interference between the degrading enzyme and the enzyme binding site. In the latter case, however, we need to know something about the bonds and the reactions involved. For halogenated compounds, the relative strength of the carbon-halogen bond requires two things: (1) an enzyme that can act on the bond and (2) a large input of energy to break the bond. In general, monochlorinated alkanes are considered degradable; however, increasing halogen substitution results in increased inhibition of degradation. Halogenated aliphatics can be degraded by two types of reactions that occur under aerobic conditions. The first is substitution, which is a nucleophilic reaction (the reacting species donates an electron pair) in which the halogen is substituted by a hydroxyl group. The second is an oxidation reaction, which requires an external electron acceptor. These two reactions are compared in Figure 7-10. Although increasing halogenation generally slows degradation, aerobic oxidation of highly chlorinated aliphatics can occur cometabolically.

7.4.1.2 Alicyclic Hydrocarbons

Alicyclic hydrocarbons are saturated carbon chains that form ring structures (Figure 7-6 on page 81). Naturally occurring alicyclic hydrocarbons are common. For example, alicyclic hydrocarbons are a major component of crude oil, comprising 20 to 67% by volume. Other examples of complex, naturally occurring alicyclic hydrocarbons include camphor, which is a plant oil; cyclohexyl fatty acids, which are components of microbial lipids; and the paraffins from leaf waxes. Anthropogenic sources of alicyclic hydrocarbons to the environment include fossil-fuel processing and oil spills, as well as the use of such agrochemicals as the pyrethrin insecticides.

It is very difficult to isolate pure cultures of bacteria that can degrade alicyclic hydrocarbons. For this reason, biodegradation of an alicyclic hydrocarbon is thought to take place as a result of teamwork among mixed populations of microorganisms. Such a team is commonly referred to as a **microbial consortium**. For example, in the degradation of cyclohexane, one population in the consortium performs the first two degradation steps, cyclohexane to cyclohexanone via cyclohexanol, but is unable to lactonize and open the ring. Subsequently, a second population in the consortium, which cannot oxidize cyclohexane to cyclohexanone, performs the lacton-

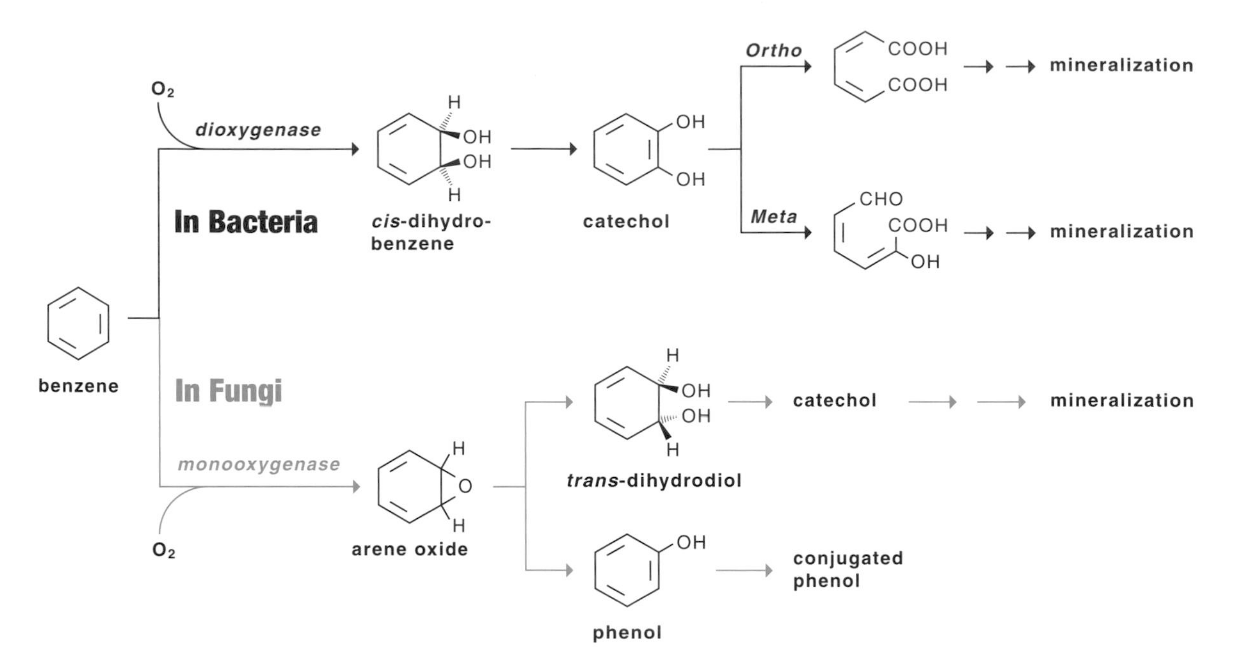

Figure 7-12 Aerobic biodegradation pathways of aromatic compounds in bacteria and fungi. Reprinted from *Microbial Degradation of Organic Compounds*, pp. 188 and 220 by courtesy of Marcel Dekker, Inc.

ization and ring-opening steps, then degrades the compound completely (Figure 7-11 on page 85).

Interestingly, cyclopentane and cyclohexane derivatives, which contain one or two hydroxyl, carbonyl, or carboxyl groups, degrade more readily in the environment than do their parent compounds. In fact, microorganisms capable of degrading of cycloalkanols and cycloalkanones are ubiquitous in environmental samples.

7.4.1.3 Aromatic Hydrocarbons

The **aromatic hydrocarbons** contain at least one unsaturated ring system with the general structure C_6R_6, where R is any functional group (see Figure 7-6 on page 81). The parent hydrocarbon of this class of compounds is benzene (C_6H_6), which exhibits the **resonance**, or delocalization of electrons, typical of unsaturated cyclic structures. Owing to its resonance energy, benzene is remarkably inert. [*Note*: As a group, the benzene-like "aromatics" tend to have characteristic aromas–hence the name.]

Aromatic compounds—including **polyaromatic hydrocarbons (PAHs)**, which contain two or more fused benzene rings—are synthesized naturally by plants. For example, they serve as a major component of the common plant polymer, lignin. Release of aromatic compounds into the environment occurs as a result of such natural processes as forest and grass fires. The major anthropogenic sources of aromatic compounds are fossil-fuel processing and utilization (burning). For example, benzene is one component of gasoline that is often released into the environment; it is of particular concern because it is a carcinogen.

Aromatic compounds, especially PAHs, are characterized by low water solubility and are therefore very hydrophobic. As is common with hydrophobic compounds, aromatics are often found sorbed to soil and sediment particles. The combination of low solubility and high sorption results in low substrate bioavailability and slow biodegradation rates. This is particularly true for PAHs having three or more rings because water solubility decreases as the number of rings increases.

A wide variety of bacteria and fungi can degrade aromatic compounds. Under aerobic conditions, the most common initial transformation is a hydroxylation, which involves the incorporation of molecular oxygen (O_2). The enzymes involved in these initial

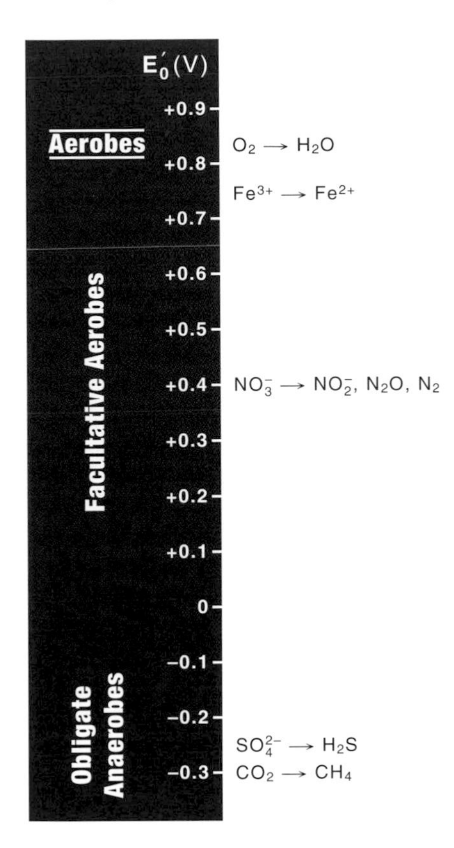

Figure 7-13 The range of redox potentials found in environments commonly inhabited by actively metabolizing microorganisms. While aerobic conditions refer to a specific potential, facultative and obligate anaerobes metabolize over spectrums of redox potential.

transformations fall into two groups: **dioxygenases**, which incorporate both atoms of molecular oxygen into the PAH, and **monooxygenases**, which incorporate only one atom of molecular oxygen. In general, bacteria transform PAHs by an initial dioxygenase attack at a *cis*-dihydrodiol, which is subsequently rearomatized to afford a dihydroxylated intermediate called catechol. The ring is then cleaved by a second dioxygenase, as shown in Figure 7-12, using either an *ortho*- or a *meta*-pathway, and further degraded.

Fungi transform PAHs by an initial monooxygenase attack. This enzyme incorporates one atom of molecular oxygen into the PAH and reduces the second oxygen to water. The result is the formation of an arene oxide, followed by the enzymatic addition of water to yield a *trans*-dihydrodiol (Figure 7-12). Alternatively, the arene oxide can be isomerized to form phenols, which can be conjugated with sulfate, glucuronic acid, and glutathione. These conjugates are similar to those formed in higher organisms and seem to aid in detoxification and elimination of PAH. In general, PAHs having two or three condensed rings are transformed rapidly, often mineralizing completely, whereas PAHs with four or more condensed rings are transformed much more slowly, often as a result of cometabolic attack.

7.4.2 Biodegradation Pathways under Anaerobic Conditions

Anaerobic conditions are not uncommon in the environment. Most often, such conditions develop in water or saturated sediment environments. But even in well-aerated soils there are microenvironments with little or no oxygen. In all of these environments, **anaerobiosis** occurs when the rate of oxygen consumption by microorganisms is greater than the rate of oxygen diffusion through either air or water. In the absence of oxygen, organic compounds can be mineralized through **anaerobic respiration**, in which an electron acceptor other than oxygen is used (see Section 7.3.1.2 on page 82). The series of alternative electron acceptors in the environment includes iron, nitrate, manganese, sulfate, and carbonate, which are listed in order from most oxidizing to most reducing. This progression means they are usually utilized in this order because the amount of energy generated for growth depends on the oxidation potential of the electron acceptor. Since none of these electron acceptors are as oxidizing as oxygen, growth under anaerobic conditions is never as efficient as growth under aerobic conditions (Figure 7-13). (See also *Respiration* on page 39 in Chapter 4.)

Anaerobic degradation pathways have not been as extensively studied as the pathways for aerobic degradation of organic compounds. Interestingly, many compounds that are easily degraded aerobically, such as saturated aliphatics, are far more difficult to degrade anaerobically. However, in at least one group of compounds—those that are highly halogenated—the halogen substituents are removed more rapidly under anaerobic conditions. But once dehalogenation has occurred, the remaining molecule behaves more typically; that is, it is generally degraded

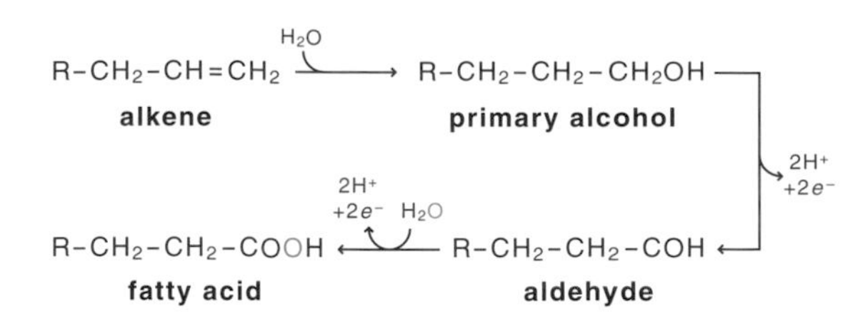

Figure 7-14 General anaerobic biodegradation pathway of an alkene.

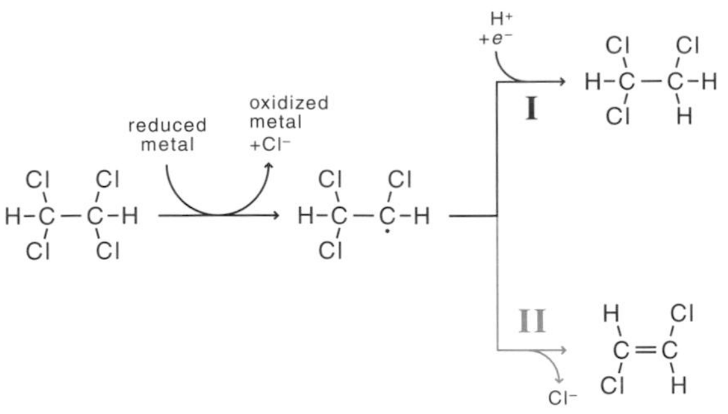

Figure 7-15 Reductive dehalogenation of a chlorinated hydrocarbon in the presence of a metal to form an alkyl radical. I) The alkyl radical scavenges a hydrogen atom. II) The alkyl radical loses a second halogen to form an alkene.

more rapidly and extensively aerobically than anaerobically. As a consequence of this sequential process, technologies have been developed that utilize sequential anaerobic-aerobic treatments to optimize degradation of highly halogenated compounds.

7.4.2.1 Aliphatic Hydrocarbons

Saturated aliphatic hydrocarbons are degraded slowly, if at all, under anaerobic conditions. We see evidence of this slow to nonexistent degradation in nature; for example, hydrocarbons in natural underground reservoirs of oil (which are under anaerobic conditions) are not degraded, despite the presence of microorganisms. However, both unsaturated aliphatics and oxygen-containing aliphatics (aliphatic alcohols and ketones) are readily biodegraded anaerobically. The suggested pathway of biodegradation for unsaturated hydrocarbons is the hydration of the double bond to an alcohol, with further oxidation to a ketone or aldehyde, followed finally by formation of a fatty acid (Figure 7-14).

Halogenated aliphatics can be partially or completely degraded under anaerobic conditions through a transformation reaction called **reductive dehalogenation**. Often a cometabolic degradation step, reductive dehalogenation may be mediated by reduced transition-metal/metal complexes. The steps in this transformation are shown in Figure 7-15. In the first step, electrons are transferred from the reduced metal to the halogenated aliphatic, resulting in an alkyl radical and free halogen. Then, the alkyl radical can either scavenge a hydrogen atom (I), or lose a second halogen to form an alkene (II). In general, anaerobic conditions favor the degradation of highly halogenated compounds, while aerobic conditions favor the degradation of mono- and disubstituted halogenated compounds.

7.4.2.2 Aromatic Hydrocarbons

Like aliphatic hydrocarbons, aromatic compounds can be completely degraded under anaerobic conditions if the aromatic is oxygenated. Recent evidence also indicates that even nonsubstituted aromatics are degraded slowly under anaerobic conditions. Anaerobic mineralization of aromatics often requires a mixed microbial community whose populations work together under different redox potentials. For example, mineralization of benzoate can be achieved by growing an anaerobic benzoate degrader in coculture with an aerobic methanogen or sulfate reducer. In this consortium, benzoate is transformed by one or more anaerobes to yield aromatic acids, which in turn are transformed to methanogenic precursors such as acetate, carbon dioxide, or formate. These small molecules can then be utilized by methanogens (Figure 7-16). This process can be described as an anaerobic food chain because the organisms higher in the food chain cannot utilize acetate or other methanogenic precursors, while the methanogens cannot utilize larger molecules such as benzoate. Methanogens utilize carbon dioxide as a terminal electron acceptor, thereby forming methane. [*Note*: Methanogens should not be confused with methanotrophic bacteria, which aerobically oxidize methane to carbon dioxide.]

7.5 Transformation of Metal Pollutants

Metals comprise a second important class of pollutants. In fact, a 1984 U.S. Environmental Protection Agency (EPA) survey of 395 remedial action sites revealed that metals (*e.g.*, lead, cadmium, zinc, copper) were the single most prevalent class of contaminants. Metals are essential components of microbial cells. For example, sodium and potassium regulate gradients across the cell membrane, while copper, iron, and manganese provide metalloenzymes for photosynthesis and electron transport. On the other hand, metals can also be extremely toxic to microorganisms. Although the most toxic metals are the nonessential metals such as cadmium, lead, and mercury, even essential metals can become toxic in high concentrations.

What is the fate of metals in the environment? Metals and metal-containing contaminants are not degradable in the same sense that carbon-based molecules are for two reasons. (1) Unlike carbon, the metal atom is not the major building block for new cellular components. (2) While a significant amount of carbon is released to the atmosphere in gaseous form as carbon dioxide, the metals rarely enter the environment in a volatile, or gaseous, phase. (There are some exceptions to this—most notably mercury and selenium, which can be transformed and volatilized by microorganisms under certain conditions.) In general, however, the nondegradability of metals means that it is difficult to eliminate metal atoms from the environment. Therefore, localized, elevated levels of metals are common, especially in industrially developed countries. Consequently, these metals can accumulate in biological systems, where their toxicity poses serious threats to human and environmental health.

Metals and metal-containing molecules can undergo transformation reactions, many of which are mediated by microorganisms. The nature of these reactions is important for consideration of metal toxicity in the environment because toxic effects more often depend on the form of the metal than on the total metal concentration. In general, the most active

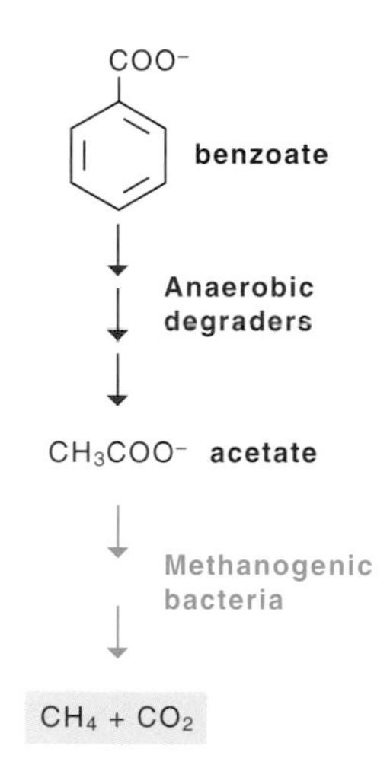

Figure 7-16 An example of an anaerobic food chain showing the formation of simple compounds from benzoate by a population of anaerobic bacteria and the subsequent utilization of the newly available substrate by a second anaerobic population, the methanogenic bacteria.

forms of added metals are free metal ions. The metals having the highest toxicity are the cations of mercury (Hg^{2+}) and lead (Pb^{2+}), although other metallic cations (arsenic, beryllium, boron, cadmium, chromium, copper, nickel, manganese, selenium, silver, tin, and zinc) also exhibit significant toxic effects. (The specific toxicities of selected metals to humans are discussed in Section 21.8.1, beginning on page 338.)

7.5.1 Effects of Metals on Microbial Metabolism

The nature of the interaction between heavy metals and microorganisms is complex. Metal toxicity requires uptake of the metal by a cell, which is dependent on many factors, such as pH, soil type, and temperature. For example, accumulation of metals by cell-surface binding increases with increasing pH. Transport of metal into a cell is also pH-dependent, with maximal transport rates in the pH range 6.0–7.0. Once a metal is taken up by a cell, toxicity can result. After an initial period in which the toxic effects of the metals are evident, microorganisms often acquire tolerance mechanisms that enable them to repair metal toxicity damage, after which they can start metabolizing and growing again at a nearly normal rate. The length of time required to develop tolerance mechanisms is influenced both by biotic

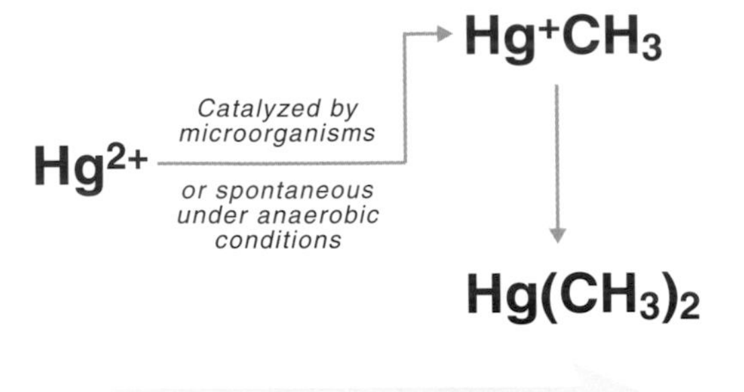

Figure 7-17 Microbial alkylation of mercury.

and abiotic factors. The biotic factors of importance may involve the physiological state of the organism in question, such as the nutritional level, or genetic adaptations that result in metal resistance. Abiotic factors include the physicochemical characteristics of the environment such as pH, temperature, and redox potential, all of which affect the precipitation and complexation of metals.

The specific toxic effects of heavy metals on microorganisms are caused by the binding of the metal to cellular ligands such as proteins or nucleic acids. This metal-ligand binding leads to conformational changes and loss of normal ligand activity. For example, the particularly strong affinity of cationic metals for protein sulfhydryl groups can lead to alterations in protein folding. Both the ligand structure and the size of the metal affect the type of binding. Large metal ions such as copper, silver, gold, mercury, and cadmium preferentially form covalent bindings with sulfhydryl groups. In contrast, small, highly electropositive metal ions such as aluminum, chromium, cobalt, iron, titanium, zinc, and tin preferentially complex with carboxyl, hydroxyl, phosphate, and amino groups.

7.5.2 Microbial Transformations of Metals

Microorganisms have developed various resistance mechanisms to prevent metal toxicity: among these mechanisms we will briefly discuss metal oxidation/reduction, metal complexation, and alkylation of metals.

Oxidation/Reduction: Metal oxidation enhances metal mobility, stimulating metal movement away from the cell. For example, the Gram-positive bacterium *Bacillus megaterium* can oxidize elemental selenium to selenite, a reaction that increases selenium mobility. Alternatively, some microorganisms reduce metals such as chromium, causing them to precipitate and become immobilized.

Complexation: Other microorganisms can effectively complex metals to polymeric materials either internal or external to the cell. For example, uranium has been found to accumulate extracellularly as needle-like fibrils in a layer approximately 2 μm thick on the surface of a yeast, *Saccharomyces cerevisiae*. In contrast, uranium accumulates as dense intracellular deposits in *Pseudomonas aeruginosa*.

Alkylation: Some microorganisms can transform metals by alkylation, which involves the transfer of one or more organic ligand groups (*e.g.*, methyl groups) to the metal, thus affording stable organometallic compounds. One such metal is mercury, as shown in Figure 7-17. The physical and chemical properties of organometals are different from those of pure metals. For example, volatility is increased, thus facilitating the movement of the organometal through the soil solution and, ultimately, into the atmosphere. In general, the extent of organometallic mobility depends on the nature and number of the organic ligands involved. But the presence of even a single methyl group can significantly increase both the volatility and **lipophilicity** (the affinity for lipids) of a metal. Transformations that increase volatility allow microorganisms to help remove a toxic metal from its environment; however, the concomitant increase in lipophilicity can cause biomagnification of the metal, resulting in toxic effects on other members of the ecosystem.

References and Recommended Reading

Atlas R.M. and Bartha R. (1993) *Microbial Ecology*, 3rd Edition. Benjamin Cummings, Menlo Park, California.

Chapelle F.H. (1993) *Ground Water Microbiology and Geochemistry*. John Wiley and Sons, New York.

Gibson D.T. and Subramian V. (1984) *Microbial Degradation of Organic Compounds*. Marcel Dekker, New York.

Hughes M.N. and Pook R.K. (1989) *Metals and Microorganisms*. Chapman and Hill, New York.

Leahy J.G. and Colwell R.R. (1990) Microbial degradation of hydrocarbons in the environment. *Microbiological Reviews*. **54**, 305–315.

Pitter P. and Chudoba J. (1990) *Biodegradability of Organic Substances in the Aquatic Environment*. CRC Press, Boca Raton, Florida.

Problems and Questions

1. Define biodegradation, making reference to the terms transformation, mineralization, and cometabolism.

2. (a) Consider *n*-octane, an eight-carbon, straight-chain aliphatic compound. Draw its structure.

   (b) Is this compound biodegradable?

   (c) Beginning with the structure you have drawn, show how you can alter it to make it less biodegradable.

   (d) Compare the structure in part (a) with Figure 7-1 on page 78, and consider the following situation: A site is contaminated by a leaking underground gasoline storage tank. The remediation firm that you work for would like to use bioremediation to clean the site. Given the structure of gasoline components (which typically include simple aliphatic, alicyclic, and aromatic compounds), what other nutrients may be required to complete bioremediation? Explain.

3. List and explain the factors that determine bioavailability of an organic compound.

4. Compare the following structures:

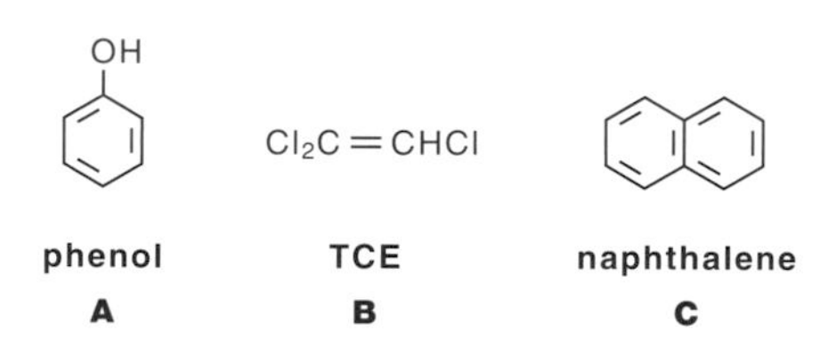

   (a) Predict the order of bioavailability.

   (b) Predict the order of biodegradability.

   (c) What does a comparison of (a) and (b) tell you about the relationship between bioavailability and biodegradability?

   (d) Which is the most likely type of biodegradation (aerobic or anaerobic) for each of the above compounds?

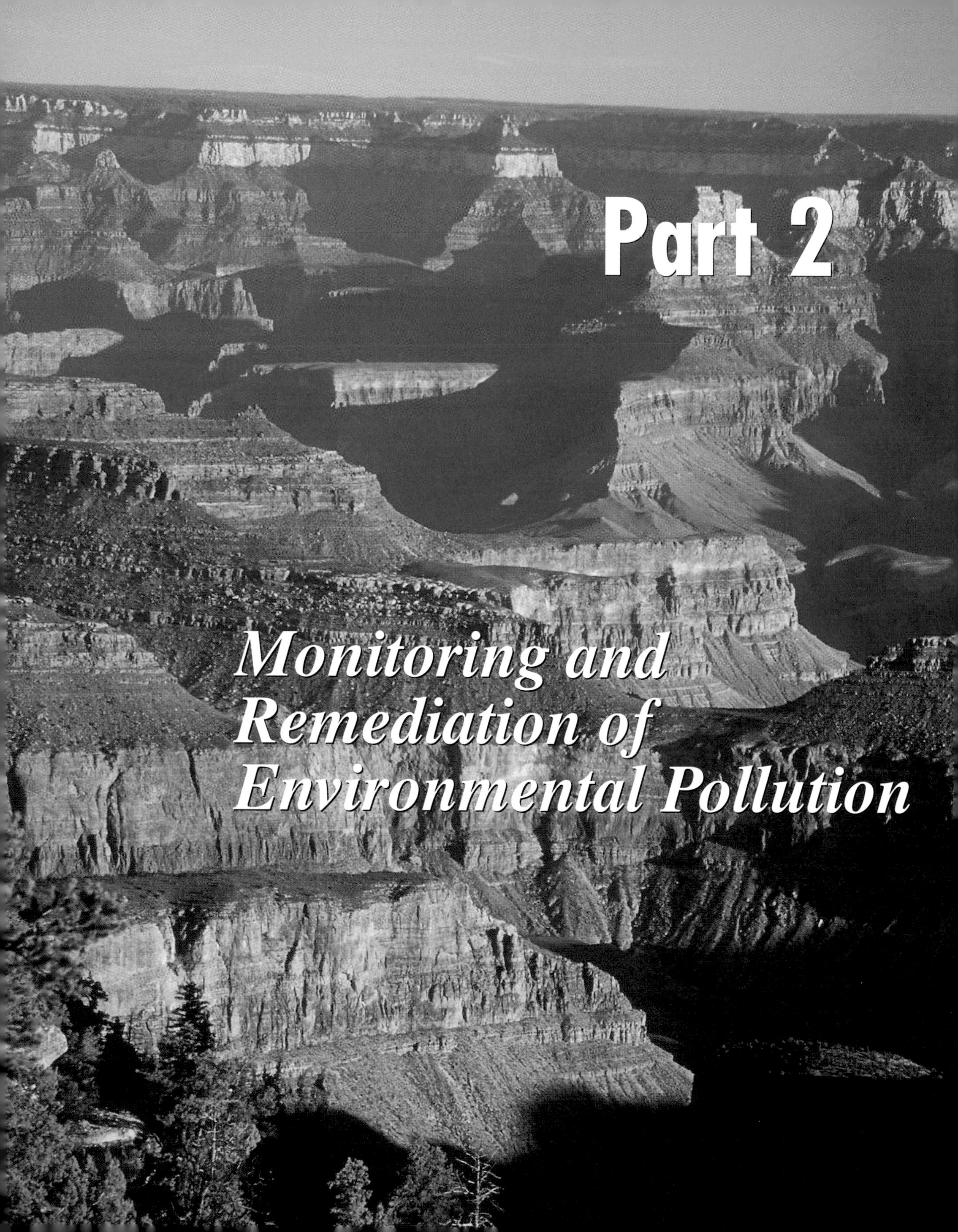

Part 2

Monitoring and Remediation of Environmental Pollution

On Reverse *Not even such majestic natural wonders as the Grand Canyon in Arizona are immune to the effects of pollution. Air quality in the canyon is constantly monitored as pollution from a coal-burning electrical generation plant located in the Four Corners area 400 km away may be creating a haze that has been obscuring the view of the canyon.*

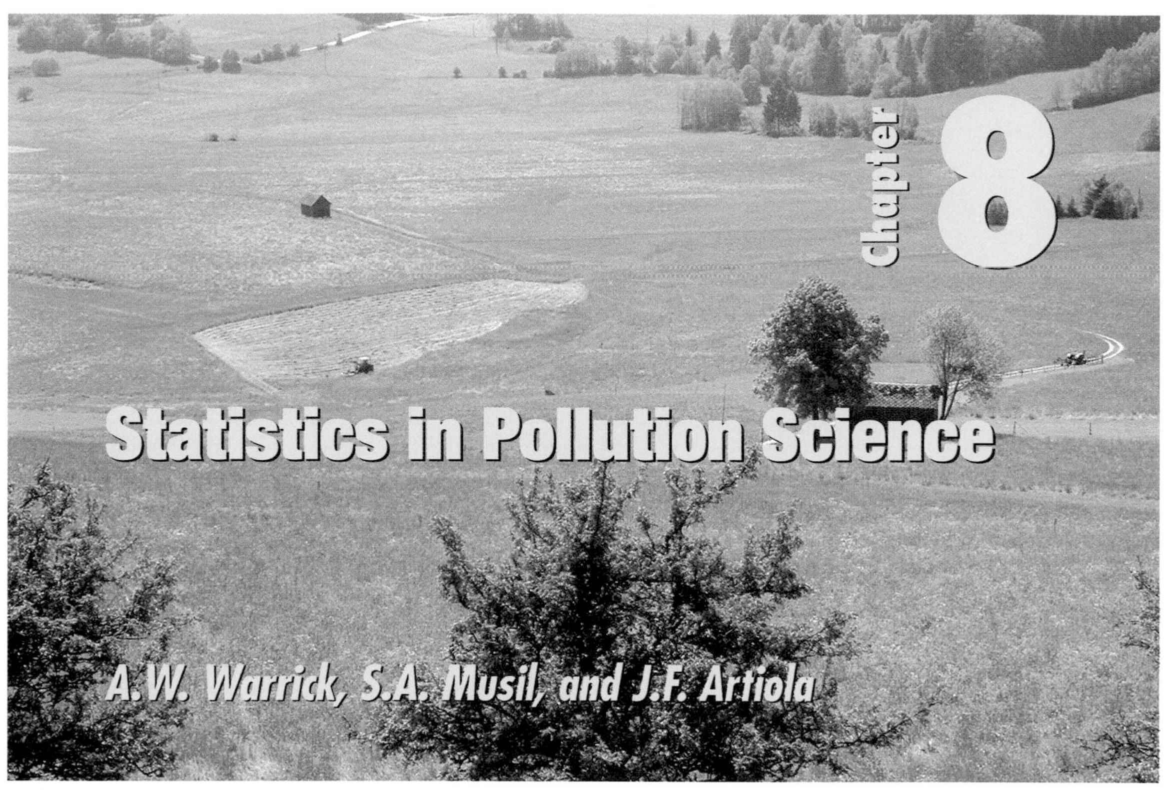

Reliable statistical analysis of environmental data is possible only through an understanding of proper sampling numbers and inherent variability of the sample as it may effect the measured value. Photograph: J.W. Brendecke.

8.1 Statistical Concepts

Statistics is a branch of mathematics that deals with the collection, analysis, and presentation of observations expressed as numbers. After an analysis, we often present numerical data in tables, graphs, and even cartoons (pictograms), which may appear in specialized scientific journals, widely read newspapers, popular magazines, or on television programs. In pollution science, as in most areas of modern life, statistical methods are applied extensively to quantify and evaluate data derived from observations. Specifically, we use statistics as a framework for reaching conclusions about environmental factors on the basis of field observations taken from a sampled site. In this chapter, we'll look at the basic principles of setting up measurement schemes, making statistical analyses, and presenting results.

8.1.1 The Statistical Sample

In pollution science, we often use the noun **sample** to denote a physical specimen or a measurement. Thus, we may speak of a vial containing a water "sample." As a verb, sample generally means to obtain a physical specimen or make a measurement. In statistics, however, the term sample often denotes the subset of the whole population that is being investigated. In this sense, a sample would be an entire set of vials collected for a particular study, rather than a single vial.

Proper sampling techniques are the foundation of acceptable statistical results. Generally, sampling involves collecting data either by physically removing specimens from a study area or by making a series of on-site measurements. In some cases, however, sampling may involve the systematic collection of opinions, as in a poll. In either case, acceptable sampling techniques and protocols require that the samples collected adequately reflect the whole **target population**. That is, our sampling techniques must result in good *statistical* samples, so that the numbers we collect represent what we're trying to measure. Only by using appropriate sampling techniques—and obtaining good statistical samples—will we be able to

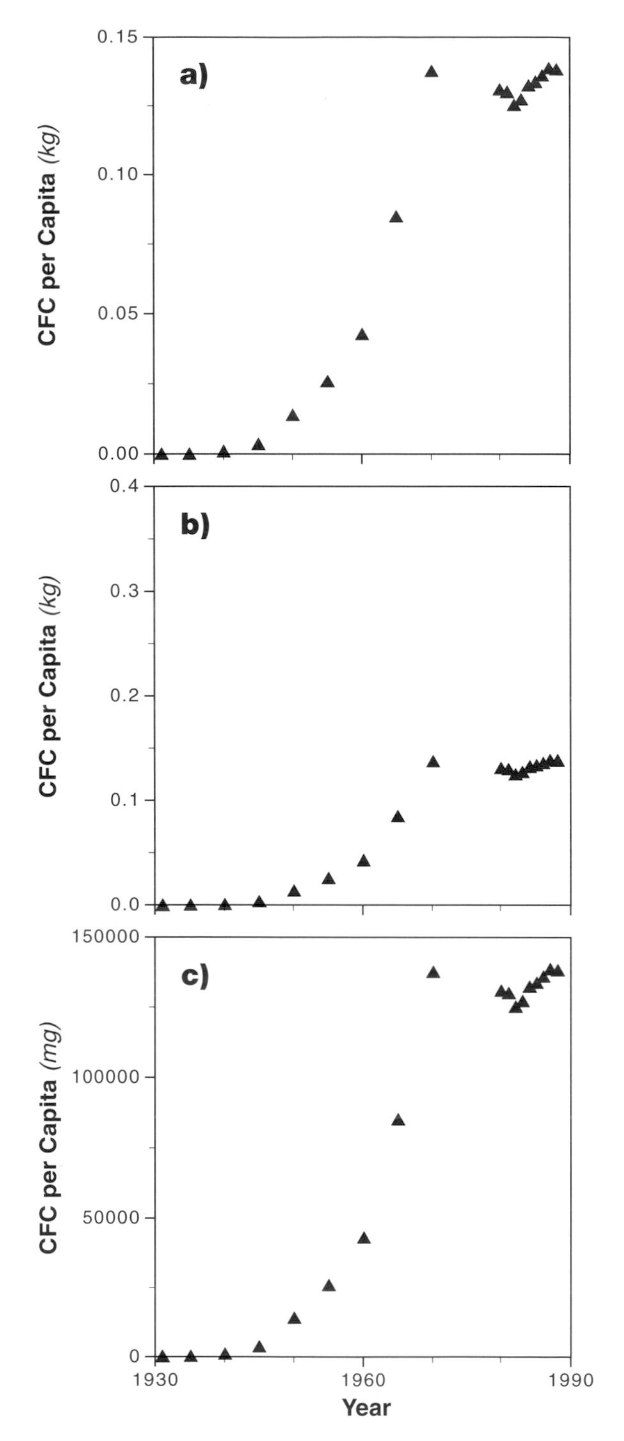

Figure 8-1 Annual chlorofluorocarbon (CFC) releases per capita. All three plots are for the same data but have different vertical scales or units. Data: Council on Environmental Quality, 1992; U.S. Department of Commerce, 1990.

analyze data accurately and thereby draw defensible, objective conclusions. (Specific sampling techniques for environmental monitoring are described in more detail in Chapter 9, beginning on page 115.)

8.1.2 Visual Presentation of Data

Numerical data can be presented in a variety of visual ways, including line graphs, bar charts, histograms, pie charts, and pictograms. Each of these methods provides a "picture" of data that can influence both the first impressions and the ultimate interpretation of someone trying to understand them. Generally, the most meaningful data presentations are graphs, which can have a variety of units, scales, legends, and colors. In any case, all presentations should be as simple as possible and clearly labeled.

Graphical displays also make it easy to abuse data. Consider, for example, the effect of scale on a graph (Figure 8-1) of the estimated annual global release per capita of chlorofluorocarbons, or CFCs. These compounds, which are commonly used in refrigerators and air conditioners, have been implicated in stratospheric ozone depletion (see Section 3.1, beginning on page 19, and Section 12.5.2, beginning on page 184). The plots in Figure 8-1a–c all show the annual CFC releases from 1930 to 1990. In Figure 8-1a the vertical (y) axis is from 0 to 0.15 kg, while the vertical axis in Figure 8-1b is from 0 to 0.4 kg. These two plots show a difference in *scale*. Our natural tendency is to think that Figure 8-1a shows a larger increase and Figure 8-1b a smaller increase simply because the results "look higher" in the first figure. Thus the very same data that seem alarming in Figure 8-1a appear to be innocuous in Figure 8-1b. Similarly, in Figure 8-1c the same CFC data are again plotted, but this time the vertical axis is measured in milligrams rather than kilograms. This figure therefore shows a difference in *units* from the other two plots. Here again, the difference influences the interpretation. The increase is apparently a huge number—from about 0 to 140,000, whereas before the maximum was a mere 0.14!

None of the plots in Figure 8-1 is "wrong" in any way—each may be useful for a variety of people for different reasons. How, then, are we to go about presenting our data? Although there are no hard and fast

Table 8-1 Statistical symbols and their definitions.

Term	Symbol	Definition
Sample mean or estimated mean	$\bar{x} = \frac{\sum_{i=1}^{n} x_i}{n} = \frac{x_1 + x_2 + \ldots + x_n}{n}$	Mean estimated from the n values $x_1, x_2, \ldots, x_n$. (Specific cases are identified by $\bar{x}$, $\bar{Y}$, $\bar{y}$.)
Mean	μ	Population or true mean. Limiting value of $\bar{x}$ as n becomes large or if all possible values of x_i are included.
Sample variance	$s^2 = \frac{\sum_{i=1}^{n} (x_i - \bar{x})^2}{n-1}$	Estimate of variance based on n values given by $x_1, x_2, \ldots, x_n$. (If including all possible values of n, use n in place of $n-1$ for denominator).
Variance	σ^2	Population or true variance for the population.
Standard deviation	s, σ	Defined for s^2 and σ^2 above. (Specific cases are identified by subscripts, such as s_x and s_y).
Probability	$P(X < x)$	The probability that a randomly selected value X is less than a specified value x.
Probability density function	$f(x)$	Function which gives frequency distribution.
Cumulative probability function	$F(x)$	Integral of $f(x)$ from $-\infty$ to x and gives P above.
Coefficient of variation	$CV = \frac{\sigma}{\mu}$ or $\frac{s}{\bar{x}}$	A relative standard deviation. Can also be expressed as a percentage.
Maximum allowable error	d	Specified error used in calculating necessary number of samples.
Slope	$b = \frac{\sum_{i=1}^{n} (X_i - \bar{X})(Y_i - \bar{Y})}{(n-1)s_X^2}$	Slope for linear regression for n data pairs. The data pairs are (X_1, Y_1), (X_2, Y_2), … (X_n, Y_n).
Intercept	$a = \bar{Y} - b\bar{X}$	Y-intercept for linear regression for n data pairs.
Correlation coefficient	$r = \frac{\sum_{i=1}^{n} (X_i - \bar{X})(Y_i - \bar{Y})}{(n-1)s_X s_Y} = \frac{b s_X}{s_Y}$	Linear relationship between two variables sampled n times. The data pairs are (X_1, Y_1), (X_2, Y_2), … (X_n, Y_n). Range of r is –1 to +1.
Coefficient of determination	r^2	Square of r (above) for linear correlation. Range is 0 to 1.
Predicted value	$Y = a + bX$	Predicted value of dependent variable Y.

Table 8-2 Thirty-six values of clay (%) taken over a 90-hectare area at a 30 cm depth. The sample mean ($\bar{x}$) and standard deviation (s) are 35.3 and 6.38, respectively. Data: Coelho, 1974.

% Clay	% Clay	% Clay	% Clay
34.7	38.8	45.5	36.1
29.5	38.5	27.0	40.1
43.8	42.6	42.5	37.4
33.3	36.3	27.2	45.3
37.5	32.1	27.2	28.5
33.3	53.2	24.9	30.2
29.5	39.9	33.6	33.6
36.1	34.7	37.4	39.9
25.4	30.5	31.3	32.3

Table 8-3 Plotted points for Figure 8-2 using clay data from Table 8-2.

Interval (x for Figure 8-2a)	Number of Observations (y for Figure 8-2a)	Maximum Value of Interval (x for Figure 8-2b)	Cumulative Number of Observations (y for Figure 8-2b)
0–5	0	5	0
5–10	0	10	0
10–15	0	15	0
15–20	0	20	0
20–25	1	25	1
25–30	7	30	8
30–35	11	35	19
35–40	10	40	29
40–45	4	45	33
45–50	2	50	35
50–55	1	55	36
55–60	0	60	36

rules, we can use a number of practical guidelines to answer this question. One general rule for a two-dimensional (2-D) plot of data is to use "convenient" numbers and employ most of the vertical range. By *convenient numbers* we mean numbers that are in commonly used units and are easy to read. In the case of the CFC production, for example, the best scale and units are probably those in Figure 8-1a on page 96. As to the significance of the data and interpretations, this determination should be made with supporting arguments, not simply by the size of the numbers or length of the axes.

8.1.3 Mathematical Terms in Statistics

In addition to graphical displays, there are a large number of mathematical ways of examining, analyzing, and presenting data. These mathematical expressions and their symbols are used to define statistics. In this chapter, the terms and symbols we will use are listed in Table 8-1 on page 97. If you are unfamiliar with these terms you may need to refer to a statistics text (see *References and Recommended Reading* on page 112).

8.2 Descriptive Measures

Descriptive measures are numbers that characterize overall measurements. That is, they refer to the set of numbers as a whole. Examples of descriptive measures are the mean and variance, which are defined in Table 8-1 and discussed in the following sections.

8.2.1 Measured Frequency Distributions

Measured values can be either discrete or continuous. **Discrete values** are countable numbers. Thus, if we count the number of mosquitoes in a trap, the results are discrete; that is, they consist solely of whole numbers, 0, 1, 2, …, and only whole numbers are possible. Similarly, if our observations are simply Yes or No, or On and Off, we can convert these observations to numerical values of 0 and 1. On the other hand, continuous values can be any real number within reasonable limits, and our observations are not limited to whole numbers. For example, if we measure the length of a frog, the temperature of a lake, or the concentration of oxygen, the values are continuous; that is, they include fractions, such as 1.23 or 0.459, along with appropriate units. **Continuous values** may involve an absolute range as well as an expected range. For example, the absolute range is 0 to 100 if we are looking at a percentage volume of, say, a gaseous constituent, but the expected range is likely to be much narrower. (In this chapter, we will emphasize continuous variables, such as might be measured with an analytical technique.)

Suppose, for example, we measure the clay content of soil at many locations within a field. (Clay content is a very important indicator of the active nature of the soil; see Section 2.2, beginning on page

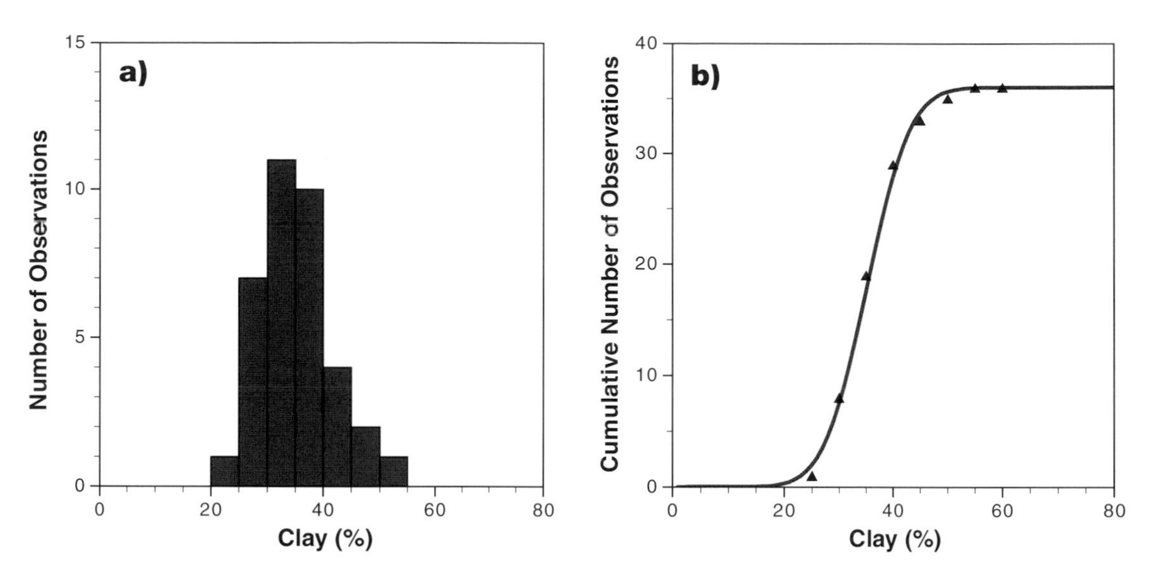

Figure 8-2 (a) Measured frequency histogram and (b) cumulative frequency distribution for the clay from Table 8-2. The solid line is a theoretical curve based on a normal distribution with the same mean and variance as those estimated from the data.

10.) Table 8-2 lists 36 values of clay percentages from soil samples collected over a 90-ha area, each at a depth of 30 cm. The measured values range from a minimum of 24.9% to a maximum of 53.2%. The trouble is, the raw values listed in the table don't really tell us much. These numbers need to be arranged—usually in ascending order—so that we can manipulate them. And once we've got our data in ordered arrays, we can draw some conclusions from them.

One way to arrange these data is to construct a **frequency histogram**, a graph in which we plot the measured values against numbers of observations; this graph lets us see how often the measured values occur. That is, the frequency histogram is a specialized bar chart that illustrates the **distribution** of values. Figure 8-2a is the frequency histogram of the clay data in Table 8-2. To prepare the frequency histogram of the clay data, we first have to break the range of measured values, from 24.9 to 53.2, into smaller, nonoverlapping, exclusive ranges called **class intervals**. These class intervals are shown on the horizontal (x) axis, while the number of observations are shown on the vertical (y) axis. Then we count the number of observations in each class interval so we can plot that number against the class interval. In this case, we chose class intervals with a width of 5. Here the intervals starting from (but not including) 0 and extending through 20 show no observations, but those from >20 to 25 and ending with >50 through 55 do show data that we can plot. Thus, we show (>20–25, 1) as the first bar, with the bar extending from 20 through 25 on either side. The second bar shows seven observations in the range from >25 through 30. And so on (see Table 8-3).

The resulting histogram shows that most of the values tend to cluster near the center *e.g.*, 30 to 35, and the bar heights decrease to either side from the center. So where did the class width of 5 come from? The fact is, we used 5 because it worked. The choice of class intervals in constructing a frequency histogram is often somewhat arbitrary. The problem is that if we choose too few classes (*i.e.*, make the class interval bigger), we lose information and the frequency is not clear; but if we choose too many classes (*i.e.*, make the class intervals smaller), the histogram tends to be too rough. Thus we must look for a compromise between information contained in the class interval and smoothness of the histogram. We can often achieve this compromise by picking an appropriate number of classes, say between five and fifteen, and then dividing that number into the range between the maximum and minimum.

There is a second method of plotting the distribution of clay values. Instead of arranging the data in histogram form as in Figure 8-2a on page 99, the results can also be plotted as a cumulative frequency distribution, as shown in Figure 8-2b. For the cumulative frequency plot, the y-values increase from 0 to $n = 36$, and the curve shows the total number of each observation, or the **cumulative frequency**—that is, each number on the y-axis represents the sum of all the observations at or below that value. Thus, each point on the plot shows the clay percentage that is smaller than the percentage on the corresponding abscissa (x-axis). The **endpoint** of each class interval is shown as a discrete point in Figure 8-2b (see Table 8-3 on page 98). Thus, we can see that there are 8 values less than 30% clay in the data set.

The information in the frequency histogram is the same as that used in the cumulative frequency plot. But even though these diagrams are just alternative ways of plotting the same data, they tell us different things. The histogram lets us visualize the **distribution**, or shape, of the data, while the cumulative frequency distribution is useful for making estimates from the data. Suppose, for example, we were interested in finding the proportion of samples that contained less than 40% clay. The graph in Figure 8-2b shows us that 29 out of 36 samples have less than 40% clay, so it's easy to estimate that proportion as 80.6%.

There are also ways to describe statistical data by using a single numerical value. One of these is the **experimental** (or **sample**) **mean**, $\bar{x}$, which is simply the arithmetic average defined by

$$\bar{x} = \frac{\sum_{i=1}^{n} x_i}{n} = \frac{x_1 + x_2 + \ldots + x_n}{n} \tag{8-1}$$

where x_i is the ith (1st, 2nd, 3rd, *etc.*) observation, $x_1, x_2, \ldots, x_n$ are the observed values, and the number (n) of values is 36. For the clay, $\bar{x}$ is the sum of all the observed values (x) divided by n, resulting in $\bar{x} = 35.3$. The mean, then, is a measure of the *center* of the values; that is, it is a measure of **central tendency**.

We are often interested in how much the sample is spread about the mean. That is, we want to know whether most of the values are close to $\bar{x}$ or are spread over a large range. The spread, or **dispersion**, from the average is given by the **experimental** (or **sample**) **variance** s^2 defined by

$$s^2 = \frac{\sum_{i=1}^{n} (x_i - \bar{x})^2}{n-1} \tag{8-2}$$

For the clay example mentioned earlier, the result is $s^2 = 1424/35 = 40.7$.

A high value of s^2 indicates that the measured values are more dispersed relative to their average. Conversely, a small value of s^2 characterizes measurements that are tightly clustered near the average.

> *Note:* Sometimes, n–1 is replaced by n for the definition of s^2 in Eq. (8-2). In fact, this is the case for many hand-held calculators and popular software packages. So long as $\bar{x}$ is only an estimate of the true mean and each measured x_i is independently chosen, the definition in Eq. (8-2) is correct and s^2 is an unbiased estimate of the "true variance." But if the exact mean (the population mean μ discussed in Section 8.2.2) is substituted for $\bar{x}$, then n should be used. Of course, if n is large, there is only a slight difference between using n and n–1, so the distinction is insignificant!

8.2.2 Populations and Theoretical Frequency Distributions

A population is a collection of all the possible values that we can measure. Suppose, for example, we're interested in the number of cars passing through an intersection each day in January. In this case, defining the relevant population is straightforward—it's the collection of thirty-one whole numbers corresponding to the (discrete) vehicles passing through the intersection each day. Better yet, we can actually measure the entire population.

However, more often than not, we cannot measure the entire population, so we have to obtain a **statistical sample** that adequately represents the population. Say, for instance, our real interest in all those cars is finding the average amount of carbon monoxide (CO) in the air over the same intersection, also in January. In this case, we could collect a series of air

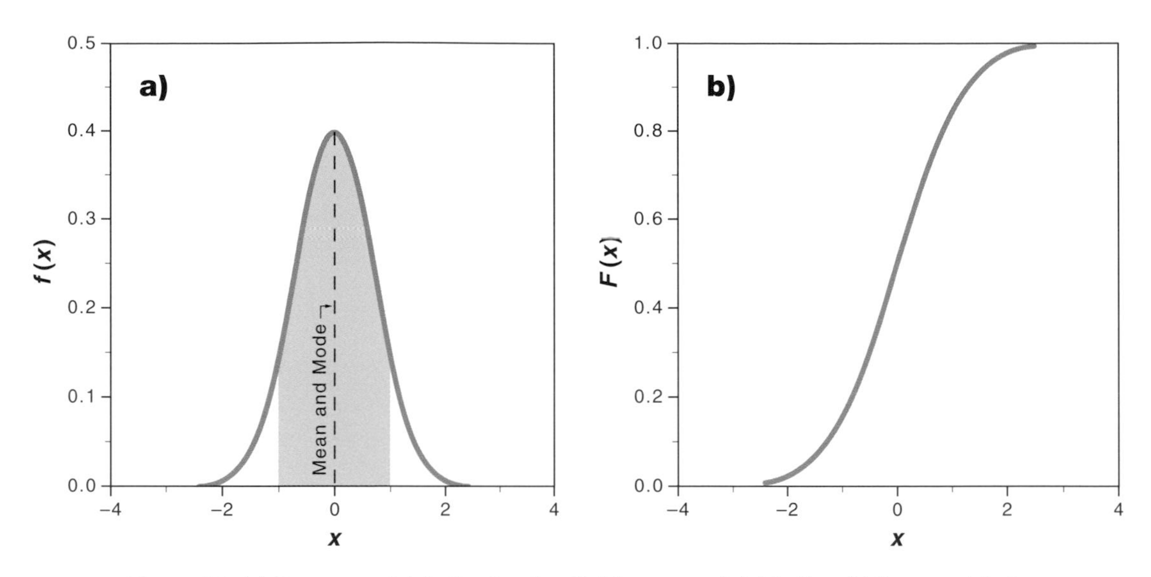

Figure 8-3 (a) Frequency distribution function $f(x)$ for a normal distribution; (b) the associated cumulative distribution function for $\mu = 0$ and $\sigma = 1$. The shaded area in (a) is ±1 standard deviation from the mean.

samples in 100-mL vials throughout the thirty-one days in January. But this isn't as simple as it sounds. In the first place, the population is defined over time, so the time of day when the samples are collected is critical. Similarly, the points to be measured are equally complex in space. We might be interested in CO concentrations from the surface to a 10- or 20-m elevation, or higher or lower. Obviously, it's neither practical nor possible to sample *all* of the air—the "true" or complete population. Consequently, we have to make decisions about such factors as times, heights, and positions within the intersection. Only after we make these decisions can we make the actual measurements that comprise our statistical sample, from which we can calculate the mean and variance of the carbon monoxide levels. Thus, statistical calculations are influenced by the definition of the true population and the manner by which estimates of this population are made. And these decisions are heavily influenced by the assumptions we make about the population we're trying to measure.

So how do we start making assumptions? In many cases, it's easiest to begin with an assumption about the shape of the data. Thus, for data analysis, it is often expedient to assume an experimental distribution similar to the theoretical distribution shown in Figure 8-3a. [*Note:* This is like the shape we saw in Figure 8-2a on page 99.] This gaussian or "bell-shaped" distribution, which is the most widely used theoretical distribution, is known as the **normal distribution**. The shape of the normal distribution can be defined as a function $f(x)$. Each normal distribution has a **population mean** (μ) and a **population variance** (σ^2): the mean μ corresponds to the center of a normal population as shown in Figure 8-3a, while the variance σ^2 is a measure of the spread or dispersion. The **mode** is the value of x corresponding to the greatest value of $f(x)$. Notice that the mode and mean are the same for a normal distribution.

Note: The population mean μ is to the population what the experimental mean $\bar{x}$ is to the sample. Similarly, the population variance σ^2 is to the population what the experimental variance s^2 is to the sample.

We define the shape of the normal distribution as a function $f(x)$ because this lets us compute theoretical results to compare with real sample distributions. It also lets us calculate appropriate probabilities. In

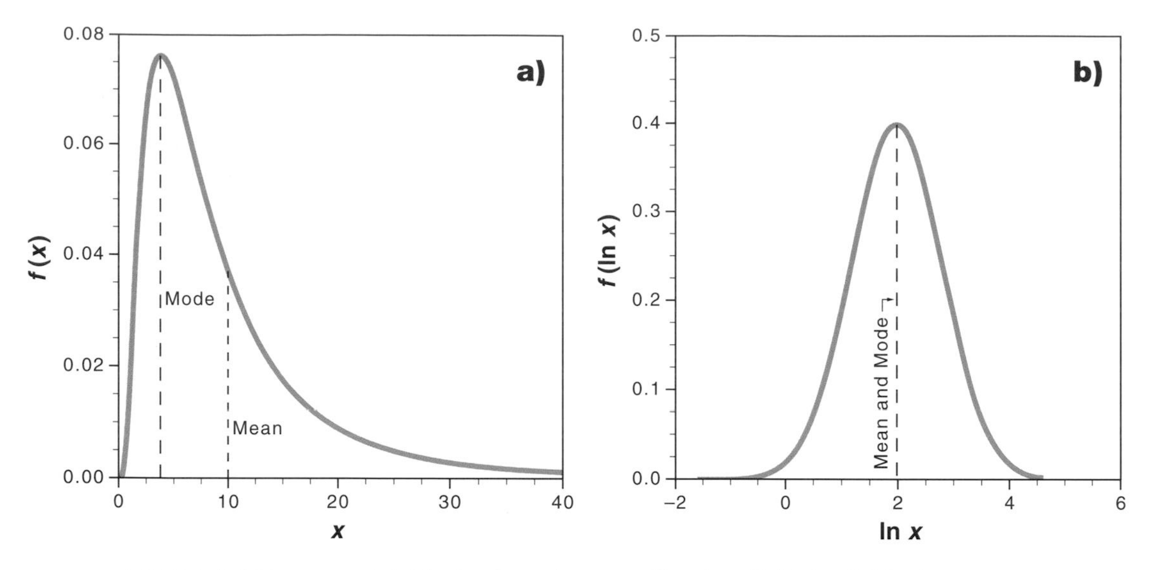

Figure 8-4 (a) Log normal distribution for x where $\mu = 10$ and $\sigma = 5$; (b) the distribution transformed by using $y = \log x$.

short, this mathematical definition is generally useful for quantifying concepts. The equation for $f(x)$ is

$$f(x) = \frac{1}{(2\pi)^{0.5}\sigma} \exp\left[\frac{-(x-\mu)^2}{2\sigma^2}\right] \tag{8-3}$$

In Eq. (8-3), exp is the exponential function; that is, it is the natural log base e ($e \cong 2.72$) raised to the power given in the argument (*e.g.*, $\exp 2 = e^2$). This definition also includes μ, which is the center or mean value, and σ, which is the standard deviation. The standard deviation is related to the dispersion or spread of the curve; in fact, it's the square root of the variance σ^2. In the plot (Figure 8-3a on page 101), we chose values of $\mu = 0$ and $\sigma = 1$. This choice lets us align the center of the theoretical normal distribution at the origin. For other values of μ, the center would be shifted (left or right) accordingly, and for other values of σ, the spread would change.

Although the μ and σ defined for the population are almost the same as $\bar{x}$ and s defined for sampled values, there is an important difference. For the theoretical distribution, the mean μ and the standard deviation σ are assumed to be known *exactly*. For a real distribution, the mean $\bar{x}$ and standard deviation s are dependent on the samples taken and only approximate a theoretical μ and σ (unless, of course, we've measured the entire population, in which case $\bar{x} = \mu$ and $s = \sigma$). This means that as the number of samples increases, $\bar{x}$ and s should get closer and closer to μ and σ, provided each member of the population has an equal chance of being selected.

8.2.3 Significance of the Distribution Function

The distribution function $f(x)$ also lets us predict some physical significance of future samples based on the numerical range of values that have been sampled. When we choose a sample, we don't know what its value will be ahead of time; but we do know that its range, average, and frequency will generally fall within the confines given by the distribution function. We can express this prediction in terms of a *probability statement* to the effect that a randomly selected value X is less than a specified value x; that is

$$P(X < x) = F(x) = \int_{-\infty}^{x} f(x)\,dx \tag{8-4}$$

We read the left-hand side of this expression as "the probability that X is less than x." The function $F(x)$, called the **cumulative frequency function**, is plotted in Figure 8-3b on page 101. For x well below zero, the probability is small that a randomly chosen value will be less than that value, so $F(x)$ is nearly 0. As x gets closer to the mean (which is 0 here), the probability is nearly 0.5 (50%) that the value will be less than 0, so $F(0) = 0.5$. For large x, the corresponding probability is close to the maximum, $F(x) = 1$ (100%). Another way of saying this is that the probability is 1 that all of the values must be included as x approaches infinity.

Now look at Figure 8-3a, where the shaded area is for $x = -\sigma = -1$ to $x = \sigma = 1$. This shaded area comprises 0.693 units and is exactly the probability that a randomly selected value will fall between $-\sigma$ and σ. Similarly, the value of an area for -2σ and 2σ is 0.954, and that for -3σ and 3σ is 0.997.

A theoretical cumulative frequency distribution from Eq. (8-4) may be compared to the experimental cumulative frequency distribution for the clay in Figure 8-2b on page 99. To do this, we substitute $\mu \cong \bar{x}$ and $\sigma \cong s$ in Eq. (8-3) and calculate $F(x)$ from Eq. (8-4). This gives us a smooth curve. To get a corresponding number of values less than x, we multiply $F(x)$ by n. This multiplication ensures that the total number of samples corresponds to a large value of x (hence ensuring that the probability is 1 that all of the values will be attained as x becomes large). Of course, the measured and theoretical distributions do not match exactly. This is a consequence of two factors: a population subset was measured rather than the total population, and the underlying distribution is not exactly normal but is slightly skewed.

8.2.4 Other Distributions

There are many other theoretical distribution functions that we can consider. One choice is the log normal function, in which the natural log of the random variable x is normally distributed. Figure 8-4a shows such a distribution. Here, the mean and standard deviation are $\mu = 10$ and $\sigma = 5$. As we can see, the center of the plot is no longer obvious since there is a long "tail" to the right. We say that this distribution is positively skewed; that is, it is no longer symmetric. This skewness tells us that the mean (given by Figure 8-1 on page 100) no longer corresponds to the maximum value of the function; instead, it falls to the right of the mode. Thus the mode—the value of x for which $f(x)$ is a maximum—is less than the mean.

So why did we call this distribution a log *normal* distribution? What's normal about it? The fact is, this distribution is normal if we transform it. To do this, we use a transform $y = \ln x$ so that we plot not the data themselves but the natural log of the data x versus the function $f(\ln x)$. The result is that the y-values are normally distributed as in Figure 8-4b. This transformation is what makes a log normal distribution look "normal" and allows us to say that x is log normally distributed. The mean and standard deviation for the transformed data $y = \ln x$ in Figure 8-4b are $\mu_y = 1.98$ and $\sigma_y = 0.805$. The relations for the mean and variance of y are well known in terms of the mean and variance of x. For example, the mean value of x is μ given by

$$\mu = \exp\left(\frac{\mu_y + \sigma_y^2}{2}\right) \tag{8-5}$$

where μ_y and σ_y^2 are the mean and variance of the transformed variable y. When should we use such a transformation? The answer is that we use it when it leads to a normal distribution. Suppose, for instance, we had an advanced statistical test that is valid only if the values are normally distributed, but the distribution of the data was skewed. If we can find a transformation that lets us obtain a normal distribution from the skewed distribution, the test will be valid on those transformed values. We can see an illustration of a transformed skewed distribution in the following example.

A study of the amount of lead in soil gave the 60 values of soil lead (in mg kg^{-1}) listed in Table 8-4 on page 104. These values are shown in the scatterplot in Figure 8-5a on page 105, where we see many small values of less than 50 mg kg^{-1} as well as several values above 100 mg kg^{-1} and two values above 250 mg kg^{-1}. The experimental mean and standard deviation of these data are $\bar{x} = 48.9$ and $s = 56.9$. The frequency distribution is given by the histogram shown in Figure 8-5b, which was obtained by counting the number of values in the class intervals between >0 through 20, >20 through 40, and so on (listed in Table 8-5). As we can see, the frequency

Table 8-4 Soil lead data showing untransformed and log-transformed values. Data: Englund and Sparks, 1988.

ID[a]	[Lead] (mg kg^{-1})	ln [Lead]	ID[a]	[Lead] (mg kg^{-1})	ln [Lead]
01	18.25	2.904	31	19.75	2.983
02	30.25	3.410	32	4.50	1.504
03	20.00	2.996	33	14.50	2.674
04	19.25	2.958	34	25.50	3.239
05	151.5	5.020	35	36.25	3.590
06	37.50	3.624	36	37.50	3.624
07	80.00	4.382	37	36.00	3.584
08	46.00	3.829	38	32.25	3.474
09	10.00	2.302	39	16.50	2.803
10	13.00	2.565	40	48.50	3.882
11	21.25	3.056	41	49.75	3.907
12	16.75	2.818	42	14.25	2.657
13	55.00	4.007	43	23.50	3.157
14	122.2	4.806	44	302.50	5.712
15	127.7	4.850	45	42.50	3.750
16	25.75	3.248	46	56.50	4.034
17	21.50	3.068	47	12.25	2.506
18	4.00	1.386	48	33.25	3.504
19	4.25	1.447	49	59.00	4.078
20	9.50	2.251	50	147.00	4.988
21	24.00	3.178	51	268.00	5.591
22	9.50	2.251	52	98.00	4.585
23	3.50	1.253	53	44.00	3.784
24	16.25	2.788	54	94.25	4.546
25	18.00	2.890	55	68.00	4.220
26	56.50	4.034	56	60.75	4.107
27	118.00	4.771	57	70.00	4.248
28	31.00	3.434	58	25.00	3.219
29	12.25	2.506	59	33.00	3.496
30	1.00	0.000	60	40.75	3.707

[a] Sample identification number

Table 8-5 Tally table showing class intervals and corresponding number of observations used to create Figure 8-5.

Normal Distribution		Log Normal Distribution	
Class Interval	Number of Observations	Class Interval	Number of Observations
0–20	20	0.0–0.6	1
20–40	17	0.6–1.2	0
40–60	10	1.2–1.8	4
60–80	3	1.8–2.4	3
80–100	3	2.4–3.0	13
100–120	1	3.0–3.6	14
120–140	2	3.6–4.2	13
140–160	2	4.2–4.8	6
160–180	0	4.8–5.4	4
180–200	0	5.4–6.0	2
200–220	0	6.0–6.6	0
220–240	0		
240–260	0		
260–280	1		
280–300	0		
300–320	1		

histogram is positively skewed—there are many small values clustered at the left and a tail of smaller bars to the right.

But what happens if we consider the logarithms of the lead concentration? Using the transform $y = \ln x$, where x (the lead content) is the random variable, we see that there are many y-values between "2" and "5," as given in Figure 8-5c. The frequency distribution of the transformed values is plotted by choosing intervals whose width is 0.6, as Figure 8-5d shows. (Remember that the choice of the intervals is somewhat arbitrary. We chose this one because it gives us classes that are large enough to be meaningful but small enough to show a distribution. Often, we can ensure such a happy compromise by deciding to choose between 5 to 15 classes.). The resulting frequency distribution for the transformed values is much more symmetric than the original distribution.

What are we to make of those two extreme values—the ones at 268 and 302.5 mg kg^{-1}—that are so far above the other points in Figure 8-5a but less so after the transformation in Figure 8-5c? These two data points may be **outliers**, that is, points that fall out of the defined target population. The problem is, we don't know whether these extremes are "for real." Sometimes, we encounter a faulty lab analysis or an otherwise obviously biased value, in which case we should remove the outliers from the analysis. But we should remove extremes only when we have a sound reason (other than opinion) to think they are outliers. If there is uncertainty, the extremes should be included in the analysis. In the case illustrated by this example, we don't have any reason to think that the values are incorrect, so they should be included in the analysis.

Raw reality is rarely normal. Many distributions of natural (and unnatural) phenomena will be

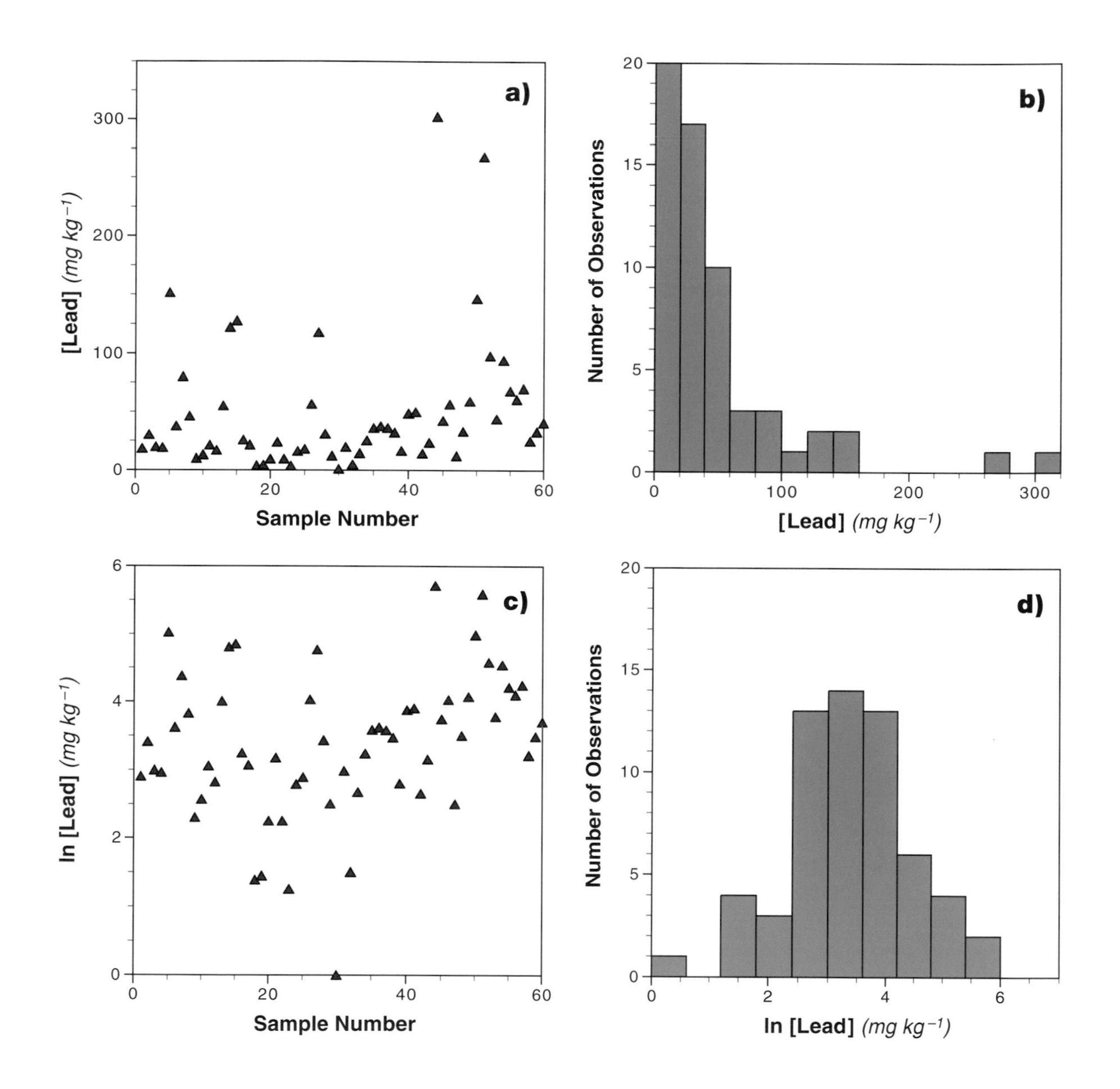

Figure 8-5 (a) Scatterplot, (b) frequency distribution, (c) transformed distribution, and (d) frequency distribution for the transformed distribution for lead in soil as given in Tables 8-4 and 8-5. Data: Englund and Sparks, 1988.

skewed, sometimes resembling the positively skewed theoretical distribution shown in Figure 8-4a on page 102 or sometimes resembling its negatively skewed counterpart. We can identify certain processes with given distributions, but these are mostly hypothetical. For example, sums of random numbers are normally distributed and random numbers multiplied together are log normally distributed. Of course, the normal and log normal are just two convenient models of actual distributions and may not provide a very good match to any real data. In fact, some data may not fit any known distribution, and assumption of an improper distribution could lead to erroneous conclusions. Many statistical tests assume a normal distribution for the population considered, which may use either the "original" values or "transformed" values. Ideally, a test/analysis will not rely heavily on the exact form of the distribution.

For all distributions, $\bar{x}$ obtained from Eq. (8-1) on page 100 and the standard deviation $\sqrt{s^2} = s$ obtained from Eq. (8-2) on page 100 are estimates of the population mean μ and standard deviation $\sqrt{\sigma^2} = \sigma$. Thus the value of $\sqrt{\sigma^2}$ is the theoretical (or population) variance and s^2 is the *estimate* of the variance[1]. The **coefficient of variation** (CV) is defined by

$$CV = \frac{\sigma}{\mu} \cong \frac{s}{\bar{x}} \tag{8-6}$$

[Alternatively, we can multiply the above by 100% to get $CV = (\sigma/\mu) \times 100\%$ or $(s/\bar{x}) \times 100\%$]. The CV is a relative measure of variability within a data subset. Since it is a ratio without dimensions, it is a convenient way to compare the variability of different properties or measurements. A low CV indicates relatively low variability. For example, the CV for the clay data was $6.38/35.3 \cong 0.18$ (or 18%) whereas the lead tended be much more variable with a CV of $56.9/48.9 \cong 1.2$ (or 120%).

[1] There are alternative estimators that are "more efficient" for known specialized distributions. For example, for a log normal distribution, Eq. (8-5) on page 103 can be used to estimate μ based on estimates for μ_y and σ_y.

8.3 Estimates of Sample Numbers

How many points must be sampled to estimate a mean value that provides an acceptably accurate result? This is a very significant problem in sampling and monitoring. Generally, we know that the more sample points we collect, the more reliable our estimate will be, so extensive sampling seems to be a good idea. However, collecting data at more points costs more in labor and analytical expenses. (*Note:* Here, we are using the word *points* to refer either to points in space with samples collected at one time or to points in time with samples collected at one location.) Thus, we need a way to tackle this problem that gives us a good compromise between the number of points we determine and the amount of money and time we have.

In fact, it is the concept of theoretical frequency distributions that helps us determine the number of samples necessary to estimate a population mean, that is, to find a suitable approximation to the average value of what we're measuring. Often, we begin by assuming a normal distribution.

8.3.1 Confidence Limits for the Mean

One classical solution for finding the number of points needed involves confidence limits, which are the endpoints of the range of values expected for $\bar{x}$. The confidence limits are measures of how closely we expect the sample mean $\bar{x}$ to approximate the population mean μ. To do this, we assume that the sample mean is itself drawn from a normal distribution having mean μ and variance σ^2/n, where n is the number of points sampled. That is, if we were to repeat the entire exercise of collecting a set of values over and over, then each time we did it we'd find an additional $\bar{x}$ from the theoretical distribution. (This is always true for random values from an underlying normal distribution; it is also true for other distributions when n is sufficiently large.) For a normal distribution, the probability that a randomly selected value falls within ±2 standard deviations is 0.954 (or 95.4%). By using a mean μ and variance σ^2/n appropriate for the distribution of $\bar{x}$, we can express this in a probability statement similar to Eq. (8-4) on page 102, where $\bar{x}$ is the estimated mean:

$$P\left(\mu - \frac{2\sigma}{n^{0.5}} < \bar{x} < \mu + \frac{2\sigma}{n^{0.5}}\right) = 0.954 \qquad (8\text{-}7)$$

This equality restates the probability that the sample mean falls between $\mu - 2\sigma/n^{0.5}$ and $\mu + 2\sigma/n^{0.5}$ and is about 0.95 (95%). The argument of P is the confidence limits; that is the limits are $\mu - 2\sigma/n^{0.5}$ and $\mu + 2\sigma/n^{0.5}$.

Of course, we could use other probability values. For instance, if a probability of 0.997 is chosen, the corresponding interval will be within ±3 standard deviations of the mean and Eq. (8-7) would become

$$P\left(\mu - \frac{3\sigma}{n^{0.5}} < \bar{x} < \mu + \frac{3\sigma}{n^{0.5}}\right) = 0.997 \qquad (8\text{-}8)$$

The 0.954 value in Eq. (8-7) and the 0.997 in Eq. (8-8) are called the **confidence levels**, which are often applied as percentages, 95.4% and 99.7%. As the confidence level increases, the probability that we have captured the mean within the stated interval also increases. So then, why don't we simply increase the confidence level to include 0.9999 or a number even closer to 1? Since the confidence interval is a function of n, the number of samples can get out of hand as the probability level gets close to 1. The sample number becomes so large that it is unrealistic in terms of time, effort, and budget. And huge numbers may add nothing to the utility of the results.

Now, we specify a **maximum allowable error** d. This is equivalent to requiring that $d \leq |\bar{x} - \mu|$. Using the 95% confidence level, the equality in Eq. (8-7) is satisfied provided that

$$d \cong \frac{2\sigma}{n^{0.5}} \qquad (8\text{-}9)$$

or

$$n \cong \frac{4\sigma^2}{d^2} \qquad (8\text{-}10)$$

There is a caveat that goes along with the maximum error d corresponding to Eqs. (8-9) and (8-10). If the value of d is determined over and over, the estimate

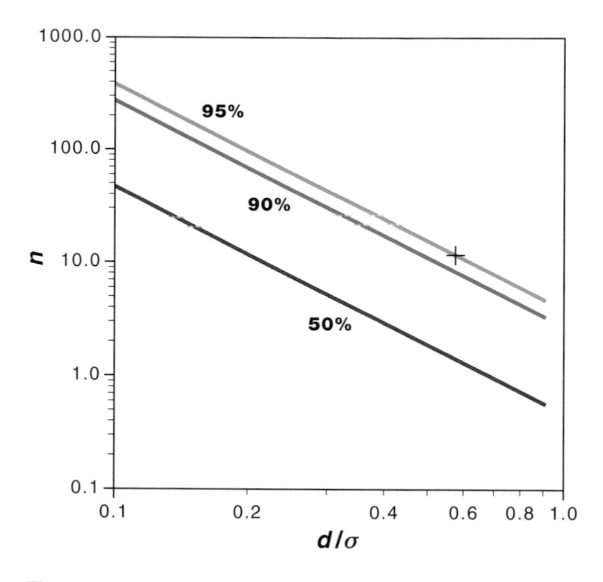

Figure 8-6 Relationship of sample numbers necessary to reach given accuracy for estimating the mean (for confidence levels of 50, 90, and 95%).

will fall within ±2 deviations of the true mean about 95% of the time. For the remaining 5%, it will be outside of that range. Of course for a single determination of n, we don't know that the value will be within the stated tolerance. What we know is that as the confidence level increases, the likelihood of being inside the stated range is higher.

Figure 8-6 is a summary that relates sample number n to σ/d for probability levels of 50%, 90%, and 95% (note the log scales). We can see that the necessary number of samples must increase if d is to be smaller, thereby giving a tighter range on the estimate. Likewise, a larger number of samples is necessary if either σ is larger or the probability that the limit is reached is higher. This is demonstrated in the following example.

Consider the clay percentages found in Table 8-2 on page 98 and then assume that $\mu = \bar{x} = 35.3$, $\sigma^2 = s^2 = (6.38)^2 = 40.7$ and $n = 36$. We are interested in the possible range of means at the 95% confidence level. Using Eq. (8-7), we find that there is approximately a 95% chance that $\bar{x}$ will fall between 33.2 and 37.4. How many sample points are needed for the estimate $\bar{x}$ to be within 10% of the true mean with a 95% confidence level?

The value of d above is 10% of the mean or $d = 0.1\mu$, giving $d/\sigma = 0.1\mu/\sigma = 3.53/6.38 = 0.553$. In Figure 8-6 on page 107, the cross hair marks $d/\sigma = 0.553$ and corresponds to $n \cong 12$. We can confirm this value by using Eq. (8-10) on page 107:

$$n \cong \frac{4\sigma^2}{d^2} = \frac{4}{(0.553)^2} \cong 13.1 \cong 13 \qquad \text{(8-11)}$$

A more exact relationship for n is given by substituting 1.96^2 for $2^2 = 4$ in Eq. (8-11), giving $n = 12.6$.

If the value of the $CV = \sigma/\mu$ is smaller, then n will be smaller and fewer points will be necessary. However, the decision gets harder if the CV is very large. We have already considered lead concentrations with a CV above 1 in the example of the lead-contaminated soil on page 103. It is common to have a CV of 1 or larger for such cases including soil transport properties (*e.g.*, hydraulic conductivity or apparent diffusion coefficient). If $CV = 1$ and the estimate needs to be within 10% of the true mean with a 95% probability, d/σ would be 0.1 and the value of n would be close to 400! Generally, it makes intuitive sense that more points are needed to estimate a mean value for a more highly variable property. The calculated sample number may be impractical. However, in this case, the tolerance level or the probability level would need to be relaxed. Perhaps $d = 20\%$ of the mean and/or a probability level of 50% is more reasonable. For the clay example with a $CV \cong 6.38/35.3$, a 50% probability level would lead to an n value of 3.

We can make an additional refinement to our estimate of the confidence of the mean value after it has been determined. The estimate of the value of $d = |\bar{x} - \mu|$ from the example above assumed that the variance σ^2 was known exactly; but, in fact, it can only be estimated. Thus the estimated value will have a slightly higher range than that calculated above. But the estimate can be calculated using the ***t*-statistic**, which takes into account the uncertainty in estimating σ by s including the number of values used in computing s. (The t-statistic is provided in elementary statistics texts.)

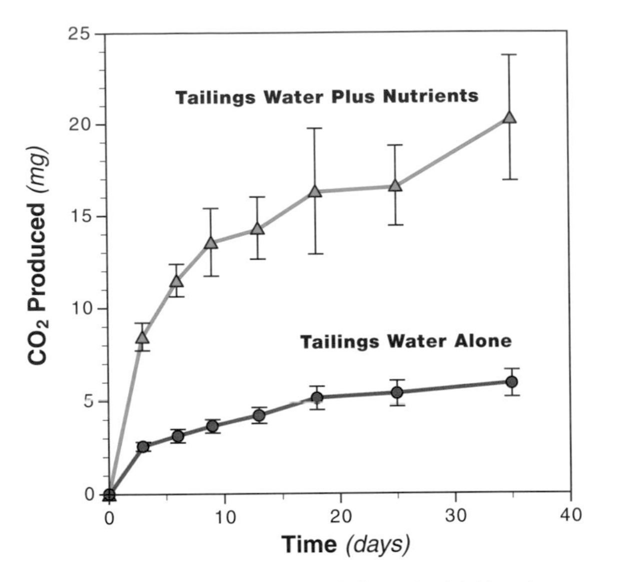

Figure 8-7 Plot of CO_2 evolved during microbial breakdown of crude oil components. The error bars show the mean $\pm s$ based on three replications. The upper curve is for tailings water plus nutrients; the lower curve is for tailings water only. Data: Herman *et al.*, 1994.

8.3.2 Expressions of Reliability of Estimates

Another common method of expressing the reliability of an estimated average value is to use the standard deviation s rather than a probability statement such as Eq. (8-7) on page 107. For example, we can give an average and standard deviation as $\bar{x} \pm s$ or we can plot this expression graphically as a bar showing the calculated range as illustrated in the following example.

Carbon dioxide (CO_2) is one of the products evolved during microbial breakdown of the tailings wastewater produced when crude oil is mined from sand deposits. The amount of CO_2 evolved is a measure of the amount of organic material metabolized by microorganisms—the higher the CO_2 values, the greater the degradation of the organic compounds. (See Chapter 7, beginning on page 77, for more information on microbial degradation of organic contaminants.) In one study, the amounts of CO_2 produced by such a breakdown process were measured over a period of 35 days. The values obtained (in milligrams) are plotted over time in Figure 8-7. Here, we see a lower curve, which shows microbial activity in unamended tailings water, and a higher

Table 8-6 Data used to derive the plot and regression line shown in Figure 8-8. The lettuce yield is fresh head weight per plant in kg/plant. Source: Heck *et al.*, 1982.

Average [Ozone] ($\mu L\ L^{-1}$)	Lettuce Yield (kg)
0.106	0.414
0.043	0.857
0.060	0.751
0.068	0.657
0.098	0.437
0.149	0.251
—Statistical Summary—	
n	6
$\overline{X}$	0.087
$\overline{Y}$	0.561
s_X	0.038
s_Y	0.231
b	–5.90
a	1.08
r^2	0.96
r	–0.98

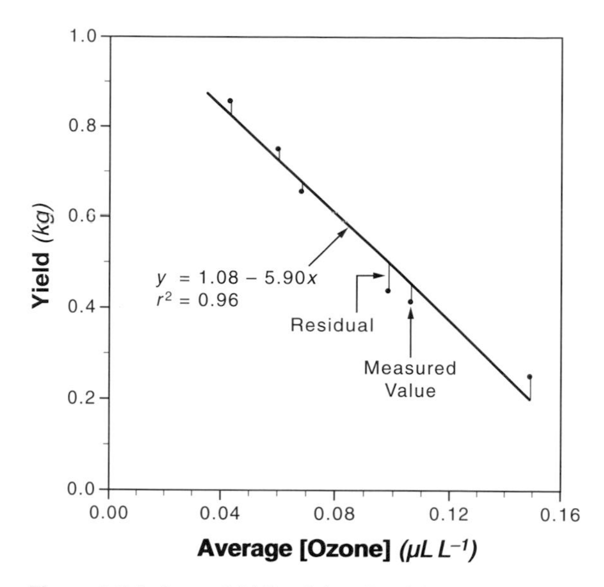

Figure 8-8 Lettuce yield (fresh head weight per plant) vs. ozone concentration and deviations of experimental points from regression line. Data: Heck *et al.,* 1982.

curve, which shows microbial activity in tailings water enriched with nutrients. This top curve reveals an increase in microbial activity with the addition of nutrients—nitrogen and phosphorus plus an organic compound similar in structure to the toxic components within oil sands tailings. The error bars at each point plotted are for the mean $\pm s$, where s is calculated from three replicates. The wider bars indicate more variation in the replications while the shorter bars represent a correspondingly narrow difference in the replications.

8.4 Regression Analysis

In the previous examples, the CFC concentration, the clay percentage, and the soil lead concentration, we examined one variable at a time. What happens if we have two variables—one dependent on the other? In that case, we might use regression analysis, which is a way of fitting an equation to a set of data to describe the relationship between the variables. This method also allows us to make some predictions. We can see how regression analysis works in the following example.

Consider the measurements listed in Table 8-6 for lettuce yield corresponding to different ozone (O_3) concentrations in the atmosphere. As we know, excessive concentrations of ozone in the lower atmosphere can be responsible for reductions in plant growth (and hazardous to human health as well; see Section 12.2.2, beginning on page 175). Suppose we wanted to use these data to predict yield at an ozone level that hasn't been measured. Can we develop an equation that relates the lettuce yield to the ozone concentration in such a way that we can make this prediction?

A logical first step in the analysis of the data is to plot one variable against the other—yield vs. ozone concentration—as shown by the points in Figure 8-8. Now we have a visual demonstration that the higher yields tend to occur at lower ozone concentrations while the lower yields occur at higher ozone levels. Happily, we also see that we can draw a straight line such that the points are all reasonably close to the line. One way to draw this line is simply to lay a ruler across the points and choose where it looks the best. As a first step, such "eyeballing" is fine, but it leaves open the question of exactly where to place the line. Fortunately, there is a more unbiased method.

The fact that we can see a straight line amongst the plotted points suggests that the relationship between the data is linear—which means that a simple linear equation will meet our need for an equation that fits the data. Thus, from simple algebra, we know that the equation of a straight line is

$$Y = a + bX \tag{8-12}$$

where Y is the yield and X is the ozone concentration. Now we can see that **linear regression** is the study of the relationship in Eq. (8-12) between a dependent variable Y and an independent variable X. The value of a is the y-intercept (*i.e.*, the value of Y corresponding to $X = 0$) and b is the slope of the line.

Next, we have to deal with a and b, whose values must be chosen so that the distances between the points and the fitted line are small. The distances are shown by the short vertical lines at each point in Figure 8-8 on page 109. Fortunately, calculating a and b is straightforward. The solutions for a and b are based on minimizing the sum of the squares of the distance between each point and the fitted line (the short vertical lines, commonly known as residuals). The solutions are

$$b = \frac{\sum_{i=1}^{n}(X_i - \overline{X})(Y_i - \overline{Y})}{(n-1)s_X^2} \tag{8-13}$$

$$a = \overline{Y} - b\overline{X} \tag{8-14}$$

where the s_X is the standard deviation of X (the ozone concentration). The values of the slope and intercept are $b = -5.90$ kg L μL^{-1} and $a = 1.08$ kg. The negative sign for the slope indicates what we already observed: the yield decreases as ozone concentration increases.

The value of the intercept can be verified by extending the line up and to the left until the level of ozone reaches a corresponding yield of $a = 1.08$ kg. Notice that, once we've established that a linear relationship exists, we can predict what other, unmeasured values might be.

There is an easy way to quantify how closely the straight line fits the data points. We do this by calculating the linear **correlation coefficient** r, which is defined by

$$r = \frac{\sum_{i=1}^{n}(X_i - \overline{X})(Y_i - \overline{Y})}{(n-1)s_X s_Y} = \frac{b s_X}{s_Y} \tag{8-15}$$

where s_X and s_Y are the standard deviations for X and for Y. The square r^2 is called the **coefficient of determination**, and it does exactly what its name implies: If X and Y follow a perfect linear relationship, r^2 will be exactly 1 with larger values indicating a better fit. For the lettuce yield plot, $r = -0.98$ and r^2 is 0.96, so the fit is very good.

The limiting values of r^2 are 0 to 1 and r has limiting values of -1 to 1. The sign of r, which is always the same as that for b, indicates whether the two variables increase together or are inversely related as in Figure 8-8. In addition, there are tests for significance of a, b, and r. These tests address such questions as "Is the slope b significantly different than zero?" which is the logical next step in a regression analysis. However, such tests are beyond the scope of this chapter.

> *Caution*: When interpreting the correlation coefficient, note carefully whether r or r^2 is presented, either on a graph or calculated in a spreadsheet. An r value of 0.85 appears to indicate an excellent fit, but it gives $r^2 = 0.72$, which is not as impressive. Additionally, you may encounter cases in which apparently low values of r^2 indicate a significant correlation, particularly if the data set is extensive.

There are some common pitfalls and misuses associated with linear regression. The first is the temptation to extrapolate beyond the range of the data points. For example, suppose we wanted to estimate the ozone concentration for which a yield of zero could be expected. We could use the regression equation $Y = 1.08 - 5.90\,X$, take $Y = 0$, and find $X = 1.08/5.90 = 0.18$. So, would this be the ozone concentration for a yield of zero? Possibly it is, but this a dangerous extrapolation beyond the original data set, and we don't really know. Even though we often need to make predictions from our data, ex-

Let the Reader Beware

To round or not to round? Is your body temperature 98.6°F?

Recent investigations involving millions of measurements have revealed that this number is wrong; normal average human body temperature is actually 98.2°F (Fahrenheit). The fault, however, lies not with Dr. Wunderlich's original measurements—they were averaged and sensibly rounded to the nearest degree: 37°C. When this temperature was converted to Fahrenheit, however, the rounding was forgotten and 98.6°F was taken to be accurate to the nearest tenth of degree. Had the original interval between 36.5°C and 37.5°C been translated, the equivalent Fahrenheit temperatures would have ranged from 97.7°F and 99.5°F (Paulos, 1995).

❖

Even a nonrandom large sample of 250,000,000 can have problems. The U.S. Census Bureau attempts to count every person in the United States; in other words, it tries to measure the true population (in a statistical sense) rather than a subset. However, the Census Bureau doesn't supplement its count with standard sampling techniques. The result is a serious underestimate of the total population, especially of the urban poor (Paulos, 1995).

❖

Cause and effect: "There are two clocks which keep perfect time. When "a" points to the hour, "b" strikes. Does "a" cause "b" to strike?" (Huff, 1954)

Cause and effect: "Wind is caused by the trees waving their branches" (Ogden Nash).

Extrapolation of numbers of TVs: "The number of sets in American homes increased around 10,000% from 1947 to 1952. Project this for the next five years and you find that there'll soon be a couple of billion of the things, Heaven forbid, or forty sets per family" (Huff, 1954).

trapolating beyond the measured data can lead to real trouble. A second misuse is to infer—or worse, imply—a causal relationship between two variables. For example, it is natural to look at the lettuce-ozone plot in Figure 8-8 on page 109 and solemnly pronounce that ozone concentration *causes* a decrease in yield. This may be correct, but the data themselves are not sufficient to prove such a statement. What if, for instance, an unidentified toxin were present in the experiment—a toxin that was supplied in increasing amounts with higher levels of ozone? The presence of such a toxin could explain the yield decrease, so ozone might not be the culprit. As to cause and effect, that must be decided by other methods, not by simply looking at one data set.

Of course, there are many data sets that have more than two dependent variables. In fact, for the study previously mentioned in this section, lettuce yields were measured for mixtures of ozone, SO_2, and NO_2, so the values in Table 8-6 on page 109 include only a subset of the data. Whereas linear regression is for one dependent variable, multiple regression deals with more than one dependent variable, such as ozone and the other two factors. A multiple regression analysis would follow the same general steps used in the linear analysis.

References and Recommended Reading

Coelho M.A. (1974) *Spatial Variability of Water Related Soil Physical Properties*. Ph.D. Thesis, The University of Arizona.

Council on Environmental Quality (1992) *Environmental Quality, 22nd Annual Report*. U.S. Government Printing Office, Washington, D.C.

Englund E. and Sparks A. (1988) *Geo-Eas (Geostatistical Environmental Software) User's Guide*. EPA600/4-88/033. U. S. Environmental Protection Agency, Las Vegas, Nevada.

Gilbert R.O. (1987) *Statistical Methods for Environmental Pollution Monitoring*. Van Nostrand Reinhold, New York.

Heck W.W., Taylor O.C., Adams R., Bingham G., Miller J., Preston E., and Weinstein L. (1982) Assessment of crop loss from ozone. *Journal of the Air Pollution Control Association*. **32**, 353–361.

Herman D.C., Fedorak P.M., MacKinnon M.D., and Costerton J.W. (1994) Biodegradation of naphthenic acids by microbial populations indigenous to oil sands tailings. *Canadian Journal of Microbiology*. **40**, 467–477.

Huff D. (1954) *How to Lie with Statistics*. W.W. Norton, New York.

Huntsberger D.V. and Billingsley P. (1987) *Elements of Statistical Inference*. 6th Edition. William C. Brown Publishers, New York.

Little T.M. and Hills F.J. (1978) *Agricultural Experimentation*. John Wiley and Sons, New York.

Matthias A.D., Artiola J.F., and Musil S.A. (1993) Preliminary study of N_2O flux over irrigated bermudagrass in a desert environment. *Agricultural and Forest Meteorology*. **64**, 29–45.

Ott W.R. (1995) *Environmental Statistics and Data Analysis*. Lewis Publishers, Boca Raton, Florida.

Paulos J.A. (1995) *A Mathematician Reads the Newspaper*. Basic Books, New York.

U.S. Department of Commerce (1990) *Statistical Abstract of the United States: 1990*, 110th Edition. U.S. Government Printing Office, Washington, D.C.

Problems and Questions

1. Consider the first 9 clay values in Table 8-2 on page 98. Calculate $\bar{x}$ and s^2. What will be the difference in the value of s^2 if n rather than $n-1$ is used in the denominator of Eq. (8-2) on page 100? Would the difference be more significant for a large or a small number of samples? (If you have a hand-held calculator that calculates $\bar{x}$ and s^2, which form does it use?)

2. Enter the values for clay percentage in Table 8-2 in a spreadsheet, calculate $\bar{x}$, s, and CV. Plot a frequency histogram if the function is available in your spreadsheet.

3. Consider the first 30 lead values in Table 8-4 on page 104. Calculate $\bar{x}$ using Eq. (8-1) on page 100. Take the transformation $y = \ln(x)$ for each value and find the mean and variance for y. How does the estimate of the population mean μ based on Eq. (8-5) on page 103 compare with $\bar{x}$? Why are they different?

4. Calculate the mean and standard deviation for each of the times shown in Table 8-7. Plot the mean and include error bars based on $\pm s$ for each.

Table 8-7 N_2O concentrations (nL L^{-1}) collected 0.05 m above a bermudagrass surface from 18–19 July 1991. Source: Matthias *et al.*, 1993

Time of Day	Replicate 1	Replicate 2	Replicate 3
19:37	325	304	308
06:37	307	305	303
08:37	309	323	336
10:37	316	322	331
12:37	321	323	319
16:12	312	309	318

5. The normal distribution Eq. (8-3) on page 102 assumes values from $x = -\infty$ to $x = \infty$. Suppose the variable of interest goes from 0 to 100. Discuss whether this automatically implies that a normal distribution is an inappropriate model for the data.

6. Estimate the number of sampling points necessary to estimate the mean value of lead concentrations based on Table 8-4 on page 104 and Figure 8-6 on page 107. Assume you want the estimate to be within 15% of the mean with a 90% probability level. How would n change if the estimate is within 50% of the mean and has a 50% probability level?

7. Plot the peanut yield vs. ozone concentration from data shown in Table 8-8. Calculate the mean and standard deviation for each variable. Calculate the slope and y-intercept of a line fitted through the points. Draw the line on your plot. Calculate r and r^2. Write the equation of the fitted line, r, and r^2 on your plot.

Table 8-8 Peanut yield (weight of marketable pods/plant in g) and ozone concentration (μL L^{-1}). Source: Heck *et al.*, 1982.

Peanut Yield (g)	[Ozone] (μL L^{-1})
157.8	0.056
142.3	0.025
122.4	0.056
92.0	0.076
68.9	0.101
40.0	0.125

8. Refer to Figure 8-7 on page 108. Fit a line to the data for the first 10 days only, using the data shown in Table 8-9. Report the calculated slope, *y*-intercept, and the coefficient of determination. Use the fitted line to predict the amount of CO_2 (mg) after 35 days. Discuss the difference, if any, between the predicted and the measured values. Is the predicted value reasonable? Why or why not?

Table 8-9 Partial data set for CO_2 evolved (mg) and time (days) for mine tailings water plus nutrients. Source: Herman *et al.*, 1994.

CO_2 (mg)	Time (days)	CO_2 (mg)	Time (days)
8.702	0	14.93	9
7.631	0	12.39	9
9.072	0	15.59	9
11.67	3	17.46	35
10.54	3	24.08	35
12.26	3	19.30	35
11.55	6		
13.96	6		
15.14	6		

9. Plot the data in Table 8-10. Find the linear regression and draw it on the plot. Calculate the coefficient of regression. Does this indicate that *X* is a good predictor of *Y*? *X* is the number of cigarettes (billions) used annually in the United States from 1944 to 1958. *Y* is an index number of production per man-hour for hay and forage crops from the same time period. Do you still think that *X* is a reasonable predictor for *Y*? Why or why not?

Table 8-10 Fifteen pairs of highly correlated data. Source: Little and Hills, 1978.

X	Y
295	73
339	78
343	85
344	91
357	100
359	109
368	119
395	125
414	129
406	135
385	142
394	139
404	140
420	147
446	156

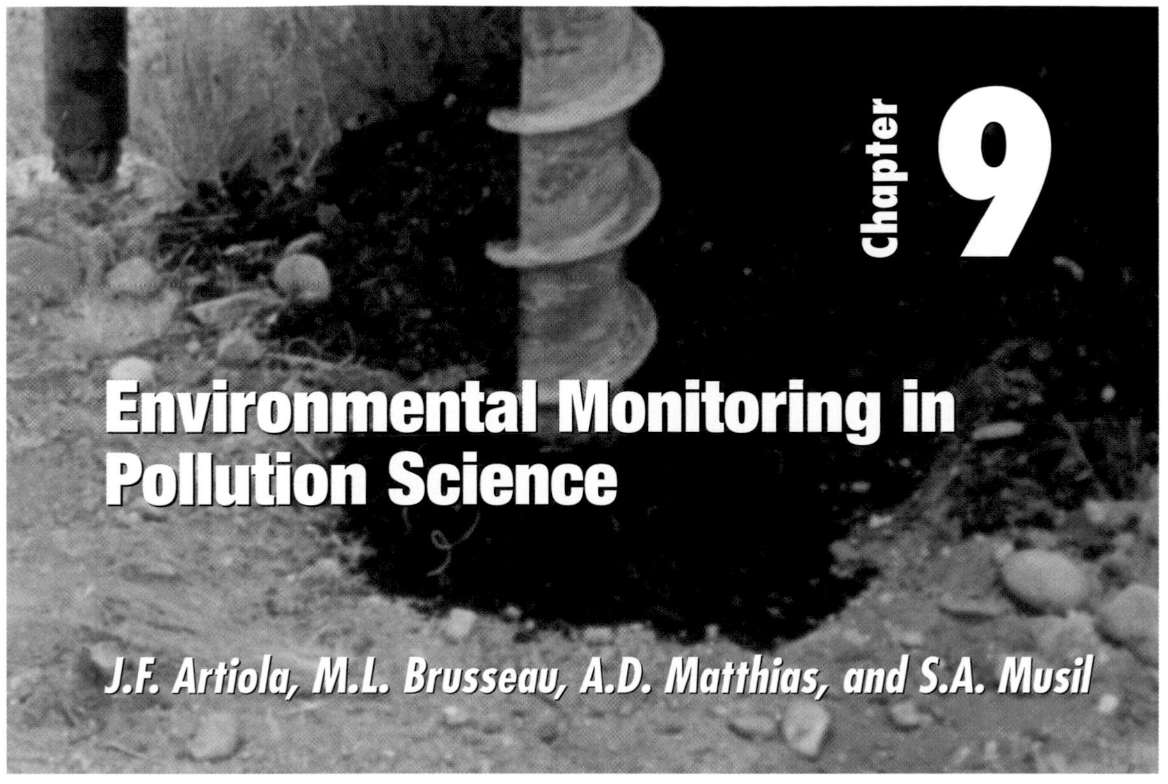

This screw auger is used to drill well bores, take disturbed soil samples and reach a specified depth before taking undisturbed soil samples using the less sturdy split-spoon auger shown in Figure 9-3 on page 119. Photograph courtesy of J.F. Artiola.

9.1 Environmental Monitoring

Everything we know about environmental systems—the chemical, biological, and physical properties of the environment as well as the many processes that influence the behavior of pollutants—is based on the collection and analysis of data. For example, our knowledge of the existence and behavior of particular bacteria in a particular soil comes from collecting soil samples and analyzing their microbiological properties. Similarly, everything we need to know about managing environmental pollution—its sources, cleanup, and regulation—depends on our ability to determine that pollution exists at a site and that it may be harmful to humans or the environment. But how can the existence, extent, and behavior of a pollutant be measured to provide this determination?

In most cases, it all starts with the collection and analysis of environmental samples, a process known as **environmental monitoring**. In this chapter, we'll briefly review some of the major components of environmental monitoring, including the development of sampling plans and the methods of sample collection for atmosphere, soil, and water environments. In so doing, we'll be looking at some important questions. For example: What, exactly, is to be sampled? Where will the points in space and time be located? and What special sample handling requirements are needed? We'll also see that, because each pollutant affects each environment differently, we need to adapt our sampling techniques and strategies to suit the unique physical, chemical, and biological properties of specific pollutants and environments.

9.2 Sampling and Sampling Plans

The word *sampling* implies the collection or measurement of a subset of specimens taken from a large population located in space and time. For example, when you taste stewed squid, you need to eat just a small amount—a *sample*—of what's on the plate to decide whether or not you like the dish. You don't have to eat every tentacle on the plate—the *population*—to arrive at your conclusion; instead, you assume that the sample bites you took truly represented the taste, flavor, and consistency of the rest of the dish. The problem is that your assumption may or may not be true: your sampling technique may have given you the wrong information or biased data.

Environmental scientists also have to deal with this fundamental assumption: the samples and/or numbers that we collect during a sampling program must reflect what we're trying to measure. To achieve a degree of sampling objectivity, we use accepted sampling techniques and protocols that allow us to collect samples that truly represent the target environment or population. Environmental sampling requires that we carefully consider the constraints and limitations imposed by each environment and use these to develop a sampling plan that is bias-free. Only when we apply unbiased sampling techniques can we adequately analyze the data and draw objective, defensible conclusions. And to do this, we need specific methods for sampling air, surface water, groundwater, and soil media.

Before embarking on a sampling exercise, we must state our objectives clearly in order to develop a realistic plan. Environmental sampling is labor- and capital-intensive, and careful planning can save on both. Some projects, for example, may have multiple objectives, which need to be considered simultaneously and planned accordingly. Other projects may be very complex, requiring contributions from several different disciplines. In this case, all pertinent investigators should contribute to the planning—the biologists, the chemists, and the engineers, as well as the requisite statistician, who can make sure that the statistical needs of the analysis are considered before the samples are taken.

The first consideration is the identification of the population of interest. That is, what is to be sampled, and where does it exist in time and space? Next, we have to decide how to sample that population. Generally, it is desirable that each specimen (member or part) of the study population be equally likely to be sampled. We may want to take samples at equal intervals (in time and space), for instance, or we may take them randomly, depending on how we plan to use and interpret the data. Further considerations include choices about the types of measurements to make and analytical techniques to use, as well as questions about methods of specimen handling and data analysis. Also important is the degree of precision required to accomplish the objective. Generally, greater precision requires more work and more resources—that is, more samples to be collected and analyzed.

A good sampling plan also provides information about the allocation of resources. If the objective is clearly stated, alternative methods can be evaluated with respect to their efficiency and cost-effectiveness. For example, is it better to sample at 50 random sites or at 50 sites on a grid? Is it better to analyze each sample separately or to mix them all together and do a single analytical determination? Is it better to find one value for each sample or to separate the sample into replicates (*i.e.*, analyze several subsets from each specimen)? Is it better to sample at 30-second intervals or daily? Each plan requires a different allocation of resources, which should be tailored to maximize the chance of obtaining the desired objectives.

When we develop and evaluate a sampling plan, we rely on previous experience with similar measurements and reviews of existing data available for the study area of interest. These provide a background for expected ranges of measurements and variability. Detailed written protocols designed to support good techniques and comparability of final results exist for many common sampling situations. Many of these protocols have been developed for such government agencies as the U.S. EPA. These agencies often expect a **quality assurance project plan (QAPP)**, previously known as the quality assurance/quality control plan (QA/QC). At a minimum, the QAPP should specify the following:

1. *Sampling strategies:* Number and types of samples, locations, depths, times, intervals
2. *Sampling methods:* Specific techniques and equipment to be used
3. *Sample storage:* Types of containers, preservation methods, maximum holding times
4. *Analytes:* A list of all parameters to be measured and detection limits
5. *Analytical methods:* A list of acceptable field and/or laboratory analytical procedures

More information about these topics can be obtained from materials listed in *References and Recommended Reading* on page 133.

The overall goal of a well designed sampling plan is to produce data with the following characteristics:

1. *High quality:* The data have a known degree of accuracy and precision.
2. *Defensible:* Documentation is available to validate the sampling plan.
3. *Reproducible:* The data can be reproduced by following the sampling plan.
4. *Representative:* The data truly represent the environment sampled.
5. *Useful:* The data can be used to meet the objectives of the monitoring plan.

It is unrealistic to expect that a sampling plan will fulfill all of these requirements all of the time. However, current regulatory guidelines demand that a good sampling plan meet these criteria at least 90% of the time.

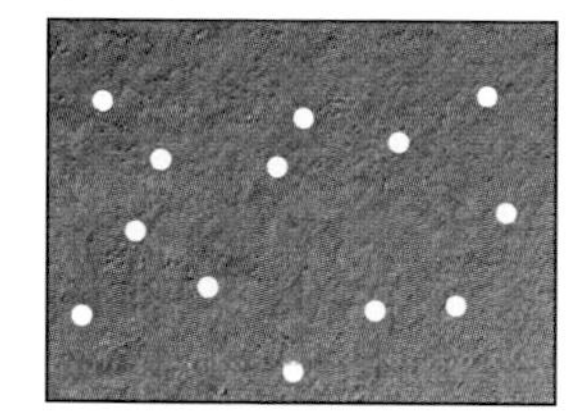

b) Transect

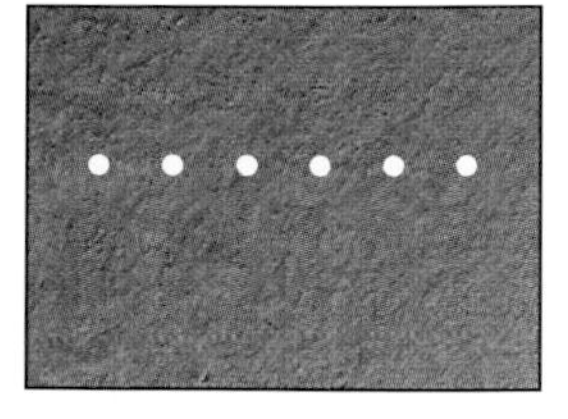

c) Two-Stage

d) Systematic Grid

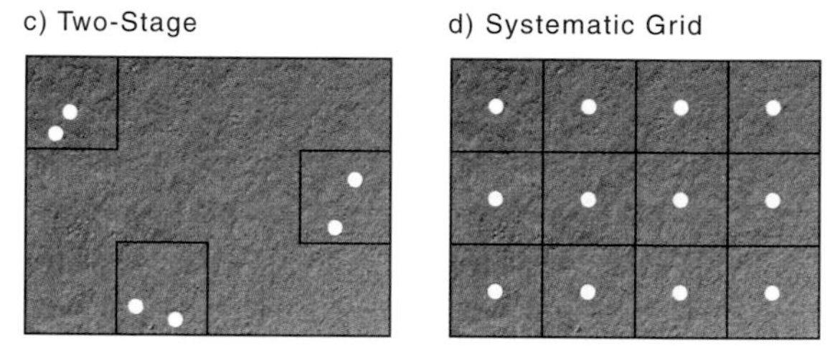

Figure 9-1 Alternative spatial sampling patterns. Adapted with permission from *Statistical Methods for Environmental Pollution Monitoring* by R.O. Gilbert; copyright © 1987 by Van Nostrand Reinhold.

9.3 Sampling Patterns

In developing a sampling plan, we have to decide where and when to take samples. Moreover, the number, location, and timing of sampling must be sufficient to obtain a statistically valid sample without exceeding the sampling budget. We do this by establishing sampling patterns.

9.3.1 Two-Dimensional Sampling Patterns

In two-dimensional sampling, we focus on where to take samples within an area, or field. That is, we have to select spatial coordinates of points that can make up a statistically valid sample within the field. Essentially, we can choose among the following sample patterns, which are illustrated in Figure 9-1.

- **Random sampling** (Figure 9-1a). In this case, we sample by choosing points randomly over the field. A reasonable way to select these positions is to divide the entire area into some meaningful minimum area to be sampled (say 1-m^2 parcels) and then choose n positions randomly. (Of course, we have to be sure our random points are truly random. Most computer spreadsheets have random-number generators that can be used to choose random coordinates.) The selected coordinates must then be accurately located in the physical field. Thus, depending on the size of the field, we might use a simple measuring tape or surveying equipment; or we might even locate sites by using global positioning satellites (GPS), depending on what maps and equipment are available.
- **Transect sampling** (Figure 9-1b). A direction, starting position, and transect length are chosen and then the measurements are taken along the transect. In some cases the sampling occurs at specified intervals along the transect. In other cases all occurrences of the parameter of interest are used, *e.g.*, when counting all members of an indicator plant species occurring along the transect. Another variation of this pattern is to choose several transects based on observed differences at the site. For instance, if sampling a riparian area, a transect along the length of the streambed and another at right angles to the streambed extending out to nonriparian habitat is useful. Sometimes a transect can supply the information needed to decide what other sampling pattern should be used for the main sampling effort. For example, a preliminary sampling along a transect might indi-

cate that a two-stage sampling pattern is the best choice for a particular study.

- **Two-stage sampling** (Figure 9-1c on page 117). Here, we stratify samples by breaking the field into regular subunits, called primary units. (Or we can stratify by slope, soil type, or some other physical property.) Within each primary unit, subsamples are taken, either randomly or systematically. Generally, we assume that there is a variability between primary units and a common random variability that is the same within each primary unit. For example, an environmental site might consist of a hillside slope and a level plain. These two primary units are likely to have different soil characteristics. In addition, each primary unit will have internal random differences of soil type.

- **Grid sampling** (Figure 9-1d). In this case, we collect samples at regular intervals at a fixed spacing. Sometimes, the spacing is the same in both directions, as shown in the figure; but the horizontal spacing may also be different from the vertical. This method is particularly useful for obtaining a map of an area, especially if little is known about the variation of the measurement from spot to spot in the overall area.

The appropriate sampling pattern depends on the site and the objectives of the study. In pollution science, transects are most often used for limited, preliminary sampling of the site. The random sampling pattern, which is often the simplest, may be appropriate for an area believed to be without trends or cycles with respect to the variable of interest. The two-stage sampling pattern offers advantages with respect to evaluating overall variability, particularly when the variability within the subunits is much smaller than that for the entire field. The grid sampling pattern is most useful if spatial distributions are to be deduced, as is the case when we need to make maps.

9.3.2 Three-Dimensional and Temporal Sampling Considerations

Three-dimensional sampling patterns, which include the vertical dimension, may involve samples at depth, as in soils or the ocean, or height, as in the atmosphere. Suppose, for example, we have to track

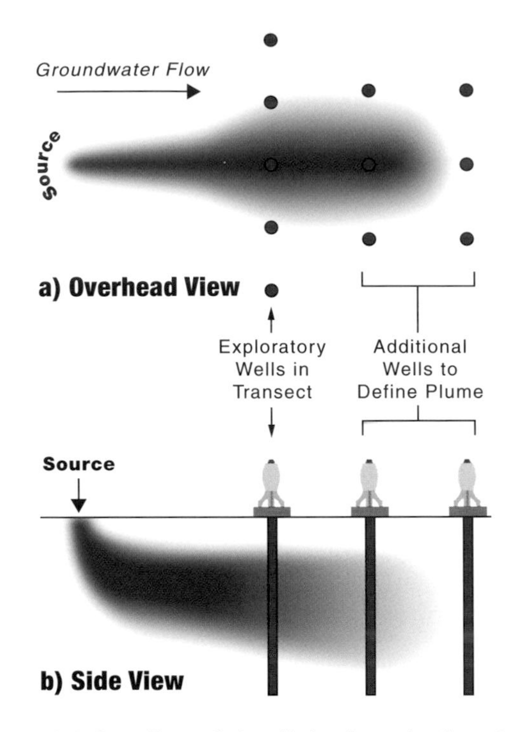

Figure 9-2 Sampling wells installed to determine the extent of a contaminant plume in groundwater: (a) overhead view, (b) side view.

the plume of a contaminant in an aquifer whose direction of movement is poorly defined. We might start by drilling several monitoring wells in a transect (crosscut) at right angles to the expected direction of flow. Once we had the initial results, we could develop a modified sampling design to minimize the total number of wells needed. In this example, the item of interest (the contaminant) must be tracked over three dimensions, as shown in Figure 9-2.

Temporal distributions also offer numerous choices for sampling. In this case, we have to make choices about sampling intervals, frequency of sample collection, and the time of sampling for which values are meaningful. Many phenomena have cyclical characteristics, such as daily or annual temperature changes. Moreover, any sample that is affected by biological activity may exhibit large temporal changes. For instance, the amount of nitrate in soil is influenced by plant nutrient uptake and by microbiological activity. However, both nutrient uptake and microbiological activity are, in turn, affected by temperature and moisture content, both of which can fluctuate. Thus, sampling under just one set of conditions may strongly bias the conclusions we draw.

9.4 Sampling the Soil Environment

Soils are anything but uniform—and this extreme heterogeneity is the single most challenging aspect of sampling in the environment. Thus, a common complication to sampling soil pollution is our limited knowledge about the composition of the soil environment. Since each soil sample collected is part of a larger, highly diverse population of components present in the soil, a valid and defensible estimate of soil pollution levels requires: (1) *a priori* knowledge of the soil heterogeneity and pollutant variance of the soil system; (2) careful use and implementation of soil sampling equipment, sampling protocols, and analytical techniques; and (3) a rigorous statistical approach to determining the appropriate soil sampling scheme and to analyzing the data.

9.4.1 Destructive Soil Sampling

The most common approach to measuring soil pollution involves removal of soil samples from the field and subsequent analysis. That is, we measure the concentrations of pollutants in the soil samples either by analyzing the soil itself or by analyzing soil-liquid extracts or acid digests. In either case, we have to use disruptive sampling methods and destructive analytical techniques to obtain our data. However, environmental soil sampling requires that soil samples be collected with minimal physical disturbance and that they be free of cross-contamination. Consequently, special sampling equipment has been designed to minimize field contamination and physical disturbance.

We have at our disposal many methods of collecting samples that can be analyzed for target pollutants. For example, solid-phase soil samples may be collected by using devices as simple as a hand shovel and as sophisticated as a mechanical split-spoon auger mounted on a truck (see Figure 9-3a). On the one hand, screw augers are suitable for taking disturbed soil samples (see the photograph at the top of page

Figure 9-3 (a) A portable drill rig equipped with a split-spoon auger for taking soil samples at depth. Notice the protective equipment worn at this hazardous waste site. (b) Workers in protective suits and respirators take samples from a split-spoon auger core extracted from the soil in (a). The worker on the right is holding a photoionization detector, which detects volatile organic compounds. The worker on the left is technically in noncompliance with rules regarding protective wear since the suit sleeves are not taped down over his plastic gloves. Photographs courtesy of M. Young.

115). On the other hand, split-spoon and hollow-stem augers are particularly useful because they allow us to take sample soil cores without disturbing the surrounding field. Moreover, a variety of hand-held and mechanical augers have been designed to use new inserts, or sleeves, for each sample collected. These sleeves may be either plastic or metal: we use plastic sleeves for routine soil inorganic parameters, and metal sleeves for possible organic contaminants. In either case, the result is a set of sealed soil cores that can be stored and transported to the laboratory for analysis without risk of cross-contamination.

9.4.2 Nondestructive Soil Sampling

Nondestructive soil sampling is performed *in situ* (on site). Such testing, which does not involve soil removal, usually focuses on liquid or gaseous components of the soil.

Soil contains water within which soluble contaminants can reside (see Chapter 6, beginning on page 63). Soil-pore water sampling devices are used to collect samples of water-soluble pollutants from the soil environment. These devices do not directly measure any pollutant parameters, but allow the *in situ* collection of samples of soil-pore water. These samples may then be analyzed for compounds of concern. Such devices also require some soil disturbance as they must be permanently inserted into the soil profile.

There are two major types of soil-pore water collection devices: *passive* or gravity-fed and *active* or suction devices. Passive devices include hollow glass brick **lysimeters** or sealed boxes. These boxes are inserted into the soil where water percolates into them under the force of gravity. Then this water is withdrawn under suction via an access tube, and preserved for analysis. Active devices include numerous types of porous cups that are used to collect soil-pore water under suction. These cups, which may be constructed of ceramic, Teflon®[1], or stainless steel, are cylindrical devices that are inserted into the soil. Then a vacuum is applied to create suction across the porous material in contact with the soil solution. This process draws soil-pore liquid into the cup. The liquid is subsequently withdrawn via an access tube and subjected to chemical analysis. Since such passive and active devices intercept and remove only a very small fraction of the soil-pore water, they can be considered nondestructive samplers. [*Note:* See the description of tensiometers in Section 5.1.3, beginning on page 48, and Figure 5-4 on page 49, as they are similar in construction and operating principles.]

Truly nondestructive or *in situ* soil monitoring is performed primarily for two soil parameters: soil moisture and salinity. These two parameters are linked to changes in physical and chemical properties of the soil—changes that may be induced by the presence of such pollutants as soluble salts. Additionally, **hydraulic conductivity**—the ability of a porous medium (soil) to transmit water (see Section 5.2, beginning on page 50)—is generally the single most important soil physical parameter that can be linked to the transport and fate of pollutants in the soil environment. To obtain this parameter, we use **infiltrometers**, which are field devices placed on the surface of the soil where they measure water infiltration rates. Two devices are commonly used for measuring infiltration: the double-ring infiltrometer and the tension disk infiltrometer. For more information see Wilson *et al.* (1995).

Neutron activation probes, time domain reflectometry (TDR), and electrical conductivity (EC) soil probes are also used for nondestructive measurements of soil water content. Both TDR and EC probes can also be used to measure salinity in soils. However, these devices require some disturbance of the soil environment because an access tube or metal probes must be inserted into the profile. The use of these devices is somewhat limited, owing largely to high costs. In addition, such devices are frequently difficult to use for various reasons. The neutron probe, for example, requires a low-energy radioactive source, while both TDR and neutron probes require the use of sophisticated power sources and measuring devices as well as frequent calibrations. For further details about the design and operation of these devices see Wilson *et al.* (1995) (see also Section 5.1.3, beginning on page 48).

Nondestructive methods are also available for collecting samples of the soil gas. Such collections are of special interest for monitoring levels of important gases such as oxygen and carbon dioxide, and for determining gas-phase concentrations of volatile organic pollutants (such as gasoline components or cleaning solvents). Active gas-sampling devices are conceptually similar to those used for pore-water sampling; thus, volumes of the fluid (in this case, gas) are extracted from the soil and then removed to the surface for subsequent analysis for the target pollutants. Passive gas-sampling devices, however, are

[1] Teflon: Registered trademark for DuPont fluorocarbon resins.

different. Such devices are usually based on placing a sorbing material, such as an activated charcoal capsule, in the soil. The capsule may be recovered after the gas-phase pollutant is sorbed into charcoal and trapped. Then the pollutants can be volatilized (thermally desorbed) and subjected to appropriate analysis.

9.4.3 Additional Soil Sampling Considerations and Constraints

Other important aspects of pollution sampling involve methods of sample collection and methods of sample preparation, some of which can be difficult to implement. One significant problem is the question of how much to sample. Because soil systems have microscopic as well as macroscopic heterogeneities, sample size can affect the precision of analyses. Theoretically, the larger the sample the better. However, it is impractical and unrealistic to collect large volumes of soil—the costs of massive collections would be prohibitively high and the large disturbance created would damage the soil system. Thus we must agree on some acceptable guidelines. As a general rule, we assume that the mass of a soil sample should be no less than 200 g. Typical soil samples range from 300 to 1000 g, depending on the type of equipment used, the depth of the soil, and the number of pollutants that must be analyzed. In addition, field samples should be thoroughly mixed and subsampled as necessary prior to analysis. And when sample bulking (composites of two or more samples) is appropriate, equal amounts of each sample must be thoroughly mixed prior to analysis.

Another important consideration is the nature of the target pollutants, which will affect the collection technique. Some samples, such as those with radioactive constituents, may be hazardous to handle. If the material to be collected is hazardous, the person doing the collecting will need special equipment, from simple gloves to respirators and protective suits (see Figure 9-3 on page 119). The nature of the sampled material—and the target pollutant—also affects the way in which the samples are handled. For example, soil samples are usually collected in plastic or glass jars, sealed, and kept cool at 4°C until analyzed. Soil-pore water extracts, which may be collected in either glass or plastic bottles, are preserved and stored using protocols similar to those used for water samples (these are discussed below). The following example illustrates an application of sampling principles and sample collection techniques.

Because it is a necessary plant nutrient, nitrate-nitrogen (NO_3-N) is often added to croplands, sometimes as chemical fertilizer and sometimes as nitrate-rich municipal sludge. Ideally, any nitrates added to the soil in a field will be consumed by the crop. Inevitably, however, some nitrates leach below the root zone, so that nitrates are the most prevalent pollutant found in groundwater in the United States (see Section 14.2, beginning on page 212, and Section 16.5.1, beginning on page 241). Thus, it is important to monitor the presence and fate of nitrates in the soil over time. Consider the case of a 28-hectare irrigated cotton field treated with a nitrate-rich municipal sludge.

Collection: Soil cores were collected in a grid pattern (see Figure 9-1d on page 117), so that for each core, a total of 9 soil samples was collected at 30-cm vertical intervals. These samples were then placed in dedicated containers, and the sample for each 30-cm interval was thoroughly mixed and subsampled. One subsample was placed in a sealable metal can for field moisture determination. A second subsample of about 300 g was collected into a sealable plastic bag for chemical and physical analyses. All subsamples were immediately stored in an ice chest and kept refrigerated until extraction and analysis. In the lab, soil-water extracts were analyzed for NO_3-N using ion chromatography.

The spatial sampling pattern described above was repeated three times during the year:

1. March, after the sludge was applied but before the cotton was planted.
2. May, ten days following preplant irrigation and during germination of the planted cotton.
3. September, when plants had reached maturity and were in full bloom.

Data Analysis: Following chemical analysis of NO_3-N concentrations, the data for each location were summed over the 2.7-m profile and the cumulative concentration reported in kilograms of NO_3-N per hectare. To facilitate nonstatistical data evaluation, an interpolation scheme was used to estimate the concentration of NO_3-N for the entire field. These data were used to plot two-dimensional (2-D) contour maps, as shown in Figure 9-4. [*Note*: Interpolation schemes tend to be more accurate as the number of data points increase; this case represents close to the bare minimum of points needed.]

In Figure 9-4, we can see that the overall concentrations of NO_3-N in the root zone of this field increased significantly following the preplant irrigation (the May contour map is a deeper shade of red). This observation is consistent with significant increases in sludge-N mineralization rates following water addition to the soil. Because sludge behaves as a slow-release N fertilizer, the concentration of N in this particular field remained fairly constant until harvest time.

The contour maps also let us make observations about significant soil textural changes in the field, which were not readily apparent from the surface. For example, the southeast (lower right side) of the field exhibits N concentrations that are consistently lower than those in the rest of the field. Soil particle size analysis of the core samples (not presented here) later showed that this location was sandier. Thus, nitrates were quickly lost in this area during irrigation events.

Ideally, more measurement points would be made and the maps would be more reliable. On the other hand, the presented data represent many hours of labor. Even with such a limited number of points, the visual display (Figure 9-4) helped in the interpretation of the seasonal and spatial variability of the measurements.

9.5 Sampling the Water Environment

Because of its dynamic (liquid) nature, the water environment is especially vulnerable to pollutant inputs. That is, the water environment is much more rapidly and extensively affected by pollution events than is the soil environment. For example, a train derailment that spills a large quantity of pesticide on land is bad enough, but its effects will be fairly well confined within a limited area. In addition, because soil is a porous solid, the pollutants will not spread quickly, thus allowing sufficient time for containment and cleanup. But the same derailment on a bridge over a river will be much more serious. This spill will inevitably result in a fast spread of the pesticide in the water, and this spread will inevitably affect, in some degree or another, the entire river downstream as well as its associated environments.

Water environments, which are generally classified as oceans, streams, lakes, reservoirs, and groundwater, are apparently more homogeneous in composition than soils. But this apparent homogeneity can be misleading. In fact, water systems are often stratified: they have different temperatures and densities, gases, chemical compositions, and, by inference, different pollutant distributions. (See also Chapter 13, beginning on page 189.) Thus, because pollutant concentrations are seldom constant throughout a body of water, water environments must be sampled with care. In the United States extensive water monitoring began as a result of the Clean Water Act of 1948 and the Safe Drinking Water Act of 1974. See Section 23.2 and Section 23.3, beginning on page 366, for a description.

9.5.1 Surface Water Sampling

In sampling bodies of water, we must consider the surface, volume, and dynamics of these systems. For example, flowing bodies of water should be sampled across the entire width of the channel flow because water moves faster in the center than at the edges. In cross-sectional sampling, flow velocities should be measured and recorded at each sampling location. In relatively stagnant enclosed bodies of water, such as lakes and lagoons, it is also very important to sample by depth, particularly where sharp temperature changes, or **thermoclines**, occur. Temporal sampling for the analysis of dynamic (high volume turnover) water systems is common because the chemical (pollutant) compositions of these bodies can change quickly. Moreover, frequent sampling intervals are warranted to detect influxes of pollutants. Less frequent sampling is usually done in water systems having no obvious point and nonpoint sources of pollution.

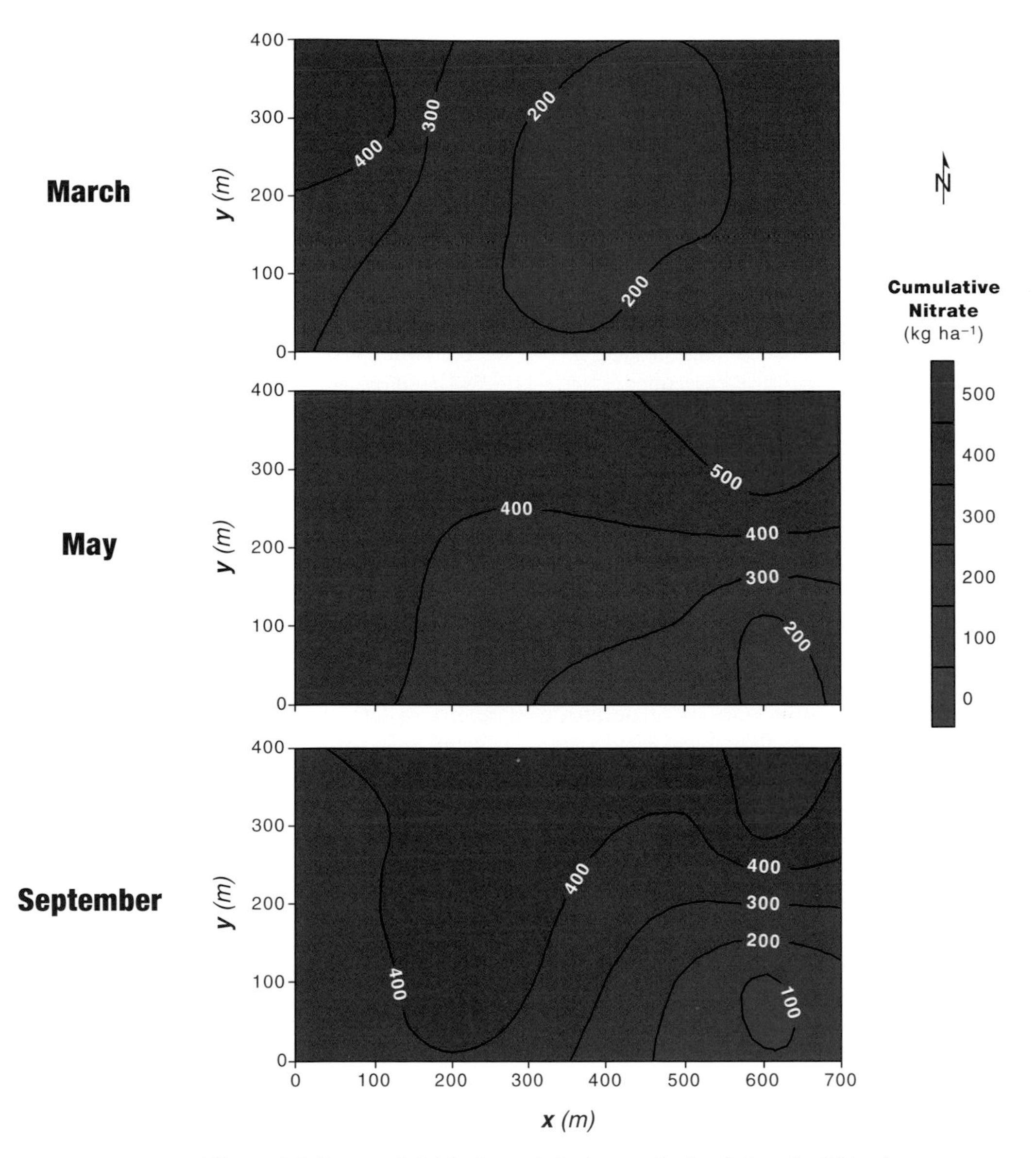

Figure 9-4 Seasonal distributions of nitrate over the top 2.7 m of a 28-hectare cotton field. Data: Artiola, 1995.

Nonpoint sources of pollution typically affect water systems slowly and evenly. In some cases, the impacts of nonpoint source pollution are mitigated by natural mechanisms; in other cases, however, their cumulative effects can lead to a quick collapse of one more parts of the aquatic system. For example, the effects of acid rain may be neutralized by the alkalinity of lakes with measurable concentrations of carbonates and bicarbonates. The buffering effect of these minerals may maintain the pH within a tolerable range (~6–8) following acid rain events for a long time (years). Then, as the alkalinity is exhausted, the pH may drop quickly below 6 after just one or two acid rain events. Such a drop in pH may trigger a series of environmental impacts to the aquatic system that are hard to reverse. These impacts may include death of fish, reduction in photosynthesis by marine algae, reduction of natural bio-

Table 9-1 Common parameters measured directly in water with field portable equipment.

Parameter	Associated Parameters	Method of Analysis/Probe
pH	High/low $[H^+]$	Hydrogen-ion electrode
Salinity	Na^+, Cl^-, plus other soluble ions	Electrical conductivity
Dissolved oxygen	Lack of O_2	Gas-sensitive membrane electrode
Temperature	High/low temperatures	Thermocouple

logical cycles that control micro- and macronutrients, and increased concentrations of toxic metals. In this case, both carbonates and bicarbonates as well as pH should be monitored.

On the other hand, point sources of pollution, which include natural drainage pathways and waterways such as rivers and creeks, and industrial and municipal discharge points or outfalls (see Section 13.3, beginning on page 192), can affect water systems much more quickly and unevenly. In this case, monitoring at or near the point of discharge is the single most important way to limit and/or quantify pollutant impacts to water systems. In these instances, it is usually more important to prepare and implement a monitoring plan with frequent sampling intervals.

We do not distinguish between destructive and nondestructive methods of sampling and analysis in water environments. Such a distinction is unnecessary because removing a sample (small volume) of water from a lake or a river, for example, does not disrupt or disturb these environments in a measurable or lasting manner. Nonetheless, we do recognize two types of water sampling: real time, *in-situ* analysis, which is done on location; and sample removal (*e.g.*, from rivers), which is followed by analysis in a laboratory. We will discuss these two approaches separately—the first with surface water sampling and the second with groundwater sampling.

With the advent of portable field analytical equipment, several surface water quality parameters or indicators can now be quantified on location and in real time. Some pollution-related parameters that can be measured with field equipment by inserting a probe directly into the water are listed in Table 9-1. Although the parameters listed in Table 9-1 are not necessarily pollutants, an imbalance or deviation from their "natural" range of values usually signifies that the water system is being stressed by one or more pollutants. For example, low concentrations of dissolved oxygen usually indicate excessive biological activity due to unusually large inputs of nutrients and organic matter. Thus, in lakes and streams, initial signs of eutrophication can be detected by measuring changes in oxygen concentrations.

Additionally, there are now numerous other tests that can be performed in the field but not directly in the water. Such tests allow us to measure the concentration of specific anions (found at concentrations >1 mg L^{-1}), such as NO_3^-, NO_2^-, PO_4^{3-}, and Cl^-, and/or to determine the alkalinity or hardness of the water, by using premixed chemicals (specific reagents that react with these ions and produce color) and portable color tables and/or colorimeters. Field tests for specific pollutants that are usually present in very low concentrations (μg L^{-1}) are also being developed. However, the accuracy of such field tests should always be questioned, as it is very difficult to remove interferences from other ions present in waters. Some additional pollutants that can be measured by using portable field equipment include metals such as Cr, Cu, Zn, Ni, and Pb, and organic chemicals such as cyanide, phenolics, oil, and tannins and lignins.

9.5.2 Groundwater Sampling

Many of the considerations discussed for surface water sampling also apply for collecting groundwater samples. However, a primary difference between the two is the method of sample collection. The majority of groundwater samples are collected from wells drilled into the subsurface. The appropriate location, drilling, and construction of monitoring wells is critical to a successful groundwater sampling program. The method used to drill the well and the materials used to construct the well should be compatible with the subsurface environment and with the pollutants for which the plan is designed. (Specifics on drilling and constructing wells can be found in the literature and in *References and Recommended Reading* on page 133.)

There are many factors to consider when developing a method to collect samples from monitoring wells. One major factor is **well purging**, which is the withdrawal of a specified volume of groundwater prior to the collection of a sample. The goal of purging a well prior to sample collection is to ensure that the water sample truly represents the properties and conditions of the subsurface environment, and not those of the well itself. The problem is that the properties of the water residing in a well can be quite different from those of the water in the porous medium. The water in a well is exposed to the atmosphere, which can lead to changes in temperature, pH, and oxygen content. Such changes can cause changes in the properties and concentrations of pollutants. Pollutants can also interact with well materials. For these and other reasons, it is desirable to discard water that has been sitting in the well and take samples from water that has been freshly drawn from the porous medium.

Another major factor to consider for groundwater sampling is the device that will be used to collect the sample. There are several criteria for selecting the appropriate sampling device, including the preservation of sample integrity, ease of use, and economics. Obviously, it is critical that the sampling device permit neither a loss of pollutant nor a change in the properties of the pollutant during collection. For example, a device used to collect samples containing volatile pollutants should not allow volatilization. Thus, a device such as a suction-lift pump, which creates a vacuum, would not be suitable for collecting samples of volatile pollutants.

9.5.3 Water Analyses—Additional Sampling Considerations and Constraints

Samples of water and nonaqueous liquids offer specific collection problems. Water samples to be analyzed for trace and ultra-trace metals, organic pollutants, and microbes pose particular difficulties with contamination. Typical problems encountered during sample collection include external contamination of samples and pollutant losses by volatilization in the case of volatile organic chemicals. In this section we will review special sample collection, handling, and preservation techniques that should be followed for proper monitoring of water environments.

As a general rule, accurate and precise analysis of low-level (trace) pollutants in water can be done only in a laboratory environment using modern sample-preparation methods and analytical equipment operated by skilled technicians. There are two very good reasons for following this general rule. The first is that the analytical instruments needed to achieve regulatory detection limits of individual pollutants are very expensive and so delicate that they require constant calibration and maintenance. The second is that laboratory procedures minimize the risk of contamination during sample preparation and analysis, thereby reducing the chances of false positive results. Moreover, water analyses cannot be done quickly; so storing samples in a controlled laboratory environment is necessary to ensure that specimens do not degrade (change chemical forms) and thus yield false negative results.

In order to minimize the problem of contamination/degradation, we pay careful attention to the type of containers used to collect water samples. Several broad categories of sample containers are available to collect water samples, depending on the parameters to be analyzed. Table 9-2 on page 126 lists these major categories together with the types of sample preservation techniques used to ensure that the pollutants in the water sample will not sorb, precipitate, or change form while in storage. Table 9-2 does not include every storage and preservation technique used to collect water samples. However, the majority of water samples are collected and stored following one or more of the methods listed in the table, some with only minor modifications. (For a comprehensive list of all parameters, see U.S. EPA, 1983.)

A further consideration is the issue of maximum holding times. Current guidelines developed by the U.S. EPA require that water samples be analyzed within a specified time interval following collection. These "maximum holding times" are quite parameter-specific and can vary greatly. For example,

Table 9-2 Major types of sample containers and methods of preservation.

Pollutant Class	Sample Container	Preservation/Storage
Metal cations (*e.g.*, Pb^{2+}, Ni^{2+}, Zn^{2+})	Plastic (HDPE[a]) bottle	Add nitric acid (final pH < 2)
Major anions (*e.g.*, Cl^-, F^-, NO_3^-)	Plastic (HDPE[a]) bottle	Keep cool at 4°C
Major cations (*e.g.*, Na^+, K^+, Ca^{2+}, Mg^{2+})	Plastic (HDPE[a]) bottle	Add nitric/sulfuric acid (final pH < 2)
Purgeable hydrocarbons	Glass vial, clear or amber	Add reducing reagent.[b] Keep cool at 4°C
Pesticides, herbicides, nonvolatile hydrocarbons, dioxins, and furans	Amber glass bottle	Add reducing reagent.[b] Keep cool at 4°C
Surfactants (from soaps), plasticizers (from plastics), and PCBs (from transformers)	Amber glass bottle	Keep cool at 4°C

[a] HDPE = high-density polyethylene plastic.

[b] Low concentrations of sodium thiosulfate or ascorbic acid are added to the water sample to minimize oxidation by oxidizing agents such as chlorine.

chemicals that degrade or change form quickly and cannot be preserved must be analyzed between 8 and 48 hours after collection: these include oxygen, phosphates, cyanide, and nitrate. On the other hand, most metals and salts, if properly preserved, can be kept for up to 6 months prior to chemical analysis. Water samples containing organic pollutants must generally be extracted (using solvents) within 7 days; but thereafter, the solvent extract containing the pollutants may be kept for up to 40 days prior to actual analysis. (See U.S. EPA, 1983, for a complete list of holding times.)

Aeration or degassing of water samples is a critical consideration in the development of sampling plans for both surface and groundwater systems. Aeration or degassing of water samples must often be prevented for several interconnected reasons. In the first place, many groundwaters have concentrations of CO_2 that are higher than those in the atmosphere. Exposure of a groundwater sample to the atmosphere during or after collection therefore results in loss of CO_2, which leads to an increase in pH. In addition, changes in pH can influence the properties of many contaminants. For example, the solubility of many metals is higher at lower pH values. Thus, the increase in pH due to a loss of CO_2 could induce precipitation of metals so that the aqueous concentrations of the metals obtained by laboratory analysis would be lower than the true values. Ultimately, we could wind up with the erroneous conclusion that metal pollution was not a problem for that system. Aeration of water samples also changes the oxidation/reduction (redox) status of the system. Changes in redox can affect many properties and processes. For example, increases in the redox status (oxygen concentration increases) can induce precipitation of iron hydroxide $Fe(OH)_3$ in the following manner:

$$\begin{aligned} Fe^{2+} &\longrightarrow Fe^{3+} + e^- \\ 2e^- + 2H^+ + \tfrac{1}{2}O_2 &\longrightarrow H_2O \\ Fe^{3+} + 3H_2O &\longrightarrow Fe(OH)_3\downarrow + 3H^+ \end{aligned} \quad (9\text{-}1)$$

Metals such as Cd, Pb, and Zn can sorb to iron hydroxide, which would reduce their aqueous-phase concentration in the sample vial. Addition of oxygen can also influence organic pollutants, for example, by inducing aerobic biodegradation.

Finally, in developing sampling plans for surface and groundwater systems, we need to consider filtration of water samples. "To filter or not to filter" is a question that must be addressed for every specific site and for each type of pollutant. A major reason to filter water samples is to remove suspended solids, allowing collection of only dissolved pollutants. Filtering, however, raises the specter of aeration or degassing during the filtration, as outlined above.

9.6 Monitoring Air Pollution

Air pollution is monitored within all large urban areas in the United States and in most other large cities worldwide. Monitoring in the United States is a direct result of the provisions of the Clean Air Act of 1970 (see Section 23.6, beginning on page 369). This Act requires that the EPA formulate, enforce, and revise (at five-year intervals) the National Ambient Air Quality Standards (NAAQS), which apply to the outdoor air pollutants known to be most harmful to human health. These pollutants, which are commonly found in urban air, include carbon monoxide, nitrogen oxides, sulfur dioxide, nonmethane hydrocarbons (NMHCs), particulate matter, and ozone. These and other pollutants are discussed in Section 12.2, beginning on page 172. To help the EPA meet the requirements of the Clean Air Act, an extensive set of data has been and is being collected by monitoring networks throughout the nation.

In addition, concerns about the impacts of air pollution on the global environment are promoting research directed toward monitoring pollution throughout the whole atmosphere. There are several ongoing national and international research programs involving environmental data collection from towers, aircraft, balloons (Figure 9-5), sounding rockets, and satellites. Satellite-borne radiometric sensors offer the means to monitor pollutants on a global scale. They provide needed data on impacts of air pollution on the earth's radiation balance, climate change, and stratospheric ozone depletion. One such effort—the Earth Observing System (EOS), which deploys satellite-borne instruments contributed by the National Aeronautics and Space Administration, by the National Oceanic and Atmospheric Administration, and by other nations—will provide atmospheric composition (and other) data well into the twenty-first century.

Figure 9-5 Tethered balloons, such as the one pictured here, are used for temperature, pressure, and humidity measurements. Photograph: A.D. Matthias.

9.6.1 Monitoring Urban Air Pollution and Sources

As set forth under Title 40 of the Code of Federal Regulations, the general goals of *in situ* urban air pollution monitoring networks are to determine the:

1. highest pollutant concentrations expected to occur in an urban area
2. representative pollution concentrations in highly populated areas
3. effects of major sources on pollution concentrations
4. background pollution concentrations.

Many questions must be considered in establishing and maintaining a network of air-quality monitoring stations. For example, we need to decide what pollutants are to be monitored, where they should be monitored, and what instruments are to be used for that purpose. We also need to know what kind of weather data needs to be collected. Finally, we have to figure out how many stations are necessary to meet the above-stated goals.

The general answer to this last question is that networks must include a "sufficient" number of urban and surrounding rural stations. Since such general answers don't help us much, let's look at a specific metropolitan area for some specific answers to this and other questions.

Metropolitan Tucson, Arizona (population about 700,000 people, area about 500 km^2), has about 33 stations operated and maintained by the Pima County Department of Environmental Quality. Most of these stations are within the city but some are in the rural outskirts of the city.

Depending on the strength and proximity of their sources, different pollutants may be monitored at different stations. Thus, the Tucson stations sited in the south part of the city mostly monitor particulate matter because they receive the dust generated by nearby mining operations. Stations sited near busy intersections and in neighborhood locations usually monitor motor vehicle pollutants, such as CO and NO_x ($NO + NO_2$). Several stations monitor ozone levels, but do so primarily during the summer months when photochemical production of ozone is highest.

Most of these air-quality monitoring stations consist of well instrumented laboratories housed in climate-controlled (air-conditioned or heated) trailers or vans. The station may contain various automated analytic instruments, recorders, calibration materials, and other laboratory equipment needed to ensure data reliability and validity. Outside air for pollutant analysis is brought into the lab via a teflon tube from a level several meters above ground, thereby ensuring sampling of well mixed boundary-layer air (see Section 3.2.5, beginning on page 28). The incoming air is diverted to an analytic instrument in separate air streams at flow rates of 0.5–1.0 liters per minute. These analytic instruments, most of which use optical properties to determine pollutant concentration, are described below, together with the pollutants they measure.

1. Carbon monoxide (CO) is typically measured by using an **infrared gas analyzer**. With this instrument the absorption of infrared radiation (see Section 3.2.4, beginning on page 26) by CO in the sample air stream is compared with absorption in a reference gas of known CO concentration. This method allows continuous nondestructive measurement of CO in the sampled air.

2. Sulfur dioxide (SO_2) is generally measured by **ultraviolet (UV) emission spectrometers**. This approach is based on the principle that SO_2 emits a measurable flux of radiation when irradiated with intense UV from a light source in the spectrometer.

3. Nitrogen oxides (NO_x) are measured by **chemiluminescence**. Two sequential chemical reactions involving ozone are used. First, NO is measured, then NO_2. Infrared radiation is emitted during oxidation of NO to NO_2 by ozone introduced into the instrument. The amount of radiation (chemiluminescence) produced is proportional to the NO concentration in the air stream. To measure NO_2, a catalyst is used to reduce all NO_2 in the air stream to NO, whose subsequent reaction with ozone permits the indirect determination of NO_2.

4. Ozone (O_3) concentrations are generally measured by using a **UV absorption spectrophotometer**, although chemiluminescent-type instruments are also used.

5. Various nonmethane hydrocarbons (NMHCs) are measured using such instruments as a **gas chromatograph**. Hydrocarbons are generally more difficult to measure than most other pollutants, and often require greater operator involvement in the measurement process.

6. Particulates that are less than 10 μm in diameter (PM_{10}) are measured using a **high-volume air sampler** situated outside the trailer or van. In this type of instrument, a known volume of air is continuously drawn through several paper filters. The filters collect particles of different size ranges. The amount of particulate matter in each size range is determined by measuring the weight change of the filter over a specified period of time, such as 24 hours. Chemical analysis, such as determination of lead, may also be done.

Data from the instruments are recorded continuously on a data logger at the station. Typically, average concentrations are calculated for each pollutant each hour of the day. The data are checked and sent to the EPA for further processing and analysis. Also, all instruments are frequently checked and calibrated using reference gas sources of known concentration. (Detailed information about these instruments may be found in Mead, 1993.)

The EPA checks the data collected by the Tucson stations and others to make sure pollutant levels meet National Ambient Air Quality Standards (NAAQS). Table 9-3 summarizes standards set by

Table 9-3 EPA monitoring standards for outdoor pollutants. Source: Mead, 1993.

Pollutant	Period of Averaging for Mean Concentrations	NAAQS[a]
CO	8 hours	9 μL L^{-1}
	1 hour	35 μL L^{-1}
NO_2	Annual	0.05 μL L^{-1}
O_3	1 hour	0.120 μL L^{-1}
SO_2	Annual	0.03 μL L^{-1}
	24 hours	0.14 μL L^{-1}
PM_{10}	Annual	50 μg m^{-3}
	24 hours	150 μg m^{-3}

[a] National Ambient Air Quality Standard.

the EPA for selected common outdoor pollutants. Pollutant concentrations that are greater than these values when averaged over a specified time period (*e.g.*, 1 hour, 8 hours, 1 year) are said to exceed the standard. For example, if the average CO concentration taken at any 1-hour period during a year at a monitoring station exceeds 35 μL L^{-1}, the city may be in violation of federal safeguards of air quality.

Standard meteorological data may also be collected at air-monitoring stations and sent to the EPA. The meteorological data generally include total and diffuse solar radiation, wind speed and direction, relative humidity, and air temperature. The EPA uses these data to interpret pollution results. These data may also serve as inputs to numerical models that relate the concentration distribution to emission rates of important pollutants. Modeling is a tool that can facilitate the assessment of potential problems, such as the potential effects of new or additional sources of pollution on air quality.

Environmental scientists in some urban areas use standard instruments such as 3-cup anemometers as well as specialized acoustic sounder systems to monitor wind, turbulence, and temperature conditions within the full depth of the urban boundary layer. These acoustic systems are especially useful for monitoring such meteorological conditions as temperature inversions (see Section 3.2.2, beginning on page 24, and Section 12.4, beginning on page 177), which affect the dispersal of pollutants.

In addition to monitoring ambient air pollution, several states now require annual direct inspection of pollutant emissions from motor vehicles. Owners of vehicles failing to meet emission standards are required to tune their cars and/or make necessary repairs to emission control devices. Motor vehicle inspection programs have been very successful in helping to reduce the emission inventories of most pollutants. Some argue, however, that a better alternative might be to use roadside sensors that can monitor exhaust pollutants directly on cars as they drive past (see Cadle and Stephens, 1994). In such remote sensing systems, a beam of infrared radiation is directed across a single lane of traffic to a detector that captures the beam after it has been attenuated by the presence of vehicle exhaust pollutants such as carbon monoxide and hydrocarbons. The degree of attenuation is then used to infer the amount of pollution emitted by the vehicle. Developed in 1989, this type of approach, which can quickly identify vehicles in noncompliance with emission standards, offers great promise for the ultimate reduction of total emissions. Commercial systems for remote sensing of vehicle exhaust have been recently developed to fulfill this promise.

9.6.2 Monitoring Gaseous Emissions from Soils and Landfills

Soils play an important role in controlling background concentrations of most air pollutants. Soils can either emit or take up from the atmosphere many trace gases, including NO_x, N_2O, CO_2, and CH_4. Similarly, waste storage areas such as municipal landfills also emit and take up such gases, and may be important localized sources of atmospheric CH_4.

Measurement of soil gas fluxes may be difficult, depending upon the location and the gas(es) monitored. For example, it is hard to obtain highly accurate flux measurements when there are large uncertainties in determining gas concentrations, which is often the case. Owing to this and other problems, the methods we use to measure gas fluxes must be adapted to suit several situations. In general, we use three different approaches to measure gas fluxes between soil and the atmosphere: chamber approaches, micrometeorological approaches, and soil profile approaches.

9.6.2.1 Chamber Approaches

Both closed and open chambers are common. Each of these approaches entails placement of an enclosure directly over the soil surface to collect emitted soil gases within the air space of the enclosure. In the **closed chamber approach**, we determine the flux by measuring the rate of increase of gas concentration within the enclosure. In the **open chamber approach**, outside air is continuously pumped through the enclosure. The flux is then calculated from the measured difference between the gas (pollutant) concentration in air entering the enclosure and that in air leaving the enclosure. Although both chamber approaches are relatively simple to use, it is still difficult to obtain samples that reflect the "real" or actual gas concentrations.

We generally use closed chambers to measure fluxes of nonreactive gases, such as N_2O and CO_2 (Figure 9-6), and open chambers to measure fluxes of reactive species, such as NO_x. For nonreactive gases, air samples from the enclosure are collected in gas-tight containers such as syringes or evacuated bottles. The air samples are then taken to a laboratory for analysis, often by gas chromatography. The problem is, no container is truly gas-tight; loss rates of N_2O from "gas-tight" glass syringes, for example, can be as much as 1 to 2% per day. Thus it is important to do the analysis as soon as possible in order to minimize such inevitable gas losses. For reactive gases such as NO_x, container losses are no problem because concentrations are measured directly at the measurement site, often by chemiluminescence. However, gas concentrations within outside air can vary spatially and temporally.

The chamber approach also presents some methodological difficulties. Chambers alter the soil environment, which can affect production and diffusion of gas from the soil to the air. In closed chambers, for example, the relatively large buildup of gas concentration in the enclosure may markedly reduce the concentration gradient between the soil and air, thereby reducing the calculated flux. Moreover, gas fluxes from soil are often highly spatially and temporally variable, and chambers can measure flux only over a relatively small soil area (*e.g.*, $<1\ m^2$). Thus, measurements at many sites are often required in order to obtain a truly spatially integrated flux value.

Figure 9-6 Closed chamber for measuring gas flux to the atmosphere. Air samples from the enclosure are collected in gas-tight syringes. Photograph: A.D. Matthias.

Figure 9-7 Micrometeorological (flux-profile) instruments used to measure N_2O flux over a large turf grass field. Photograph: A.D. Matthias.

9.6.2.2 Micrometeorological Approaches

There are several micrometeorological approaches, all of which involve the continuous measurement of gas concentration in the atmospheric surface layer (see Section 3.2.5, beginning on page 28). These approaches include **eddy-correlation** and **flux-profile** methods, which are used to measure trace gas fluxes over relatively large areas. For the eddy-correlation approach, rapid sampling (~1 sample per second) of gas concentration and wind speed at one

Figure 9-8 (a) View of a landfill site in Tucson, Arizona showing gas sampling wells. A special bag located in the desiccator at the worker's feet is used to collect the gas sample. Gas sampling bags exhibit low permeability to liquids and gasses, with some types offering additional protection of the gas sample from ultraviolet degradation. (b) A gas sample is syringe-extracted using an innovative technique that allows the sample to be analyzed on-site with a portable gas chromatograph or stored temporarily in metal canisters prior to off-site laboratory analysis. Photographs: G. Monger.

height is required. Rapid sampling of concentration is difficult to achieve for most trace gas species, although fast-response instruments such as infrared gas analyzers are commercially available for use with CO_2 and some other gases. Sampling of wind is done with fast-response sonic anemometers. In flux-profile approaches, vertical concentration gradients are measured in the surface layer (Figure 9-7). Generally, the sampling requirements for the flux-profile approach are not as demanding as those of the eddy-correlation approach. However, it is very difficult to resolve the often small gas concentration gradients, especially when the atmosphere is well mixed. (See, for example, Table 8-7 on page 113, which lists N_2O concentration data collected within the surface layer above bermuda grass.) Micrometeorological approaches have an advantage over chamber approaches because they do not disturb the soil environment during the measurement process and because they allow us to make valid flux measurements over relatively large areas (*e.g.*, 100 to 1000 m^2).

9.6.2.3 Soil-Profile Approaches

Soil-profile approaches provide useful data about sources and sinks of gas contaminants in soil. In some applications, the measured gas content of soils may be used to estimate gas fluxes from areas such as landfills. Soil gas samples are collected by various means, including active and passive sampling. In active sampling, we collect gas samples by inserting stainless steel sampling probes (tubes) into the soil to several depths. Then we use a vacuum pump or gas-tight syringe to extract the soil gas samples, typically storing them in gas sampling bags, glass syringes, or evacuated containers for chemical analysis (Figure 9-8). In passive sampling, a sorbent material such as charcoal may be used to collect the soil gas. In this case, the sorbent is housed in a container placed in the soil so that gas contaminants passively move through the soil to the collection device. Depending upon gas production or loss rates in the soil, fluxes to the atmosphere may be inferred from the measurement of vertical concentration gradients and estimation of soil gas diffusion coefficients.

9.7 Accuracy, Precision, and Errors in Environmental Measurements

It is important to understand that the results of all our measurements are only as good as our instruments, and our instruments give us good results only if they are properly calibrated. That is, we can't make measurements without a known point of reference. In general, a **reference** is a standard—such as a fixed point, a length, a mass, or a cycle in time or space—that we trust not to change. Specifically, then, we must constantly calibrate our instruments against a reference, or known standard, to ensure their reliability. Thus, when we are making measurements of pollutants in environmental samples, we often use a "certified" standard solution traceable to the National Institute of Standards and Technology to calibrate our instrument. For example, to calibrate a spectrophotometer for aluminum measurement, we would first dilute a certified standard several times to make a series of known solutions, as listed in Table 9-4. When we run these solutions on the spectrophotometer, we get a reading that corresponds to the concentration of aluminum in each solution. We can then plot the solution concentration (x-axis) against the instrument response (y-axis) and perform a regression as detailed in Section 8.4, beginning on page 109. The resulting equation defines the best fit among the (x,y) points. Thus, we have a curve of the true values for each concentration.

Table 9-4 Calibration of a spectrophotometer to measure aluminum in an unknown sample.

[Al] in Standard Solution (mg L^{-1})	Spectrophotometer Response (intensity units)
0.000	4,142
0.500	17,315
2.00	63,305
5.00	161,486
10.00	320,087
Unknown sample	250,090
Statistical Evaluation	
Linear regression equation	$y = a + bx$
x	Instrument response
y	[Al]
a	0.00
b	3.206×10^4
r^2	0.9997
[Al] in the unknown sample	7.8 mg L^{-1}

Following this calibration procedure, we can obtain the response of our sample concentration and use the calibration curve equation to compute the actual concentration in the sample. This process is repeated for each unknown sample. If we make repeat measurements of the same sample, we can check the precision and accuracy of the instrument. We first met these concepts in Section 8.2.1, beginning on page 98, where we looked at the statistical definitions of sample mean and variance. Now, however, we can define **accuracy** as a measure of proximity to the true or reference value, and **precision** as a measure of the reproducibility of a method repeated on a homogeneous sample under controlled conditions. Note that these two measures are not related. It is possible, for example, to have replicate analyses whose values are very precise (close together), but inaccurate (far from the true value). Replicate analysis of reference materials can be evaluated in terms of precision by computing the percent of deviation between the measured and true values, and by computing confidence limits when many subsamples have been analyzed (>30) (see Section 8.3.1, beginning on page 106). An accepted way of reporting precision is by computing the standard deviation or the coefficient of variation (CV) (Eq. 8-6 on page 106).

Instruments with poor precision and/or accuracy produce biased measurements. Biased data are the result of either systematic or random errors in the procedure, which lead to a deviation from the true value. A **systematic error** is usually caused by poor equipment calibration or by artifact(s) within the procedure. Contamination of control samples, poor reagent quality, and extraction inefficiencies are examples of conditions that cause these errors. Systematic errors are relatively easy to correct because they tend to affect only the accuracy of the measurement. **Random errors**, however, are hard to detect (and hence harder to correct) since they are usually caused by random events or conditions that occur during the execution of a procedure. These errors typically affect the precision (random increases in the variance) of measurements. Both types of errors may produce a positive or negative bias.

We further define the errors in measurements as follows: **Type I errors (false positive)** which result from overestimating the amount of a component in a sample; and **Type II errors (false negatives)** which result from underestimating the amount of a component in a sample. Both types of errors are common in the analysis of soil and waste samples. Usually, these are related to matrix effects (enhancement/reduction of a measurement signal) and poor sample preparation (extraction/digestion), resulting in low pollutant recoveries.

References and Recommended Reading

Arizona Water Resources Research Center (1995) *Field Manual for Water Quality Sampling.* The University of Arizona, Tucson.

Artiola J.F. (1995) Unpublished data. Department of Soil, Water and Environmental Science, University of Arizona.

Boulding J.R. (1994) *Description and Sampling of Contaminated Soils: A Field Guide.* 2nd Edition. Lewis Publishers, Boca Raton, Florida.

Cadle S.H. and Stephens R.D. (1994) Remote sensing of vehicle exhaust emissions. *Environmental Science and Technology.* **28**, 258A–264A.

Carter M.R. (1993) *Soil Sampling and Methods of Analysis.* Canadian Society of Soil Science. Lewis Publishers, Boca Raton, Florida.

Csuros M. (1994) *Environmental Sampling and Analysis for Technicians.* Lewis Publishers, Boca Raton, Florida.

Denmead O.T. (1983) Micrometeorological methods for measuring gaseous losses of nitrogen in the field. In *Gaseous Loss of Nitrogen from Plant-Soil Systems* (J.R. Freney and J.R. Simpson, Editors), pp. 133–157. Marinus Nijhoff/Dr. W. Junk Publishers, Netherlands.

Gilbert R.O. (1987) *Statistical Methods for Environmental Pollution Monitoring.* Van Nostrand Reinhold, New York.

Greensberg A.E., Trussell R.R., and Clesceri L.S. (1992) *Standard Methods for the Examination of Water and Wastewater*, 18th Edition. APHA/AWWA/WPCF. American Public Health Association, Washington, D.C.

Keith L.H. (1991) *Environmental Sampling and Analysis: A Practical Guide.* American Chemical Society, Washington, D.C.

Mead S.C. (1993) *Annual Data Summary, Air Quality in Tucson, Arizona 1993.* Pima County Department of Environmental Quality, Arizona.

NASA (1988) *Earth System Science, A Closer View: A Report to the Earth System Sciences Committee, NASA Advisory Council.* National Aeronautics and Space Administration, Washington, DC.

Nielsen D.M. (1991) *A Practical Handbook of Ground-Water Monitoring.* Lewis Publishers, Boca Raton, Florida.

Rolston D.E. (1986) Gas flux. In *Methods of Soil Analysis, Part 1, Physical and Mineralogical Methods* (A. Klute, Editor), 2nd Edition, pp. 1103–1119. American Society of Agronomy, Madison, Wisconsin.

Stedman D.H. (1989) Automobile carbon monoxide emission. *Environmental Science and Technology.* **23**, 147–149.

Taylor J.K. (1987) *Quality Assurance of Chemical Measurements.* Lewis Publishers, Boca Raton, Florida.

Ullom W.L. (1995) Soil gas sampling. In *Handbook of Vadose Zone Characterization and Monitoring* (L.G. Wilson, L.G. Everett, and S.J. Cullen, Editors), pp. 555–567. Lewis Publishers, Boca Raton, Florida.

U.S. EPA (1983) *Methods for Chemical Analysis of Water and Wastes.* Revised Edition, March 1983. EPA-600/4-79-020. U.S. Environmental Protection Agency, Environmental Monitoring and Support Laboratory. Research and Development. Cincinnati, Ohio.

U.S. EPA (1986) *Methods of Analysis of Hazardous Solid Wastes*, 3rd Edition. SW-846. U.S. Environmental Protection Agency, Office of Solid Waste. Washington, D.C.

Wilson L.G., Everett L.G., and Cullen S.J. (1995) *Handbook of Vadose Zone Characterization and Monitoring*. Lewis Publishers, Boca Raton, Florida.

Problems and Questions

1. Suppose you want to sample 12 random locations from a field that is 100 × 100 m (see Figure 9-1a on page 117). How would you find 20 random positions? [*Hint:* Divide the field into a grid that has at least 12 squares. Use a random-number generator from any spreadsheet.]

2. An unknown sample of aluminum in water has a spectrophotometer response of 51,003. Assuming the sample was diluted 1:5 before it was analyzed, use the equation parameters for the aluminum calibration shown in Table 9-4 on page 132 to compute final concentration of aluminum.

3. How many and what types of sample bottles would you need to take to the field if you wanted to collect six water samples and have them analyzed in a laboratory for lead, zinc, 2,4-D (an organic chemical pesticide), pH, and salinity? Explain your answer.

4. You are preparing a standard solution of sodium from a NaCl standard that is only 93% pure. How many grams of this chemical do you need to make 1 liter of a 2000 mg L^{-1} standard solution?

5. You have already sampled a field and collected 50 soil samples to be analyzed for NO_3-N, salinity, and pH. But your budget has been suddenly reduced; now it allows for the analysis of just 20 samples. How would you proceed? Would you (a) composite samples or (b) discard some and analyze others? Explain your answer. [*Hint*: Remember there are three parameters that must be analyzed.]

6. What are the main pollutants monitored in urban air? Why are different pollutants often monitored at different locations within the city?

7. Why are weather data collected together with air pollution data in a city? Explain their important relationships.

8. Address the following factors for the design of a sampling program that will monitor two pollutants—nonvolatile hydrocarbons and metals such as lead. What type(s) of sample container(s) would you choose and what sample preservation would you use prior to analysis?

Land application of municipal sewage sludge from the city of Tucson to cotton fields. Photograph adapted by permission from page 1 of Environmental Microbiology: A Laboratory Manual, *by I.L. Pepper, C.P. Gerba, and J.W. Brendecke; copyright © 1995 by Academic Press, Inc. All rights reserved.*

10.1 Waste Disposal in Perspective

Human activity inevitably produces waste materials that must be managed. Some wastes, such as plastic and glass, can be recycled while others, such as municipal wastes, can be reused. However, many wastes cannot be used beneficially and must be disposed of by special technologies to ensure that pollution does not result. In this chapter we will survey the techniques used to treat most municipal wastes and some industrial wastes. Although new waste-treatment techniques are being developed every day, the majority of wastes produced by industrial societies are being treated and/or disposed of using the methods described below.

10.1.1 Types and Classes of Wastes

Wastes may be classified by their physical, chemical, and biological characteristics. One important classification criterion is their consistency. **Solid wastes** are waste materials having less than ~70% water. This class includes municipal solid wastes such as household garbage, industrial wastes such as power plant fly ash and flue gas desulfurization wastes, mining wastes such as mine spoils, and oilfield wastes such as drill cuttings. **Liquid wastes** are usually wastewaters, including municipal and industrial wastewaters, that contain less than 1% suspended solids. Such wastes may contain high concentrations (>1%) of dissolved species, such as salts and metals. For example, oilfield liquid wastes often have salt (NaCl) concentrations greater than that found in seawater. **Sludge** is a class of wastes intermediate to solid and liquid wastes. Sludges usually contain between 3% and 25% solids, while the rest of the material is water dissolved species. These materials, which have a slurry-like consistency, include municipal sludges, which are produced during secondary treatment of wastewaters, and sediments found in storage tanks and lagoons.

Federal regulations classify wastes into three different categories; these categories are based on hazard criteria. **Non hazardous** wastes are those that pose no immediate threat to human health and/or the

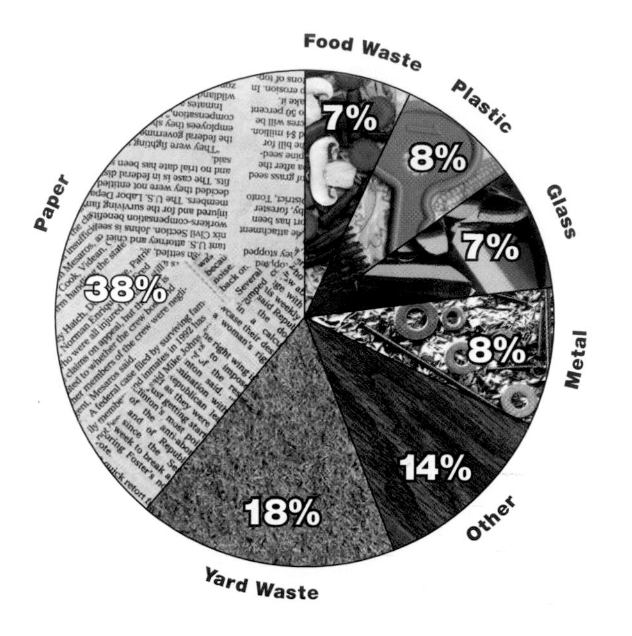

Figure 10-1 The nature of household trash (MSW) in the United States (on a volume basis). Data: U.S. EPA, 1994.

environment. Examples include municipal wastes such as household garbage and many high-volume industrial wastes such as NaCl-laden effluents. **Hazardous wastes** are of two types: those that have "characteristic" hazardous properties (such as ignitability, corrosivity, or reactivity) and those that contain leachable toxic constituents. Other hazardous wastes include listed wastes, which are identified with a particular industry or industrial activity. The third category from industry is classified generically as **special wastes** by origin, and are regulated with waste-specific guidelines. Examples include mine spoils, oilfield wastes, spent oils, radioactive wastes, medical wastes, and nonhazardous industrial wastes. All hazardous wastes are regulated under Subtitle C of the Resource Conservation and Recovery Act (RCRA) (see Table 23-1 on page 367).

10.2 Landfills

Landfilling is the disposal of waste materials at a specific site, usually in the soil or shallow vadose zone. The waste materials thus disposed of may be municipal solid waste (MSW) or industrial hazardous wastes. Most state and/or federal rules permit landfills to contain either hazardous or nonhazardous wastes, but more than 95% of landfilling involves MSW. Municipal solid waste consists of food, animal, and plant residues, as well as miscellaneous household nontoxic wastes such as paper, cans, plastic, metals, and bottles (Figure 10-1). Until the mid 1980s, these materials were buried or dumped haphazardly within soil or the vadose zone, then simply covered with a layer of soil as a cap (Figure 10-2a). There was no regulation of these landfills, nor were protective mechanisms in place to prevent or minimize releases of pollutants from the landfills.

The Environmental Protection Agency (EPA) estimates that the United States currently produces more than 200 million metric tons of MSW each year, or about 2 kg of trash per person per day. Developed nations, such as the United States and Canada, lead the world in the production of trash, producing much more discardable material than developing countries (Figure 10-3 on page 138). Currently, about 17% of the MSW is recycled, while the other 83% is mostly disposed of in landfills or incinerated (Figure 10-4 on page 138). People hoped that much of the landfilled material would biodegrade. In reality, however, most of the waste landfilled is still in landfill sites 30–40 years later. Surveys of landfills have shown that conditions at those sites are often not favorable for biodegradation. Such conditions include various factors: low moisture, low oxygen concentration, and high heterogeneity of materials, many of whose components are nondegradable or very slow to degrade. Thus, many old landfills serve as "waste repositories," releasing pollutants to the groundwater and the atmosphere.

Despite the fact that many landfills are nearly full, the number of landfills in the United States has been decreasing, from 18,500 in 1979 to 6,500 in 1988. In fact, the EPA predicts that by the year 2000, there will only be 3,250 active landfills. This decrease is partially attributable to the fact that many full landfills have been closed. Moreover, new regulations and landfill designs are increasing the costs of land-

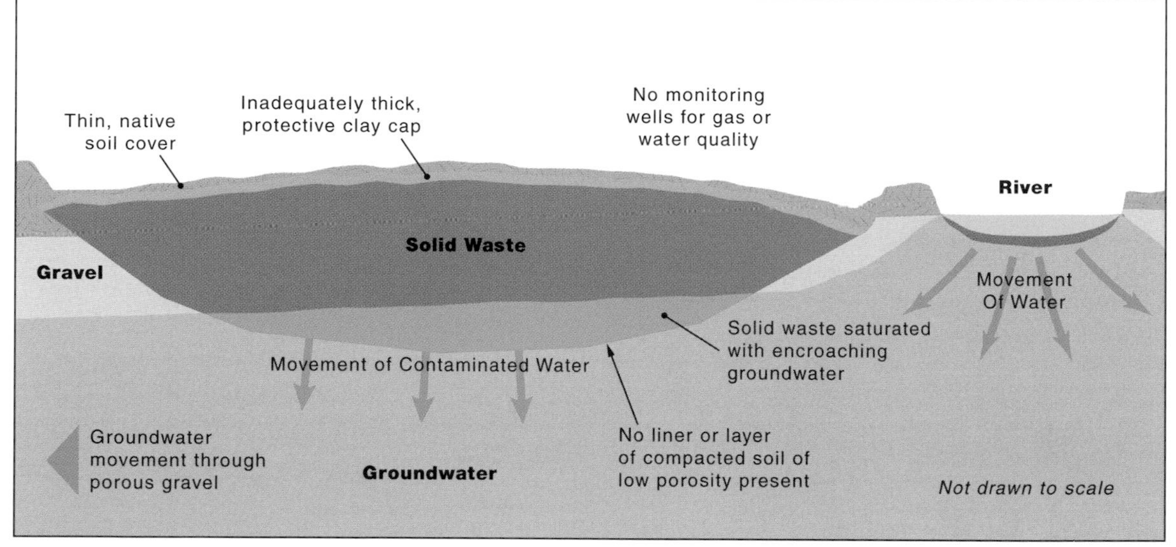

a) Old-Style Sanitary Landfill

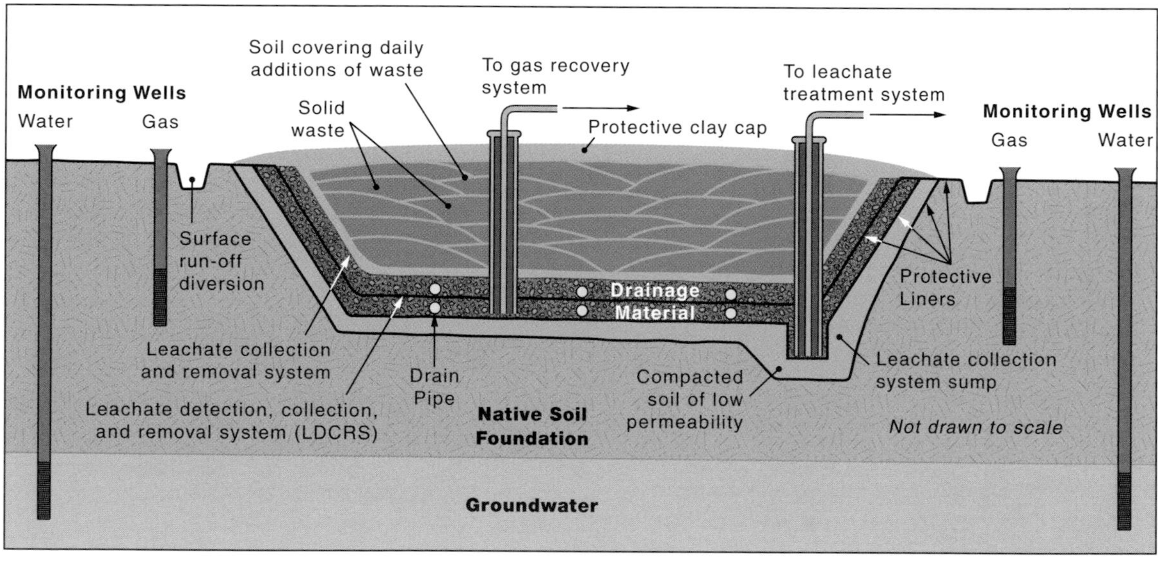

b) Modern Sanitary Landfill

Figure 10-2 Landfills: (a) Old-style sanitary landfill where location was chosen more out of convenience or budgetary concerns than out of any environmental considerations. Here, an abandoned gravel pit located near a river exemplifies an all too common siting arrangement. (b) Cross-section of a modern sanitary landfill showing pollutant monitoring wells, a leachate management system, and leachate barriers. In contrast to the old-style site above, the modern sanitary landfill emphasizes long-term environmental protection. Additionally, whereas landfills have formerly been abandoned when full, a modern landfill is monitored long after closure.

Figure 10-3 Fresh Kills Landfill in New York, the largest human-made entity in the world. Here, garbage is being analyzed for research purposes. Photograph: C.P. Gerba.

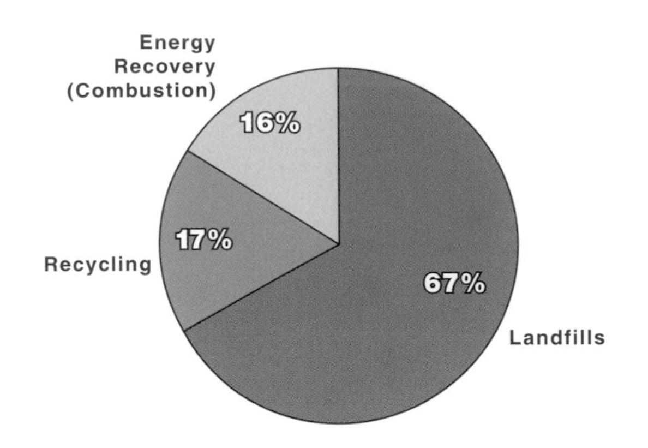

Figure 10-4 Breakdown of the major avenues of trash (MSW) disposal in the USA (on a volume basis). Data: Miller, 1992.

filling while recycling and energy recovery have decreased the overall rate of production of MSW.

10.2.1 Modern Sanitary Landfills

Old landfills were usually located in old quarries, mines, natural depressions, or excavated holes in abandoned land. Conditions in these landfills are conducive to pollution, as these sites are often hydrologically connected to surface streams or groundwater sources. Water pollution is caused by the formation and movement of leachate, which is produced by the infiltration of water through the waste material. As the water moves through the waste, it dissolves components of the waste; thus, landfill leachate consists of water containing dissolved chemicals such as salts, heavy metals, and often synthetic organic compounds. Such water-solute transport processes are typically slow, so the effects of pollution from old landfills may take years to show up. In addition to contaminating water, landfills release pollutants into the air. Anaerobic microbial processes generate greenhouse gases, such as nitrous oxide (N_2O), methane (CH_4), and carbon dioxide (CO_2).

A modern **sanitary landfill** is designed to meet exact standards with respect to containment of all materials, including leachates and gases (Figure 10-2a on page 137). New design specifications are predicated on minimal impact to the environment, both short- and long-term, with particular emphasis on groundwater protection. Landfill site selection is based on geology and soil type, together with such groundwater considerations as depth of the water table and use. Figure 10-2a shows a schematic drawing of a typical modern landfill. In this design, the new landfill is located in an excavated depression, and fresh garbage is covered daily with a layer of soil. The bottom of the landfill is lined with low-permeability liners made out of high-density plastic or clay. In addition, provisions are made to collect and analyze leachate and gases that emanate from the MSW.

New landfills currently cost up to $1 million per hectare to construct. Additionally, they require costly permanent monitoring for potential pollutant releases to the surrounding environment. Thus, many communities are faced with difficult decisions with respect to disposal of MSW—despite the fact that landfill design has improved efficiency with respect to pollution prevention and innovative strategies have decreased the rate of MSW production.

There are other problems, as well. Perhaps the most difficult problem associated with construction of new landfills is that of locating or siting a landfill. Nobody wants to live next door to a "dump," however sanitary. Real or perceived local community concerns about potential health effects associated with living near a landfill can result in controversy, often delaying the siting of a landfill for months or years. Thus, on the one hand, we can expect the costs of MSW disposal to rise as new landfills are located

Table 10-1 Percentage (on a volume basis) of wastes being recycled by commodity. Source: Miller, 1992.

Commodity	Recycled (%)
Plastics	1
Rubber and leather	2
Metals	4
Glass	8
Paper	23
Aluminum	25

in increasingly remote areas. On the other hand, these costs may force communities to look for alternative forms of MSW disposal, such as the development of new, more cost-effective recycling and waste-minimization strategies.

10.2.2 Reduction of MSW

The primary methods used to reduce MSW are combustion (or incineration), source reduction, and recycling. These approaches are widely used in densely populated areas where land scarcity limits the use of landfills. In Europe and Japan, for example, less than 15% of MSW is sent to landfills.

Combustion or incineration today does more than burn trash: much of the heat derived from burning is converted to other forms of energy. **Incineration** usually involves combustion of unprocessed solid waste; however, in some instances nearly 25% of the waste is processed into pellets, termed **refuse-derived fuel**, prior to burning. A new technique called **mass burning** can burn MSW at temperatures up to 1130°C, trapping the resultant heat to generate steam and electricity. Incineration is a relatively efficient method of reducing MSW, often reducing it as much as 90% by volume and 75% by weight. But this technique is not without drawbacks. One of the primary problems associated with incineration involves disposal of the ash, which can contain toxic metals and refractory organics. Thus, incinerator ash itself often needs to be treated as a hazardous waste. Moreover, even modern incinerators can contribute to air pollution. Finally, incinerators are expensive and the costs of incineration are increasing each year.

In contrast to incineration, **source reduction** is aimed at prevention; thus, it is a fundamental way of reducing MSW because it tends to eliminate the need for landfills. Source reduction can be accomplished by a variety of methods, including the use of less material for packaging and the practice of municipal composting. For example, plastic containers may be reduced in size or eliminated altogether, and yard waste may be separated from other sources of MSW and used as compost. Municipalities often sell the composted material for use as a soil amendment or fertilizer.

Recycling involves collecting certain types of trash, breaking them down into their components, and using those components to make new products. Materials that can be recycled include aluminum cans, plastics, glass, paper, cardboard, and metal. In many municipalities, recycling is mandatory; residents must separate recyclables from the rest of their trash to avoid increased payments for trash collection. However, even with the recent growth of recycling programs, only a small percentage of the total MSW generated in the United States is recycled (see Table 10-1). About 17% of the total household waste generated in the United States is recycled.

10.3 Wastewater Treatment

Most of the high-volume liquid waste in the United States is treated in wastewater treatment plants prior to ultimate disposal. Most United States municipalities have one or more wastewater treatment plants, and large industrial complexes may also maintain their own treatment plants if they cannot discharge their wastewater into municipal sewage systems. In the United States, more than 90% of all municipal wastewater and nonhazardous industrial wastewaters, as well as more than 10% of all hazardous wastewaters, are treated in wastewater treatment plants prior to ultimate disposal.

Basic treatment processes include:

1. *Primary treatment*, or separation of solids from liquids;

2. *Secondary treatment*, or biological oxidation of the most soluble organic constituents into CO_2 and microbial biomass, resulting in the production of sludges;

Table 10-2 Maximum (lifetime) cumulative metal-loading rates within the state of Arizona. Source: Arizona Department of Environmental Quality, 1995.

Metal	Concentration in Soil ($kg\ ha^{-1}$)
Arsenic	41
Cadmium	39
Chromium	3000
Copper	280
Lead	300
Mercury	17
Molybdenum	41
Nickel	110
Selenium	100
Zinc	560

3. *Tertiary treatment*, or chemical stabilization of the wastewater for safe reuse and/or disposal into waterways.

This last process usually targets the chemical oxidation of biological toxins (bacteria and viruses) that can cause human health problems (see Section 19.3, beginning on page 282). Strong oxidizing agents are used in this process, such as chlorine gas, perchlorate salts, ozone gas, and UV radiation. (See Section 20.3, beginning on page 302, for details on these three types of treatment.)

Additionally, most inorganic industrial liquid hazardous wastes are treated chemically to oxidize, precipitate, or otherwise alter the composition of toxic constituents prior to discharge into the environment. For example, wastewaters having high metal concentrations can be treated with alkaline solutions to precipitate the metals out. This process, which can be called chemical secondary treatment, is responsible for the bulk of inorganic sludges that have high concentrations of metals and salts (see Section 18.3.2 on page 273).

Ultimately, wastewater treatment plants produce reclaimed water, which is usually discharged into the environment or is used to irrigate farmland, parks, or golf courses. Wastewater-treatment plants also produce large volumes of sludges, which present a challenge in their ultimate safe disposal. In some cities, such as Honolulu, Hawaii, primary treated effluent is disposed of through ocean outfalls (see Section 13.2.1, beginning on page 190).

Figure 10-5 Municipal sewage sludge from the city of Tucson is pumped into a tanker truck for transport to agricultural land where it will be injected into the soil. Photograph: I.L. Pepper.

10.4 Landfarming

Landfarming is the practice of disposing of sludges produced at wastewater treatment plants on agricultural land. These sludges may contain either solid wastes or liquids that result from municipal waste treatment. The amount of material that can be deposited in a given area depends on the chemical and physical composition of the sludge as well as the characteristics of the soil. In 1993, new regulations were issued by the EPA that require the monitoring of certain potential pollutants. Of particular importance are metals, such as lead and mercury, which cannot be destroyed and therefore accumulate in the soil over time after multiple sludge applications (see also Section 7.5, beginning on page 89). Total metal loadings are regulated for these metals, and may not be exceeded during the lifetime of landfarming in a particular region. Lifetime maximums for metals in Arizona are shown in Table 10-2.

10.4.1 Methods of Sludge Disposal

Solid municipal wastes are generated when treated wastes are dewatered, dried, or (in some cases) composted. These solid materials are normally trucked from the wastewater treatment plant to the disposal site, where they are then disked into the land.

Liquid sludges, which contain between 1 and 8% solids by dry weight, can also be trucked to the disposal site in tankers (Figure 10-5). The liquid sludge can then be pumped into **terrigators** (Figure 10-6),

Figure 10-6 Land application of municipal sludge is one of the most successful methods of treating waste as a resource. Here, a terrigator (a) injects sludge beneath the soil surface for subsequent growth of furrow-irrigated cotton (b). Currently all sludge from Tucson, Arizona is utilized via land application. Photographs: J.F. Artiola.

which inject the sludge into the soil to a depth of 30 cm. This method of disposal is advantageous because it limits odor problems; in addition, sludges can be applied at known rates. The primary disadvantage of this practice is the extra cost associated with it.

10.4.2 Benefits of Landfarming

Nutrients: All municipal sludges contain essential plant nutrients, present either as available inorganic forms or as unavailable organic constituents. The organic forms, particularly those of nitrogen, are often quickly mineralized to inorganic forms (see Section 14.2.2, beginning on page 212). Thus, the application of sludges can reduce the need for fertilizer applications to crops. An analysis of a municipal sludge is shown in Table 10-3 on page 142. In terms of essential plant nutrients, this sludge is high in nitrogen, phosphorus, calcium, and other macro elements. It also contains variable amounts of trace elements, as well as some heavy metals.

Water: Liquid sludges contain up to 99% water. In arid lands such sludges are even more valuable as they can help reduce pre-plant irrigation needs.

Sludge as a Soil Amendment: Since many sludges contain large amounts of organic matter (up to 40% on a dry weight basis), they can act as a beneficial soil amendment. Additional organic matter helps in the formation of soil aggregates and increases soil microbial activity. This in turn improves air and water permeability (better soil structure) and increases plant nutrient availability.

Reduced Pollution: In most instances, the broad land application of municipal wastes at appropriate loading rates results in far less pollution than if the material had been concentrated by disposal at a single site. When loading rates are controlled, the soil has a chance to transform many waste components into plant-available nutrients. Thus, plants are able to take up these nutrients and complete their natural cycle. These include carbon, nitrogen, sulfur, and phosphorus. Other important sludge components that are recycled include micronutrients such as zinc, iron, and copper. However, when found in high concentrations these metals can be toxic to plants. Additionally, the soil environment helps stabilize other potential pollutants found in sludges such as lead, cadmium, zinc, arsenic, *etc.* by trapping them into their solid phases. When this happens the pollutants do not leach into the groundwater and are much less likely to be taken up by plants. However, it is important to note that the metal contaminant itself is still

Table 10-3 Analysis of anaerobically digested sludge from Tucson, Arizona. Source: I.L. Pepper, 1991.

Element	Concentration (dry weight basis)
Metals	———($mg\ kg^{-1}$)———
Copper	520
Nickel	13
Lead	59
Chromium	29
Cadmium	3.5
Zinc	1900
Silver	4.7
Arsenic	ND[a]
Mercury	0.51
Molybdenum	12
Selenium	ND[a]
Other Elements	———($g\ 100\ g^{-1}$)———
Phosphorus	3.3
Calcium	3.6
Magnesium	0.45
Sodium	0.4
Organic carbon	16.6
Nitrogen	
total Kjeldahl N	3.4
inorganic N	0.16
Total solids	2.8

[a] ND = Below detection limits.

retained in the soil and remains a potential source of pollution.

10.4.3 Potential Hazards of Landfarming

10.4.3.1 Nitrates

Sludges frequently contain large amounts of nitrogen (N) as nitrate (NO_3^-) and ammonium (NH_4^+), as well as organic forms of nitrogen that are readily transformed microbially to NO_3^- within soil (see Section 14.2.2, beginning on page 212). Nitrate ions are very soluble in water and are also electrostatically repelled by the likewise negatively charged soil colloids. Consequently, this chemical form of nitrogen can be readily leached from the soil profile below the root zone (see Section 6.3, beginning on page 67). Excessive applications of wastes containing high concentrations of nitrogen can lead to significant NO_3^- losses from the soil root zone because plants cannot remove all the NO_3^- before percolating water carries it beyond the root zone. Therefore, waste loading rates must be limited to those sufficient to provide adequate and timely plant nitrogen needs while minimizing leaching losses. To date, we have no reliable monitoring technology to quantify NO_3^- leaching losses below the plant root zone. However, a combination of soil, plant, and waste monitoring, together with high irrigation efficiency, can limit NO_3^- losses below the root zone. At present, the water that is recharged into an aquifer cannot exceed the EPA potable drinking water standard of 10 mg L^{-1} of nitrogen in the form of nitrate (NO_3-N).

10.4.3.2 Metals

All sludges have small amounts of essential trace metals needed for plant growth, but they also contain variable amounts of potentially toxic heavy metals. Moreover, even essential trace elements can be present in such high concentrations that they induce toxicities to plants or microorganisms. Metals of particular concern include Zn, Cu, Cd, Ni, Pb, Hg, Mo, and As (see also Section 21.8.1, beginning on page 338). The amount of metal contaminants in a particular sludge depends on the amount of industrial inputs into the municipal sewage system. It is therefore illegal to discharge excessive amounts of metals into municipal wastewater lines, and such wastes must often be treated on site, prior to disposal (see Section 18.3, beginning on page 273). The soil environment also influences the toxicity of the metals associated with any particular sludge. Sludge-amended soils with high pH have lower plant-available metal concentrations than do sludge-amended low-pH soils. This is because the water solubility of most metals increases as pH decreases. Thus, one management strategy to reduce metal mobility and toxicity in plants is to lime soils to a neutral or alkaline pH. The organic matter content of soils also affects metal availability. In general, soils with high organic matter content (>5%) exhibit relatively low metal uptake by plants, as metals are sorbed and complexed by the polymer-like organic carbon structure of organic matter. However, when low-molecular-weight organic molecules (usually present in the early stages of plant tissue decay) form complexes with metals, their mobility and plant availability can be dramatically increased in the soil environment. Once metals are introduced into soil, their bioavailability and mo-

Table 10-4 Major groups of organic chemicals that may be monitored in MSW. Source: 40 Code of Federal Regulations, Part 257.

Pollutants by Groups[a]	Origins and Comments
Aldrin/dieldrin, heptachlor, DDT/DDE/DDD, lindane, toxaphene, malathion, hexachlorobenzene, 2,4-dichlorophenoxyacetic acid (2,4-D), hexachlorobutadiene	Pesticides and herbicides. Some are found in household chemicals. Many chlorinated pesticides have been banned from use.
Benzo[a]pyrene, benzo[a]anthracene, phenanthrene	Motor oils and diesel fuel,. These occur naturally as by-products of fuel combustion (oil, wood).
Polychlorinated biphenyls (PCBs)	Electrical and chemical manufacturing. PCBs are banned.
Dioxins, furans	By-products from the synthesis of phenol-based pesticides, such as 2,4-D.
Phenol	Household products and disinfectants.
Pentachlorophenol	Wood preservative. Pentachlorophenol is very persistent in the environment.
Benzene, methylene chloride, methyethyl ketone tetrachloroethylene, trichloroethylene, hexachlorobutadiene	Some household products such as paints. These chlorinated solvents are very volatile.
Vinyl chloride, bis(2-ethylhexyl)phthalate, tricresyl phosphate, dimethylnitrosamine, benzidine, 3-3′-dichlorobenzidine	Plastics and plasticizers.

[a] No limits have been established for use of these chemicals.

bility can be manipulated by changing the valence state of the metal or by altering soil factors that influence their solubilities. While short-term effects of metal additions to soil can be beneficial, the long-term fate of these metals is more difficult to predict. This is because metal pollutants do not biodegrade and therefore continue to accumulate in the soil environment (see also Section 7.5, beginning on page 89).

10.4.3.3 Toxic Organic Chemicals

Municipal sludges can also contain toxic organic compounds that originate from household wastes or industrial wastewaters. All municipal sludges contain trace amounts of the most common organic chemicals, such as pesticides, polynuclear aromatic hydrocarbons (PAHs), plasticizers, volatile organics, and solvents. Depending on the magnitude of the inputs, many of these organic compounds are degraded in the wastewater treatment processes. However, the more refractory and insoluble compounds, such as chlorinated pesticides, can pass through wastewater treatment plants without degrading because they are sorbed in the biosolids (see Section 7.2, beginning on page 77). New regulations for disposal of wastes on land require periodic monitoring not only of metals but also organic pollutants (Table 10-4). To date, there is no evidence that residual concentrations of any of these chemicals in MSW present an immediate pollution or health problem. As may be expected, the degradation, bioavailability, and mobility of these chemicals are largely dependent on soil type, and particularly soil organic-matter content. Other general soil physical and chemical conditions, such as particle size distribution, pH, oxygen levels, and soil moisture, also affect their degradation. Therefore, as with the metals, the long-term effects of these chemicals are difficult to predict, particularly since many can accumulate in the soil environment.

10.4.3.4 Pathogens

All municipal wastes contain bacterial and viral pathogens as well as intestinal parasites. This aspect

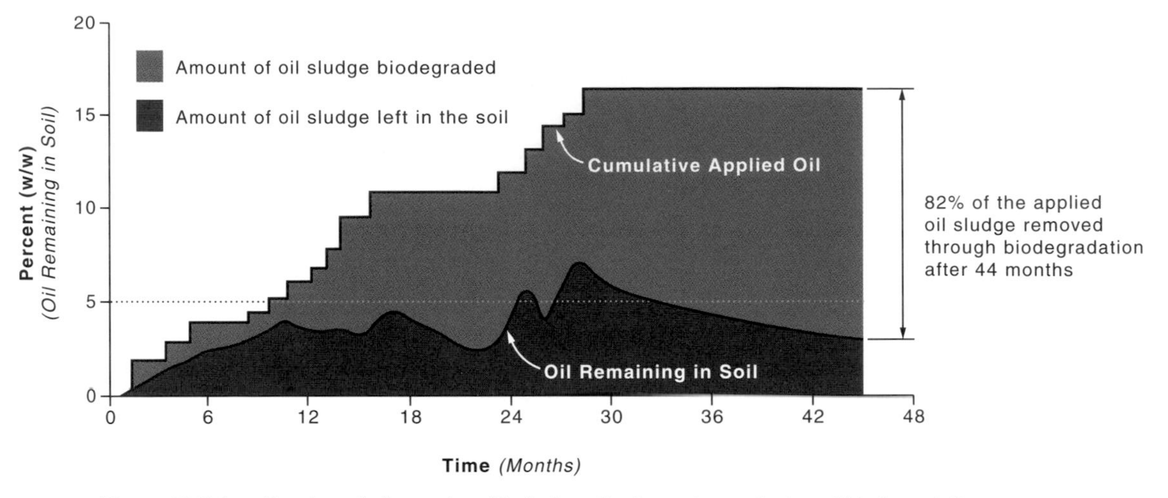

Figure 10-7 Landfarming of oily wastes. (Typical application rates and rates of biodegradation are shown.) Figure 2-6 from publication #4379, *The Land Treatability of Appendix VIII Constituents Present in Petroleum Industry Wastes*, May 1984. Courtesy of the American Petroleum Institute.

will be covered in detail in Chapter 19, beginning on page 279.

10.4.4 Landfarming of Refinery Sludges

Landfarming techniques can also be used to treat hazardous wastes, particularly oily sludges, whose major constituents are oil, sediments, and water. The rates of application of these sludges, which must be based on the properties of the particular material, need to be optimized; that is, the rates should not be so low that they require large areas of land, nor so high that soil microbes are overwhelmed and degradation rates are decreased. Typical annual applications do not exceed 5% total oil in the soil plow layer (Figure 10-7). In addition, nitrogen and phosphorus fertilizers are often added to optimize C:N and C:P ratios, and lime may be added to optimize soil pH values. Thus if waste disposal is carefully managed, landfarming is an ecologically sound practice, whether municipal or refinery wastes are utilized.

In the early 1980s, land treatment of hazardous oily wastes came under intense scrutiny by the EPA. Land disposal restrictions, begun in early 1992, now prohibit land treatment of hazardous oily wastes. These restrictions have compelled the petroleum industry to look at alternative disposal methods, including landfilling and incineration. On the positive side, petroleum refineries have also been spurred to develop waste-minimization strategies and to look aggressively for waste pretreatment methods that can render oily wastes nonhazardous enough to be landfarmed.

10.5 Deep-Well Injection of Liquid Wastes

Deep-well injection of liquid wastes into the subsurface is another waste disposal method. This method greatly reduces the potential hazard posed by the wastes by disposing of them in a place where, in principle, the chance of direct contact between the waste and humans is minimal. Deep-well injection is widely used in the south-central part of the United States and in heavily industrialized states such as California, Michigan, and New York. Examples of deep-well injection of liquid wastes include oilfield wastes (brines), metal-containing hazardous wastewaters and slurries, and wastewaters with high concentrations of toxic organic chemicals such as chlorinated hydrocarbons, pesticides, and radioactive wastes.

Deep-well injection is most suitable for handling large volumes of liquid or slurry wastes that have a water-like consistency (low viscosity). Watery liq-

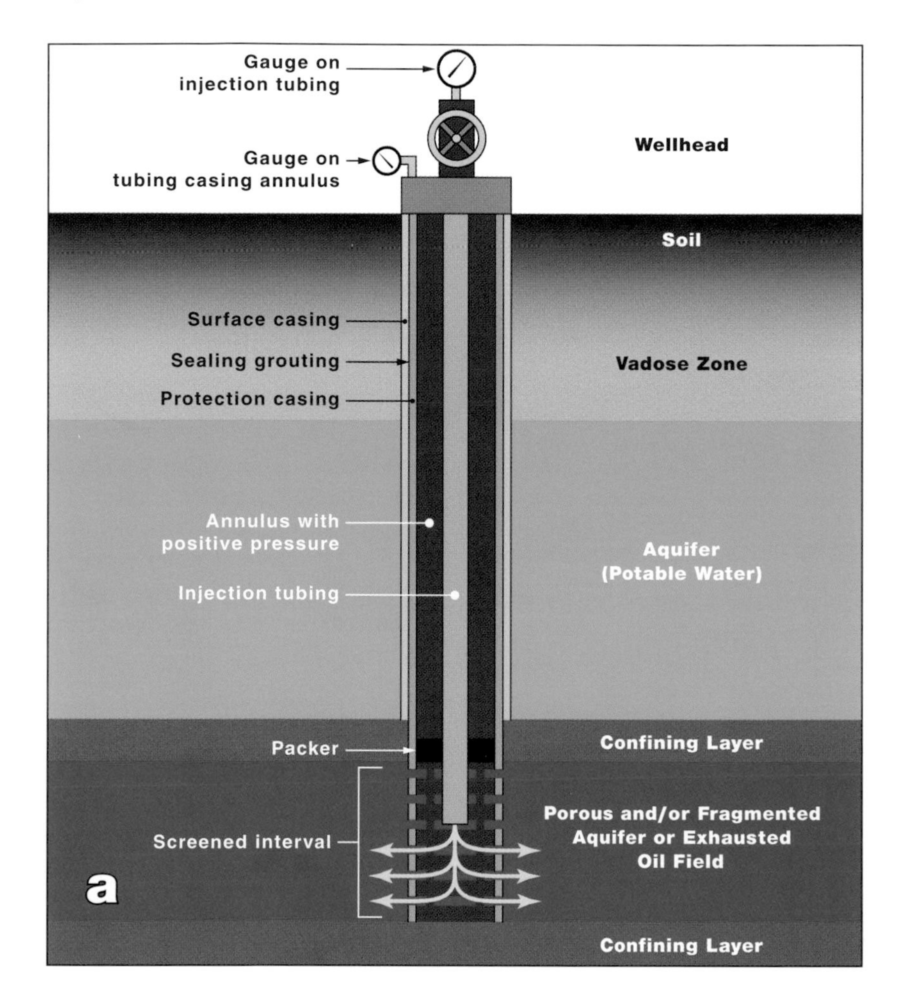

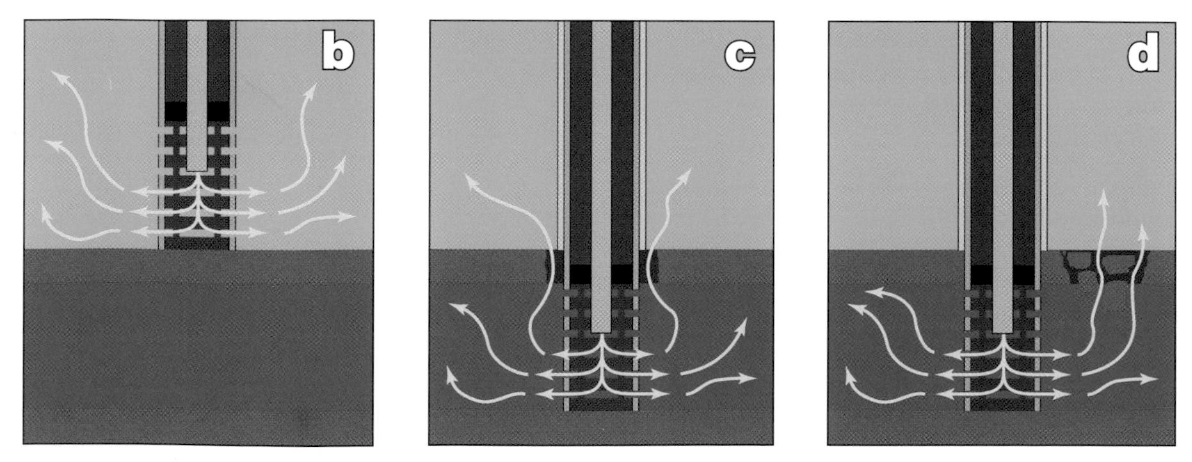

Figure 10-8 Deep-well injection design (a) and potential forms of system failures that can lead to groundwater pollution: (b) direct injection into the aquifer used for drinking water; (c) leaking well bore; (d) movement of waste to a zone that supplies drinking water. Adapted with permission from *The Technology of Injecting Wastewater into Deep Wells for Disposal* by D.L. Warner and J.H. Lehr; copyright © 1981 by Premier Press. All rights reserved.

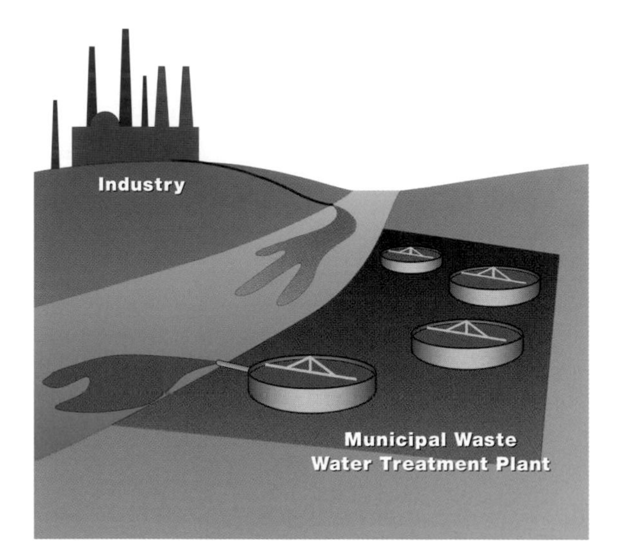

Figure 10-9 Major point-source discharges into a water stream.

uids are usually pumped down into confined, aquifer-like zones composed of highly water-permeable material, such as sandstone or limestone. Similarly, oily wastes can be deep-well injected into high-permeability subsurface zones. The depth of the injection zones, usually hydrologically confined, ranges from about 200 to 4500 m, and most are located between 700 and 2500 m (Figure 10-8a on page 145). Drilling and constructing wells for these depths can be very expensive; therefore, deep-well injection is primarily used by the petroleum industry, which already possesses in-service oilfield wells. Once constructed, these systems are comparatively cheap to operate and maintain.

Deep-well injection does, however, pose several problems. One concern is the possible clogging of the injection zone due to solid particles or bacterial growth. Also, injected waste may contaminate resident groundwater. This possibility is of major concern if the groundwater is a current or potential drinking-water source. Waste injection can cause groundwater contamination in several ways: (1) direct injection into the aquifer used for drinking water (Figure 10-8b on page 145); (2) leaking wells (waste leaks from the well bore into an aquifer used for drinking water, Figure 10-8c); and (3) movement of waste to a zone that supplies drinking water, *e.g.*, through fractures in an upper confining layer (Figure 10-8d). Deep-well injection is regulated by the EPA under the **Underground Injection Control (UIC) Program**, which permits up to five classes of injection wells.

10.6 Incineration and Immobilization

10.6.1 Incineration

The process of incineration that destroys highly toxic and hazardous organic wastes differs from MSW incineration, where energy is often produced. In general, low-temperature (up to 850°C) and high-temperature (~1200°C) incinerations use energy to oxidize carbon- and water-containing wastes completely to CO_2 and H_2O vapor. However, some incinerators can serve as heat-energy sources, especially when oily wastes, such as spent oils, are used as fuel. But these incinerators must meet stringent emissions levels. Incineration efficiencies are also closely regulated, and destruction of all organic compounds must exceed 99.99%.

Incineration cannot be used with wastes that have high concentrations of water and noncombustible solids, nor can it be used for radioactive materials. Moreover, incinerators are very expensive to build and maintain, and waste incineration expenses often exceed $500 a barrel. Thus, this technology is mostly limited to low-volume wastes. Finally, incinerators produce low volumes of highly toxic metal ash, which is usually disposed of in landfills. Thus, incinerators do not eliminate the problems of metal contamination in the environment.

But perhaps the overriding limitation of incineration in the United States is its lack of acceptance by the general populace. Public perceptions, both well-founded and misplaced, about the potential environmental threats posed by this technology have severely limited its development. Current estimates of waste incineration range from 15 to 20% of all solid wastes produced in the United States. These figures may include mass incinerators used for waste volume reduction, incinerators capable of producing usable energy in the process, and hazardous waste incinerators.

Table 10-5 NPDES water quality parameters. Source: Van der Leeden *et al.*, 1990.

Water Quality Parameter	Maximum allowable limits for discharge based on 30-day averages
Biochemical oxygen demand (BOD)	30 mg L^{-1}
Suspended solids	30 mg L^{-1}
pH	6.5–9.0
Specific organic chemicals	Variable[a]
Metals	Variable[a]

[a] Source-specific. These limits may be set by each state on the basis of aquatic and wildlife water quality standards.

10.6.2 Immobilization

Wastes high in metal and radioactive content must be physically or chemically immobilized or stabilized prior to final disposal. The process of waste solidification is a physical one that involves trapping or encapsulating the waste in a stable matrix. For example, when wet cement is mixed in with sludges, it yields a solid block that is stable after a few days of curing or drying. This kind of solidification method binds the wastes in a relatively low-porosity matrix that cannot be easily deformed or cracked under typical landfill overburden pressures. Consequently, in a landfill environment, percolating water does not readily infiltrate the matrix so metals or salts are less likely to leach out. Solidification agents must be tested for compatibility and stability when mixed with waste streams, as well as for leachability of toxic constituents under landfill conditions. Such testing is done according to the EPA's Toxicity Characteristics Leaching Procedure (TCLP) method. Solidification agents, which must have a liquid consistency initially and harden quickly, include concrete cements, lime and gypsum slurries, epoxy resins, and asphalt (bitumen). Wastes that are routinely solidified prior to landfilling include hazardous inorganic wastes, radioactive wastes, metallic wastes, dissolved (or liquid) metals, and metal-contaminated soil and sediments. Although wastes high in organic carbon are not likely candidates for this technique, PCB-contaminated soils and sediments are sometimes solidified.

Immobilization can also be accomplished through an energy-intensive process that involves the use of molten glass (amorphous silica). This process, known as **vitrification**, entraps the waste inside the silica matrix. Upon cooling, the resulting material has the consistency of glass and is impermeable to liquids for long periods of time. However, glass blocks can crack with uneven overburden pressures. Therefore, when vitrified materials (particularly radioactive wastes) are placed in landfills, a cushion of softer materials such as clay or gypsum is usually added to minimize uneven external pressures. Vitrification can also be accomplished *in situ* by inserting conducting electrodes into contaminated soils and applying high volumes of electricity, thereby melting resident silica minerals. However, this process is very expensive.

Another form of immobilization that involves a "permanent" association of the contaminant to a stable solid phase of the soil is physical or chemical fixation. These immobilization mechanisms occur naturally in soils and are an integral—and intricate—part of MSW land disposal. As a deliberate technique, fixation has great potential and promises to be very economical. However, issues of long-term stability and reversibility of the reactions involved remain unresearched, thus limiting its widespread use.

10.7 Point-Source Discharges into Open Water

Provided they meet general criteria, liquid wastes may be discharged through pipes, sewers, or ditches into an open-water environment (Figure 10-9, see also Chapter 13, beginning on page 189). In the United States, this practice requires a National Pollution Discharge Elimination System (NPDES) permit. Such wastes must be nonhazardous and meet the water-quality standards established by the EPA (Table 10-5). Limits exist on total dissolved solids (TDS), organic carbon (OC), dissolved oxygen (DO), biochemical oxygen demand (BOD), and pH levels. In addition, stringent restrictions apply to specific pollutants, such as metals and pesticides. Examples of the wastes routinely discharged under these conditions include reclaimed wastewaters from sewage treatment plants, oilfield wastewaters, and some industrial process waters. Special permits can be obtained for the discharge of liquid wastes that exceed one or more specified limits. These exceptions are typically limited to nonhazardous components that are also very common in the environment. For example, salty water used in oil drilling may be discharged directly into the ocean because the water in the ocean is already very salty. However, the same salty water may not be discharged into a riverway, since most rivers have total dissolved solids that are 10–500 times lower than that of the ocean. The discharge of wastewaters into surface waters and streams is very common, and is done primarily by municipal wastewater treatment plants and industries that use large volumes of water in their processes. Examples of these industries include power plants, paper mills, and food processing plants.

10.8 Special Wastes and Practices for Their Disposal

Hundreds of millions of tons of mine spoils and other nonhazardous solid wastes are being produced each year as a result of mining, oil and gas production, and electric power generation. Generally, these wastes are composed of relatively inert, physically or chemically modified geologic materials such as quartz, clays, carbonate and sulfate minerals, and fly ash. Because these residues are considered benign, most are stockpiled in mine tailing fields, lagoons, and natural depressions. Some of these waste materials are landfilled in special dry landfills, or they may be used as fill materials for roadbed and embankment construction (see also Section 18.4.4, beginning on page 276). Unlike other waste material that is treated and/or recycled, the ultimate disposition of these wastes will not change. Massive quantities of these wastes are being produced today, and they are likely to remain wherever they are being stockpiled. Strategies to aid in the long-term physical stabilization of these waste sites are under active consideration, including soil capping, revegetation, and water runoff control.

References and Recommended Reading

American Petroleum Institute (1984) *The Land Treatability of Appendix VIII Constituents Present in Petroleum Industry Wastes.* Publication #4379, May 1984.

Arizona Department of Environmental Quality (1995) Land Application of Biosolids. Rule 15 in *Arizona Administrative Register*, Volume 1, Issue 24, June 30, 1995. Arizona Department of Environmental Quality, Phoenix, Arizona.

Brown K.W., Carlile B.L., Miller R.H., Rutledge E.M., and Runge E.C.A. (1986) *Utilization, Treatment, and Disposal of Waste on Land.* Soil Science Society of America, Madison, Wisconsin.

Cope C.B., Fuller W.H., and Willets S.L. (1983) *The Scientific Management of Hazardous Wastes.* Cambridge University Press, Cambridge England.

Cote P., and Gilliam M. (1989) Environmental aspects of stabilization and solidification of hazardous and radioactive wastes. ASTM STP 1033. American Society of Testing and Materials, Philadelphia.

ERT (1984) Land treatability of Appendix VIII constituents present in petroleum industry wastes. Document B-974-220. Environmental Research and Technology, Inc., Houston, Texas.

Fuller H.W. and Warrick A.W. (1985) *Soils in Waste Treatment and Utilization*, Volume II. CRC Press, Boca Raton, Florida.

Miller G.T. (1992) *Environmental Science*, 4th Edition. Wadsworth, Belmont, California.

Pepper I.L. (1991) *Agricultural Sludge Utilization.* In: Annual Report to Pima County Wastewater Management Division. Department of Soil, Water and

Environmental Science, The University of Arizona, Tucson, Arizona.

U.S. EPA (1983) *Land Application of Municipal Sludge: Process Design Manual*. EPA 625/1-83-016. United States Environmental Protection Agency, Municipal Environmental Research Laboratory, Cincinnati, Ohio.

U.S. EPA (1994) *MSW Handbook, 1994*. Electronic Handbook version 1.2. United States Environmental Protection Agency, Office of Solid Waste, Washington. D.C.

Van der Leeden F., Troise F.L., and Todd D.K., Editors (1990) *The Water Encyclopedia*. Lewis Publishers, Chelsea, Michigan.

Warner D.L. and Lehr J.H. (1981) *Subsurface Wastewater Injection: The Technology of Injecting Wastewater into Deep Wells for Disposal*. Premier Press, Berkeley, California.

Problems and Questions

1. Modern landfills have venting systems. Explain why.
2. How is the bulk of municipal solid waste disposed of in the United States? In other countries? Explain your answer.
3. Explain some of the potential long-term problems to the soil that may arise from the land treatment of wastes.
4. Why can't sludge be applied to fields indefinitely? Describe a waste that could be used indefinitely as a fertilizer. Explain.
5. Table 10-3 on page 142 lists the characteristics of liquid sludge from Tucson, Arizona, but the data are reported on a dry-weight basis. How many liters of this liquid sludge (which contains 97.2% water on a mass basis) would you apply per hectare to add 50 kg of phosphorus to the soil? [*Hint*: Remember that the chemical analyses were performed after removing the water from the sludge. Thus, you must consider the dilution factor in your final answer.]
6. Each year 100 metric tons of sludge is applied to land that contains 4% total nitrogen (dry weight). Sludge is applied for 5 years, but only 80% of the nitrogen is assimilated by plants and another 10% leaches below the root zone each year. At the end of the fifth year, how many tons of nitrogen are left in the soil? [*Hint*: Plot each year's added and residual nitrogen amounts, then add up total residual nitrogen.]

Restoration of water and land contaminated by oil spills has proven to be one of the most successful applications of bioremediation to date, exhibiting both public acceptance and economical feasibility. Above, an oil tanker is moored offshore near Los Angeles, California. Photograph: A.B. Brendecke.

11.1 Basic Concepts

Public concern with polluted soil and groundwater has encouraged the development of government programs designed to control and clean up this contamination. In the United States, the first major piece of federal legislation that dealt with cleanup of contaminated environments was the Water Pollution Control Act of 1972. This act authorized funds for the cleanup of hazardous substances released into navigable waters, but did not provide for cleanup of contaminated land. Subsequently, in 1976, the first legislation to directly address the problem of contaminated land was passed—the Resource Conservation and Recovery Act (RCRA), which authorized the federal government to order operators to clean up hazardous waste emitted at the site of operation. This act did not, however, cover abandoned contaminated sites.

The **Comprehensive Environmental Response, Compensation, and Liability Act (CERCLA)** of 1980—otherwise known as the **Superfund** program—explicitly addressed the cleanup of hazardous waste sites (see Section 23.4, beginning on page 368). Together with a later amendment, the **Superfund Amendments and Reauthorization Act (SARA)** of 1986, CERCLA is the major federal act governing activities associated with the cleanup of environmental pollution. In addition, the Department of Defense has a separate cleanup program (Installation Restoration Program) for military sites, and the Department of Energy has a program designed specifically for cleanup of radioactive waste sites associated with the production of materials needed for nuclear weapons. Below the federal level, other governing bodies have also enacted legislation to regulate and control pollution. For example, some states have passed smaller versions of the federal Superfund program.

Because of its importance, we will briefly discuss the major components of Superfund. Its purpose is twofold: to respond to releases of hazardous substances on land and in navigable waters and to clean up contaminated sites. The former deals with future releases whereas the latter deals with sites of exist-

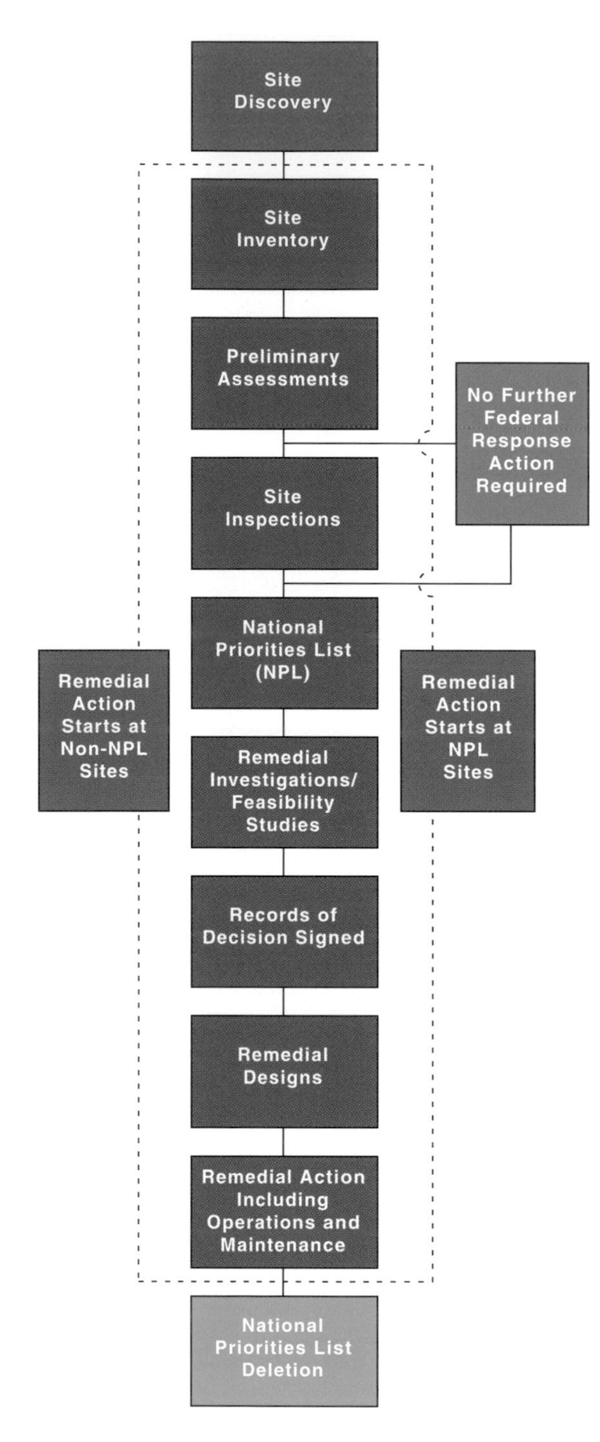

Figure 11-1 The Superfund process for treatment of a hazardous waste site. Adapted from U.S. EPA, 1988.

ing contamination. There are two types of responses available within Superfund: (1) removal actions, which are responses to immediate threats, such as leaking drums; (2) remedial actions, which involve cleanup of hazardous sites. The Superfund provisions can be used either when a hazardous substance is actually released or when the threat of such a release is substantial. They may also be used when the release of a contaminant or threat thereof poses imminent and substantial endangerment to public health and welfare. The process by which Superfund is applied to a site is illustrated in Figure 11-1.

The first step is to place the potential site in the **Superfund Site Inventory**, which is a list of sites that are candidates for investigation. The site is then subjected to a preliminary assessment and site inspection, which may be performed by a variety of local, state, federal, or even private agencies. The results of this preliminary investigation determine whether or not the site qualifies for the **National Priority List (NPL)**, which is a list of sites deemed to require remedial action by the Environmental Protection Agency (EPA). Currently, the EPA has placed more than 1200 sites on the NPL. [*Note*: Non-NPL sites may also need to be cleaned up, but their remediation is frequently handled by nonfederal agencies, with or without the help of the EPA.]

The two-component **remedial investigation/feasibility study (RI/FS)** is the next step in the process. The purpose of the RI/FS is to characterize the nature and extent of risk posed by contamination and to evaluate potential remedial options. The investigation and feasibility study components of the RI/FS are performed concurrently, using a "phased" approach that allows feedback between the two components. A diagram of the RI/FS procedure is shown in Figure 11-2.

The selection of the specific remedial action to be used at a particular site is a very complex process. The goals of the remedial action are to protect human health and the environment, to maintain protection over time, and to maximize waste treatment (as opposed to waste containment or removal). Specifically, Section 121 of CERCLA mandates a set of three categories of criteria to be used for evaluating and selecting the preferred alternative: threshold criteria, primary balancing criteria, and modifying cri-

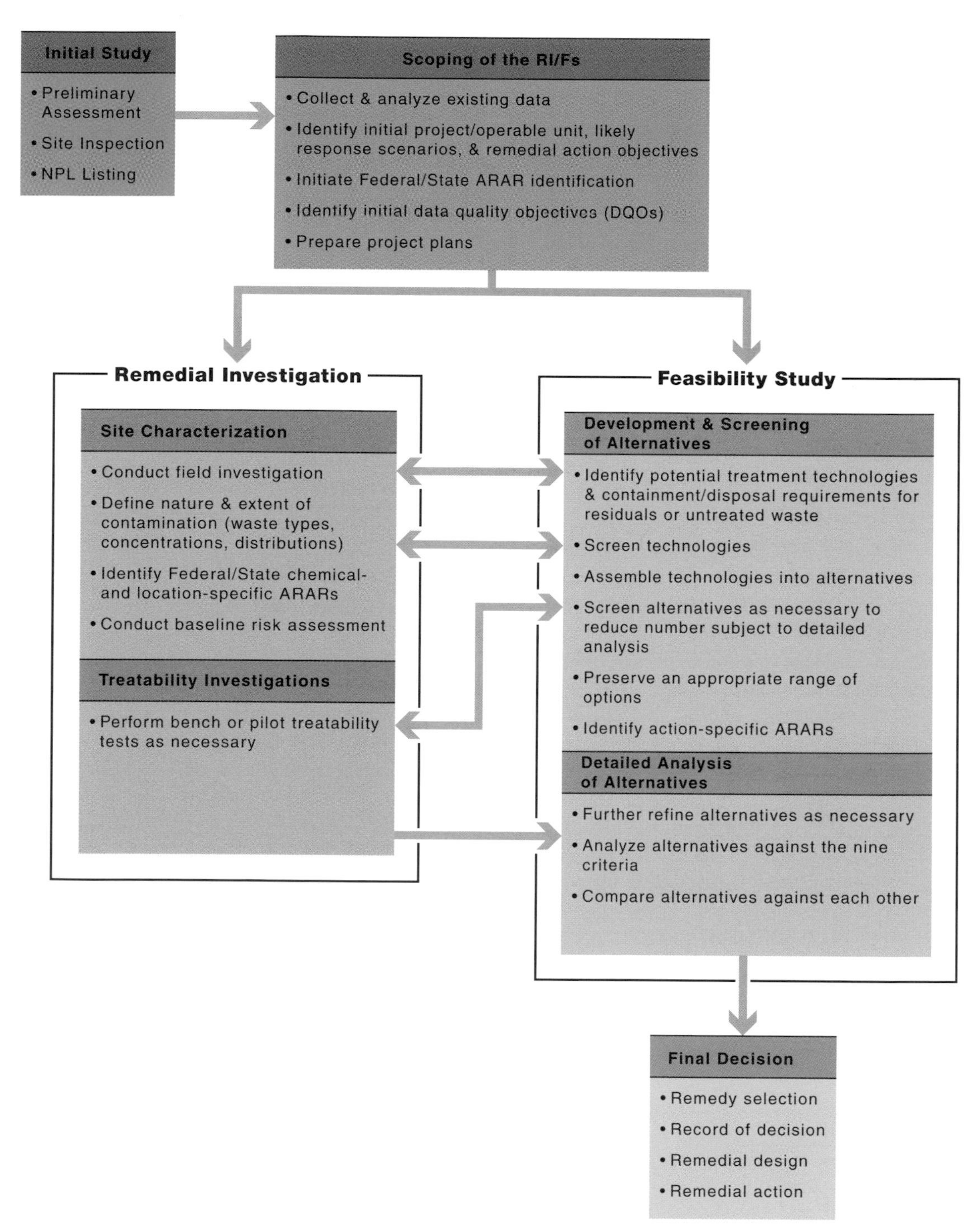

Figure 11-2 The remedial investigation/feasibility study process (RI/FS). ARAR(s) = Applicable or relevant and appropriate requirement(s). Adapted from U.S. EPA, 1988.

teria. The **threshold criteria** ensure that the remedy protects human health and the environment and is in compliance with applicable or relevant and appropriate requirements (ARARs); the **balancing criteria** ensure that such trade-off factors as cost and feasibility are considered; and the **modifying criteria** ensure that the remedy meets state and community expectations.

After a remedial action has been selected, it is designed and put into action. Some sites may require relatively simple actions, such as removal of waste-storage drums and surrounding soil. However, the sites placed on the NPL generally have complex contamination problems, and are therefore much more difficult to clean up. Only a few NPL sites have been completely cleaned up since the inception of the Superfund program.

An important component of Superfund and all other cleanup programs is the problem of deciding the target level of cleanup. Related to this problem is the question: "How clean is clean?" If, for example, the goal is to lower contamination concentration, how low does that concentration level have to be before it is "acceptable?" As you can imagine, the stricter the cleanup provision, the greater will be the attendant cleanup costs. It may require tens to hundreds of millions of dollars to return large, complex hazardous waste sites to a pristine condition. In fact, it may be impossible to completely clean many sites. However, a site need not be perfectly clean to be usable for some purposes. It is very important, therefore, that the physical feasibility of cleanup and the degree of potential risk posed by the contamination be weighed against the economic impact and the future use of the site. Consideration of the risk posed by the contamination and the future use of the site allows scarce resources to be allocated to those sites that pose the greatest current and future risk.

There are three major categories or types of remedial actions: (1) **containment**, where the contaminant is restricted to a specified domain to prevent further spreading; (2) **removal**, where the contaminant is transferred from an open to a controlled environment; and (3) **treatment**, where the contaminant is transformed to a nonhazardous substance. Since the inherent toxicity of a contaminant is eliminated only by treatment, this is the preferred approach of the three. Containment and removal techniques are very important, however, when it is not feasible to treat the contaminant. Although we will focus on each of the three types of remedial actions in turn, it is important to understand that remedial action often consists of a combination of containment, removal, and treatment.

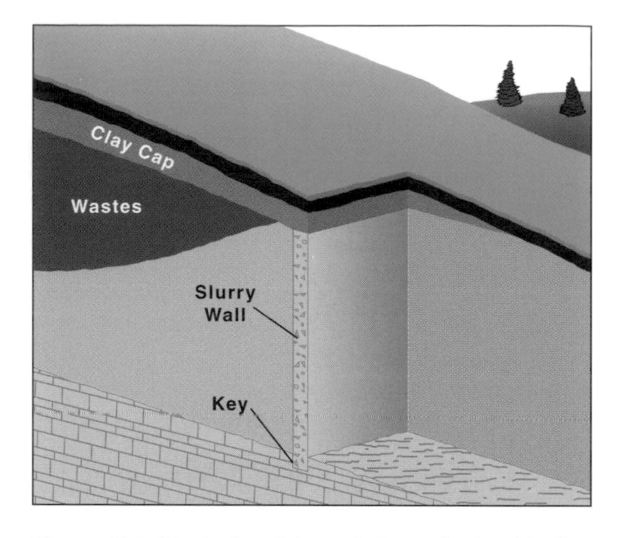

Figure 11-3 Physical containment of a contaminant by the use of a slurry wall. Adapted from U.S. EPA, 1985.

11.2 Containment

Containment can be accomplished by controlling the flow of the fluid that carries the contaminant or by directly immobilizing the contaminant. Here, we will discuss the role of physical and hydraulic barriers in containing contaminated water.

11.2.1 Physical Barriers

The principle of a physical barrier is to control the flow of water, thereby preventing the spread of contaminant. Usually, the barrier is installed in front (downgradient) of the contaminated zone (see Figure 11-3); however, barriers can also be placed upgradient or both up- and downgradient of the contamination. Physical barriers are primarily used in unconsolidated materials such as soil or sand, but they may also be used in consolidated media such as rock if special techniques are employed. In general, physical barriers may be placed to depths of about 50 meters. The horizontal extent of the barriers can vary widely, depending on the size of the site.

One important consideration in the employment of physical barriers is the presence of a zone of low permeability, into which the physical barrier can be seated or keyed. Without such a key into a low permeability zone, the contaminated water could flow underneath the barrier. Another criterion for physical barriers is the permeability of the barrier itself. Since the goal of a physical barrier is to minimize fluid flow through the target zone, the permeability of the barrier should be as low as practically possible. Another factor to consider is the potential of the contaminants to interact with the components of the barrier and degrade its performance. The properties of the barrier material should be matched to the properties of the contaminant to minimize failure of the barrier.

There are three major types of physical barriers: slurry walls, grout curtains, and sheet piling. **Slurry walls** are trenches filled with slurries of clay or mixtures of clay and soil (see Figure 11-3). **Grout curtains** are hardened matrices formed by cement-like chemicals that are injected into the ground. **Sheet piling** consists of large sheets of iron that are driven into the ground. Slurry walls are the least expensive and most widely used type of physical barrier, and they are the simplest to install. Grout curtains, which can be fairly expensive, are limited primarily to sites having consolidated subsurface environments. Sheet piles have essentially zero permeability and are generally of low reactivity. They can leak, however, because it is difficult to obtain perfect seals between individual sheets. Moreover, sheet piling is generally more expensive than slurry walls, and it is difficult to drive sheet piles into rocky ground.

11.2.2 Hydraulic Barriers

The principle behind hydraulic barriers is similar to that behind physical barriers—to manipulate and control water flow. But unlike physical barriers, which are composed of solid material, hydraulic barriers are based on fluid potentials (see Section 5.1.2, beginning on page 46). They are generated by the pressure differentials arising from the extraction or injection of water. The key performance factor of this approach is its capacity to capture the contaminant plume, that is, to limit the spread of the zone of contamination. Plume capture is a function of the number, placement, and flow rate of the wells or drains. Often, attempts are made to optimize the design and operation of the containment system so that plume capture is maximized while the volume of contaminated water removed is minimized.

The simplest hydraulic barrier is that established by a drain system. Such a system is constructed by installing a perforated pipe horizontally in a trench dug in the subsurface and placed to allow maximum capture of the contaminated water. Water can then be collected and removed by using gravity or active pumping. Drains are, however, effective containment mechanisms only for shallow contaminant zones.

Wells are more complicated—and more versatile—than drain systems. Both extraction and injection wells can be used in a containment system, as illustrated in Figure 11-4 on page 157. An extraction well removes the water entering the zone of influence of the well, creating a cone of depression. Conversely, an injection well creates a pressure ridge, or mound of water under higher pressure than the surrounding water, which prevents flow past the mound. One major advantage of using wells to control contaminant movement is that this is the only containment technique that can be used for deep systems (>50 m). In fact, wells can be used on contaminant zones of any size; the number of wells is simply increased to handle larger problems. For these and other reasons, wells are the most widely used method for containment, despite disadvantages that include the cost of long-term operation and pump maintenance, and the need to store, treat, and dispose of the large quantities of contaminated water pumped to the surface.

11.3 Removal

11.3.1 Excavation

A time-honored method for removing contaminants is excavation of the soil in which the contaminants reside. This technique has been used at many sites and is highly successful. There are, however, some disadvantages associated with excavation. First, excavation can expose site workers to hazardous compounds. Second, the contaminated soil requires treatment and/or disposal, which can be expensive. Third, excavation is usually feasible only for relatively small areas. Excavation is most often used to remediate shallow, localized, highly contaminated source zones.

11.3.2 Pump and Treat

The pump-and-treat method, currently the most widely used remediation technique for contaminated groundwater, removes contaminated water from the subsurface by using one or more wells to pump it out. Furthermore, clean water brought into the contaminated region by the pumping action removes, or "flushes," additional contamination by inducing desorption from the solid phase (see Sections 6.3–6.4, beginning on page 67). The contaminated water pumped from the subsurface is directed to some type of treatment operation, which may consist of air stripping, carbon adsorption, or perhaps an above-ground biological treatment system (see Sections 11.4.1.1 through 11.4.1.2, beginning on page 160). Illustrations of a pump-and-treat system are provided in Figure 11-5 on page 158 and Figure 11-7a on page 162.

Usually discussed in terms of its use for such saturated subsurface systems as aquifers, the pump-and-treat method can also be used to remove contaminants from the vadose zone. In this case, it is generally referred to as ***in situ* soil washing**. For this application, infiltration galleries, *i.e.*, a network of porous, underground pipes whose function is to supplement large areas of underground soil with fluid amendments, can be used in addition to wells to introduce water to the contaminated zone.

When using water flushing for contaminant removal (as in pump-and-treat), contaminant-plume capture and the effectiveness of contaminant removal are the major performance criteria. Recent studies of operating pump-and-treat systems have shown that the technique is very successful at containing contaminant plumes and, in some cases, shrinking them. However, it appears that pumping is frequently ineffective for completely removing contaminants from the subsurface. There are many factors that can limit the effectiveness of water flushing for contaminant removal.

1. **Presence of low-permeability zones**. When low-permeability zones (*e.g.*, silt/clay lenses) are present within a sandy subsurface, they create domains through which advective flow and transport are minimal in comparison to the surrounding sand. The groundwater flows preferentially around the silt/clay lenses, rather than through them. Thus, contaminant located within the silt/clay lenses is released to the flowing water primarily by pore-water diffusion, which can be a relatively slow process. Thus the mass of contaminant being removed diminishes with each volume of water pumped, thereby increasing the time required to completely remove the contaminant.
2. **Rate-limited desorption**. Research has revealed that adsorption/desorption of many solutes by porous media can be significantly rate-limiting (see Section 6.3.3, beginning on page 70). When the rate of desorption is slow enough, the concentration of contaminant in the groundwater is lower than the concentrations obtained under conditions of rapid desorption. Thus, less contaminant is removed per volume of water, and removal by flushing will therefore take longer.
3. **Presence of immiscible liquid**. In many cases, residual phases of immiscible organic liquids may exist in portions of the contaminated subsurface. Since it is very difficult to displace or push out residual saturation with water, the primary means of removal must be dissolution into water and volatilization into the soil atmosphere. It can take a very long time to completely dissolve immiscible liquid, thus greatly delaying removal. The immiscible liquid, therefore, serves as a long-term source of contamination.

Because pump-and-treat is a major remedial action technique, methods are being tested to enhance its effectiveness. One way to improve the effectiveness

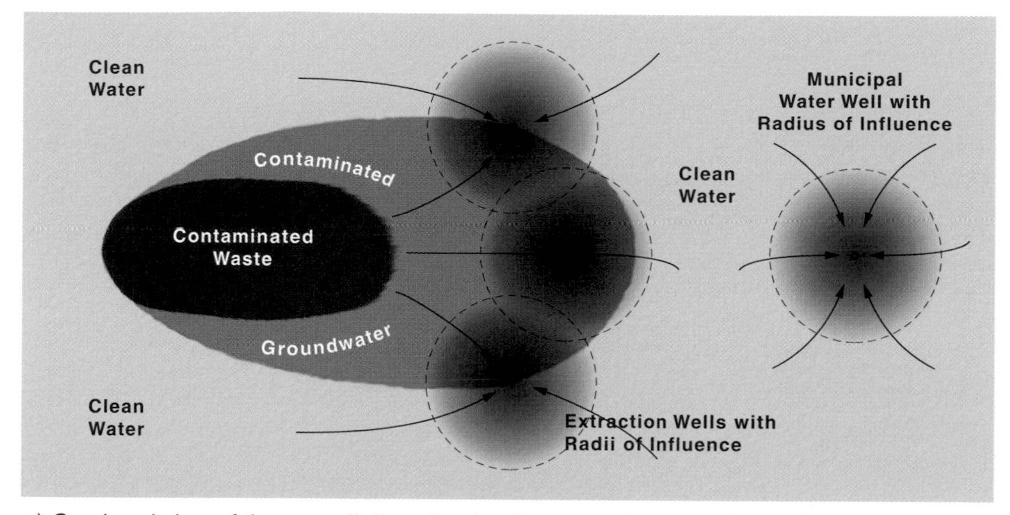

a) Overhead view of the remediation site showing groundwater surfaces, flow directions, and contaminated waste (soil material removed).

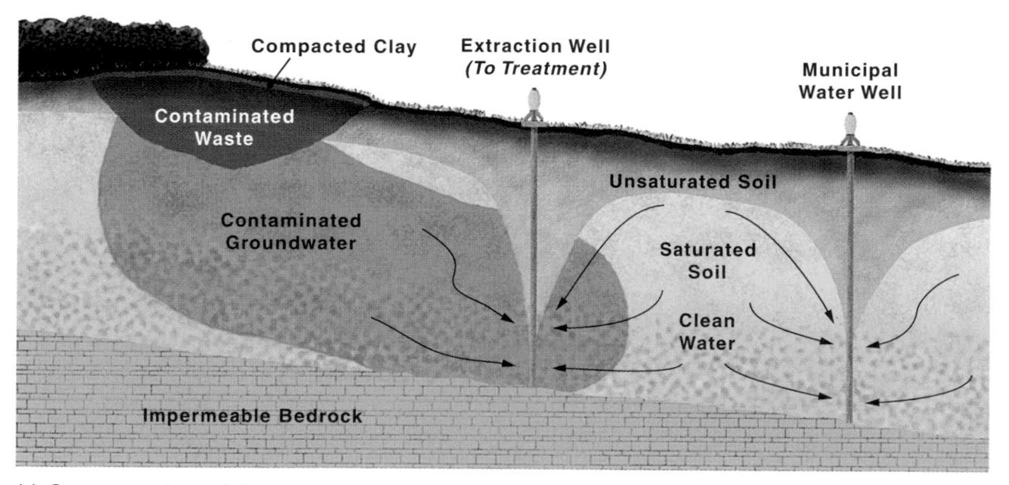

b) Cut-away view of the above site showing lateral movement of leached contaminants and groundwater.

Figure 11-4 Containment of a contaminant plume by a hydraulic barrier. Adapted from U.S. EPA, 1985.

of pump-and-treat is to alter the status of the **contaminant source zone**, that is, the area in which contaminants were disposed or spilled. If the source zone remains untreated or uncontrolled, it will serve as a continual source of contaminant requiring removal. Thus, failure to control or treat the source zone can greatly extend the time required to achieve site cleanup. It is important, therefore, that the source zone at a site be delineated and addressed in the early stages of a remedial action response. Other methods of improving this technique involve enhancing removal, as detailed below.

11.3.3 Enhanced Removal

Contaminant removal can be difficult because of such factors as low solubility, high degree of sorption, and the presence of immiscible-liquid phases, all of which limit the amount of contaminant that can be flushed by a given volume of water. Approaches

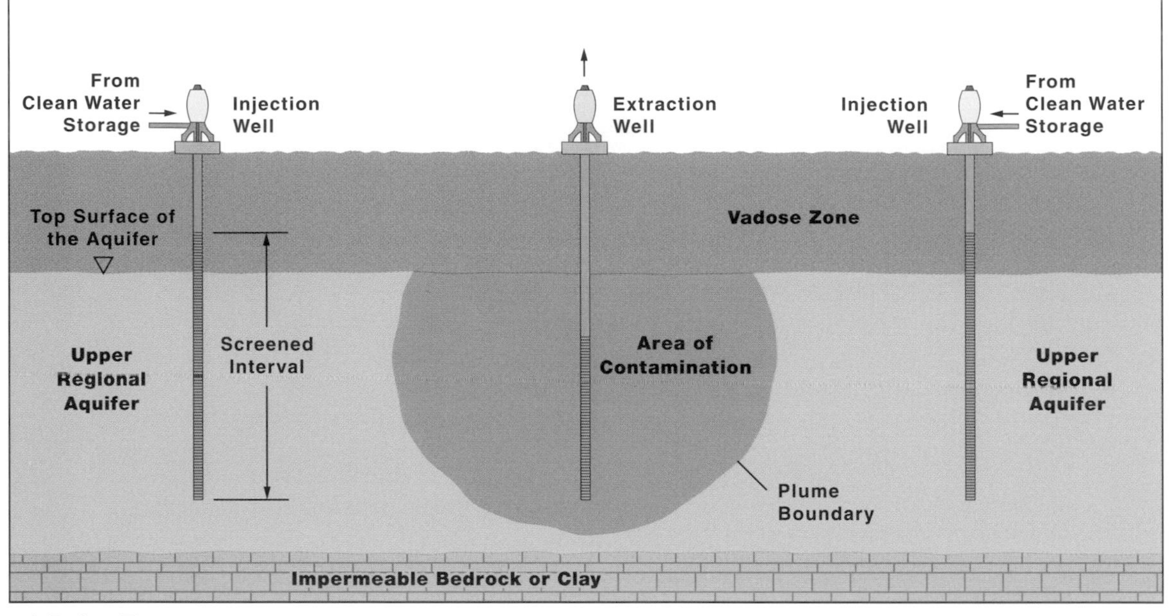

a) Before Treatment

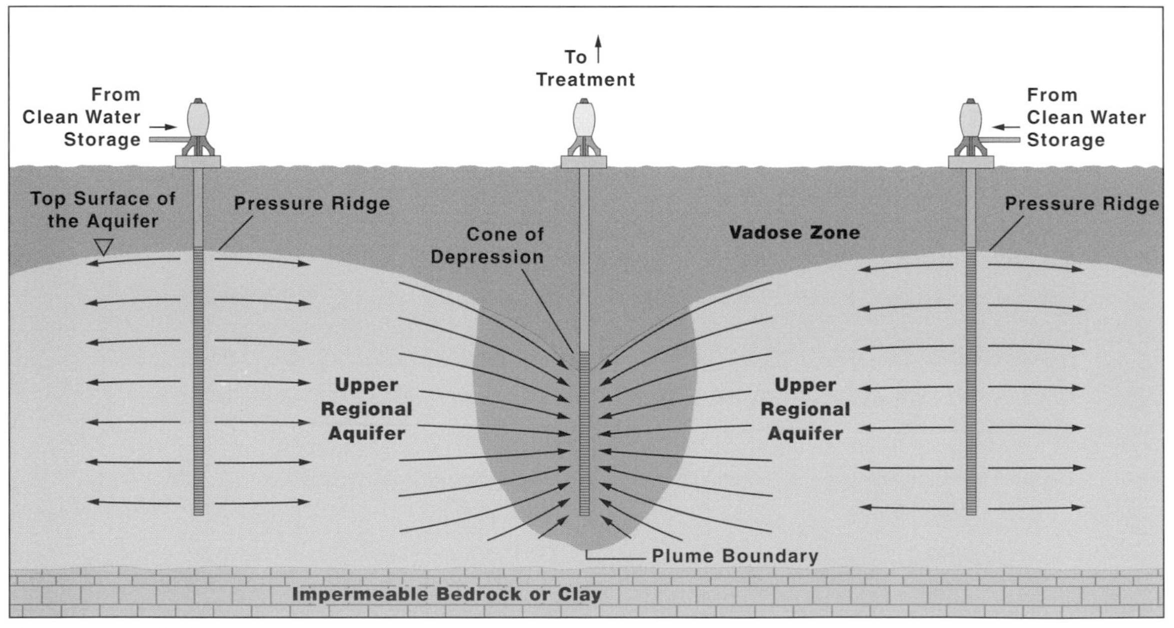

b) After Flow Initiation

Figure 11-5 Removal of a contaminant by the pump-and-treat processes.

are being developed to enhance the removal of low-solubility, high-sorption contaminants. One such approach is to inject a chemical into the aquifer, such as a surfactant (*e.g.*, detergent), that will promote dissolution and desorption of the contaminant, thus enhancing removal effectiveness (see also Section 11.4.1.4 on page 164). Such surfactants work like industrial and household detergents, which are used to remove oily residues from machinery, clothing, or dishes: individual contaminant molecules are "solubilized" inside of surfactant micelles, which are groups of individual surfactant molecules ranging from 5 to 10 nm in diameter. Alternatively, surfactant molecules can coat oil droplets and emulsify them into solution (see Figure 11-6). In laboratory tests, surfactants have been successfully used to increase the apparent aqueous solubility of organic contaminants. However, only a few field tests have been attempted, and results have been mixed. A key factor controlling the success of this approach in the field is the ability to deliver the surfactant to the places that contain the contaminant. This would depend, in part, on potential interactions between the surfactant and the soil (*e.g.*, sorption) and on properties of water flow in the soil.

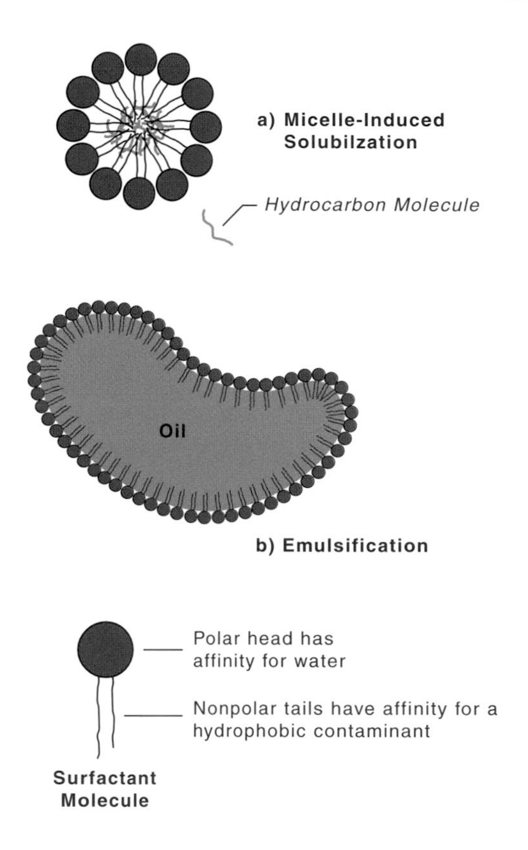

Figure 11-6 Two types of interactions between surfactants and organic contaminants. (a) Micelle-induced solubilization involves stabilization of a few molecules of the organic contaminant by the surfactant. These droplets are on the order of a few nanometers in diameter. (b) Emulsification occurs when the surfactant coats droplets of the contaminant.

11.3.4 Soil Vapor Extraction

The principle of soil vapor extraction, or soil venting, is very similar to that of pump-and-treat: a fluid is pumped through a contaminated domain to enhance removal. In the case of soil venting, however, the fluid is air rather than water. There are two key conditions for using soil venting. First, the soil must contain a gas phase through which the contaminated air can travel. This condition limits the use of soil venting to the vadose zone. Second, contaminants must be capable of transfer from other phases (solid, water, immiscible liquid) to the gas phase. This requirement limits soil venting to volatile contaminants.

Because air is much less viscous than water, much less energy is required to pump air. Thus, it is usually cheaper to use soil venting for removing volatile contaminants from the vadose zone. Once the contaminant is removed from the vadose zone, it is either released to the atmosphere or placed into a treatment system. The major performance criteria for soil venting are the effectiveness of capturing and removing the contaminant. The effectiveness of contaminant removal by soil venting can be limited by the same three factors that limit removal by water flushing (see Section 11.3.2 on page 156).

A technology somewhat related to soil vapor extraction is **air sparging**, which involves the injection of clean air into the saturated zone or aquifer. The primary purpose of this injection is to volatilize contaminants from the soil into air bubbles. Owing to buoyancy effects, the air bubbles rise and eventually make their way to the vadose zone, where they are captured with a soil vapor extraction system. In some cases, *in situ* biodegradation can be enhanced by air sparging because oxygen is added to the subsurface (see Section 11.4.1.1, beginning on page 160).

11.4 *In situ* Treatment

In situ treatment technologies are methods that allow in-place cleanup of contaminated field sites. There is great interest in these technologies because they can be used with minimum economic costs. Second, *in situ* treatment can, in some cases, eliminate the risk associated with the hazardous form of the contaminant. The two major types of *in situ* treatment are biological (*in situ* bioremediation) and chemical.

11.4.1 Bioremediation

The objective of bioremediation is to exploit the naturally occurring biodegradative processes (discussed in Chapter 7, beginning on page 77) to clean up contaminated sites. There are several types of bioremediation: *in situ* bioremediation is the in-place treatment of a contaminated site; *ex situ* bioremediation is the treatment of contaminated soil or water that is removed from a contaminated site; and intrinsic bioremediation is the indigenous level of contaminant biodegradation that occurs without any stimulation or treatment. All of these types of bioremediation are receiving increasing attention as viable remediation alternatives for several reasons. These include generally good public acceptance and support, good success rates for some applications, and the comparatively low cost of bioremediation when it is successful. As with any technology, there are also drawbacks. First, success can be unpredictable because biological systems are themselves complex and unpredictable. A second consideration is that bioremediation rarely restores an environment to its original condition. Often, the residual contamination left after treatment is strongly sorbed and not available to microorganisms for degradation. Over a long period of time (years), these residuals can be slowly released, generating additional pollution. There is little research concerning the fate and potential toxicity of such released residuals: therefore, both the public and regulatory agencies continue to be concerned about the possible deleterious effects of residual contamination.

Domestic sewage waste, which has been treated biologically for many years, is one of bioremediation's major successes. From this success, it is easy to see that biodegradation is dependent on pollutant structure and bioavailability (see Section 7.3.2 on page 83). Therefore, application of bioremediation to other pollutants depends on the type of pollutant or pollutant mixtures present, and the type of microorganisms present. The first successful application of bioremediation outside of sewage treatment was in the cleanup of oil spills, where aerobic heterotrophic bacteria were used to biodegrade hydrocarbon products. In the past few years, many new bioremediation technologies have emerged that are being used to address other types of pollutants; these are listed in Table 11-1.

Several key factors are critical to successful application of bioremediation: environmental conditions, contaminant and nutrient availability, and the presence of degrading microorganisms. Thus, when biodegradation fails to occur, it is important to isolate the "culprit"—the limiting factors of bioremediation. This task can be very complex. Initial laboratory tests on soil or water from a polluted site can usually determine the presence or absence of degrading microorganisms; such tests may also reveal an obvious environmental factor that limits biodegradation, such as extremely low or high pH. Often, however, the limiting factor is not easy to identify. For example, pollutants are often present as mixtures, and one component of the pollutant mixture can have toxic effects on the growth and activity of degrading microorganisms. Similarly, low bioavailability, which is another factor that can limit bioremediation, can be very difficult to evaluate in the environment.

Most of the developed bioremediation technologies are based on two standard practices: the addition of oxygen and the addition of other nutrients.

11.4.1.1 Addition of Oxygen or Other Gases

One of the most common limiting factors in bioremediation is the availability of oxygen. Oxygen is required for aerobic biodegradation (see Section 7.3.1.1, beginning on page 81). Moreover, oxygen is sparingly soluble in water and has a low rate of diffusion (movement) through both air and water. The

Table 11-1 Current feasibility of bioremediation.

Chemical Class	Frequency of Occurrence	Status of Bioremediation	Evidence of Future success	Limitations
Hydrocarbons and derivatives				
Gasoline, fuel oil	Very frequent	Established		Forms nonaqueous-phase liquid
Polycyclic aromatic hydrocarbons	Common	Emerging	Aerobically biodegradable under a narrow range of conditions	Sorbs strongly to subsurface solids
Creosote	Infrequent	Emerging	Readily biodegradable under aerobic conditions	Sorbs strongly to subsurface solids; forms nonaqueous-phase liquid
Alcohols, ketones, esters	Common	Established		
Ethers	Common	Emerging	Biodegradable under a narrow range of conditions using aerobic or nitrate-reducing microbes	
Halogenated aliphatics				
Highly chlorinated	Very frequent	Emerging	Cometabolized by anaerobic microbes; cometabolized by aerobes in special cases	Forms nonaqueous-phase liquid
Less chlorinated	Very frequent	Emerging	Aerobically biodegradable under a narrow range of conditions; cometabolized by anaerobic microbes	Forms nonaqueous-phase liquid
Halogenated aromatics				
Highly chlorinated	Common	Emerging	Aerobically biodegradable under a narrow range of conditions; cometabolized by anaerobic microbes	Sorbs strongly to subsurface solids; forms nonaqueous phase—solid or liquid
Less chlorinated	Common	Emerging	Readily biodegradable under aerobic conditions	Forms nonaqueous phase—solid or liquid
Polychlorinated biphenyls				
Highly chlorinated	Infrequent	Emerging	Cometabolized by anaerobic microbes	Sorbs strongly to subsurface solids
Less chlorinated	Infrequent	Emerging	Aerobically biodegradable under a narrow range of conditions	Sorbs strongly to subsurface solids
Nitroaromatics	Common	Emerging	Aerobically biodegradable; converted to innocuous volatile organic acids under anaerobic conditions	
Metals (Cr, Cu, Ni, Pb, Hg, Cd, Zn, *etc.*)	Common	Possible	Solubility and reactivity can be changed by a variety of microbial processes	Availability highly variable and controlled by solution- and solid-phase chemistry

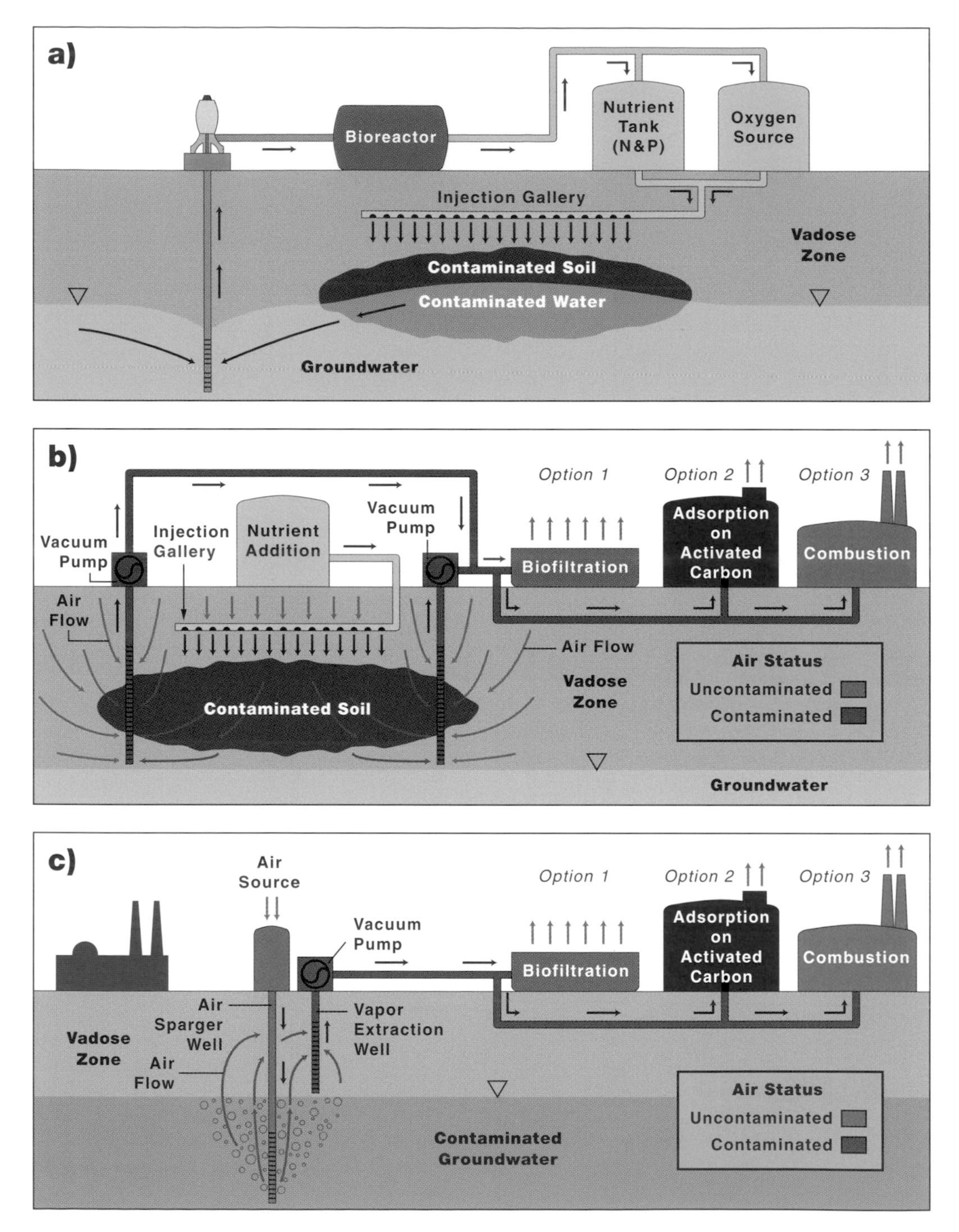

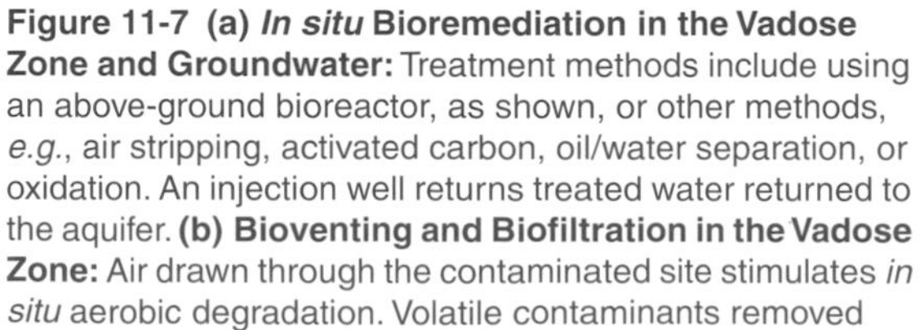

Figure 11-7 (a) *In situ* Bioremediation in the Vadose Zone and Groundwater: Treatment methods include using an above-ground bioreactor, as shown, or other methods, *e.g.*, air stripping, activated carbon, oil/water separation, or oxidation. An injection well returns treated water returned to the aquifer. **(b) Bioventing and Biofiltration in the Vadose Zone:** Air drawn through the contaminated site stimulates *in situ* aerobic degradation. Volatile contaminants removed with the air are treated in a biofilter, by adsorption on activated carbon, or by combustion. **(c) Bioremediation in the Groundwater by Air Sparging:** Air pumped into the contaminated site stimulates aerobic degradation in the saturated zone. Volatile contaminants brought to the surface are treated by biofiltration, activated carbon, or combustion. Adapted from Figures 3-1, 3-2, and 3-3 in National Research Council, 1993.

combination of these three factors makes it easy to understand why inadequate oxygen supplies often limit bioremediation.

Several technologies have been developed to overcome a lack of oxygen. Consider the typical bioremediation system shown in Figure 11-7a. This system is used to treat a contaminated aquifer, together with the contaminated zone above the water table. It contains a series of injection wells or galleries and a series of recovery wells, thus providing a two-pronged approach to bioremediation. First, the recovery wells remove contaminated groundwater, which is treated above ground, in this case using a **bioreactor** containing microorganisms that are acclimated to the contaminant. Following bioreactor treatment, the clean water is supplied with oxygen and nutrients (which may not be needed if the level of contamination is very low), and then reinjected into the subsurface. The reinjected water provides oxygen and nutrients to stimulate *in situ* biodegradation. In addition, the reinjected water flushes the vadose zone to aid in removal of the contaminant for above-ground bioreactor treatment. This remediation scheme is a very good example of a combination of physical/chemical/biological treatments that can be used to maximize the effectiveness of the remediation treatment.

Bioventing is a technique used to add oxygen directly to a site of contamination in the vadose zone (unsaturated zone). Bioventing is a combination of soil venting technology and bioremediation. The bioventing zone, which is denoted by red air flow arrows in Figure 11-7b, includes the vadose zone and contaminated regions just below the water table. As shown in this figure, a series of wells have been constructed around the zone of contamination. To initiate bioventing, a vacuum is drawn on these wells to force accelerated air movement through the contamination zone. This effectively increases the supply of oxygen throughout the site, and hence the rate of contaminant biodegradation. In the case of volatile pollutants, some of the pollutants are removed as air is forced through this system (see Section 11.3.4 on page 159). This contaminated air can also be treated biologically by passing the air through above-ground soil beds (biofilters) in a process called **biofiltration**, as shown in Figure 11-7b. Figure 11-8

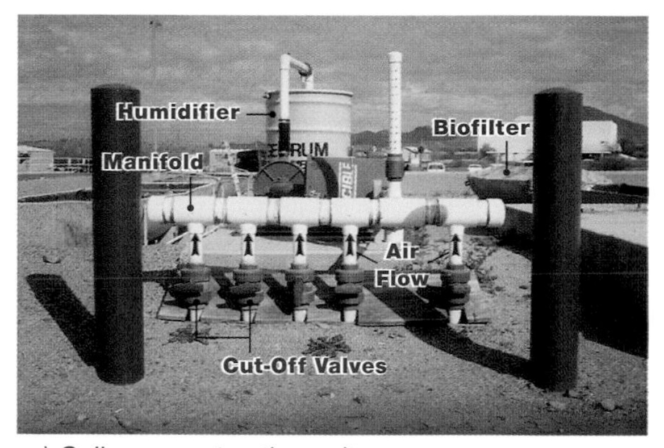

a) Soil vapor extraction unit

b) Side view of a biofilter

Figure 11-8 Bioremediation of petroleum vapors from a leaking underground storage tank. (a) Red arrows denote the direction of air flow from out of the ground and through the manifold, which controls the air flow. The air then passes through the humidifier, a tank containing water, and into the biofilter. The large red vertical cylinders serve as protective barriers. (b) Air flows into the biofilter through a pipe running lengthwise along the bottom of the filter and into the soil via orifices in the pipe. Photographs: E.M. Jutras.

shows an operational soil venting unit that utilizes biofiltration to remediate petroleum vapors from a leaking underground storage tank (LUST).

In contrast, air sparging can be used to add oxygen to the saturated zone (Figure 11-7c). In this process, an air sparger well is used to inject air under pressure below the water table. The injected air displaces water in the soil matrix, creating a temporary air-filled porosity. This causes oxygen levels to increase, thereby enhancing biodegradation rates. In addition, volatile organics that volatilize into the air stream can be removed by a vapor extraction well, as discussed in Section 11.3.4, beginning on page 159.

Methane is another gas that can be added with oxygen into extracted groundwater and reinjected into the saturated zone. Methane is used specifically to stimulate methanotrophic activity and cometabolic degradation of chlorinated solvents. As described in Section 7.2, beginning on page 77, methanotrophic organisms produce the enzyme methane monooxygenase to degrade methane, and this enzyme also cometabolically degrades several chlorinated solvents. Cometabolic degradation of chlorinated solvents is currently being tested in field trials to determine the usefulness of this technology.

11.4.1.2 Nutrient Addition

Second only to the addition of oxygen in bioremediation treatment is the addition of nutrients, nitrogen and phosphorus in particular. Many contaminated sites contain organic wastes that are rich in carbon but poor in nitrogen and phosphorus. Nutrient addition is illustrated in the bioremediation schemes shown in Figures 11-7a and 11-7b on page 162. Injection of nutrient solutions takes place from an above-ground, batch-feed system. The goal of nutrient injection is to optimize the carbon/nitrogen/phosphorus ratio (C:N:P) in the site to approximately 100:10:1. However, sorption of added nutrients can make it difficult to achieve the optimal ratio.

11.4.1.3 Stimulation of Anaerobic Degradation Using Alternative Electron Acceptors

Until recently, anaerobic degradation of many organic compounds was not considered feasible. Now, however, it is being proposed as an alternative bioremediation strategy, even though aerobic degradation is generally a much more rapid process. Because it is difficult to establish and maintain aerobic conditions in some saturated subsurface systems, several alternative electron acceptors have been proposed for use in anaerobic degradation. These acceptors include nitrate, sulfate, and iron (Fe^{3+}) ions, as well as carbon dioxide. Although limited in number, field trials using nitrate have shown promise. This relatively new area in bioremediation will undoubtedly receive increased attention in the next few years.

11.4.1.4 Addition of Surfactants

While the use of surfactants is being investigated for improving pump-and-treat, as discussed in Section 11.3.3 on page 157, surfactant addition has also been proposed as a technique to enhance the biodegradation of contaminants. Surfactants can be synthesized chemically, but they are also produced by many microorganisms. Such microbially produced surfactants are known as **biosurfactants** (see Figure 11-9). Like synthetic surfactants, biosurfactants increase the solubility and decrease the sorption of contaminants, thus increasing their bioavailability and enhancing the rates of biodegradation.

11.4.1.5 Addition of Microorganisms

If appropriate biodegrading microorganisms are not present in soil or if microbial populations have been reduced because of contaminant toxicity, specific microorganisms can be added as "introduced organisms" to enhance the existing populations. This process is known as **bioaugmentation**. Scientists are now capable of creating superbugs—organisms that can degrade pollutants at extremely rapid rates. Such organisms can be developed through successive adaptations under laboratory conditions, or they can be genetically engineered. In terms of biodegradation, these superbugs are far superior to organisms found in the environment. One way to take advantage of the superbugs is to use them in bioreactor systems under controlled conditions. Extremely efficient biodegradation rates can be achieved in bioreactors that are used in above-ground treatment systems.

The problem with use of superbugs is that introduction of a microorganism to a contaminated site may fail for two reasons. First, the introduced microbe often cannot establish a niche in the environment; in fact, these introduced organisms rarely survive in a new environment beyond a few weeks (see Section 4.2.1 on page 41). Second, microorganisms, like contaminants, can be strongly sorbed by solid surfaces, so there are difficulties in delivering the introduced organisms to the site of contamination. Currently, very little is known about the establishment of environmental niches or about microbial transport; these are areas of active research. Perhaps in the next few years scientists will gain further un-

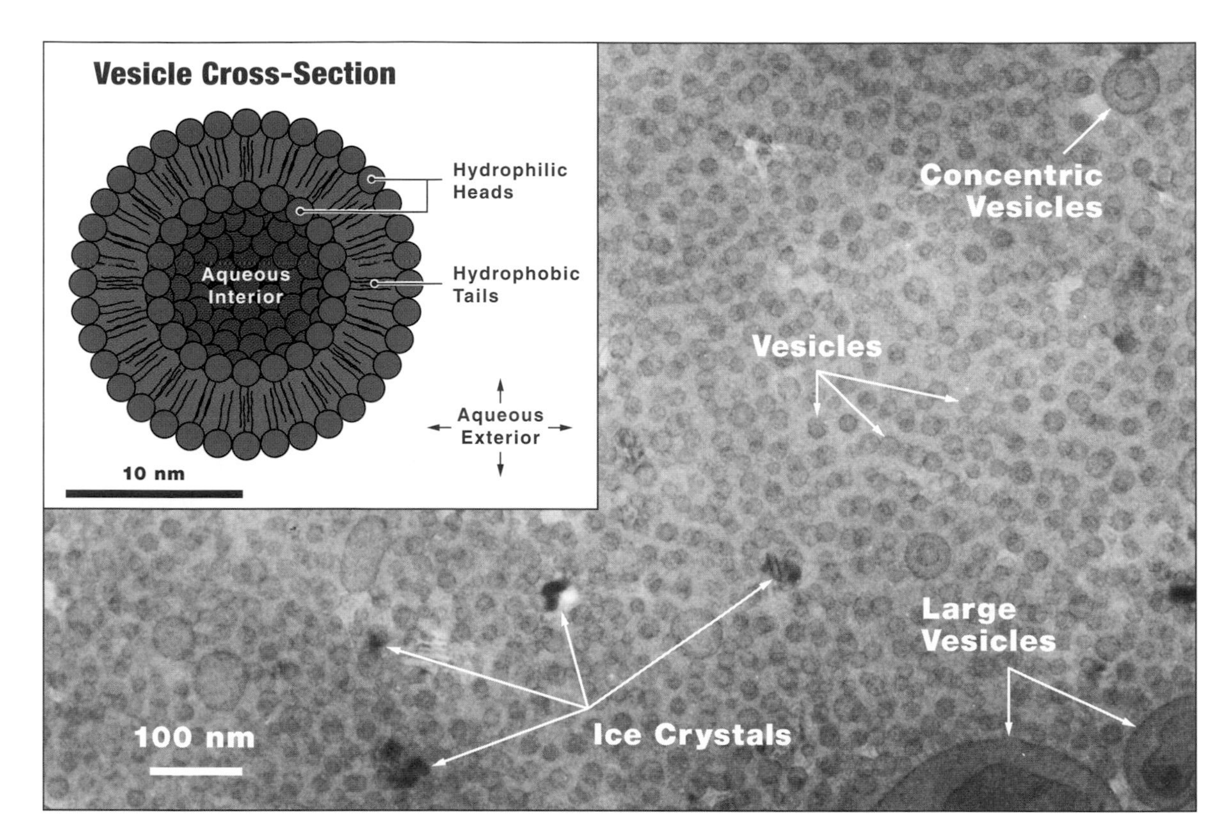

Figure 11-9 Cryo-transmission electron micrograph of a microbially produced surfactant, rhamnolipid. This compound acts as a "biological soap" and aids in solubilization of organic contaminants with low water solubility. Inset illustration based on Figure 1 on page 570 of Champion *et al.*, 1994. Photomicrograph courtesy of R.M. Miller.

derstanding of microbial behavior in soil ecosystems. However, until we discover how to successfully deliver and establish introduced microorganisms, their addition to contaminated sites will not be a feasible method of bioremediation.

11.4.1.6 Metal Contaminants

Current approaches to bioremediation of metals are based upon the complexation, oxidation-reduction (redox), and alkylation reactions introduced in Section 7.5, beginning on page 89. Microbial leaching, microbial surfactants (biosurfactants), volatilization, and bioaccumulation/complexation are all strategies that have been suggested for removal of metals from contaminated environments. Unfortunately, the number of accompanying field-based studies has, thus far, been small.

Bioleaching: *Ex situ* removal of metals from soil can sometimes be accomplished by microbial leaching, or **bioleaching**. This technique has been used in mining to remove metals such as copper, lead, and zinc from low-grade ores. In bioleaching, metals are solubilized as a result of acid production by specific microorganisms such as *Thiobacillus ferrooxidans* and *T. thiooxidans*. Analogously, in bioremediation, this process has been used to leach uranium from nuclear-waste-contaminated soils and to remove copper from copper tailings. Another potential application is the treatment of sewage sludge earmarked for disposal in soil (see also Section 10.4, beginning on page 140). Sludge-amended soils exhibit improved productivity, but also show increased metal

Case Study: The *Exxon Valdez*

In March 1989, 40.9 million liters of crude oil was spilled from a large tanker—the *Exxon Valdez*—in Prince William Sound, Alaska. Subsequently spread by a storm, this oil eventually coated onto the shores of the islands in the Sound. Initially, more than 2,200 kilometers of Alaska coastline was polluted. A resultant court settlement included $900 million to be paid by the Exxon Corporation for damage assessment and restoration of natural resources. A good portion of this sum was spent on conventional cleanup, primarily by physical methods. But these methods failed to remove all of the oil on the beaches, particularly that under rocks and in the beach sediments. In late May 1989, Exxon reached a cooperative agreement with the Environmental Protection Agency (EPA) to test bioremediation as a cleanup strategy. The approach followed was to capitalize on existing conditions as much as possible. For example, scientists quickly realized that the beaches most likely contained indigenous, oil-degrading organisms that were already adapted to the cold climate. Preliminary studies revealed that, indeed, the polluted beaches contained such organisms, together with a plentiful carbon source (spilled oil) and sufficient oxygen. In fact, natural biodegradation, also called intrinsic bioremediation, was clearly already occurring in the area. Further studies showed that intrinsic degradation rates were limited by the availability of nutrients such as nitrogen, phosphorus, and certain trace elements. At this point, time was of the essence in treating the contaminated sites because the relatively temperate Alaskan season was rapidly advancing. Therefore, it was decided to attempt to enhance the intrinsic rates of biodegradation by amendment of the contaminated areas with nitrogen and phosphorus.

In early June 1989, field demonstration work was started. Several different fertilizer nutrient formulations and application procedures were tested. One problem encountered was the tidal action, which tended to quickly wash away added nutrients. Therefore a liquid oleophilic fertilizer (Inipol EAP-22), which adhered to the oil-covered surfaces, and a slow-release water-soluble fertilizer (Customblen) were tested as nutrient sources. Within about two weeks after application of the fertilizers, rock surfaces treated with Inipol EAP-22 showed a visible decrease in the amount of oil. Subsequent independent scientific studies on the effectiveness of bioremediation in Prince William Sound concluded that bioremediation enhanced removal from three- to eightfold over the intrinsic rate of biodegradation without any adverse effects to the environment.

Bioremediation in Prince William Sound was considered a success not only because contaminated areas were restored but because the rapid rate of removal of the petroleum contamination prevented further spread of the petroleum to uncontaminated areas. This well-documented and highly visible case has helped gain attention for the potential of bioremediation.

content. The use of *T. ferrooxidans* and *T. thiooxidans* has been demonstrated on the laboratory scale for leaching metals from contaminated sludge before soil application. Bacterial surfactants can also be used for removal of metals from contaminated soils and water.

Volatilization: Although alkylation is not a desirable reaction for most metals, particularly arsenic or mercury, alkylation with subsequent volatilization has been proposed as a technique for remediation of selenium-contaminated soils and sediments. For example, selenium-contaminated soils from San Joaquin, California, were remediated using selenium volatilization stimulated by the addition of pectin in the form of orange peels. It was found that the addition of pectin enhanced the rate of selenium alkylation, ranging from 11.3 to 51.4% of added selenium; this result suggests that volatilization is a feasible approach to treating selenium-contaminated soils.

Bioaccumulation/Complexation: Technologies for metal removal from solution are based on several specific microbial-metal interactions: the binding of metal ions to microbial cell surfaces, the intracellular uptake of metals, and the precipitation of metals via complexation with microbially produced ligands. These interactions lead to the **bioaccumulation** of metals within cells or on the outside of the cell. Treatment of metal-containing waste streams generally involves the use of **biofilms**, which are concentrated films of microorganisms that may be composed of bacteria, fungi, or algae. Some systems utilize viable organisms, while other systems utilize processed, nonviable cells. In either case, the biofilm essentially traps metals as the contaminated water is pumped through. Bioaccumulation has been used to treat acid mine drainage, mining effluent waters, and waste streams from nuclear processing. Metals removed include zinc, copper, iron, manganese, lead, cadmium, arsenic, and uranium.

11.4.2 *In situ* Chemical Treatment

In situ chemical remediation is a process in which the contaminant is degraded by promoting a transformation reaction, such as hydrolysis or oxidation/reduction, within the soil or vadose zone. Although this approach has been used much less frequently than *in situ* bioremediation, it has begun to receive increasing attention. It can be accomplished by injecting a reagent into the contaminated zone of the subsurface or by placing a permeable treatment barrier downgradient of the contamination. The key factor controlling the performance of the injection technique is the delivery of the reagent to the desired locations.

In situ treatment walls or barriers are of particular interest for chlorinated solvents such as trichloroethene. This approach involves digging a trench downgradient of the contaminant plume and filling that trench with a wall of permeable material that can degrade the contaminant to nontoxic by-products. The wall is permeable so that the water from which the contaminant has been removed can pass through. For example, iron filings can degrade compounds such as trichloroethene by reacting with chlorine atoms while permitting water to pass through freely.

References and Recommended Reading

Champion J.T., Gilkey J.C., Lamparski H., Retterer J., and Miller R.M. (1995) Electronmicroscopy of rhamnolipid (biosurfactant) morphology: Effects of pH, cadmium, and octadecane. *Journal of Colloid and Interface Science*. **170,** 569–574.

National Research Council (1993) *In Situ Bioremediation, When Does It Work?* National Academy Press, Washington, DC.

National Research Council (1994) *Alternatives for Ground Water Cleanup*. National Academy Press, Washington, D.C.

Norris R.D. and Matthews J.E. (1994) *Handbook of Bioremediation*. Lewis Publishers, Boca Raton, Florida.

U.S. EPA (1985) *Remedial Action at Waste Disposal Sites*. EPA/625/6-85/006. U.S. Environmental Protection Agency, Washington, D.C.

U.S. EPA (1988) *Guidance for Conducting Remedial Investigations and Feasibility Studies under CERCLA*. EPA 540/G-89/004. United States Environmental Protection Agency, Washington, D.C.

Problems and Questions

1. A small volume of soil, 10 m by 20 m by 5 m, is contaminated with a radioactive material. What remediative action would you recommend? Explain.
2. Groundwater is contaminated with trichloroethene (TCE); what remediative action would you recommend (a) to contain the contaminant plume and (b) to clean up the contaminated water?
3. Identify four factors that can limit the effectiveness of soil venting as a remediative action.
4. What bioremediative process has been successfully practiced widely throughout the United States for the past several decades? Why has it been successful?
5. What are the three entities that are routinely added to contaminated sites to enhance biodegradation? Why?
6. What are the major factors to consider when evaluating the performance of a physical barrier? A hydraulic barrier? Compare the two.

Part 3

Sources, Extent, and Characteristics of Pollution

On Reverse *A sewage outfall dumps raw sewage from Mexico City into the Tula River, which transports it to agricultural regions outside the city. Farming communities which receive the sewage have prospered from the nutrients it delivers to the soil while avoiding significant contamination of groundwater supplies due to an impermeable subsurface layer protecting the aquifer. Infections from helminths, however, have been a problem among workers coming in contact with the sewage.*

Photograph *C.P. Gerba*

Recent research has determined that some stands of trees showing signs of damage from acid rain, such as those in the Harz Mountains of Germany (above), have been dying from other causes, including old age. Photograph: J.W. Brendecke.

12.1 Air Pollution Concepts

In this chapter we discuss air pollutants, including their sources and effects on human activity, as well as their transport to, and fate in, the atmosphere. We also describe the role of air pollution in such major environmental issues as potential global warming and ozone depletion.

An **air pollutant** is any gas or particulate that, at high enough concentration, may be harmful to life and/or property. A pollutant may originate from natural or anthropogenic sources, or both. Pollutants occur throughout much of the troposphere; however, pollution close to the earth's surface within the boundary layer is of most concern because of the relatively high concentrations resulting from sources at the surface (see Section 3.2.5, beginning on page 28).

Atmospheric pollutant concentrations depend mainly on the total mass of pollution emitted into the atmosphere, together with the atmospheric conditions that affect its fate and transport. Obviously, air pollution has many and varied sources, including cars, smokestacks, and other industrial inputs into the atmosphere as well as wind erosion of soil. Large emissions from both anthropogenic and natural sources over long periods enhance concentrations, as do the chemical and physical properties of these pollutants. For example, when nitrogen oxides and hydrocarbons in car exhaust are emitted into warm, sunlit air, they readily form ozone molecules (O_3). Similarly, the solubility of a pollutant affects how efficiently it is removed by rain.

Atmospheric conditions have a major effect upon pollutants once these pollutants are emitted to (*e.g.*, nitrogen oxides from car exhaust) or formed within (*e.g.*, O_3) the atmosphere. Pollution dispersal is controlled by atmospheric motion, which is affected by wind, stability, and the vertical temperature variation within the boundary layer. Stability, in turn, influences both air turbulence and the depth at which mixing of polluted air takes place.

Wind determines the horizontal movement of pollution in the atmosphere. Pollution emitted from a point source, such as a smokestack, is generally dispersed downwind in the form of a "**plume**." Wind speed establishes the rate at which the plume contents are transported. Strong winds flowing over rough land surfaces enhance mixing of air by producing shear stress (mechanical mixing) much like that created when an electric fan circulates air in a room. Also, wind direction establishes the path followed by the pollution.

Once present in the atmospheric boundary layer, a pollutant may undergo a series of complex transformations leading to new pollutants, such as O_3. Also, the removal of pollutants from air by rain and snow, gravity, or by surface deposition is influenced by boundary-layer conditions. These removal processes, in turn, are also affected by the type and roughness of the underlying ground surface.

Even when emissions are relatively constant, pollutant concentrations can quickly change owing to variations in atmospheric conditions. When atmospheric conditions are stable, relatively low emissions can cause buildup of pollution to hazardous levels—a situation that occurs during radiation inversions at night. In contrast, unstable conditions, such as wind and rain, may effectively dilute pollution to relatively "safe" concentrations despite a fairly high rate of emissions.

Air pollution, which is of major public concern, is currently the object of extensive scientific research. Its effects on life, including human health, productivity, and property, are not yet fully understood, even though exposure to high levels of pollution is a daily experience for many people. The cost of such pollution, whether expressed in terms of direct biological consequences or in terms of economic impact, is enormous. Worldwide, urban air pollution affects nearly a billion people, exposing them to possible health hazards. In the United States alone, billions of dollars are spent annually to prevent, control, and clean up air pollution; other developed nations are incurring similar costs. The United Nations considers air pollution to be a major global problem.

Most commonly, air pollution can and does harm life. It harms the human respiratory system. Emphysema, asthma, and other respiratory illnesses may result from or be aggravated by chronic exposure to

Figure 12-1 Atmospheric deposition from urban air pollution is contributing to long-term degradation of marble monuments, such as this Roman-era temple in Athens, Greece. Photograph printed by permission; copyright © 1995 by Jeffrey W. Brendecke. All rights reserved.

certain pollutants, such as O_3. Vegetation, too, can be harmed by uptake of pollutants through the leaf stomates or by deposition of pollutants on the leaf surfaces. Sufficiently high concentrations of sulfur dioxide, for example, may cause leaf lesions in susceptible plants. Chronic exposure to relatively low levels of pollution can harm plants by reducing their resistance to disease and insect predators.

Air pollution can also damage property. It can erode the exterior surfaces of buildings, particularly those constructed of limestone materials that react with acids in precipitation (see Figure 12-1). Further evidence of the deleterious effects of air pollution on property can be seen in the damaged paint finishes on cars regularly parked downwind from ore smelters.

12.2 Sources of Air Pollution

Air pollution is not a new problem. Lead in Swedish lake sediments indicates that air pollution produced from lead mining and silver production in ancient Greece and Rome affected air quality throughout Europe. Early written accounts of air pollution refer mainly to smoke from burning wood and coal. For example, in the thirteenth century, King Edward I of England prohibited the use of sea coal, the burning of which produced large amounts of soot and sulfur dioxide (SO_2) in the atmosphere over London (see Figure 23-3 on page 369 for a contemporary example). The industrial revolution increased pollution so markedly that air quality deteriorated significantly in Europe and North America. By the mid-nineteenth century, many cities in the United States and Europe were experiencing the consequences of air pollution. By the early twentieth century, the term "smog" was coined to describe the adverse combination of smoke and fog in London. In Los Angeles, photochemical smog alerts became common by the mid-1940s. The first major air pollution disaster in the United States occurred in 1948, when approximately twenty lives were lost as a result of industrial pollutants trapped in very stable air over Donora, Pennsylvania, in the Monongahela River Valley. In December 1952, stagnant air and coal burning caused severe smog conditions in London that ultimately took the lives of nearly 4000 people.

Virtually all metropolitan areas are affected by air pollution, especially those situated in valleys surrounded by mountains (*e.g.*, Mexico City) or along coastal mountain ranges (*e.g.*, Los Angeles). But even unpopulated areas far from cities may be affected by long-range transport of pollution either from urban areas or from such rural sources as ore smelters or coal-burning power plants. For example, pollution from a coal-burning power plant in the Four Corners area of Arizona may be reducing visibility in Grand Canyon National Park 400 kilometers west of the plant.

Most of the air we breathe is elemental oxygen (O_2) and nitrogen (N_2). About 1% is composed of naturally occurring constituents such as carbon dioxide (CO_2) and water vapor. A small part of this 1% may, however, be air pollutants, including gases and **particulate matter** (*i.e.*, suspended aerosols composed of solids and liquids). Anthropogenic air pollution enters the atmosphere from both fixed and mobile sources. Fixed sources include factories, electrical power plants, ore smelters, and farms, while mobile sources include all forms of transportation that burn fossil fuels. Mobile sources account for nearly 50% of the pollutants emitted to the atmosphere in the United States (see Figure 12-2a). Fuel combustion from stationary sources accounts for nearly 33%, and industrial processes account for about 16% of emissions in the United States. Natural sources of air pollution include winds eroding dust from cultivated farm fields, smoke from forest fires (Figure 12-3 on page 174), and volcanic ash that is emitted into the troposphere and stratosphere.

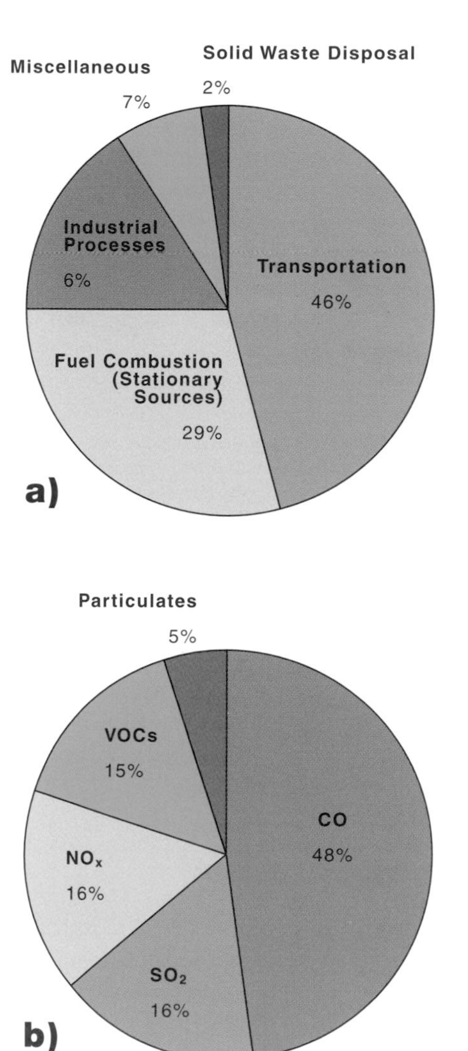

Figure 12-2 (a) Source contributions to air pollution in the United States; (b) primary pollutant emissions in the United States. Reprinted by permission from page 298 of *Essentials of Meteorology: An Invitation to the Atmosphere*, by C. Donald Ahrens; copyright © 1993 by West Publishing Company. All rights reserved.

There are many types of air pollutants. Some gases, such as CO_2, although produced by burning fossil fuels, are generally not considered pollutants be-

Figure 12-3 Wildland fires are a natural source of atmospheric pollution. The "Rincon Fire" in mountains east of Tucson, Arizona. Photograph: J.W. Brendecke.

cause they are essential to plant life. Many pollutants, such as dust particles, exist naturally in the atmosphere and become hazardous only when their concentrations exceed air-quality standards set by such regulatory agencies as the U.S. Environmental Protection Agency (EPA). The EPA classifies air pollutants according to two broad categories: primary and secondary air pollutants.

12.2.1 Primary Pollutants

Primary air pollutants enter the atmosphere directly from various sources. The EPA designates five types of primary air pollutants for regulatory purposes: carbon monoxide, hydrocarbons, particulate matter, sulfur dioxide, and nitrogen oxides. Each type is described below.

12.2.1.1 Carbon Monoxide

Carbon monoxide (CO), which is the major pollutant in urban air, is a product of incomplete combustion of fossil fuels. Carbon monoxide has relatively few natural sources. It is a part of cigarette smoke, but the internal combustion engine is the major source, with about 50% of all CO emissions in the United States originating from cars and trucks. Emissions, therefore, are highest along heavily traveled highways and streets. Of the EPA-designated primary pollutants in the United States, CO emissions currently contribute about 50% of the total emissions (see Figure 12-2b on page 173). Fortunately, CO concentrations are decreasing in the United States owing to improved fuel efficiencies in cars.

Carbon monoxide is highly poisonous to most animals. The EPA standards currently limit human exposure to a 24-hour average of 9 nL L^{-1} or a 1-hour average of 35 nL L^{-1}. [*Note*: The alternative unit of parts per billion (ppb) is also commonly used. See Footnote 1 on page 19 for more information on units for describing gas concentrations.] When inhaled, CO reduces the ability of blood hemoglobin to attach oxygen. Although relatively stable, it is short-lived in the atmosphere because it is quickly oxidized to CO_2 by reaction with hydroxide radicals. Some atmospheric CO may be removed by soil microbes. In order to increase the oxidation of CO to CO_2 during fuel combustion, some cities require the use of oxygenated gasoline containing ethanol or other additives during winter months.

12.2.1.2 Hydrocarbons

Hydrocarbons (HCs), or volatile organic carbons (VOCs), are compounds composed of hydrogen and carbon. **Methane** (CH_4), the most abundant hydrocarbon in the atmosphere, is an active greenhouse gas. Volatile organics include the **nonmethane hydrocarbons** (NMHCs), such as benzene, and their derivatives, such as formaldehyde. Some of these compounds (*e.g.*, benzene) are carcinogenic, and some relatively reactive HCs contribute to ozone production in photochemical smog.

Hydrocarbons are produced naturally from decomposition of organic matter and by certain types of plants (*e.g.*, pine trees, creosote bushes). In fact, HCs emitted from vegetation may be a major factor in smog formation in some cities, particularly those near forested areas of the southeastern United States. A large proportion of HCs and NMHCs is generated by human activity. Some NMHCs, including formaldehyde, are readily emitted from indoor sources, such as newly manufactured carpeting. Hydrocarbons are also emitted into the atmosphere by fossil-fuel combustion and by evaporation of gasoline during fueling of cars. To mitigate this latter source, some cities in southern California require that service-station gasoline pumps be fitted with a special trap to collect HC vapors emitted during fueling of vehicles. Because transportation is the primary source of HCs, concentrations tend to be highest near heavily traveled roadways.

12.2.1.3 Particulate Matter

The category of particulate matter comprises solid particles or liquid droplets (aerosols) small enough to remain suspended in air. Such particles have no general chemical composition and may, in fact, be very complex. Examples include soot, smoke, dust, asbestos fibers, and pesticides, as well as some metals (including Hg, Fe, Cu, and Pb). We can characterize particulate matter by size. Particles whose diameters are 10 μm or larger generally settle out of the atmosphere in less than a day, whereas particles whose diameters are 1 μm or less can remain suspended in air for weeks. Smaller particulate matter, whose particles are 10 μm or less, have come to be known as PM_{10}.

The effects of particulates in the air are various. Some particulates, especially those containing sulfur compounds, are emitted by volcanoes. These can reach the stratosphere, where they may significantly alter the radiation and thermal budgets of the atmosphere and thus produce cooler temperatures at the earth's surface. Tropospheric particulates may cause or exacerbate human respiratory illnesses. Especially harmful to the human respiratory system is the fraction of mid-sized particles, PM_{10}. In large cities particulates also reduce visibility. In the United States about 40% of particulates come from industrial processes, and about 17% come from highway vehicles.

12.2.1.4 Sulfur Dioxide

About 67% of **sulfur dioxide** (SO_2) emissions come from burning sulfur-containing fossil fuels, such as coal, which may contain up to 6% sulfur. Ore smelters and oil refineries also emit significant amounts of SO_2. At relatively high concentrations, SO_2 causes severe respiratory problems. Sulfur dioxide is also a source of acid rain, which is produced when SO_2 combines with water droplets to form sulfuric acid (H_2SO_4). At sufficiently high concentrations, SO_2 exposure is harmful to susceptible plant tissue. Sulfur dioxide and other tropospheric aerosols containing sulfur are believed to affect the radiation balance of the atmosphere, which may cause cooling in certain regions.

12.2.1.5 Nitrogen Oxides

Nitrogen oxides (NO_x stands for an indeterminate mixture of NO and NO_2) are formed mainly from N_2 and O_2 during high-temperature combustion of fuel in cars. Catalytic converters are used to reduce emissions. Nevertheless, NO_x causes the reddish-brown haze in city air, which contributes to heart and lung problems and may be carcinogenic. Nitrogen oxides also contribute to acid rain because they combine with water to produce nitric acid (HNO_3) and other acids. Natural sources of nitrogen oxides include certain soil bacteria.

12.2.2 Secondary Pollutants

The second EPA category of air pollutants is known as "**secondary air pollutants**", which are formed during chemical reactions between primary air pollutants and other atmospheric constituents, such as water vapor. Generally, these reactions must occur in sunlight; thus, they ultimately produce **photochemical smog**. Photochemical smog is most common in the urban areas where solar radiation is very intense.

A simplified set of some of the reactions involved in photochemical smog formation is given here:

$$\begin{array}{lll}
N_2 + O_2 & \longrightarrow 2NO & \textit{(inside engine)} \\
2NO + O_2 & \longrightarrow 2NO_2 & \textit{(in atmosphere)} \\
NO_2 + h\nu & \longrightarrow NO + O & \downarrow \\
O + O_2 & \longrightarrow O_3 & \downarrow \\
NO + O_3 & \longrightarrow NO_2 + O_2 & \downarrow \\
HC + NO + O_2 & \longrightarrow NO_2 + PAN & \downarrow
\end{array} \qquad (12\text{-}1)$$

As indicated by the reactions in Eq. (12-1), photochemical smog is composed mainly of O_3, **peroxyacetyl nitrate (PAN)**, and other oxidants. Ozone formation is closely tied to weather conditions. Favorable conditions for O_3 formation include air temperatures exceeding 32°C, low winds, intense radiation, and low precipitation. Unfortunately, many major U.S. cities exceed the current federal air-quality standard for O_3 (an average O_3 concentration

Figure 12-4 Hydrocarbons interact with nitrogen oxides under the influence of ultraviolet light, resulting in photochemical smog. In urban centers, such as Los Angeles, pictured above, atmospheric pollutants can concentrate and pose severe health hazards. Photograph printed by permission; copyright © 1995 by Alyson B. Brendecke. All rights reserved.

>120 nL L^{-1} for 1 hour 1 day per year averaged over a 3-year period).

As the reactions indicate, HCs are necessary for ozone buildup in the atmosphere. In the absence of HCs, solar UV (photons, $h\nu$) breaks down the NO_2 into NO and O. Next the O atom combines with O_2 to form O_3, which then combines with the NO to reform NO_2 and O_2. Ozone would not accumulate in the atmosphere if it weren't for the fact that HCs disrupt the reaction cycle by reacting with NO to form more NO_2. Hydrocarbons from car emissions and other sources, therefore, play an important part in O_3 formation in urban environments. But not all O_3 in the lower atmosphere results from human activities; natural sources include lightning and diffusion of some O_3 downward from the stratosphere.

In most western U.S. cities, photochemical smog is often referred to as **brown air** (O_3 + PAN + NO_x). Industrial eastern and midwestern U.S. cities generally receive less intense sunlight than western cities; thus, smog in those cities is sometimes referred to as **gray air** because of smoke and SO_2 emanating from burning coal.

Ozone may be either hazardous or beneficial, depending largely on where it is. For example, it is hazardous as an oxidant in smog, but in the O_3 layer it is beneficial because it absorbs UV radiation. Smog ozone reduces the normal functioning of lungs because it inflames the cells that line the respiratory tract. Other health effects include an increase in asthma attacks, increased risk of infection, and reductions in heart and circulatory functions.

Smog O_3 may damage plant as well as animal life. In vegetation the main damage occurs in the foliage, with smaller effects on growth and yield. In the United States, it has been implicated in the loss of conifer trees near Los Angeles, and is suspected of doing similar damage to trees in the Appalachian Mountains. It may also have a harmful effect on saguaro cacti in the desert downwind from Tucson and Phoenix in Arizona. Several plant species are also very susceptible to PAN in smog, particularly those in the Los Angeles area.

Photochemical smog is generally most severe in the Los Angeles basin of the California coast. Los Angeles has many cars and hence high emissions of NO_x and HCs. At certain times of the year, particularly spring and fall, weather conditions in this area are dominated by subtropical high pressure with clear, calm air—conditions that exacerbate air stagnation. The factors influencing smog formation in the Los Angeles basin can be summarized as follows:

1. Numerous sources of primary pollutants.
2. Inversions that inhibit turbulent mixing of air.
3. Few clouds, which permit high UV intensity.
4. Light winds unable to disperse pollutants.
5. Complex coastal mountain terrain to accumulate pollutants.

Ozone and NO_x pollution in the troposphere is not limited to urban areas. In the mid-1990s, the EPA reported that O_3 and NO_x concentrations were increasing in rural areas in the southeastern and midwestern United States. Most of this increase is probably attributable to upwind urban sources; however, some rural soils may be a greater NO_x source than is fossil-fuel combustion. In fact, some estimates indicate that soils may emit as much as 40% of the total amount of NO_x emitted into the atmosphere. This percentage is very uncertain as measurements of NO_x fluxes from soils are scanty. Soil NO_x fluxes may also be highly spatially and temporally variable.

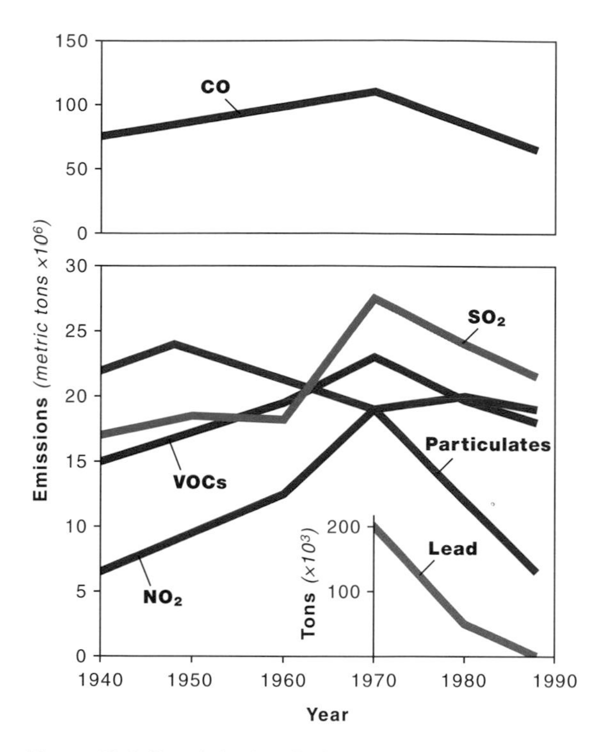

Figure 12-5 Trends in air pollution in the United States. Reprinted by permission from page 302 of *Essentials of Meteorology: An Invitation to the Atmosphere*, by C. Donald Ahrens; copyright © 1993 by West Publishing Company. All rights reserved.

Tropospheric NO_x strongly controls the concentrations of oxidants such as OH and O_3, which may affect the health of about a quarter of the U.S. population. As in urban areas, O_3 in the rural atmosphere is controlled by reactions involving NO_x, HCs, OH, and other tropospheric species. The study of the production and destruction processes of O_3 in the rural troposphere is currently an area of active research by the EPA and other government agencies. Many questions remain unanswered concerning the reasons why O_3 concentrations continue to be high in both urban and rural areas despite recent efforts to curb emissions of the O_3 precursors, NO_x and HCs.

12.3 Pollution Trends in the United States

Concentrations of nearly all types of air pollutants have generally declined or held steady in the United States since about 1970 (Figure 12-5). This decline is mostly attributable to general compliance with the federal air quality regulations set forth in response to the U.S. Clean Air Act of 1970. Although air quality is improving overall, many specific urban areas fail to meet the air-quality standards set for some pollutants. Poor air quality is estimated to affect the lives of about 100 million people in the United States alone.

Despite the fact that transportation continues to be the major source of pollution in the United States, the proportion of its contribution is diminishing. While the number of cars is increasing in most urban areas, fuel efficiencies have increased and pollutant emissions per vehicle are declining owing to improvements in technology such as catalytic converters and other pollution control devices. We can find evidence of improved air quality in the marked decline of atmospheric lead (Pb) concentrations since 1970. Atmospheric lead comes mainly from the burning of lead-containing gasoline in cars and trucks. Thus the introduction of unleaded gasoline was a significant factor in this decline. Now required for cars in the United States because of environmental health concerns, unleaded gasoline is also used because leaded fuels deactivate catalytic convertors.

It is generally recognized that reducing air pollution through control of emissions at the source is the best approach, which is the goal of the EPA and other regulatory agencies. Total control of pollutant emissions is certainly not feasible for various economic and technological reasons, but efforts at reducing emissions are helping to improve air quality in most locations.

12.4 Weather and Pollutants

What happens to pollutants in the atmosphere? The answer depends on several factors. Pollutants are transported by wind and turbulence, and they may undergo chemical transformations before being deposited on the earth's surface. Thus, weather conditions strongly affect the fate of air pollutants.

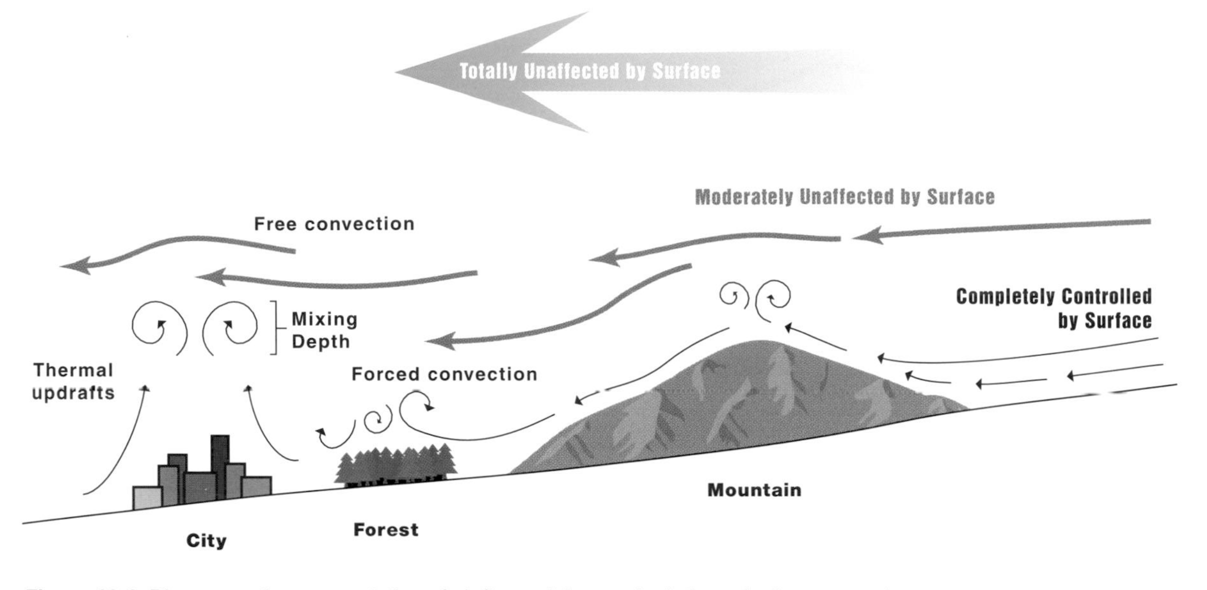

Figure 12-6 Diagrammatic representation of air flow, mixing, and relative velocity over varying terrain as affected by height.

12.4.1 Stability and Inversions

The stability of boundary-layer air (see Figure 3-4 on page 28) largely determines how quickly pollutants are moved upward from their ground sources. Stability is primarily a function of the vertical air temperature gradient relative to the adiabatic lapse rate (see Section 3.2.2, beginning on page 24). Strong instability associated with buoyancy causes efficient air mixing (and pollution dispersal) over a large mixing depth of the boundary layer (from 100 to 1000 m). Good mixing often occurs on warm days when the ground is heated by sunlight. In contrast, pollution is poorly dispersed on days or nights when the atmosphere is stable. At those times, turbulent movement of pollution upward is slow or nearly nonexistent.

We know from Section 3.2.2 that temperature inversions influence atmospheric stability; thus they play an important role in determining the concentrations of air pollutants. The effects of inversions are intensified by limited air drainage out of enclosed valleys, as is the case in Los Angeles (Figure 12-4 on page 176) and Mexico City (Figure 12-10 on page 183). Various processes may generate inversions, including surface cooling caused by loss of infrared radiation or by evaporation, atmospheric subsidence, and topographic effects.

Caused mainly by infrared radiation emission from the surface to the sky, ground-surface cooling generally occurs during clear, calm nights, with inversion heights extending about 100 m above the ground. Such inversions commonly occur throughout the western United States during fall, winter, and spring when the air is relatively dry and skies are clear. Such conditions readily permit cooling by longwave loss of energy from the ground. Tucson, Arizona, for example, often experiences radiation inversions in the cooler months, so that wintertime pollution problems are exacerbated in the area. Cooling of the ground by evaporation of water from soil and plants may also establish inversions. Evaporative cooling can occur during the day or night, particularly over irrigated fields. This type of inversion may be important in relation to certain agricultural activities, such as the aerial application of pesticides over large irrigated fields. The depth of an evaporatively cooled inversion layer is usually just a few meters.

Warming of the atmosphere by subsidence causes inversions over regions that have semipermanent

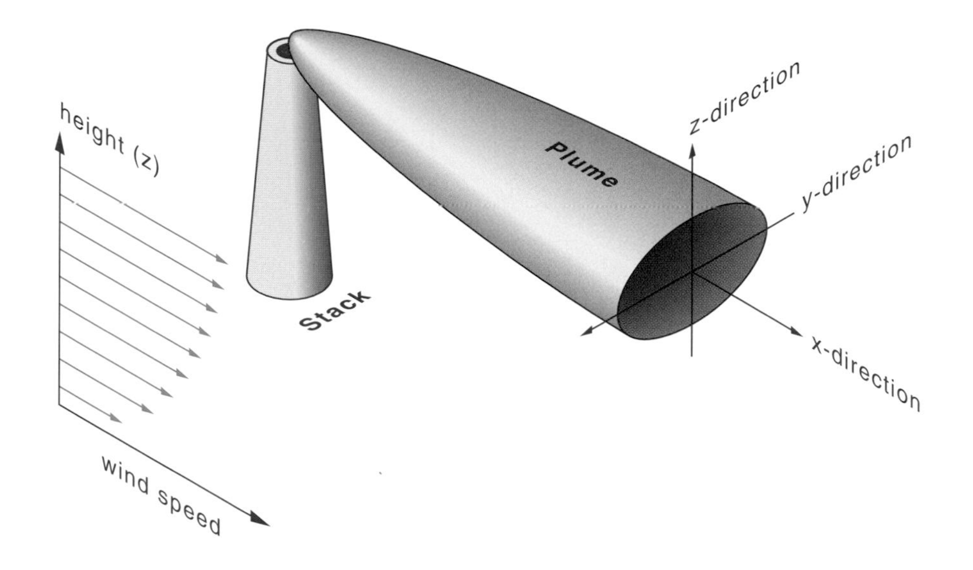

Figure 12-7 Plume pattern resulting from continuous stack emission into a near-neutral stable boundary layer under moderate winds.

high pressure (anti-cyclonic flow), such as the southwestern United States. As air subsides (sinks), it encounters higher pressure and thus is warmed (see Section 3.2.2, beginning on page 24, and Figure 3-3 on page 27). Within regions of high pressure, an inversion height may be several hundred meters above ground; thus the air may be very stable over a large depth of the atmosphere. Because subsidence inversions may last from several days to weeks, the result is highly polluted conditions at ground level. For example, subsidence inversions are a major factor in reducing air quality in the Los Angeles basin (see Figure 12-4 on page 176).

Inversions associated with topography result from adiabatic warming of air as it flows downslope over mountainous terrain. These inversions may exacerbate air pollution problems in populated, mountainous areas such as Denver, Colorado.

12.4.2 Wind and Turbulence in Relation to Air Pollution

Wind affects turbulence near the ground, thus affecting the dispersion of pollutants released into the air. Turbulence (largely up-and-down motion of air) is generated in part by air flow over rough ground. The greater the wind speed, the greater the turbulence, and hence the greater the dispersion of pollutants that are near the ground (see Figure 12-6).

We can visualize the dispersion of pollutants in air by looking at the familiar cloud or "plume" of pollution emitted continuously by a smokestack (see Figure 12-7). As the plume contents are carried away from the stack by the wind, the size of the plume increases owing to dispersion. Because of dispersion, the pollutant concentration within the plume decreases with increasing distance from the source.

Dispersion of pollution downwind from a smokestack is affected by the roughness of the ground surface. Because of friction between the atmosphere and the ground, wind speed is slowed markedly near the ground. If the surface is relatively rough, as it is when trees and buildings are present, the air flow tends to be turbulent and the increase of wind speed with increasing height is relatively small. Greater surface roughness increases turbulence, which helps disperse pollutants. Air flow over a smooth surface, such as a large mowed lawn, tends to be less turbulent and the wind speed variation with height is relatively large (Figure 12-6).

$$\chi_{(x,\,y,\,z,\,H)} = \frac{Q}{2\pi\,\sigma_y\,\sigma_z\,\bar{u}}\exp\left(-\frac{y^2}{2\sigma_y^2}\right)\left[\exp\left(-\frac{(z-H)^2}{2\sigma_z^2}\right)+\exp\left(-\frac{(z+H)^2}{2\sigma_z^2}\right)\right] \tag{12-2}$$

The cone-shaped plume in Figure 12-7 on page 179 illustrates the pattern of pollutant dispersal downwind of a point source. Several factors affect the plume, including the effective height (H) of emission, which is a measure of how high the pollutants are emitted into the atmosphere directly above the source. The height is dependent upon source characteristics and atmospheric conditions. Generally, a tall smokestack produces relatively low ground-level pollutant concentrations, because turbulence tends to dilute the pollution before reaching the ground. Driven by buoyancy, fast-moving pollutants are initially transported high up into the atmospheric boundary layer because they are warmer than the surrounding air. But as the pollutants cool and merge with the ambient air, the plume begins to move sideways with the wind. Then turbulence caused by the air flow over the surface and by possible instability governs the diffusion of the plume contents.

Usually, turbulence helps mix plume contents uniformly in such a way that the concentration follows a gaussian distribution about the plume's central axis. Mathematically, pollutant concentration χ (kg m^{-3}) at any point in the plume is described by Eq. (12-2) where Q is the rate of emission of pollution from the source (kg s^{-1}), σ_y and σ_z are the horizontal and vertical standard deviations of the pollutant concentration distributions in the y and z directions, and $\bar{u}$ is the mean horizontal wind speed within the plume (m s^{-1}). This model, which is applicable to continuous sources of gases and particulates less than about 10 μm in diameter (larger particles quickly settle to the ground), can be used to model plume concentrations over horizontal distances of 10^2 to 10^4 m. With this gaussian plume model it is assumed that no deposition of plume contents to the ground surface takes place. In fact, it is assumed that plume contents are "reflected" from the ground back to the air. The values of σ in the equation are estimated from any one of several empirical formulas that relate σ to downwind distance (x) and stability conditions. These formulas include the following equations, which were developed by the **Brookhaven National Laboratory (BNL)**.

$$\sigma_y = ax^b \quad \text{and} \quad \sigma_z = cx^d \tag{12-3}$$

where a, b, c, and d are parameters dependent upon stability. [See Hanna *et al.* (1982) for a summary discussion of BNL equations as well as other approaches.] At ground level $z = 0$, and along the plume centerline $y = 0$. Thus, from Eq. (12-2), the concentration can be simply calculated by

$$\begin{aligned}\chi_{(x,\,H)} &= \frac{Q}{\pi\,\sigma_y\,\sigma_z\,\bar{u}}\exp\left(-\frac{H^2}{2\sigma_z^2}\right)\\ &= \frac{Q}{\pi\,ax^b\,cx^d\,\bar{u}}\exp\left(-\frac{H^2}{2(cx^d)^2}\right)\end{aligned} \tag{12-4}$$

One type of plume (shown in Figure 12-7 on page 179) typically occurs under windy conditions with stability conditions at or near neutral. Within such a plume, mixing occurs mainly by frictionally generated turbulence, and pollutant diffusion is nearly equal in all directions (*i.e.*, the σ values are nearly equal and the plume spreads out in the familiar cone pattern, known as **coning**). Coning can occur day or night and is often seen during cloudy and windy conditions. Depending upon effective source height and atmospheric conditions, the plume may reach the ground close to the source. Using Eqs. (12-3) and (12-4), we can estimate the ground-level ($z = 0$) concentration of the plume composed of a pollutant, say SO_2, emitted into the atmosphere at a known effective height. Suppose we have the following: $Q = 0.5$ kg s^{-1}, $H = 25$ m; $\bar{u} = 2$ m s^{-1}; near neutral stability, and BNL parameters $a = 0.32$, $b = 0.78$, $c = 0.22$, and $d = 0.78$. Then the ground-level SO_2 concentration along the plume centerline at an arbitrary distance of $x = 500$ m from the source will be 4.7×10^{-5} kg m^{-3} (or 47 mg m^{-3}).

Plumes may change because of changes in the wind velocity and boundary layer stability. When the atmospheric boundary layer is strongly stable, such as during radiation inversions at night or during subsidence inversions, a **fanning** pattern may be evident in the plume, as illustrated in Figure 12-8a. Under these conditions, there is almost no vertical motion and the BNL parameters are $a = 0.31$, $b = 0.71$, $c = 0.06$, and $d = 0.71$. Lack of vertical motion thus effectively forces the plume into a relatively narrow layer, while changes in wind direction may spread the plume out laterally, resulting in a V or fan pattern—hence the term. A constant wind direction, however, forces the plume into a tightly closed fan pattern, which follows a relatively straight and narrow path. Over flat terrain the plume in Figure 12-8a may be unchanged for very long distances. If there is no vertical air movement, ground-level concentrations downwind of a tall smokestack can be nearly zero. However, if the source is close to the ground (*i.e.*, H is small), or if changes in topography cause the plume to intercept the ground, the ground-level concentrations can be very large.

By midmorning, surface heating by solar radiation typically begins to break down the inversion developed during the previous night, as illustrated in Figure 12-8b. Unstable conditions develop near the ground, resulting in vertical mixing of the air. With moderately unstable conditions, $a = 0.36$, $b = 0.86$, $c = 0.22$, and $d = 0.86$. In this situation, pollution is transported downward toward ground level. Stable conditions above, however, limit dispersion of pollutants upward. Thus the remaining inversion effectively puts a "lid" over the ground-level pollution. This situation is known as **fumigation** and generally

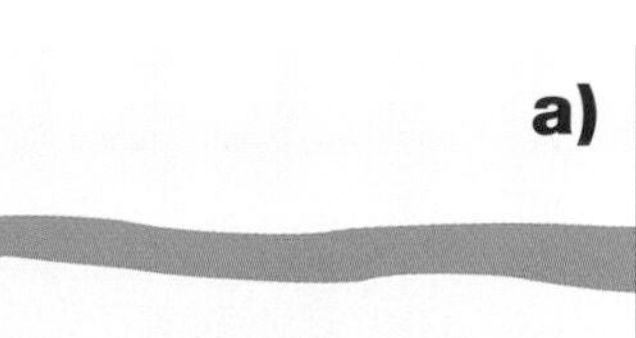

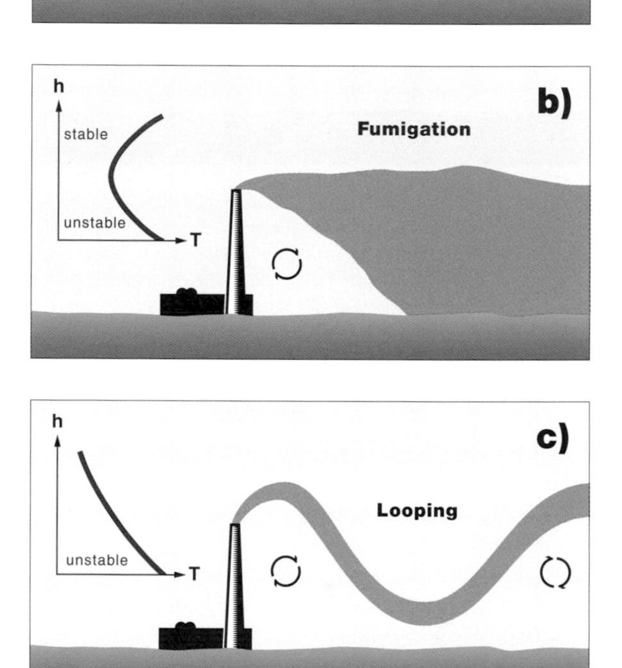

Figure 12-8 Effect of atmospheric stability upon resultant stack plume pattern during (a) inversion (fanning pattern) (b) dissipation of inversion near ground (fumigation pattern) and (c) lapse conditions (looping pattern). Adapted by permission from page 305 of *Essentials of Meteorology: An Invitation to the Atmosphere*, by C. Donald Ahrens; copyright © 1993 by West Publishing Company. All rights reserved.

lasts for periods of an hour or less. Fumigation is highly conducive to enhanced ground-level pollutant concentrations.

By early afternoon, lapse conditions (*i.e.*, negative vertical temperature gradient) generally become fully established within the boundary layer due to strong surface heating by the sun. During much of the afternoon, air motion mainly exhibits the large turbulent eddies associated with buoyancy. These eddies are generally larger than the plume diameter,

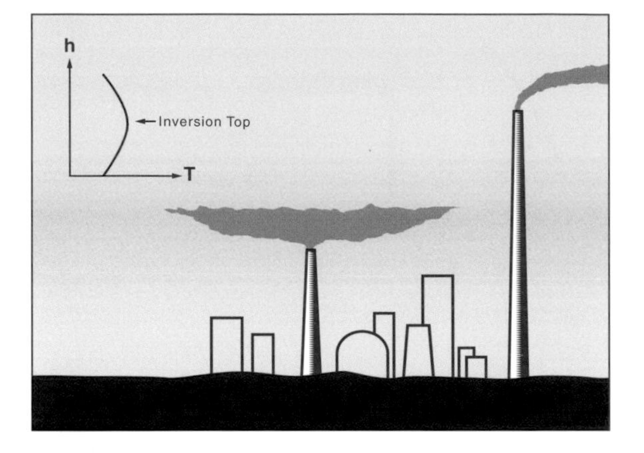

Figure 12-9 Plume pattern resulting from stack emissions above and below the inversion height. Adapted by permission from page 304 of *Essentials of Meteorology: An Invitation to the Atmosphere*, by C. Donald Ahrens; copyright © 1993 by West Publishing Company. All rights reserved.

and thus transport the plume upward and downward in a sinusoidal path or **looping** pattern, as illustrated in Figure 12-8c. The loops are carried with the overall wind pattern and generally increase in size with increasing distance downwind from the source. The motion may bring the plume contents to ground level quite close to the source. Because of turbulence, however, the plume eventually becomes dispersed at relatively large distances.

By early evening, a radiation inversion often rebuilds from ground level upward. Stable conditions near the ground inhibit transport of plume contents downward, but unstable air aloft (above the inversion height) allows dispersal upward. This upward transport, known as **lofting**, is highly favorable for dispersing pollutants. Lofting is most effective when the source is above the inversion height. Figure 12-9 illustrates how the plume contents emitted above the inversion height are dispersed upward. It also shows, however, that plume contents emitted below the inversion height are essentially trapped in a fan-type plume configuration.

Topography downwind from pollution sources affects air quality, especially in mountainous areas. For example, air drainage into relatively enclosed valleys during winter and/or inversion conditions can cause accumulation of pollutants within the valleys. Thus urban areas in valleys with restricted air flow are particularly prone to high pollution levels. In addition, in coastal areas, air flow from the ocean (sea breezes) can be blocked or channeled by mountain ranges. This situation is common in the Los Angeles basin, which is surrounded by mountains that restrict air flow from the Pacific Ocean. Thus dispersal of pollutants from sources in the basin is inhibited.

12.4.3 Pollutant Transformation and Removal

As pollutants move with the wind, chemical reactions often occur between the pollutants and other atmospheric chemical species. Although the pathways and rates of many of these chemical reactions are poorly understood, they are an important factor affecting the fates of many air pollutants.

Most pollutants, such as CO, remain in the atmosphere for relatively brief periods, lasting only a few days or weeks. Thus, if emissions were completely curtailed, the lower atmosphere would quickly lose nearly all of its pollutants. However, some pollutants—volcanic ash and sulfur-containing aerosols, for example—emitted high into the stratosphere could remain there for months before settling back to the surface. These long-lasting upper-atmospheric pollutants can alter the earth's climate, as evidenced by lower air temperatures resulting from such volcanic eruptions as that of Mt. Pinatubo in the Philippines. In addition, synthetic **chlorofluorocarbon** (CFC) compounds can remain in the atmosphere for many years before they break down.

Pollution leaves the atmosphere by gravitational settling, dry deposition, and wet deposition.

Gravitational settling: Gravitational settling removes most particles whose diameters are greater than about 1 μm. Particles less than 1 μm in diameter are often small enough to stay in the atmosphere for long periods. Particles greater than about 10 μm in diameter quickly settle out.

Dry deposition: Dry deposition is a mass-transfer process that results in adsorption of gaseous pollutants by plants and soil. Dry deposition to plants is dependent upon uptake of the pollution through sto-

Figure 12-10 Increased industrialization and subsequent affluence in developing countries has dramatically escalated carbon dioxide emissions into the atmosphere. Above, traffic in Mexico City. Photograph printed by permission; copyright © 1995 by Jeffrey W. Brendecke. All rights reserved.

matal openings in plant leaves and upon turbulent transport in the air. Dry deposition to bare soil involves not only turbulent transport of pollutants in air above the soil, but also soil microorganisms that take up such pollutants as CO.

Wet deposition: Rain is a very effective remover of gases and small particulates. Raindrops increase in size as they fall toward the ground, and thus they increasingly capture more pollutants. Raindrops, in effect, "sweep up" pollution as it falls through the air. The ability of the rain to remove pollutants depends upon the rainfall intensity, the size and electrical properties of the drops, and the solubility of the polluting species.

> *Note*: Rain is naturally acidic (pH 5–6) because it absorbs atmospheric carbon dioxide to form carbonic acid. Downwind from major sources of the oxides of nitrogen (NO_x) and sulfur (SO_2), rain becomes even more acidic as it absorbs these oxides and forms nitric and sulfuric acids. The effects of this pollutant absorption process are particularly evident in the low pH of rain downwind of the major industrial centers of North America and Europe. The low pH of rain has been implicated as a cause of pH decreases in lake and forest ecosystems.

12.5 Major Air Pollution Issues

12.5.1 Global Warming

Carbon dioxide is generally not considered to be an air pollutant because it is not hazardous to human health at ambient atmospheric concentrations; moreover, it is essential for carbon fixation by plants. It is, however, an important greenhouse gas and a major by-product of fossil-fuel burning, which steadily increases the atmospheric concentration of carbon dioxide (Figure 12-10).

Numerical climate model predictions indicate that increased warming by greenhouse gases (mainly CO_2) may add about 1°C to the global mean temperature by the year 2025, and about 3°C by the end of the twenty-first century. Although there is much uncertainty in the predicted warming, many scientists believe a temperature increase is inevitable. The possible warming could have major long-term consequences for life on earth.

Carbon dioxide is by far the most abundant and important atmospheric trace gas contributing to the natural **greenhouse effect** (see also Section 3.1, beginning on page 19, and Section 3.2.4, beginning on page 26). It is released to the atmosphere by various processes including deforestation and land clearing, fossil-fuel combustion, and respiration from living organisms. Carbon dioxide is readily absorbed by water, with warm water absorbing more than cold water, so it is removed from the atmosphere by the oceans and other bodies of water. Photosynthesis by land and water plants (phytoplankton) also removes significant amounts of CO_2 from the atmosphere. Removal by plants is particularly apparent during the summer when average CO_2 concentrations decrease. In addition, large amounts of CO_2 may eventually be fixed as limestone. The atmosphere currently contains about 720 billion metric tons (BMT) of carbon in the form of CO_2, and a 3-BMT excess enters the atmosphere each year. Research indicates that this excess gives rise to about 1.5 $\mu L\ L^{-1}$ mean annual increase in the global concentration of CO_2.

In addition to CO_2, the other main greenhouse gases are CH_4, N_2O, CFCs, and O_3. Water vapor is also an important, but variable, greenhouse species.

All of these species are, as the term "greenhouse" implies, efficient absorbers of longwave radiation. This absorption helps maintain a relatively warm climate on earth. However, because greenhouse gas concentrations continue to increase, it is conceivable that the earth's climate may be altered and become much warmer. Therefore, much scientific research is currently being directed toward improving our understanding of the atmospheric budgets of these trace gases and their role in the greenhouse effect.

Atmospheric CH_4 concentrations have also steadily increased at a rate of about 1% per year in recent decades. This rapid increase is commonly attributed to increased worldwide rice and livestock production. Increased mining of natural gas resources for energy production may also be an important factor.

Synthetic CFCs are also significant contributors to the greenhouse effect. Chlorofluorocarbons are used in refrigerators, air conditioners, foam insulation, and in industrial processes. Very efficient longwave absorbers, they are also involved in depleting stratospheric O_3.

Nitrous oxide (N_2O) is an especially good absorber of longwave radiation; one molecule of N_2O is equivalent to about 200 CO_2 molecules in terms of its ability to absorb longwave radiation. Currently, atmospheric N_2O accounts for only about 5% of the greenhouse effect, but this percentage is expected to increase in coming years. A 25% N_2O increase in atmospheric concentration may, according to numerical model predictions, increase global mean temperature by about 0.1 K. Worse, N_2O has a very long atmospheric lifetime, estimated to be about 150 years, which is far longer than the atmospheric lifetime of any other nitrogen oxide. Thus, the current buildup of N_2O could affect the earth's climate well beyond the twenty-first century.

Global warming is a poorly understood, but much discussed subject. Many scientific, political, and economic questions remain unresolved. Some of these questions include the following:

1. If the predicted warming occurs as a result of increased CO_2 concentrations, how will vegetation distributions and plant growth rates be affected?
2. How quickly will warming occur?
3. How will the oceans affect and be affected by the rate of warming?
4. How will clouds and aerosols affect the warming?
5. How will rainfall patterns change?
6. Will the temperature changes occur uniformly over the land surfaces?
7. How can greenhouse gas emissions be reduced worldwide, yet allow economic development to continue?
8. Are current numerical models of global warming accurate and reliable?

12.5.2 Ozone Depletion

Stratospheric O_3 depletion is another global environmental concern related to pollution. Concern about O_3 first emerged in the early 1970s when modeling studies indicated that a proposed fleet of supersonic transport (SST) aircraft could emit enough NO_x to damage the O_3 layer. The results from the modeling studies helped put an end to plans to build the fleet, but such considerations remain major factors in plans for aircraft development.

In the mid-1970s, concern shifted to the possible O_3-depletion effects of manufactured CFCs used as refrigerants, propellants, cleaning compounds, and foam insulation. Intensive study of the effects of CFCs on stratospheric O_3 led to a 1979 U.S. ban of the use of CFC propellants in aerosol spray cans. Growing worldwide use of CFCs, together with evidence of CFC-induced decline of stratospheric O_3 concentrations over Antarctica, convinced 24 industrialized nations to sign the Montreal Protocol (1987 and 1990), which calls for a complete phaseout of the production of CFCs by the year 2000.

Stratospheric O_3 absorbs UV light, decreasing the amount of UV striking living organisms on the earth's surface. Satellite and ground-based measurements have shown that there is a temporary decrease in O_3 concentrations over Antarctica each year. In recent years, O_3 depletion has engendered serious concerns about the causes and possible ecological

Table 12-1 Chemical species that are believed to catalyze the destruction of O_3 molecules in the atmosphere.

Cycle	X	XO
NO_x	NO	NO_2
Water	HO·	HO_2·
CFC	Cl·	ClO·

and human-health consequences if this trend continues. In humans, increased UV would probably increase the incidence of skin cancer, including melanoma. Other organisms are also vulnerable to UV; phytoplankton, for example, has declined by 6–12% in areas near Antarctica. The decline in this one-celled organism is thought to be due to increased amounts of UV that are reaching surface waters.

The chemical pathways leading to the formation of stratospheric O_3 start with the photodissociation of molecular oxygen (O_2) by solar UV radiation (photons of energy $h\nu$, where h is Planck's constant and ν is the frequency). The UV photon splits O_2 into two oxygen atoms (O), each of which recombines with undissociated O_2 (in the presence of another chemical species, M) to form two O_3 molecules. These two reactions, which result in a net formation of O_3 are given here:

$$\begin{array}{lll} & O_2 + h\nu & \longrightarrow O + O \\ & 2O_2 + 2O + M & \longrightarrow 2O_3 + M \\ \hline \text{Net Reaction:} & 3O_2 + h\nu & \longrightarrow 2O_3 \end{array} \quad (12\text{-}5)$$

The two O_3 molecules quickly convert back to molecular oxygen via

$$\begin{array}{lll} & O_3 + h\nu & \longrightarrow O + O_2 \\ & O_3 + O & \longrightarrow 2O_2 \\ \hline \text{Net Reaction:} & 2O_3 + h\nu & \longrightarrow 3O_2 \end{array} \quad (12\text{-}6)$$

The process of production and loss of the O_3 molecules by photodissociation is very important because, overall, it helps prevent harmful UV from reaching the earth's surface. These production/destruction schemes indicate that the chemistry of stratospheric O_3 would be straightforward if there were no other reactive chemical species in the stratosphere. However, other chemicals, such as CFCs and NO_x, are present and play an important destructive role. This is indicated by the following general catalytic cycle:

$$\begin{array}{lll} & X + O_3 & \longrightarrow XO + O_2 \\ & O_3 + h\nu & \longrightarrow O + O_2 \\ & XO + O & \longrightarrow X + O_2 \\ \hline \text{Net Reaction:} & 2O_3 + h\nu & \longrightarrow 3O_2 \end{array} \quad (12\text{-}7)$$

where X and XO represent the compounds or free radicals that catalyze the destruction of O_3 molecules. Mainly NO_x, water vapor, and CFCs, these species are summarized in Table 12-1.

Of these catalysts, CFCs are entirely anthropogenic, whereas nitrogen oxides come from both natural and synthetic sources. Stratospheric water vapor comes mainly from natural processes. In addition to the three main catalysts, other chemicals may play a role in controlling stratospheric O_3 levels. For example, recent evidence indicates that methyl bromide, which is used as a soil fumigant, may reach the stratosphere, where it can undergo a catalytic reaction sequence with O_3 similar to those of the three main chemical species.

The catalytic reactions do not destroy all the O_3 present in the stratosphere. The reason they don't is that reactions also occur between the catalysts, and these result in chemicals that do not deplete O_3. Some of the chemicals eventually return to the earth's surface (*e.g.*, HNO_3 in rain).

Further aspects of each of these catalytic cycles are described in Graedel and Crutzen (1993) and are briefly discussed in the following paragraphs.

12.5.2.1 NO_x/O_3 Destruction Cycle

The nitrogen oxides in the stratosphere come mainly from photodissociation of nitrous oxide, which originates mostly from microbial processes at the earth's surface. Nitrous oxide is also a major greenhouse gas. It is produced mainly within moist soils by microbial denitrification of nitrate fertilizer, but it can also be biologically produced in oceans. Since the 1970s there has been concern that increasing agricultural use of nitrogen fertilizers could increase the amount of nitrous oxide reaching the stratosphere, ultimately depleting O_3. Measurements of atmospheric N_2O indicate that its concentration is increasing by about 0.25% per year. A 25% increase in N_2O by the late twenty-first century could reduce total stratospheric O_3 by 3–4%, which could increase the incidence of skin cancer by 2–10%.

Nitrous oxide is not known to be lost within the troposphere; however, it is converted to NO in the stratosphere mainly by the following reactions:

$$\begin{aligned} N_2O + h\nu &\longrightarrow N_2 + O(^1D) \\ N_2O + O(^1D) &\longrightarrow 2NO \end{aligned} \qquad (12\text{-}8)$$

The two NO molecules formed initiate the O_3 destruction reactions described previously. Note that $O(^1D)$ in Eq. (12-8) denotes atomic oxygen in an electronically excited state.

12.5.2.2 H_2O/O_3 Destruction Cycle

The stratosphere is generally very dry. However, enough water vapor is present to react with electronically excited atomic oxygen to produce the free radical HO· via

$$H_2O + O(^1D) \longrightarrow 2HO\cdot \qquad (12\text{-}9)$$

The catalytic water cycle has less influence upon O_3 concentrations than do the other reaction cycles, but it can be significant when sufficient water vapor is present.

12.5.2.3 CFC/O_3 Destruction Cycle

Chlorofluorocarbons (*e.g.*, $CFCl_3$ and CF_2Cl_2) are relatively stable in the troposphere; but once in the stratosphere, they are photodissociated by solar UV. This photodissociation produces the catalysts Cl and ClO, both of which deplete O_3. There is evidence that links the O_3 depletion in the Antarctic region to CFCs and other pollutants that carry chlorine and bromine into the stratosphere. Chlorine monoxide (ClO) has been identified as the chief cause of O_3 depletion in polar regions. Weather patterns and volcanic eruptions may also play a part.

Chlorofluorocarbons are also implicated in possible global warming; because many are extremely efficient absorbers of longwave radiation, they contribute to the earth's greenhouse effect.

References and Recommended Reading

Ahrens C.D. (1993) *Essentials of Meteorology: An Invitation to the Atmosphere.* West Publishing Minneapolis/St. Paul, Minnesota.

Albritton D.L., Monastersky R., Eddy J.A., Hall J.M., and Shea E. (1992) *Our Ozone Shield: Reports to the Nation on Our Changing Planet.* Fall 1992. University Cooperation for Atmospheric Research, Office for Interdisciplinary Studies, Boulder, Colorado.

Dickinson R., Monastersky R., Eddy J., Bryan K., and Matthews S. (1991) *The Climate System: Reports to the Nation on Our Changing Planet.* Winter 1991. University Cooperation for Atmospheric Research, Office for Interdisciplinary Studies, Boulder, Colorado.

Graedel T.E. and Crutzen P.J. (1993) *Atmospheric Change: An Earth System Perspective.* W.H.Freeman, New York.

Hanna S.R., Briggs G.A., and Hosker, Jr. R.P. (1982) *Handbook on Atmospheric Diffusion.* U.S. Department of Energy, Washington, D.C.

Mitchell J.F.B. (1989) The "greenhouse" effect and climate change. *Reviews of Geophysics.* **27**, 115–139.

Oke T.R. (1987) *Boundary Layer Climates.* Routledge, New York.

Schlesinger W.H. (1991) *Biogeochemistry: An Analysis of Global Change.* Academic Press, San Diego, California.

Problems and Questions

1. Describe how surface air temperature inversions form. Why are air-temperature inversions important relative to air pollution in urban areas?
2. What factors affect atmospheric stability? Explain.
3. Describe how atmospheric stability affects the dispersion of SO_2 pollution emitted from a smokestack.
4. Describe the processes that remove air pollution.
5. What is the difference between EPA-designated primary and secondary air pollution? Give an example of each type of pollutant.
6. What is photochemical smog? Explain how it is formed.
7. Explain how O_3 in the stratosphere is beneficial, whereas O_3 in the troposphere is harmful.
8. Explain how anthropogenic chlorofluorocarbons destroy stratospheric O_3.

Turbid river waters, such as these in the Potomac River in Maryland, USA, present technical challenges for the municipal treatment plants that must produce clean, clear drinking water for their customers. Photograph: J.W. Brendecke.

13.1 Surface Freshwater Resources

Freshwater is a scarce and valuable resource—one that is easily contaminated. And, once contaminated to the extent it can be considered "polluted," freshwater quality is difficult and expensive to restore. Thus the study of water pollution has focused primarily on streams and lakes, and most of the scientific tools developed by such regulatory agencies as the U.S. Environmental Protection Agency (EPA) have been applied to protecting water quality in this segment of earth's surface waters.

Fresh surface water accounts for just 1/10,000 of the total water available on the planet; yet this quantity seems immense when the volume is expressed as 125,000 km^3. Such a volume of water could cover the contiguous United States to a depth of about 12 m. On a global scale, this amount is quite constant year to year, being constantly replenished by precipitation of water previously evaporated from the ocean (350,000 km^3) and from land areas (70,000 km^3). Unfortunately, most of the precipitation falls back into the ocean, and only 110,000 km^3 falls on the land. More than half of the 40,000 km^3 of water that doesn't evaporate runs off to the ocean in flood events and is not available for use throughout the year.

Lakes contain almost all of the fresh surface water on the planet. The water in rivers and streams makes up less than one percent of the volume in lakes. This fact alone suggests that lakes require special protection from contamination. Another fact to consider is that it takes many years to replenish lakes owing to the relatively small amount of precipitation that falls on the lake and the small amount of stream water that runs directly into lakes. On average, lake replenishment takes 100 years, whereas the replacement time for water in streams and rivers is 11 days. Thus, if contaminants are distributed throughout the average lake, the incoming water cannot restore the lake to its initial quality for a long time.

The water stored in reservoirs and lakes, together with the water that flows perennially in streams, is subject to heavy stress; because it's used for water supplies, agriculture, industry, and recreation, it can be easily misused. Most cities and industries in the United States discharge wastewaters to streams and rivers, rather than to lakes and reservoirs. Even though the wastewaters are treated, large quantities of contaminants flow downstream on the way to the ocean as the water is used over and over again.

13.2 Marine Water Resources

Oceans contain most of the water of the planet. Yet, even with the phenomenal volume of water in which contaminants may be dispersed, marine resources can be polluted. Using various biological and physical parameters, we usually classify the ocean environment as three components: the coastal zone, the upper mixed layer, and the abyssal ocean. Several regulatory agencies and international organizations share different responsibilities for those components. The coastal zone, which is most susceptible to the day-to-day kinds of contamination found in freshwater lakes and rivers, is often the province of water quality regulatory agencies established by individual nation states. International organizations have traditionally dealt with pollution concerns of the open ocean and its seabed, which roughly includes the other two components. In addition to these physically described components of the sea, there are legally defined (and disputed) zones, sometimes overlapping, that influence regulatory practices, as indicated in Figure 13-1.

13.2.1 The Coastal Zone

The coastal zone extends from the low-tide line to the 200-meter depth contour, tending to match the geophysical demarcation of the continental shelf. The coastal zone can be as wide as 1400 km along some coasts and less than a kilometer along others. The average width of the zone worldwide is about 50 kilometers, comprising about 8% of the surface of the ocean. (The coastal zone of Alaska is larger than that of the rest of the United States.) Within the coastal zone definition, the difference between estuaries and the open coast is important in considering the disposal of wastewater and the potential for pollution problems.

Almost all of the water-carried wastes of a continent enter the coastal zone through an estuary. **Estuaries** are bodies of water with a free connection to the sea whose salinity is measurably diluted with fresh water, as from a river. Because estuaries provide critical and limited habitat for marine organisms to rear and feed their young, water quality is of special concern. Species that inhabit the coastal zone, and especially the estuaries, have to be very resilient to such natural environmental stresses as wide daily variations in salinity, turbidity, temperature, and UV radiation. Owing to this natural resiliency, coastal organisms may be able to tolerate contaminants associated with industrial and municipal wastes better than residents of the continental shelf, where the natural environment is quite stable. The estuarine habitat must be maintained primarily because of its limited extent, as distinguished from the shelf habitat, which is enormous. For this reason, treated wastewater effluents are usually discharged offshore rather than into estuaries in coastal regions. A large pipeline or tunnel, called an **outfall**, is used to transport the effluent to the disposal site (see Figure 13-1 and Figure 13-2 on page 193).

We also need to take the physical features of the coastal zone into account in disposing of treated wastes offshore. For example, continental headlands that protrude into the sea can impede both circulation of water and exchange of nearshore water with open ocean water. Outfalls are therefore best located far offshore rather than inside the region of headland influence. Similarly, outfalls should not be located close to shore in the vicinity of estuaries or bays because tidal incursions can carry diluted wastes into the estuary, thereby eroding one of the advantages offered by offshore disposal.

13.2.2 Open Ocean Waters

A variety of conditions and circumstances can contribute to the contamination of the open ocean waters beyond the coastal zone, including atmospheric fallout, oil spills, and dumping of hazardous wastes and sewage sludge as practiced by some countries of the world. Floatable and soluble materials tend to stay in the **upper mixed layer** of the ocean, where they may be decomposed. This upper layer is also the most active photosynthetic zone of the ocean, where plant—and hence animal—life can thrive. The depth of this layer, which varies between 100 and 1000 meters,

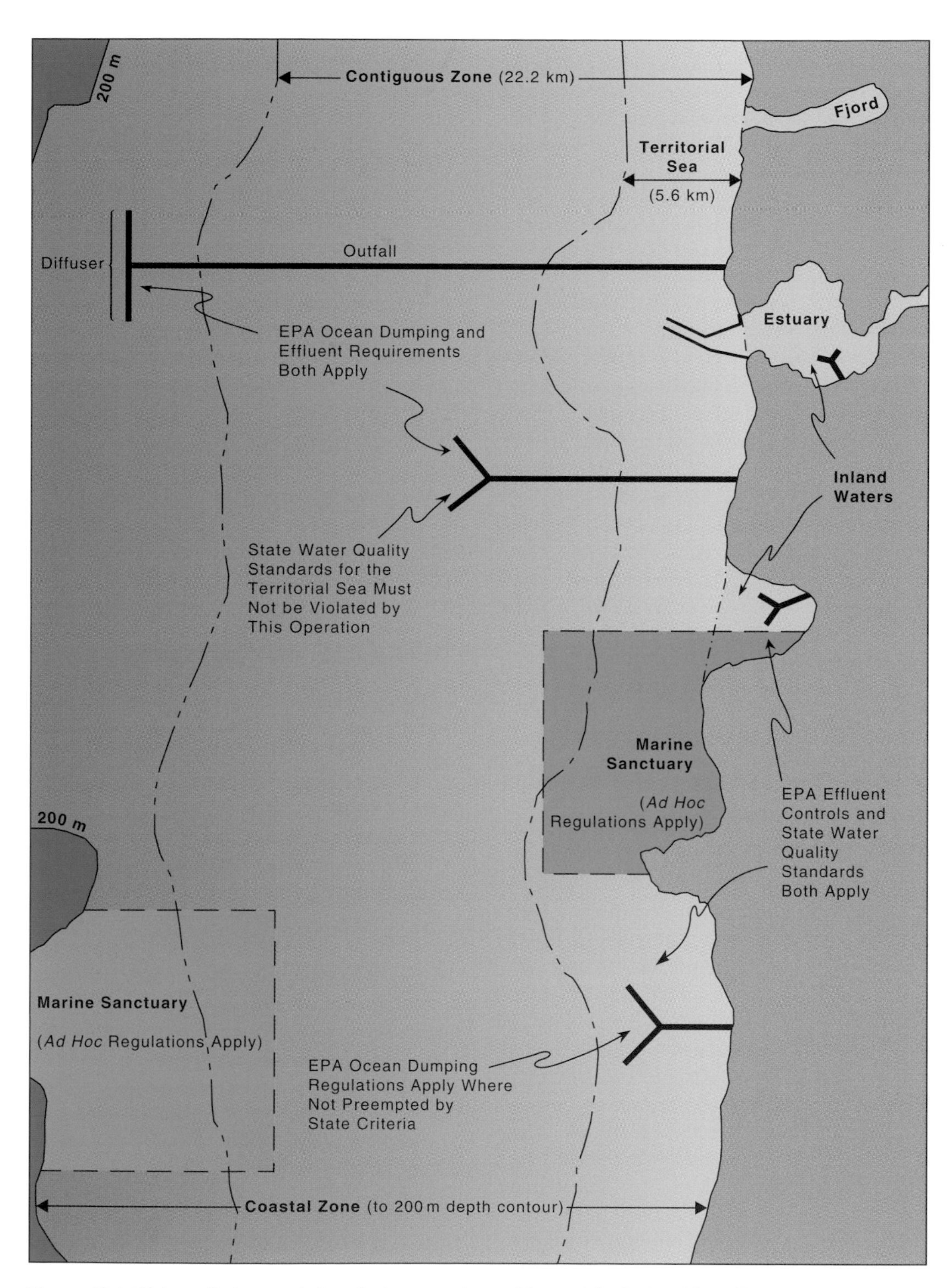

Figure 13-1 Water pollution regulations in the coastal zone. The outfalls depicted (both T- and Y-shaped) all use diffusers (see Section 13.6.1, beginning on page 200).

changes with season and geographic location. Although mixing between the upper and deeper layers of the ocean is impeded by strong density gradients, particles formed in the upper mixed layer, or discharged there, may eventually settle so far down that they can no longer be resuspended by surface generated turbulence, and thus become part of the natural detrital sediment load of the deep ocean waters.

Because the quality of the water of the upper mixed layer can significantly affect all life there, it is important to take precautions with waste disposal operations. When ocean disposal of certain materials is justified, we can use technologies to avoid contamination of the upper mixed layer and facilitate transit and long-term retention of the material in the deep waters of the open ocean, that is, the **abyssal ocean** (see also Section 13.6.2, beginning on page 201). For example, containers have been proposed for disposal of such materials as xenobiotic chemicals or radioactive wastes. Pipelines can also be used to carry liquid carbon dioxide to the seabed, where it can be retained for a long time—conceivably long enough to help reduce the rate of global warming (Section 12.5.1, beginning on page 183).

13.3 Sources of Surface Water Pollution

Pollution is a qualitative term. It describes the situation that occurs when the level of contaminants is such that intended water use is impaired. It takes just a small amount of contaminant to pollute a water body intended for a drinking-water supply. But the same water might not be considered polluted if the water were to be used, for example, for agriculture. Nor is pollution restricted to contaminants. Physical factors of the environment can also contribute to pollution. For example, heated water discharged from a power plant can change the temperature of an aquatic environment. It might not be a problem in a lake or a river during the winter, but it can certainly be a problem in the summertime. Moreover, heated water or water containing some contaminant may not be a problem at any time of the year, provided it is rapidly mixed with the surface water, and the diluted material doesn't accumulate over time. There are also many kinds of contaminants that can usually be accommodated by the natural environment without resulting pollution; but in many situations, these same contaminants (sometimes in conjunction with other contaminants) can cause pollution even in well mixed water bodies.

The major sources of surface water contamination are construction, municipalities, agriculture, and industries. However, the water delivered to earth in the form of precipitation is not necessarily pure to begin with. Near the coast, it may contain particulate and dissolved sea salts; further inland, it may contain organic compounds and acids scrubbed from contaminants added to the atmosphere both by natural processes and by anthropogenic (human) activities. Gases from plant growth and decay and gases from geological activity are examples of naturally derived atmospheric contaminants that can be returned to earth via precipitation. The acid rain problem of the New England states is a classic example of anthropogenically derived atmospheric contaminants that contribute to surface water pollution.

13.3.1 Construction and Land Use

Because construction involves land clearing and grading, this activity is a major source of sediment contamination of streams. Agriculture and forestry are also operations traditionally responsible for pollution of streams with particulate matter. In addition, other land use activities such as recreational off-road vehicle use can destabilize soil formations, thus increasing erosion. The runoff of fine-grained soil particles, especially in hilly terrain with high rainfall, can smother gravel beds that serve as spawning habitats for fish. In addition, suspended particulates and colloidal material can harbor microorganisms, necessitating increased treatment for municipal water supply facilities (see Section 20.5, beginning on page 315). Nor is the damage associated with surface water pollution limited to aquatic resources: erosion also decreases the topsoil resource needed for sustainable plant growth.

Erosion, both anthropogenic and natural, is a continuous process. All things considered, tens of thousands of hectares of riparian area are lost to erosion each year in the United States, and hundreds of millions of tons of topsoil are carried by rivers to the sea per year. Natural erosion of riverbanks in some areas masks the pollution caused by human activities. For example, such rivers as the Mississippi naturally carry about 500 mg L^{-1} of suspended solids, so even many tons of solids added over a section of the river by a construction event might not adversely affect the use of the water. In other areas, the amount of riverbank erosion can be exacerbated by poor land use practices. Some streams are naturally very sediment-free, so the addition of as little as a ton of soil in a year can cause pollution of a major portion of the stream.

Beginning in the 1970s, U.S. pollution control agencies began to restrict road construction and timber harvesting operations in forest areas to reduce soil erosion. Although these restrictions were primarily intended to reduce impacts on fish habitats, their serendipitous result was to enhance retention of topsoil. The 1980s saw a great deal of attention devoted to reducing soil runoff from individual home site construction projects, as well as large home and shopping center developments. Then, in the 1990s, the promotion of "no-till" planting in the agriculture industry helped to reduce wind- and water-carried loss of topsoil, and this in turn helped to reduce potential water pollution problems in rivers and reservoirs (see Figure 14-4 on page 218). Measures designed to control soil loss to rivers may have reduced the concentrations of other contaminants, such as pesticides and metals, that can also pollute surface waters. However, the main sources of many of these contaminants are municipalities and industries as well as agriculture.

13.3.2 Domestic and Municipal Sewage

Domestic sewage includes human wastes and a variety of waste materials that arise from household products such as foods, soaps, paints, oil, and grease. Consequently, these wastes contain microorganisms, metals, plastics, decomposable organic chemicals, and xenobiotic chemicals. Municipal sewage comprises a matrix that includes all the above plus water-carried wastes from industries, schools, hospitals, and airports connected to the municipality's sewer system. In many cities, street washings and other wastes poured into streets contribute to the contaminant load in municipal sewage.

Figure 13-2 An old-style outfall (employing no diffuser) showing grit accumulation on the seabed off of southeastern Florida. Primary effluent is shown rising from the upper end of the open pipe. Photograph: R. McAllister.

Surface water pollution from many of the contaminants in domestic or municipal sewage may result when these wastes are poorly treated or when even highly treated sewage is discharged into a watercourse that is physically or biologically incapable of accommodating the waste load. Sometimes, it is the magnitude of the waste load that results in pollution. Other times, it is a matter of certain contaminants that are not efficiently removed by treatment.

Human wastes make up a considerable portion of the contaminants in sewage; thus, many of the stream water pollution problems that have occurred—and pollution control technologies that have been developed—are related to human wastes. The components of human wastes that characteristically cause stream pollution problems worldwide are intestinal microorganisms (see Section 19.3.2.4, beginning on page 287) and organic matter.

13.3.2.1 Intestinal Microorganisms

Many intestinal microorganisms, which can remain viable in surface waters for a long time, can cause illness, disease, or even death if they come in contact with humans. The primary method of defense for downstream water users against waterborne outbreaks of disease (such as typhoid and cholera) has traditionally been high-quality physical treatment of drinking water and disinfection with chemicals such as chlorine. However, disease and illness can also be a problem for other uses of the stream resource, such as recreational bathing and harvesting of shellfish. For example, bacteria, viruses, and other microorganisms can cause upper respiratory and ear infections in bathers, while fairly large disease episodes have been attributed to the ingestion of improperly cooked shellfish harvested from polluted coastal waters. Also, some recreational users, misled by the clarity and coolness of steams in remote areas, become ill following ingestion of raw or poorly disinfected water. Visibly clear and ostensibly unpolluted mountain stream water can easily be polluted by wastes from one or two humans upstream, or even by the native warm-blooded animal population. For example, the gastrointestinal disorder known as giardiasis, caused by the protozoan *Giardia*, is the result of animal-caused pollution of water and snow (see Section 19.3.1.1, beginning on page 291). It is prudent—if not entirely necessary—to assume, without qualification, that all surface waters are contaminated with microorganisms of human health significance.

To some degree, the quality of surface water for recreation is protected by the treatment of domestic and municipal sewage at the point of discharge into the watercourse, and by the natural purification processes in the water. High-quality sewage treatment processes can remove more than 99.99% of the microorganisms in sewage, although this level of treatment is not always provided. Moreover, major storm events can greatly increase the volume of water flowing through older sewage systems (combined sanitary and storm sewers), thus increasing the likelihood that untreated and poorly treated sewage will reach the watercourse. Such events frequently result in the closing of nearby bathing beaches owing to the temporarily high concentrations of bacterial counts indicative of fecal pollution. However, wastewater treatment alone is never sufficient to provide sanitary river water suitable for municipal drinking water.

13.3.2.2 Organic Matter

Of the many organic chemicals in municipal sewage, we can identify two broad groupings based on the types of surface water pollution problems they cause. One grouping contains the persistent synthetic chlorinated organic compounds that were produced by the petrochemical industry in ever increasing amounts after World War II; these include insecticides and chlorinated hydrocarbons such as polychlorinated biphenyls (PCBs) and trichloroethylene (TCE) (see Section 7.4, beginning on page 83). Such chemicals degrade very slowly in the environment and can cause long-lasting toxicity, either acutely or chronically as a result of bioaccumulation in tissues. The term **xenobiotics** was coined to represent both the foreign (*xeno*) nature of such chemical forms and their biologically active (*biotic*) character.

Other organics are grouped in a class labeled **biodegradable organic matter**, which includes natural organic compounds found in wastes, such as carbohydrates and proteins. Many industrial chemicals also fit into this category. The problem with these organics (a problem that evolved fairly early in the modern urban age) is the depression of the dissolved oxygen content of stream waters caused by their microbial decomposition. The main feature of this pollutional effect in the stream is that the loss of dissolved oxygen is detrimental to aquatic organisms. A secondary effect is that if all the dissolved oxygen is consumed by aerobic microbial oxidation of the wastes, anaerobic decomposition becomes prominent, releasing noxious septic gases from the stream. Wastewater treatment methods, including trickling filters and activated sludge aeration, were developed in the first fifty years of the twentieth century to deal with this problem.

13.4 Quantification of Surface Water Pollution

Even as treatment methods were being developed, methods were simultaneously developed to quantify and assess both the disease threat and the dissolved oxygen problem posed by the discharge of municipal wastes into water bodies. Quantitative methods are also used to calculate the degree of treatment needed. Such methods are based on an understanding of the physical and biochemical processes controlling the decay of microbes and chemicals over time.

13.4.1 Die-Off of Indicator Organisms

Early in the development of quantitative assessment methods, scientists realized that many different microorganisms that exist in human wastes could not be effectively cultured and counted. Consequently, they settled on the coliform group of organisms to serve as an indicator of fecal pollution because they could be cultured and counted easily. Known to exist in large numbers in the gut of all warm-blooded animals, the coliform group provides a good indication of fecal pollution; however, it was not very specific, so other indicators such as fecal coliforms and streptococci may also serve as indicators of the sanitary quality of water (see also Section 19.3.2.4, beginning on page 287).

Tests have shown that 99.99% of the indicator bacteria can be removed by wastewater treatment. But the residual 0.01% remains a problem, raising concerns about the quality of water for recreational use, for example. The fact is, the number of bacteria in sewage is tremendous (500 million to two billion per 100 milliliters), so the percentage of removal is not as useful as is the actual remaining concentration in assessing quality. The microorganism concentrations allowed for various uses of water are relatively small. For drinking purposes, the concentration of fecal coliforms should, of course, be zero, but a concentration of less than 1 per 100 mL may be allowed; and for bathing, a concentration of 1000 per 100 mL is frequently accepted. If, for purposes of demonstration, we assume that raw sewage has a population of one billion (10^9) per 100 mL (see Table 19-7 on page 296), a removal efficiency of 99.99% would still leave 100,000 per 100 mL. The good news is that the concentration of organisms decreases with time and distance downstream of the discharge point owing to natural processes. And, depending on the location and proximity of uses, the point of discharge and method of discharge can be designed to optimize the rate of natural purification, further reducing downstream pollution problems. The bad news is that we have only skimpy data on all the processes that affect microbial die-off.

The concentration C of bacterial indicators of fecal contamination has been observed to decrease with time t according to a first-order reaction, the equation of which is

$$\frac{dC}{dt} = -KC \qquad (13\text{-}1)$$

where K is the **die-off rate constant**. One parameter frequently employed in water pollution analyses obtained by solution of the first-order reaction equation is t_{90}, which is the time required for 90% die-off of the bacteria. This parameter is analogous to the half-life (t_{50} or $T_{1/2}$) used in radioactivity studies (see Eq. 6-17 on page 74) and is calculated in the same fashion. Thus, given the value of K

$$t_{90} = \frac{2.3}{K} \qquad (13\text{-}2)$$

The general solution of Eq. (13-1) is used to find the concentration C_t at any time t, after the initial concentration C_0 at $t = 0$ is determined:

$$C_t = C_0\, e^{-Kt} \qquad (13\text{-}3)$$

This information can be used to calculate freshwater or marine die-off. To find the concentration of bacteria after effluent has traveled in the river for, say, 8 hours, it is necessary to determine the value of K for the specific situation being studied. Results of many studies in lakes and streams have shown that K varies widely, depending on the temperature of the water, the amount of sunlight, and the depth at which the plume travels. An average value for fresh water is about $K = 0.038$ per hour; however, values from 0.02 to 0.12 per hour have been measured. Using Eq. (13-3) and $K = 0.038$ per hour, the initial concentration $C_0 = 100{,}000$ per 100 mL would be reduced to about 74,000 per 100 mL in $t = 8$ hours due to die-off alone.

In many cases, it is preferable to predict the concentration at a given distance downstream, rather than at time increments. Most U.S. rivers have a remarkably uniform, low-flow current of 1.5–2 km per hour, which can be used to convert low-flow travel time to distance. In our example, this travel distance would be between 12 and 16 km. Travel times for other flow conditions vary from river to river, so field measurements may be needed to relate bacterial concentrations to specific locations downstream.

In analyzing coliform die-off cases in the marine environment, we often use an average value of 1.2 per hour for *K*, which is nearly thirty times greater than that of freshwater. This rapid die-off rate is usually attributed to the salinity of the marine environment, although it may also be related to a greater concentration of predatory animals. In addition, the natural flocculation and sedimentation of particles that occurs in estuaries could account for removal of bacteria from the water column. There are, however, other factors that can reduce the die-off rate. For example, when an effluent is discharged at a great depth, the die-off can slow down considerably because sunlight cannot penetrate deeply enough.

Calculating seawater die-off is similar to freshwater die-off. The value of *K* for marine waters usually ranges from 0.3 to 3.8 per hour. Recently, however, *K* values as low as 0.02 per hour have been found where an effluent plume is transported in a layer far below the surface, say 40 meters, suggesting that die-off is reduced because of the low penetration of UV radiation to that depth. What difference would this low *K* value make in the concentration?

Using the average value of $K = 1.2$ per hour, the original concentration, $C_0 = 100{,}000$ per 100 mL, would die off to $C_t = 6.7$ per 100 mL in $t = 8$ hours using Eq. (13-3) on page 195. But using $K = 0.02$ per hour instead of 1.2 per hour, we get 85,000 per 100 mL. Thus, for a 60-fold reduction in *K*, the concentration is increased by a factor of 85,000/6.7 = 13,000! It is evident from this example how important it is to have accurate values for *K* and how widely the results can vary with equally good, but different, estimates for *K*.

Variations in the rates of indicator bacterial die-off are not the only problem we have to contend with. Noncoliform pathogenic microorganisms may decay at rates different from those of our coliform indicators. Therefore, as water analysis techniques become more sophisticated, we will need to conduct many field observations to establish values that can be used to predict die-off rates for specific pathogens.

In addition to the decrease of bacterial concentrations due to die-off in either fresh or marine surface waters, bacterial concentrations are decreased as the water is diluted with upstream ambient water at the point of effluent discharge and further diluted as it flows downstream. The effect of dilution may or may not be important in meeting water quality criteria, depending on the initial mixing, the nature of the subsequent flow patterns, and the distance to water use areas. But before discussing the mechanics of the dilution process, let's examine how the same first-order decay process used for assessing indicator bacteria die-off can be applied to the analysis of the fate of biodegradable organics.

13.4.2 Organic Matter and Dissolved Oxygen

We know that biodegradable organic compounds are decomposed by bacteria and other organisms that live in surface waters. While some organics are mineralized to carbon dioxide and oxides of nitrogen, others are synthesized into more microbial biomass, most of which is subsequently decomposed as well. All this decomposition consumes dissolved oxygen (DO), upon which many desirable species of fish, other aquatic organisms, and wildlife depend. Consequently, depressed dissolved oxygen concentrations are inimical to these life forms. For example, some fish can survive at concentrations near 1 mg L^{-1}, but most are adversely affected at DO concentrations below 4 mg L^{-1}. The trouble is that there is not very much DO in surface waters to begin with. The maximum amount of oxygen a pure surface water can hold is a function of salinity and temperature, but, compared to the maxima of many other substances, it is remarkably low. Freshwater contains from 8 to 15 mg L^{-1} DO over the temperature range 30 to 0°C, respectively, and seawater contains 6–11 mg L^{-1} over the same range. Note that solubility of O_2 *decreases* with increasing temperature, which is the opposite of the temperature solubility relationship observed for most substances in water.

Domestic sewage can contain about 300 to 400 mg L^{-1} of organic compounds, 60% of which is readily degradable by bacteria commonly found in nature. *Readily degradable* implies that most of the material will be decomposed within about a week in a stream or other body of water that is sufficiently large. The change in the concentration of the organic matter with time is conveniently described by the first-order decay equation used to describe bacterial die-off. However, the value of this K depends on the specific organic compounds in the sewage. For domestic sewage, an average value is about $K = 0.4$ per day, ranging from 0.1–0.7 per day. As more industrial wastes are contributed to the sewer system, the rate constant may increase or decrease. The amount of organic material discharged to a surface water depends on the population served by the municipal sewer system. Each person contributes about 90 grams per day of organics; so if the population is 100,000 people, the mass emission rate is 9 metric tons per day. The concentration of organics in the sewage depends on the amount of water added by the individual households and that added by other water uses in the community. If the average water use is 300 liters per person per day, the concentration is 300 mg L^{-1}.

13.4.3 Measurement of Potential Oxygen Demand of Organics in Sewage

13.4.3.1 Chemical Oxygen Demand (COD)

A parameter frequently used for industrial wastes, particularly where industrial waste contributes heavily to the sewer system, is the **chemical oxygen demand (COD)**, which is a measure of the amount of oxygen required to oxidize the organic matter—and possibly some inorganic materials—in a sample. Note that the method employed to obtain this parameter, which involves reflux of a sample in a strong acid with an excess of potassium dichromate, does not specifically measure the organic content in the sample, but rather the amount of oxygen required for oxidation. This approach therefore provides a direct measure of the potential impact of oxygen consumption on the oxygen content of the water body.

13.4.3.2 Biochemical Oxygen Demand (BOD)

The **biochemical oxygen demand (BOD)** is the most commonly used parameter in the analysis of oxygen resources in water. The BOD is the amount of oxygen used over the course of time, usually 5–20 days, as the organic matter is oxidized both microbially and chemically.

13.5 Determining BOD

The laboratory method used to determine BOD has changed very little since it was initiated in the 1930s. We begin by setting up many sample bottles to contain a sample of waste, either mixed in water from the disposal site or from a standard laboratory supply. Then we use standard methods to find the initial concentration of DO, after which the bottles are incubated in a dark water bath at a given temperature, usually 20°C. Every day for five or more days, we open a few of the bottles and measure the remaining DO. The difference between the initial value and the value at each time period, that is, the **demand**, is plotted as the BOD_t for the series of days, where BOD_t is the BOD at time = t.

From the data obtained in the laboratory, we construct a smooth curve that lets us calculate the reaction rate coefficient K by graphical or analytical methods. The curve we construct, whose equation is

$$BOD_t = BOD_L\,(1 - e^{-Kt}) \qquad (13\text{-}4)$$

becomes more and more horizontal as time progresses (see Figure 13-3 on page 198). By extrapolating the curve to horizontal, we can make an estimate of the ultimate value, called the limiting value of BOD, or BOD_L.

We can use the curve in Figure 13-3 on page 198 derived from the sample BOD data to compute the value of K for the wastewater sample in this laboratory test.[1] First, by extrapolating this curve to horizontal, we see that the estimated value of BOD_L is

[1] Mathematical techniques that let us simultaneously estimate K and BOD_L are available in advanced textbooks.

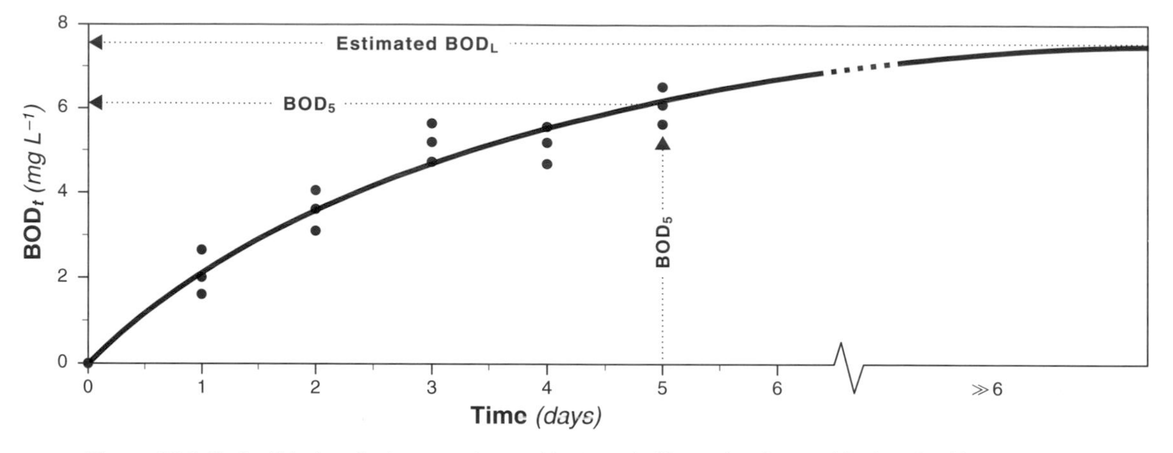

Figure 13-3 Typical biochemical oxygen demand test results illustrating the graphical method for determining BOD_L and BOD_5 for use in solving Eq. (13-4) on page 197 to find K as described in the text. Note that, in actuality, the extrapolated curve is asymptotic to $y = BOD_L$.

about 7.6 mg L^{-1} and the BOD at five days[2] (BOD_5) is equal to 6.1 mg L^{-1}. Next, by solving Eq. (13-4) on page 197 for K and substituting these values and $t = 5$, we obtain $K = 0.32$ per day.

13.5.1 Impact of BOD on Dissolved Oxygen of Receiving Waters

Typically, municipal sewage is treated to some degree to remove organics, and hence to reduce BOD, before discharge to surface waters in developed countries of the world. The amount of BOD remaining after treatment can range from 10 to 70% of the amount originally in the sewage. The impact on receiving waters depends on many environmental and waste characteristics, most of which are briefly mentioned later in this chapter. After the BOD parameters already explained, the next most important considerations are the amount of DO in the water body before the sewage is added at $t = 0$, called the initial DO (DO_0), and the rate at which additional oxygen is transferred from the atmosphere to the receiving water.

Many rivers and streams have depressed DO concentrations, that is, a **DO deficit**, because of BOD added by cities upstream. Recall that rivers can hold only a small amount of DO: 8–15 mg L^{-1} depending on temperature. This is called the **DO saturation value (DO_s)**. The local deficit—the difference between the saturation value and the observed initial value at the location of waste discharge—must be included in the computation of the downstream DO deficit caused by a new effluent discharged to the stream. But before presenting an example calculation, let's first examine how nature deals with the DO deficit.

Deficits tend to be redressed by oxygen gas derived from the atmosphere. Such replenishment occurs by a process of gas-liquid mass transfer at the surface and subsequent mixing throughout the depth of water. This overall process can be described in an approximate manner by a first-order equation, the solution of which is

$$D_t = D_0 \, e^{-Rt} \tag{13-5}$$

where D_t is the deficit at time t, D_0 is the initial deficit (*i.e.*, $DO_s - DO_0$), and R is the reaeration coefficient.

---

[2] In general, when "BOD" is referred to in water quality regulations, it is understood to mean the BOD at *five days,* or **BOD_5**. In technical writing, however, the absence of a subscript means the BOD at *any time,* or **BOD_t**. Obviously, it's a good idea to use the subscripted forms, thereby avoiding any possible confusion.

Note that the larger the magnitude of R, the quicker a given deficit is removed. The value of R depends on the degree of vertical mixing in a water body, as well as its overall depth. It varies from 0.1 per day in small ponds to above 1 in rapidly moving streams.

Most DO problem situations require that we determine the oxygen deficit resulting from the simultaneous effects of the oxygen demand of a waste and the competing restoration of oxygen from atmospheric reaeration. The combined result is termed the **self-purification capacity** of the water body. The deficit shown graphically as a function of time (or distance) is known as the **oxygen sag curve** because of its characteristic spoon shape. The equation for the curve is

$$D_t = \frac{K(\text{BOD}_\text{L})}{R-K}(e^{-Kt} - e^{-Rt}) + D_0 e^{-Rt} \quad (13\text{-}6)$$

where the terms are as defined above.

An example of the curve produced by the equation is shown in Figure 13-4. The values of K and BOD_L are taken from Figure 13-3; the value for D_0, the initial deficit, was 3 mg L^{-1} due to upstream discharges; and a reaeration coefficient R of 0.5 per day was chosen for a large, slow-flowing stream. This example might represent a case of highly treated sewage, a typical effluent plume after an initial dilution of 4, or poorly treated sewage dispersed from a diffuser (Section 13.6.1, beginning on page 200) that provides an initial dilution factor of 25.

One interesting result from the solution of the oxygen sag equation is that the deficit, irrespective of the saturation value of DO, is the same as long as the initial deficit is the same. Sometimes, water quality standards impose limits both on the amount of deficit *per se* and on the resulting DO value itself. For example, a regulation could require that a waste discharge must neither increase the deficit by more than 10% nor depress the resulting DO concentration below 5 mg L^{-1}.

When water systems are heavily used or highly valued water uses are threatened, additional factors require consideration:

1. the diurnal demands and supplies of oxygen from photosynthesizing organisms;

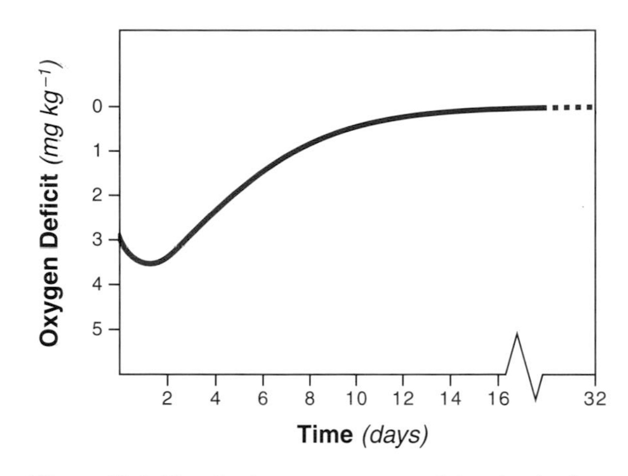

Figure 13-4 Dissolved oxygen sag curve determined using values of BOD_L and K from Figure 13-3. The equation for this curve is given in Eq. (13-6).

2. the oxygen demand of organic materials deposited in the sediment layer;
3. the oxygen demand of nitrogen compounds discharged in the effluent;
4. variations in the reaeration coefficient R with travel time due to flow conditions in the water body;
5. the wide range of sewage flows encountered over the lifetime of a river.

Comprehensive computer programs are available to describe the concentration of oxygen in large watersheds consisting of dozens of interconnecting streams and dozens of wastewater inputs; however, data on plume travel, K values, and R values still have to be obtained with time-consuming laboratory studies and physically demanding field studies. Dilution of wastes is one of the most important factors to consider in assessing impacts. The methods used to assess effects on the dissolved oxygen resource of a water body and bacterial contamination are applicable to a wide variety of toxic chemical problems.

13.6 Dilution of Effluents

Dilution can be—but is not necessarily—an effective way to prevent pollution of surface waters. Environmental scientists do not categorically embrace the old saw, "Dilution is the solution to pollution."

Instead, we recognize through the analytical process of risk assessment that certain principles underlie the utility of dilution in managing waste discharges. The first principle concerns the concentration dependence of the pollutant response mechanism. Here we want to know if the effect of the contaminant is directly related to its concentration. That is, if the concentration is reduced sufficiently, will the degree of effect be directly reduced?

Further, once the contaminant concentration is reduced, can it subsequently become more concentrated? Reconcentration is a phenomenon often associated with sediments and persistent organic chemicals such as PCBs. Dilution of waste streams containing high concentrations of suspended solids may prevent significant pollution near the discharge site, but such sediments may eventually settle out of the water column and concentrate in depressions in the stream bed. There they can cause a variety of problems. Even in the ocean, waste disposal can lead to accumulation of sediments in the seabed. With persistent organic chemicals, whose solubility in water is low and whose affinity for sorption to animal tissues is high, adverse bioaccumulation can occur even from highly diluted mixtures. Thus, dilution may or may not solve a potential concentration-related problem.

13.6.1 Dilution in Streams and Rivers

Aside from concentration-toxicity considerations, social and economic considerations enter into the decision to use dilution. In addition, its use is also dependent on the availability of a sufficient quantity of dilution water. In the arid southwestern United States, for example, many streams are ephemeral; that is, they contain water only after major rainfall events. In other streams in arid climates, the effluent from municipal treatment plants is the predominant flow for more than 50% of the year; that is, they are effluent-dominated streams. In such situations, dilution is somewhere between small and nil, except during storm runoff. For example, if the streamflow is 10 million liters per day and the effluent flow is 30 million liters per day[3], the contaminants dissolved in the effluent will be reduced in concentration by just 25%, assuming the concentration of each contaminant upstream is zero. (Generally, the effective dilution will be even less than that because there is almost always some measurable concentration of contaminants upstream.)

The general equation used in determining the concentration after dilution is

$$c_f = \frac{c_e v_e + c_a v_a}{v_e + v_a} \tag{13-7}$$

where

c_f = cross-sectional average final concentration in the stream
c_e = concentration in the effluent
v_e = volume flux of the effluent
c_a = concentration in the ambient dilution water upstream
v_a = volume flux of the ambient dilution water

In large rivers, the amount of dilution achieved depends on the method used to discharge the effluent into the river. To maximize the dilution, it is necessary to employ a diffuser, which consists of a pipeline with many exit orifices across the width of the river. For example, if the river flow is 120 million liters per day, an effluent discharge of 30 million liters per day yields an 80% reduction in concentration of contaminants if a diffuser is used to disperse the effluent across the river. However, if a diffuser is not used, the immediate dilution would be much less.

In many streams construction of diffusers may be either inappropriate or prohibited. Consequently, the amount of dilution depends on natural mixing processes that occur during streamflow. But in large streams the effluent plume may hug the bank for many miles, so it is not actively diluted with the main flow. It is not possible to estimate a range of values for the dilution rate in cases like this. It is reasonable to assume that without a physical structure, such as a diffuser, to mix the effluent into the river, we may consider only the natural die-off process in assessing the impact of bacterial contamination on downstream water uses. Similarly, if there is no dilution of the BOD, we can count only on the decom-

[3] The unit MGD (million gallons per day) is most commonly used by wastewater treatment plants in the United States. 1 MGD is equal to 3.78 million liters per day or 3.78×10^3 m^3 day^{-1}.

position of organics and reaeration to restore or maintain the DO resource of the river.

13.6.2 Dilution in Large Bodies of Water

Large bodies of water, particularly open coastal waters, offer much greater opportunity for effective dilution of waste streams. Effective initial dilution is achieved by a multiport diffuser on the end of the outfall discharge pipe. Because the density of most wastewaters is very close to that of freshwater, the discharge of effluent to deep marine waters creates a strong buoyant force. Thus, the effluent, no matter how deep the discharge, will rise to the surface of the sea if it is not trapped by density gradients below the surface. As it rises toward the surface, the effluent effectively mixes with the surrounding ambient water, resulting in more and more dilution. A good analogy is the increasing width of a smoke plume as it rises in the atmosphere with the amount of dilution being proportional to the square of the plume width.

In the effective placement of ocean outfalls, however, depth is not the only determining factor. All other things being equal, the greater the extent of vertical travel, the greater the amount of initial dilution. But those "other things" must be truly equal. If for example, a location chosen for its great depth has poor circulation, the net result may be less effective dilution of wastes than that offered by placement in shallower, but more open, water. Such considerations are a major concern in the placement of outfalls in fjords, bays, and, sometimes, estuaries (see Figure 13-1 on page 191).

Depth does not always provide the same opportunity for greater initial dilution in lakes and reservoirs because the difference in density between wastewater and receiving waters may be very small. In fact, it is not uncommon for industrial wastes to have a density greater than that of lake water, so these wastes tend to settle along the bottom rather than rise to the surface. However, cooling waters from large thermal-electric power stations have a density less than that of most lake waters because of their high temperatures. Thus, a deep discharge site can be advantageous for achieving effective reduction of thermal effects.

Many countries still allow the practice of dumping partially treated municipal sewage sludge into the ocean. For many years before being banned in the United States in the 1980s, such sludge from New York City and Philadelphia was dumped in the Atlantic Ocean. Typically a portion of the sludge is particulate matter possessing a sufficiently high density to settle to the seabed, although currents and turbulence spread the material throughout the disposal zone. Other materials disposed of in the ocean, such as dredged sediments from harbors and waterways and some industrial wastes, behave similarly to sewage sludge.

Figure 13-6 on page 203 is taken from a study to determine if pollution of the upper mixed layer caused by dumping from underway barges could be reduced by bringing the vessels to a stop before discharging their contents. This would facilitate more effective seabed accumulation (which has its own set of problems) of sludge particles while resulting in less particulate dispersion in the surface waters.

Sewage sludge dispersion is divided into three phases: convective descent, dynamic collapse, and long-term dispersion. **Convective descent** (Figure 13-6, *top*) occurs initially and is inversely analogous to the buoyant rise of municipal effluent discharged from a multiport ocean outfall diffuser on the seabed, *i.e.*, dispersion is driven by the density difference between the sludge and the less dense seawater (note the rapidity with which the plume descends towards the simulated seabed in Figure 13-6). As can be seen from the figure, there is still some material of relatively low density, even some floatables, that will eventually be carried away from the disposal site and become dispersed by surface currents. Additionally, very dense materials, such as metal objects and grit, have already descended to much greater depths towards the seabed.

After the initial convective descent, the process of **dynamic collapse** occurs (Figure 13-6, *middle*). Now, the cloud begins to visually distinguish itself into three portions, centered around the middle of the cloud (see Figure 13-5, a close-up view of Figure 13-6 at 1 minute 25 seconds), which is at buoyant equilibrium with the surrounding water. Above the middle, the coagulating effect of the salts in seawater causes some of the colloidal material to flocculate,

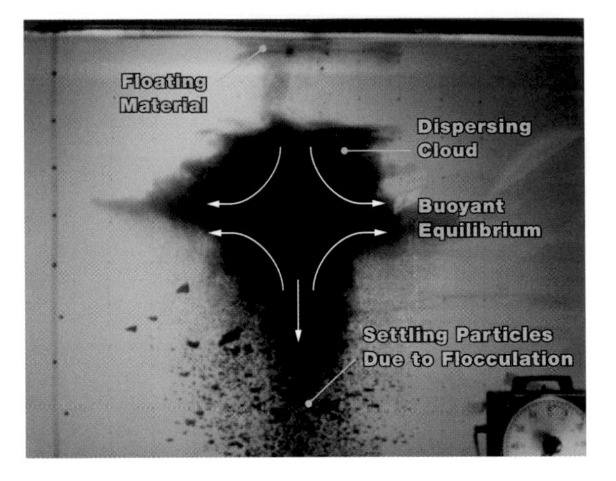

Figure 13-5 A close-up view of the photograph in the time series (Figure 13-6) taken at 1 minute, 25 seconds. The white arrows denote general directions of particle movement during dynamic collapse, with some particles flocculating and settling towards the bottom of the sedimentation tank. Photograph: D.J. Baumgartner.

become more dense, and collapse towards the middle. In contrast, other particles have descended below the zone of buoyant equilibrium to depths where the ambient density of the water is actually higher than the particles; after losing the initial momentum imparted to them from the force of dumping, these particles rise towards buoyant equilibrium in the middle of the cloud. The result is a cloud which collapses into a thin layer and slowly disperses laterally (**long-term dispersion**, Figure 13-6, *bottom*). In nature this zone of buoyant equilibrium corresponds to the strong density gradient separating the upper mixed layer from the deep ocean waters (see Section 13.2.2, beginning on page 190). Continuing flocculation forces a portion of the waste to settle out of the dispersion cloud and descend towards the seabed (Figure 13-6 at 00:50–20:00 and Figure 13-5). This phenomenon of dynamic collapse also occurs with sewage effluent plumes trapped by density gradients below the surface, as discussed further in Section 13.6.2.1.

13.6.2.1 Initial Dilution and Transport

The term **initial dilution** specifically identifies the amount of dilution achieved in a plume owing to the combined effects of momentum and buoyancy-induced mixing of the fluid discharged from the orifice. This term is used both in regulatory practice and in plume hydrodynamics. The rate of dilution caused by these forces is quite rapid in the first few minutes after exiting the orifice, then decreases markedly after the momentum and buoyancy are dissipated. Ambient currents also influence the rate of dilution during the buoyant rise of the plume irrespective of momentum and buoyancy. As current speed increases, so does initial dilution. In many cases, an initial dilution of 100:1, commonly sought in design of outfalls, is sufficient to reduce toxicity of chemical contaminants to an acceptable level. [*Note*: When bacterial die-off is an important consideration, the *distance* from a designated use area is usually a more important factor than initial dilution *per se*. In this case, the time it takes microbes to travel a long distance increases the likelihood that they will be inactivated.]

Following initial dilution, waste streams undergo additional dilution, or dispersion, as they are transported by ambient currents and mixed with the surrounding water by turbulence. The process is analogous to the dispersion that takes place when smoke plumes dissipate in the atmosphere after the smoke has risen to an equilibrium level. In some aquatic systems, however, the effluent plume is not as easily observed.

Most modern coastal cities employ multiport ocean outfalls far offshore to protect beaches and nearshore recreational areas from the effects of bacterial contamination. These outfalls are frequently designed to maintain a diluted waste stream below the surface of the sea. Such systems are especially useful during the summer recreational season because they keep the immediate area of discharge free of unsightly messes. Moreover, they reduce landward transport of the diluted waste by onshore wind currents toward peak beach activities. The disadvantage of subsurface trapping lies in the fact that initial dilution is reduced compared with plumes rising to the surface; but this disadvantage is offset by the reduced risk of onshore transport.

13.6.2.2 Measurements and Calculations

The dilution achieved in ambient transport of waste stream plumes in large water bodies can be described

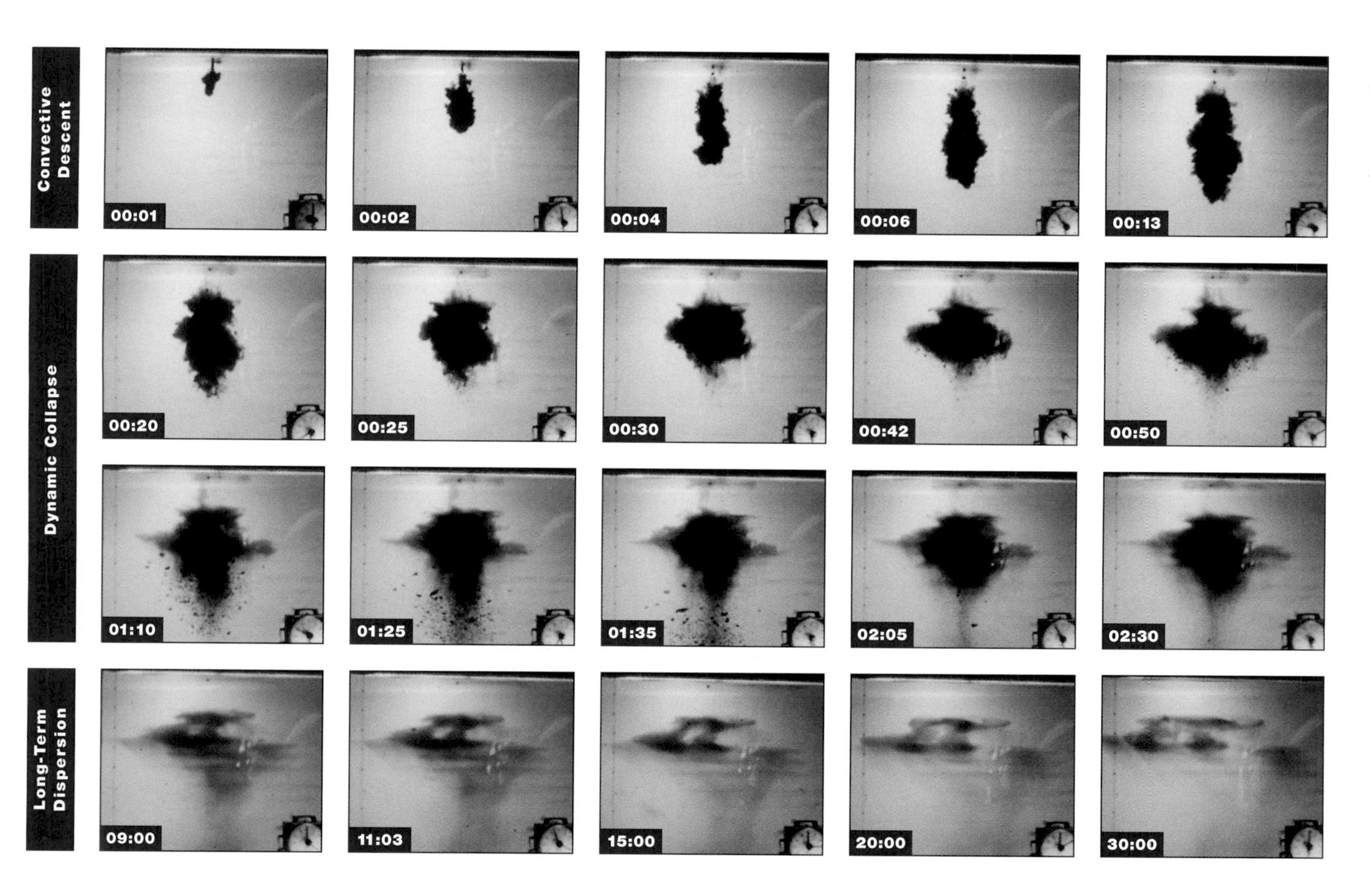

Figure 13-6 A montage of time series photographs of a hydraulic model test of dispersion and settling of New York City sludge in a simulated ocean system. Times are reported in minutes:seconds. Photographs: D.J. Baumgartner.

by physical laws—essentially the same laws used for describing contaminant flow in groundwater (Section 5.3.1, beginning on page 54) and gaseous flow in the atmosphere (Section 12.4.2, beginning on page 179). However, we cannot solve the equations completely for the general case because we lack existing data. That is, data obtained from field studies or from reports of previous studies are needed for empirical coefficients in the equations. However, many computer programs are available to obtain *approximate* solutions of the equations for complex cases involving multiple waste inputs and variable current speeds. In addition, we can sometimes use simplifications of the governing equations for many pollution assessment problems to obtain satisfactory estimates of contaminant concentration as a function of travel time or distance.

One simplified equation that has been used successfully over the past thirty years for large bodies of water gives us the maximum concentration at a distance X:

$$c_{\max} = c_{p_i} \operatorname{erf}\sqrt{\frac{Ub^2}{16\,\varepsilon_0\,X}} \qquad (13\text{-}8)$$

where

- $c_{\max}$ = centerline (maximum) concentration at distance X
- c_{p_i} = plume concentration at the end of initial dilution
- erf(*arg*) = standard error function of the given argument *arg*
- U = current speed in the X-direction
- b = width in the Y-direction (orthogonal to X) at the end of initial dilution
- ε_0 = constant horizontal (Y-direction) eddy diffusivity
- X = travel distance [*Note:* U/X can be replaced by $1/t$ (time)]

In using this equation it is important to use values that are expressed in consistent units. For example, the parameters U, b, and ε_0 contain a length unit that must be expressed consistently. Similarly, U and ε_0 also contain a time unit. Another way to appreciate this requirement is to recognize that the argument *arg* of the error function must be dimensionless. The standard error function (erf) serves here as a mathematical representation of the way contaminants are observed to vary laterally (Y-direction) as the plume is transported in the X-direction. It describes the normal distribution curve used in evaluating variance around a mean value (see Section 8.2, beginning on page 98). Values can be found from a tabular listing in a handbook, or by using the standard error function in a spreadsheet program. The **transport dilution factor** is equal to the reciprocal of the value of erf(*arg*).

In a typical problem involving finding the transport dilution factor, we might estimate the highest concentration that would occur near the beach from an outfall 8 km (*i.e.*, 8×10^5 cm) offshore when the onshore current speed U is 15 cm s^{-1}. The width b of the plume after initial dilution, which would be obtained from local observations, would likely be about 1000 m (*i.e.*, 10^5 cm). And we can use a commonly cited value for the eddy diffusivity of $\varepsilon_0 = 10^4$ cm^2 s^{-1}. Now substituting these values into Eq. (13-8) we get:

$$c_{\max} = c_{p_i} \operatorname{erf}\sqrt{1.17} = 0.87\,c_{p_i} \qquad (13\text{-}9)$$

The result is typically surprising: it shows that even with a travel distance of 8 km in a large, open body of water, the concentration is diluted to only 87%, resulting in a transport dilution factor of 1.2. (Contrast this factor of 1.2 to an initial dilution factor of about 100, which is what we expect for an ocean outfall.)

This example shows us that initial dilution is often more important in reducing harmful levels than is dilution due to transport. This is true not only for BOD and DO problems, but also for most contaminants, including heavy metals, ammonia, and toxic organics that are found in partially treated effluents. However, in the case of indicator bacteria, the value of transport distance is realized, as demonstrated in the following example.

Assume an outfall whose degree of treatment is only 90% effective, so that instead of the 100,000 coliforms per 100 mL used in the example on the freshwater die-off of indicator organisms (Section 13.4.1, beginning on page 195), the initial count of coliforms is 100,000,000 per 100 mL. The initial dilution of 100:1 would reduce the concentration to 1% of the initial count, or 1,000,000. Then, using the reciprocal of the transport dilution factor found above (0.87), we can calculate that transport to the beach would further reduce the count to C_0 = 870,000 per 100 mL. The 8-km distance would be covered in t = 14.8 hours at a speed of 15 cm s^{-1}. If the die-off rate constant K is found to be above 0.46 per hour, near the usually expected lower range of values, the beach concentration of bacteria ($C_{t=14.8}$), after substituting the values for C_0, K, and t into Eq. (13-3) on page 195, would be about 960 per 100 mL.

Even with some form of secondary treatment (see Section 20.3.2.2, beginning on page 304), some regulatory agencies require disinfection to reduce bacterial concentrations to bathing water standards. The result in the foregoing example demonstrates that chlorination of the effluent might not be necessary with 90% removal, a level typically achieved in secondary treatment.

Because ocean outfalls often provide greater travel distances, coastal communities may not have to use chlorine disinfection in the sewage treatment plant to reduce microbial contamination of human-use areas of the marine environment. This is important because use of chlorine can be hazardous. Also, chlorine may combine with organic materials in the sewage to produce compounds harmful to marine organisms and to people who consume those organisms (see Section 20.5, beginning on page 315). The distance required for indicator organisms to die off to an acceptably low level depends on current speed and direction, as well as on the bacterial concentration at the end of the rapid initial dilution process.

13.7 Dye Tracing of Plumes

Frequently, an oceanographic study is necessary to measure the bacterial concentration in the drifting plume as the current carries the water toward shore. Such studies usually employ a tracer, or dye, which is added to the sewage so that the plume can be followed for several kilometers. The data obtained can then be used to calculate a die-off rate constant, which may be useful for predicting the bacterial concentrations at different distances under a variety of current conditions.

Dye tracing (Figure 13-7a on page 207) is a well known technique commonly used in hydraulic models and prototype field settings, although in deep outfall situations tracers can be quite costly because of the large volumetric flow rates and large dilutions usually achieved within a short time frame. The rate of dye addition Q_d to the effluent flow V_e needed to provide a dye concentration of C_d following dilution of S_a is

$$Q_d = \frac{V_e C_d \alpha_a S_a}{W \alpha_d} \qquad (13\text{-}10)$$

where

α_a = specific gravity of the diluted plume
α_d = specific gravity of the dye solution
W = weight fraction of dye in stock solution

Figure 13-7b shows the required dye rate in liters per hour for various dilution factors, and effluent flows in million liters per day, to achieve an ambient dye concentration of 1 µg L^{-1} in sea water. Rhodamine WT, typically used in dye studies, is available as a 20% solution (α_d = 1.19) in small (57 L) drums. Fluorometers used in field sampling can easily detect this dye at concentrations of 0.5 to 1 µg L^{-1}.

We can use Figure 13-7b to estimate the amount of dye needed to trace an effluent flow in a water body, or any similar aquatic mixing question. If the flows marked on the abscissa of the graph and the slanted lines representing the dilution factors do not match exactly with the problem we have, Eq. (13-10) may be used to refine the estimate. Let's use both in an example problem.

Suppose we have an effluent of 330 million liters per day. A regulatory permit requires the effluent to be diluted by a factor of 200 at the end of a mixing zone. Suppose we set up our sampling boat at the mixing zone boundary and hope to measure 1 µg L^{-1} of dye as the plume passes under our boat. How much dye do we need to add to the effluent? Using the graph, estimate 330 along the abscissa and draw a line up to the dilution factor line S_a = 500. Now estimate where S_a = 200 would lie on that line. From that point draw a line horizontally to the ordinate and estimate the dye requirement as 12 L h^{-1}.

For a more precise estimate, we will use Eq. (13-10). First convert V_e = 330 million liters per day (*i.e.*, 330 × 10^6 L day^{-1}) to 13.7 × 10^6 liters per hour (L h^{-1}). For C_d use 10^{-9} g dye per gram of sea water (this is approximately equal to 1 μg L^{-1}, assuming a specific gravity of 1). The specific gravity of sewage effluent diluted 200:1 with seawater is $\alpha_a = 1.023$. (If this were a discharge to fresh water, the specific gravity would be 1.0.) Using $S_a = 200$, $W = 0.2$, and $\alpha_d = 1.19$ for rhodamine WT as cited above and substituting into Eq. (13-10), we obtain $Q_d = 12$ L h^{-1} (rounded from 11.8), verifying our estimate from Figure 13-7b.

13.8 Spatial and Temporal Variation of Plume Concentrations

The concentrations of water quality indicators, such as bacteria, are neither uniform nor steady with respect to the space and time scales involved in regulating the concentrations at the end of the mixing zone. In general, we assume that the concentrations of constituents in the horizontal extent of a plume from an outfall diffuser are uniform. But we can make no such assumption about the vertical direction. Vertical nonuniformity is commonly encountered in design, performance analysis, and compliance monitoring, although in rivers it is not nearly the problem it is in estuaries, coastal water, and some lakes and reservoirs. Generally associated with density stratification in the receiving water, vertical nonuniformity is also associated with transport of a plume in a relatively thin lens as compared to the depth of the water column. For instance, if the plume is traveling on the surface, its constituents will be dispersed downward; as these constituents disperse into the water column, the concentration of pollutants near the bottom edge of the plume gradually becomes less than that at the surface. (Thus if a permit condition requires that a maximum value be reported, sampling should be done at the surface, not at mid-depth.) Similarly, the dilution water mixed with the effluent being discharged is also vertically variable due to physical processes influencing the advection of ambient water into the region of the discharge. Dissolved oxygen (DO) is an example of one water-quality indicator that exhibits vertical nonuniformity in many riverine impoundments (reservoirs), lake, estuarine, and coastal situations.

Some transport and dispersion models produce estimates in terms of the **centerline concentration**, which is the maximum concentration for the cross section of the plume at a given distance downstream from the orifice. As the plume width expands with increasing distance, the maximum concentration progressively decreases. For example, the centerline (maximum) concentration at a distance of 60 meters from the diffuser may be 100 mg L^{-1}, while at 120 meters from the orifice the maximum concentration would be closer to 70 mg L^{-1}. Other models calculate an average concentration for the cross section of the plume, and this of course also decreases downstream: the average concentration is always smaller than the maximum concentration. Both values need to be considered in field or lab verification studies, and both values may be useful for regulatory purposes.

13.9 Compliance Monitoring

Water pollution regulatory practice in the United States is founded on a system of discharge permits—known as the National Pollution Discharge Elimination System (NPDES) permits. Holders of these permits, *e.g.*, municipal sewage treatment authorities and industries, must comply with the restrictions and requirements of their particular permit, such as limits on concentrations and mass emission rates of specific constituents. They also have to meet water quality standards established for the water body into which they discharge their effluents. Some permits, especially for coastal water discharges, require elaborate environmental monitoring projects. The permit holder is required to conduct monitoring activities and report the results to demonstrate compliance with permit conditions. On occasion, regulatory agencies conduct special studies to verify and revise monitoring programs. Monitoring data reflect the wide variations of conditions found in the natural environment, and dischargers and regulators are often challenged to reconcile monitoring results with predictions used in setting permit conditions.

13.9.1 Mixing Zones

Permit conditions of regulatory agencies usually allow exceptions to one or more of the water quality criteria within a mixing zone adjacent to the point of discharge. A mixing zone might be established by purely arbitrary considerations or by use of data and simulations with mathematical models. Many mixing zone determinations are made on the basis of the expected dilution rate that will be provided by efficient diffuser designs intended to optimize initial dilution. For large bodies of water, a common approach is to describe the width of the zone as the depth of water at the disposal site, and the length as the length of the diffuser. For large rivers, it is common to restrict a mixing zone so it does not extend completely across the river, thereby leaving a "safe passage" that lets aquatic species avoid high concentrations of wastewater constituents. But sometimes the shape of a mixing zone is entirely arbitrary, say a rectangular zone downstream from a discharge pipe equal to one-fourth the width of the stream and extending downstream for one kilometer. Frequently, there are two or three mixing zones for different groups of contaminants and degrees of toxicity.

13.9.2 Regulatory Use

Regulatory interest may be appropriately directed toward both discrete and average values of contaminants. For example, the state of California and the U.S. EPA specify maximum allowable instantaneous values for some parameters as well as several temporal average values (*e.g.*, thirty-day and six-month arithmetic means). In some cases, these regulations are based on knowledge of the effects on aquatic organisms. In other cases, these values are specified to acquire statistics on the performance of the wastewater treatment plant.

Criteria that are expressed in terms of temporal averages (daily to semiannual) suggest that plume concentrations be assessed extensively in three dimensions, both at the boundary of the mixing zone and, in some cases, at sensitive biological resource locations down-current. Current speed and direction play significant roles when assessing the concentrations at the boundary. By incorporating data on the cyclical variation of effluent composition, density profiles, and current direction, it is possible to construct a running six-month average (or median) for a number of points on the mixing zone boundary. The six-month average is expected to be quite variable at these points, and the point with the highest exposure

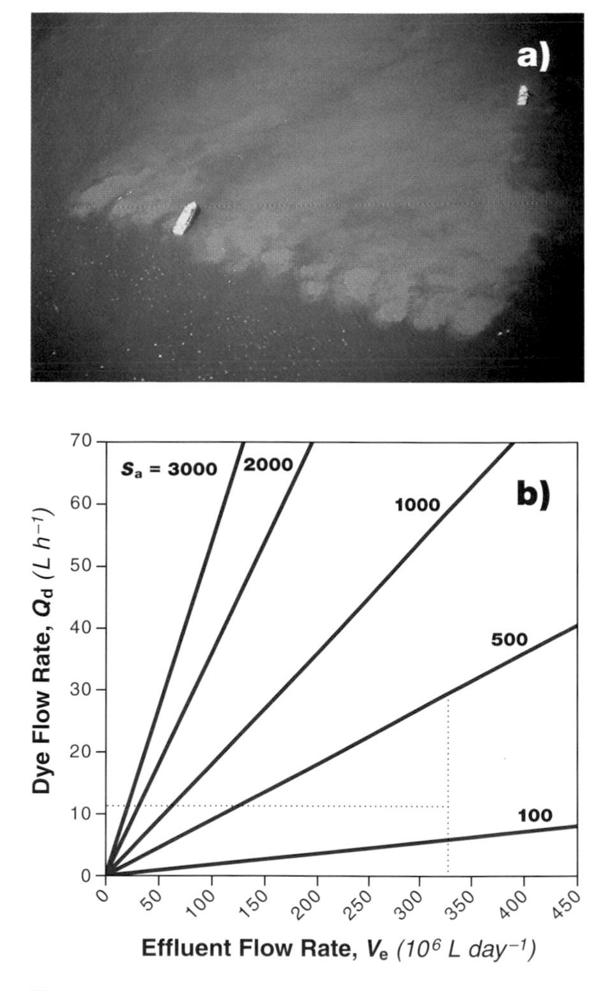

Figure 13-7 (a) A surfacing plume dyed with rhodamine WT of partially treated sewage offshore from San Francisco, California. The dye serves as a tracer for monitoring bacterial counts in drifting sewage plumes. Two monitoring vessels are visible. Photograph: W. Smith (b) A graph used to determine the dye required to provide 1 μg L^{-1} in diluted effluent as used in the study in (a). The dotted lines illustrate using the graph to solve the problem in the text.

frequency may not have the highest average concentration.

Beyond the mixing zone, there may be regions where current streams of diluted effluent, each leaving the zone at a different time in a different direction, would converge over a reef, a kelp forest, or a swimming area. In this case, the frequency and duration of exposure may be more important than the highest observed concentration in assessing the overall impact on these resources.

13.9.3 Verification Sampling

Aside from the question of whether discrete values or cross-sectional averages are used to test compliance with criteria, the way in which field samples are used to verify or compare with model results is an important consideration.

In laboratory or field verification studies of plume performance, the average value is measured or captured in a sample bottle only by chance. Characteristically, the field value measured is from a very small spatial region and represents a signal over a certain time span. Many samples are sought from the same cross section in order to arithmetically compute an average. In the laboratory, using hydraulic models, this is relatively easy to do. But in the field, where multiple plumes are usually involved, sampling is more complicated. We're usually trying to take samples from a moving flow field too deep below the surface to see, using a moving sampler mounted on a moving boat. It is therefore reasonable to assume some uncertainty as to what portion of the cross section the value represents.

For these reasons, field verification studies of submerged plumes in deep rivers, lakes, and coastal waters are best attempted for a cross section as far from the outfall as practical, as long as the region is still within the range where the plume is continuous. Nearer to the outfall, the values are changing more rapidly and the dimensions of the plume are much smaller, making it much harder to get the sampler in the right place, or even in the plume. In addition it is best to conduct the study when currents are low so that the plume rises nearest to the surface. Placement of the sampling device may be improved because it may even be possible to see the plume. Aside from the ease of sampling, samples taken during low currents may be especially useful for verification of regulatory compliance.

References and Recommended Reading

Bacow L.S. and Wheeler M. (1984) *Environmental Dispute Resolution*. Plenum Press, New York.

Baumgartner D.J., Frick W.E., and Roberts P.J.W. (1994) *Dilution Models for Effluent Discharges*, 3rd Edition. Report 600/R-94/066. U.S. Environmental Protection Agency, Washington, D.C.

Expert Panel on the Role of Science at EPA (U.S.) (1992). *Safeguarding the Future: Credible Science, Credible Decisions: The Report of the Expert Panel on the Role of Science at EPA to William K. Reilly, Administrator.* Report 600/9-91/050. U.S. Environmental Protection Agency, Washington, D.C.

National Research Council (1990) *Managing Troubled Waters: The Role of Marine Environmental Monitoring*. National Academy Press, Washington, D.C.

National Research Council (1993) *Managing Wastewater in Coastal Urban Areas*. National Academy Press, Washington, D.C.

Relative Risk Reduction Strategies Committee, United States Environmental Protection Agency. Science Advisory Board (1990) *Reducing Risk: Setting Priorities and Strategies for Environmental Protection.* Report SAB-EC-90-021. U.S. Environmental Protection Agency, Washington, D.C.

Problems and Questions

1. Calculate the time required for the coliform count to diminish to 1000 per 100 mL (the bathing water criterion) from a sewage discharge containing 10^7 per 100 mL. The die-off coefficient is expected to be 1.5 per day (base e).

2. Suppose currents in the receiving body of water average 0.8 kilometers per hour. How far away from the bathing area must the discharge be located?

3. The example BOD curve in the chapter (Figure 13-3 on page 198) shows a BOD value of 6.1 mg L^{-1} at day five. It looks like the ultimate BOD (BOD_L) is about

7.6 mg L^{-1}. We solved the BOD equation using these values to find that the value of K was about 0.32 per day. Suppose the ultimate value was 15. What would be the value at day five?

4. Suppose another waste has an ultimate value of 15, but the waste is more readily degradable. Would the value at day five be lower or higher? Why?

5. Calculate the oxygen concentration in the receiving body of water at 1, 2, 4, and 8 days caused by discharge of partially treated sewage with a BOD of 30 mg L^{-1} and a decay coefficient of 0.1 per day. The reaeration coefficient is 0.3 per day. This is a pristine body of water with no initial deficit in the region of discharge. Consequently, the ambient water is essentially saturated at 10 mg L^{-1} of dissolved oxygen.

High cash value production crops, such as lettuce, require frequent nitrogen fertilizer applications, which with irrigation can result in nitrate pollution of underground aquifers. Photograph: T.L. Thompson.

14.1 Soil as a Source of Plant Nutrition

Plants need an array of chemical elements in order to complete their life cycles. The 16 **essential elements** required for the growth of all plants are C, H, O, N, P, K, Ca, Mg, S, Fe, Mn, Zn, Cu, Mo, B, and Cl. Interestingly, soil microbes require these same elements. These elements, which usually occur in ionic form, must be taken up by plants from the soil solution. In undisturbed ecosystems, plants obtain these nutrients from the soil solution via mineral weathering, atmospheric inputs, inputs from stream deposition, or nutrient recycling due to death and decomposition of vegetation. The availability of the nutrients depends on abiotic soil factors (Chapter 2, beginning on page 9) and chemical and biological parameters (Chapters 6 and 7, beginning on pages 63 and 77, respectively). Throughout much of history, agricultural crop production has relied on these same nutrient sources, sometimes augmented by the use of animal or human wastes.

When the crops are harvested, nutrients are removed from the soil. Therefore, over long periods of time crop production may result in removal of large amounts of nutrients, with a concomitant decline in productivity. In addition to nutrient removal, long-term cropping may result in losses of soil organic matter. For example, after grassland soils at the Rothamsted Agricultural Station in England were planted to annual crops, soil organic matter content declined by 30–50% within 30 years.

Fertilizers are materials that contain essential plant nutrients and can be used to supplement, or even completely replace, the nutrients supplied by natural processes. Fertilizers are usually soluble inorganic compounds, which dissolve to give ionic nutrient forms in soil solution, but organic materials such as compost or sewage sludge may also serve as fertilizers (see Section 10.4, beginning on page 140). Because plants take up nutrients in ionic form, the source of those ions is irrelevant: plant nutritional needs are met equally well by fertilizers and nutrients produced by natural soil processes. It matters only that the ions be supplied. That the effects of fertilizers on crop productivity can be dramatic has

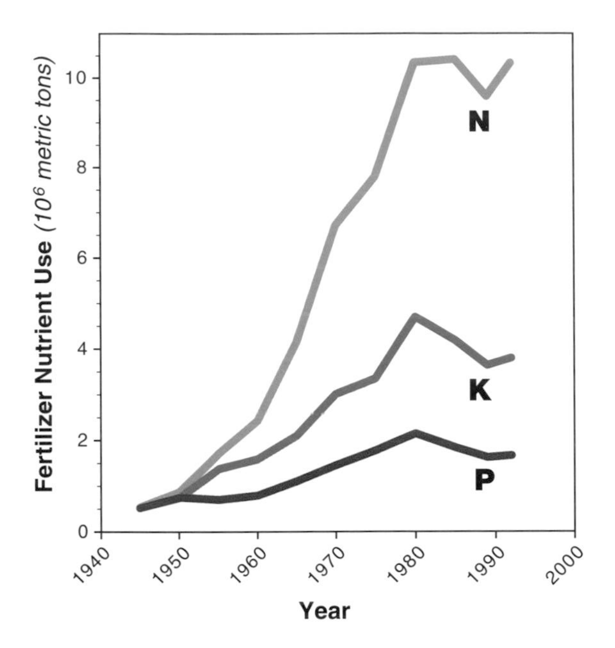

Figure 14-1 Historical trends for fertilizer nutrient use in the United States. Data: Berry, 1992.

been illustrated by research. For example, the Morrow Plots at the University of Illinois, which were established in 1876 to study the effects of long-term cropping on soil productivity, showed that between 1876 and 1954, corn yields on unfertilized plots declined drastically. But when fertilizers were added in 1955, the result was an immediate fourfold increase in crop yields.

14.2 Nitrogen as a Source of Pollution

Fertilizers may contain any of the essential nutrients, but the majority of fertilizers applied to agricultural soils contain nitrogen (N), phosphorus (P), potassium (K), or some combination thereof. These are the so-called macronutrients because plants take them up in larger amounts than the other essential nutrients. Because agricultural soils often lack sufficient amounts of one or more of these nutrients for maximum crop productivity, they are often applied in the form of fertilizers. Figure 14-1 illustrates trends for macronutrient use in the United States. Fertilizer use dramatically increased around the time of World War II, as improved crop varieties and management practices, together with increased mechanization, made fertilizer use both practicable and profitable. Since the mid 1980s, however, fertilizer use has leveled off, reflecting both lower agricultural profitability and increased environmental concerns related to fertilizer use.

14.2.1 Nitrogen Fertilizer Use in Crop Production

The nutrient most commonly applied to U.S. cropland is nitrogen. For example, in 1992 a total of 10.3 million metric tons of N was applied—an average of 83 kg for every hectare of cropland. Table 14-1 shows the most commonly used fertilizer materials. Recently, the use of high-N gas or liquid compounds, such as ammonia (NH_3) and urea ammonium nitrate, has increased, whereas the use of lower-analysis solid fertilizers has decreased. Nitrogen fertilizers may be applied by several methods, including broadcasting on the soil surface (with or without subsequent incorporation), injection beneath the soil surface, or injection into irrigation water for crops. Nitrogen fertilizers are commonly applied prior to planting (*preplant*) and may be applied to growing crops (*side-dressing*) where crop management conditions permit.

14.2.2 Nitrogen Cycling in Soils

Nitrogen is subject to a wide variety of chemical and biochemical reactions within the soil. When it is introduced into the soil in fertilizers, nitrogen is usually in the form of urea ($CO(NH_2)_2$), ammonia (NH_3), or as salts of ammonium (NH_4^+), or nitrate (NO_3^-). But regardless of which form of N fertilizer is applied to soils, the same types of reactions occur. Plants usually take up N as NH_4^+ or NO_3^-. Such ions may be obtained directly from dissolution of salts or indirectly by other processes. Urea, for example, is subject to hydrolysis in soils:

$$CO(NH_2)_2 + H_2O \xrightarrow{urease} 2NH_3 + CO_2 \qquad (14\text{-}1)$$

where urease is a microbial enzyme. This reaction can result in losses of N to the atmosphere as gaseous NH_3 when urea is applied at or near the soil surface. Losses are enhanced if soil surface pH is 7 or

Table 14-1 Commonly used N and P fertilizer materials. Data: Berry, 1992.

Nutrient	Fertilizer	Nutrient Content (% by weight)	Amount used in 1992 (million metric tons)
N	N solutions	32	7.31
	NH_3	82	4.54
	$CO(NH_2)_2$ (urea)	46	3.21
	NH_4NO_3	35	1.72
	$(NH_4)_2SO_4$	21	0.74
P	Superphosphate	10.5	0.03
	Triple superphosphate	21	0.45
N and P	Ammonium phosphate	18-N 20-P	3.11

greater. The fate of ammonium can be (1) fixation within the interlayers of soil clay minerals, (2) conversion to NH_3 and gaseous loss, (3) uptake by plants or microorganisms, or (4) **nitrification**:

$$NH_4^+ + \tfrac{3}{2}O_2 \longrightarrow NO_2^- + H_2O + 2H^+ \qquad (14\text{-}2)$$

$$NO_2^- + \tfrac{1}{2}O_2 \longrightarrow NO_3^- \qquad (14\text{-}3)$$

Nitrification is catalyzed by aerobic chemoautotrophic bacteria, which use the oxidation of N as an energy source. Equation (14-2) is catalyzed by species of *Nitrosomonas*, and Eq. (14-3) is catalyzed by species of *Nitrobacter* (see also Eqs. T4-9 and T4-10 on page 39).

Rates of nitrification are strongly influenced by soil temperature and pH. Maximum rates of nitrification are achieved when the soil is aerobic, the soil pH is between 6 and 8, and the soil temperatures are between 25 and 35°C. Nitrate is subject both to uptake by plants or microorganisms and to **denitrification** in anaerobic saturated soils with low redox potentials (see Eq. T4-14 on page 40):

$$2NO_3^- + 12H^+ + 10e^- \longrightarrow N_2 + 6H_2O \qquad (14\text{-}4)$$

This reaction illustrates a reduction half-reaction, in which nitrate is used as a terminal electron acceptor by facultative or obligate (incapable of aerobic metabolism) anaerobes. Many bacterial genera can use nitrate as a terminal electron acceptor.

The biological and nonbiological reactions governing soil N transformations are often considered collectively as the **nitrogen cycle** (Figure 14-2 on page 215). In most terrestrial ecosystems, by far the largest soil N pool is organic. This organic matter may consist of plant or animal residues in various stages of decomposition together with complex organic compounds resulting from microbial degradation. This degradation product is called **humus**. Humus, whose chemical composition is both uncertain and variable, is highly resistant to further microbial action (see Section 2.2.7, beginning on page 15). When added to soils, fresh organic material may be used by heterotrophic microorganisms as a carbon and energy source. The organic N present in the material is converted to NH_4^+ by the following process:

Aminization: Organic N to amino acids and amino sugars

Ammonification: Amino acids and amino sugar to NH_4^+

Aminization and ammonification together constitute **mineralization**. Conversely, inorganic soil nitrogen (NH_4^+, NO_3^-) may be assimilated by microorganisms in a process called **immobilization**. Soil microorganisms require specific C:N ratios in their tissues, typically 5:1 to 10:1. Relative amounts of mineralization and immobilization are determined by the C:N ratio of material added to soils. In general, addition of material with C:N < 30 promotes net mineralization while >30 promotes net immobilization.

Nitrogen fixation is the conversion of atmospheric nitrogen (N_2) to forms useful to plants and microorganisms. Nitrogen fixation may be accomplished biologically. Some bacteria have the ability to convert N_2 to NH_3 through catalysis by the enzyme nitrogenase:

$$N_2 + 10H^+ \longrightarrow 2NH_4^+ + H_2 \quad (14\text{-}5)$$

Some N-fixing bacteria live in symbiotic association with plants (*e.g.*, rhizobia with legumes, see Figure 4-2 on page 33), while others live independently of plants but within root rhizospheres. Biological N fixation, which accounts for 1.40×10^{11} kg of annual N additions to terrestrial ecosystems, is important for world food production and represents an essentially nonpolluting method of N fertilization. Unfortunately, many of the most important staple crops, such as grains, have no microbial associations allowing N fixation. Nitrogen fixation may also occur through industrial processes such as the Haber-Bosch technology and through abiotic processes such as lightning in the atmosphere (Figure 14-2).

The soil N cycle comprises both biological and nonbiological reactions. Of these, the biological reactions are the most important—both in magnitude and in the size of the pools involved—for controlling the fate and forms of soil N. These are shown in green in Figure 14-2. Mineralization and immobilization are controlled largely by the buildup and decay of soil heterotrophic biomass. Biological N fixation and denitrification are the reactions that control the relationship between the large pools of N in the atmosphere and soil.

One important nonbiological reaction in the N cycle is **nitrate leaching**. Because of its negative charge, nitrate is not strongly adsorbed to soil colloids and is highly mobile within the soil liquid phase (see Section 6.3.1, beginning on page 67). As water percolates through the soil, nitrate within soil pores is highly subject to leaching below the root zone and into the vadose zone. Once into the vadose zone, nitrate may enter the groundwater and become a pollutant (Figure 14-2). Nitrate is, in fact, the most ubiquitous of all groundwater contaminants. Since the mid 1970s, there has been a great increase in concern over the contamination of groundwater by nitrate. This increase is attributable to the proven association of nitrate with methemoglobinemia (blue-baby disease: an often-fatal disease of babies less than six months old) and its possible association with other diseases. It is still unclear how much nitrate contamination results from crop production and how much results from livestock operations or natural geological processes. Some researchers have attempted to use N isotope ratios, ($^{15}N : ^{14}N$) in groundwater to determine the source of groundwater nitrate. The reliability of this methodology is in question, however, due to the very small differences in isotope ratios as well as limited understanding of subsurface hydrology. Repeated attempts since the mid 1970s to use this method have failed to produce a reliable method for elucidating the source of groundwater nitrate.

14.2.3 Groundwater Nitrate Pollution

Nitrates may pollute both groundwater and surface water. Groundwater pollution involves the risks associated with consuming high-nitrate water, while surface water pollution can lead to **eutrophication**—increased algal growth and oxygen depletion. Most of the concern surrounding nitrate pollution has focused on groundwater pollution. Groundwater supplies approximately 20% of all water used in the United States and a disproportionate amount of this total is used for drinking water, especially in rural areas. The Environmental Protection Agency (EPA) has established a maximum contaminant level of 10 mg NO_3-N L^{-1} as the maximum safe level for drinking water, and the World Health Organization (WHO) has established a maximum safe level at 11.3 mg NO_3-N L^{-1} for drinking water. These concentrations pose a hazard primarily for infants; the safe level for adults is probably higher, but remains unknown.

It is undeniable that significant amounts of nitrate pollution of groundwater are due to crop production and that fertilizers may contribute to this problem. However, the precise origins and relative contributions of various nitrate sources have not yet been determined: that is, the variable sources of nitrates in groundwater render definitive determination an intractable problem. Besides crop production, animal confinement operations, municipalities, industries, and geologic formations are all possible sources of NO_3^- in groundwater. Furthermore, it is difficult to generalize about the risk of groundwater pollution associated with a given activity because many fac-

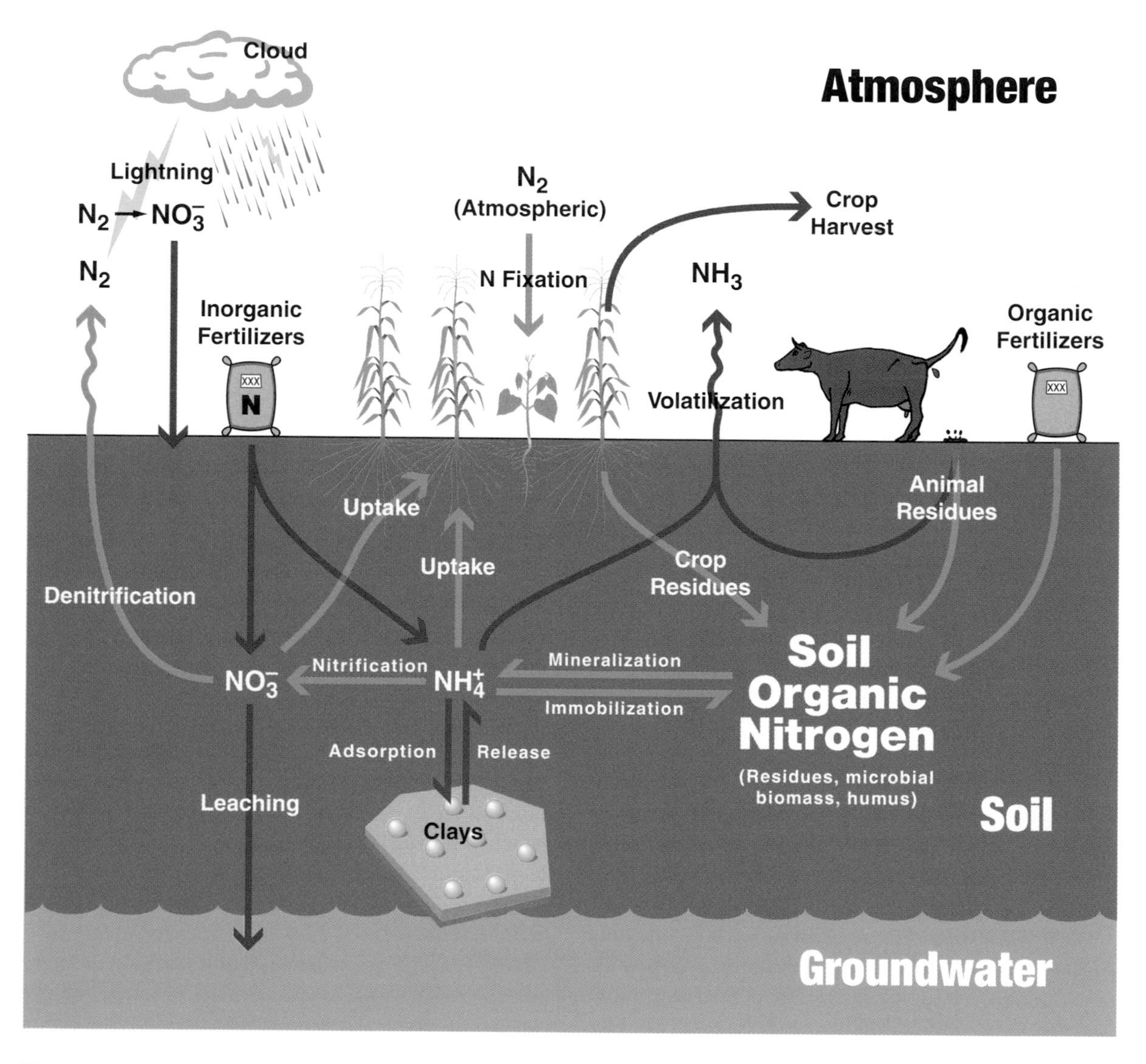

Figure 14-2 Soil nitrogen transformations.

tors, including groundwater depth, well location, and soil and vadose zone properties, affect the nitrate pollution that results from a specific activity.

In 1991 the EPA published a survey of groundwater wells within the United States that had been monitored for nitrate concentrations. Overall, most groundwater in the United States was found to be uncontaminated by nitrate. However, certain concentrated areas of contamination do exist. Data from the U.S. Geological Survey (USGS) indicate that the 20 states with the largest (total sales) agricultural marketings had a higher percentage (7.1%) of wells testing above 10 mg L^{-1} NO_3-N than did the other 30 states (3.0%). Notable areas of nitrate contamination include portions of the Great Plains overlying the Ogallala aquifer; central Arizona and central California, which are characterized by intensive irrigated agriculture; and parts of the Northeast and the Platte River Valley of Nebraska, characterized by intensive agriculture and shallow groundwater (Figure 14-3 on page 217). In general, these "hotspots" seem to be associated with areas that are now or were in the recent past important agricultural areas. In Nebraska, nitrate pollution is associated both with corn production and shallow groundwater. In central Arizona, contamination surrounding the Phoenix area seems to be related to the current or past presence of irrigated crop production.

Exceptions to this general pattern are those areas in which nitrate pollution is associated with geologic deposits rather than agricultural activities. Such areas include parts of the Las Vegas aquifer in Nevada, as well as aquifers in parts of North Dakota, Montana, and Wyoming. The nitrate in these aquifers is probably derived from exchangeable NH_4^+ that has been nitrified and leached downward from overlying sediments to the water table.

14.2.4 Factors Affecting Pollution from Nitrogen Fertilizers

Soil and climatic conditions have an important influence on how efficiently fertilizers are used. Anionic NO_3^- is generally more subject to loss than is NH_4^+. Because it is negatively charged, NO_3^- is more subject to leaching and it can also easily be lost via denitrification. Therefore, we can see that rates of nitrification in soils significantly affect the potential for nitrate loss. We know that neutral pH conditions (pH 6–8) and warm temperatures (25–35°C) promote rapid nitrification. In addition, nitrate leaching is maximized in coarse-textured soils and in soils with large continuous pores, such as those under minimum or reduced tillage (see Figure 14-4 on page 218). Much nitrate leaching may also occur during intense rainfall events when water moves rapidly through the larger pores. The loss of nitrate by denitrification is an anaerobic process. Therefore, it is promoted by saturated soil conditions where O_2 has been depleted, as well as by warm temperatures and a source of readily oxidizable carbon, such as fresh crop residues (see also the special section, *Applied Theory: Biological Generation of Energy*, beginning on page 37 in Chapter 4).

Extensive use of N fertilizers presents an interesting societal dilemma, as do many features of modern technology. On the one hand, there is no doubt that use of N fertilizers has contributed tremendously to worldwide crop production in recent decades. Consider the United States for example. Nearly 25% of all U.S. crop production is directly attributable to N fertilizers, and the resulting high levels of crop production have given U.S. consumers some of the most abundant and least expensive food in the world. Moreover, nitrogen fertilization is vital for profitability. In most areas of the United States, the cost of N fertilizer is low relative to the value of crops, thus giving crop producers an incentive to apply excess N as "insurance" against low, unprofitable yields. On the other hand, despite these benefits, excessive use of N fertilizers can pose a threat to groundwater supplies. How, then, should we balance the benefits and risks associated with modern crop production? This problem remains a significant challenge.

14.2.5 Best Management Practices

The goal of any fertilization program is to add enough nutrients to maximize plant productivity without wasting nutrients. Wasted nutrients represent an economic loss to crop producers, and, in excess, can damage the environment. Growers can manipulate several variables relating to fertilizer and water applications to maximize plant uptake while minimizing economically and environmentally detrimental nutrient losses. Methods and times of application can be varied, as can the selection of fertilizer materials. Efficient and timely use of appropriate fertilizers is the goal of Best Management Practices (BMPs), which are currently being developed by researchers for a wide range of crops in many parts of the United States. But no matter how efficient crop production is or how advanced management tools have become, a certain amount of inefficiency continues to be associated with any crop production system. In fact, some inefficiency is probably inherent because of the many chemical and biochemical reactions that compete with plant uptake for the fertilizer nutrients. For example, many reports from around the country have established that no more than 30–70% of N applied is usually recovered by crops within the first year. The fertilizer that is not utilized by plants therefore remains in the soil or is lost by various mechanisms.

Figure 14-5 on page 218 shows an idealized crop yield curve. As more fertilizer is applied, we begin to see **diminishing returns**; that is, the crop re-

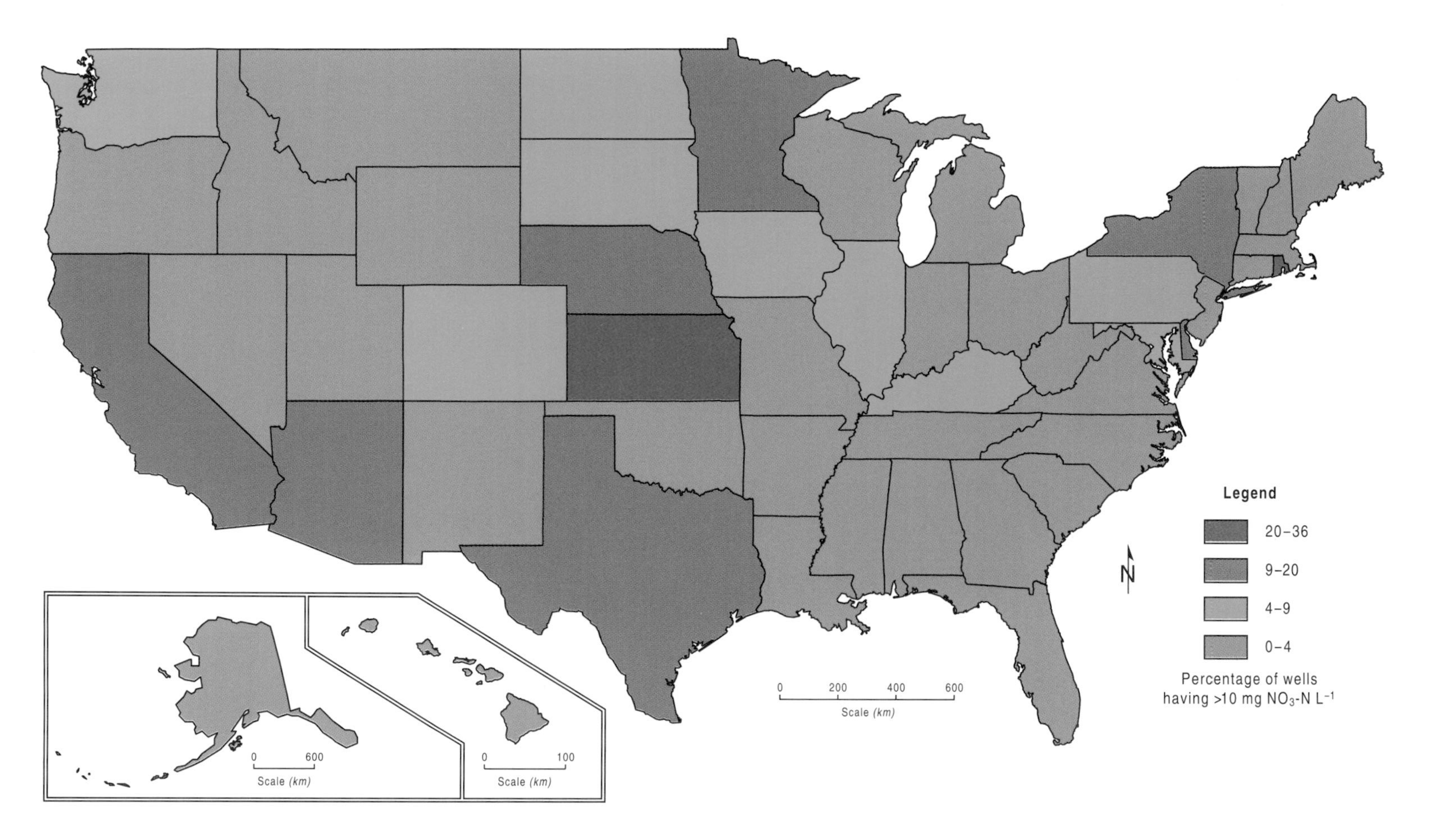

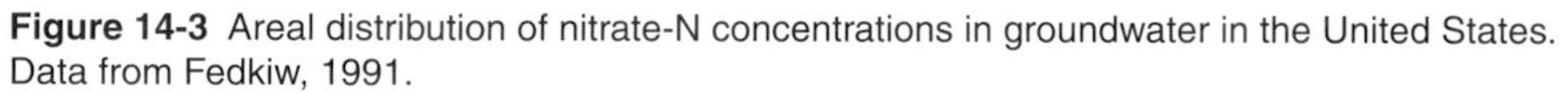

Figure 14-3 Areal distribution of nitrate-N concentrations in groundwater in the United States. Data from Fedkiw, 1991.

Figure 14-4 Minimum tillage in corn (Virginia, USA). (a) The planter sows corn seed directly in a rye cover crop that has been treated with an herbicide to kill it by the time corn starts growing. This practice aids in the control of soil erosion (see also Section 15.4 on page 234). (b) However, as the soil is neither plowed nor disked, large, continuous pores are not disturbed, allowing preferential flow to quickly carry water and dissolved substances, such as nitrate, to lower depths (see Section 5.3.3, beginning on page 58). Photographs: J.W. Brendecke.

sponse to each additional increment of fertilizer is smaller than the response to the previous increment. At low rates of application, yields may be low, but the percentage of N losses is low. As fertilizer application increases, however, N losses also increase due to leaching, denitrification, or other mechanisms. The amount of N loss can be minimized through efficient management of water and N inputs.

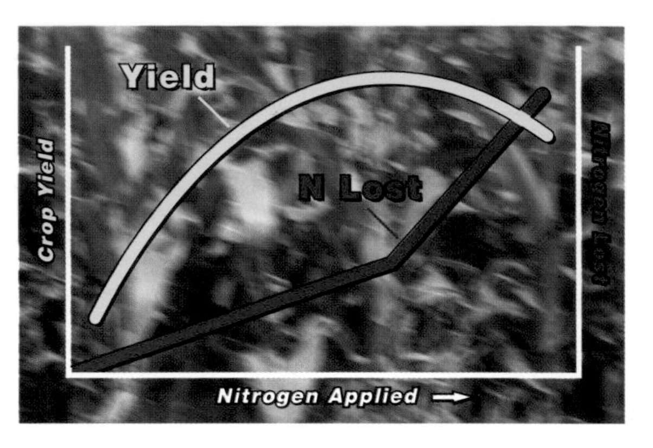

Figure 14-5 An idealized crop yield curve and the concomitant loss of nitrogen.

Nitrogen losses increase sharply when fertilizer is applied in excess of plant needs.

One BMP that can enhance the efficiency of N fertilizer involves **slow-release fertilizer** materials, which release N over a period of days or weeks. Three general types of slow-release N fertilizers are available: (1) low-solubility substances requiring decomposition, such as urea derivatives; (2) soluble fertilizers coated with insoluble materials; and (3) sparingly soluble materials. Currently, their high cost limits the use of slow-release fertilizers largely to turfgrass and high-value crops such as citrus.

Other BMPs involve controlling the amount and timing of N fertilizer applications. It is possible to apply N fertilizer when and where the plant needs it. Figure 14-6 shows the cumulative N uptake for a typical crop harvested at physiological maturity, such as corn, wheat, or cotton. Obviously, the plants need little N early in their growth. Therefore, N applied prior to or at planting has a high potential for loss. Ideally, the crop producer should apply the N exactly when the crop needs it, which can be determined by employing other management tools such as plant-tissue testing to assess the true nutritional status of the plant. This management practice is fine in principle; unfortunately, in practice it is often difficult for producers to apply N exactly when needed, because of equipment or other constraints. Application of N at the time of plant demand for N is most easily achieved in irrigated production systems, where N fertilizers can be introduced with irrigation water.

Figure 14-6 Seasonal trend for N uptake by an idealized annual crop. Based on Figure 44A on page 66 of Doerge *et al.*, 1991. Photographs: J.W. Brendecke.

Still other BMPs involve improved water management. For example, **irrigation water management** is an important component of Best Management Practices. Irrigation practices that result in large and uneven applications of water to soils, which are all too often the norm in many areas, greatly increase the risk of losses of mobile nitrate from the crop root zone. Techniques that can make water applications more uniform and efficient can help to minimize nitrate losses. For example, laser-leveling of land allows for even water distribution across a field, and pulsed rather than continuous flow of water down a furrow has resulted in more uniform water and fertilizer applications. In addition, above- or below-ground drip irrigation systems have great potential for increasing water and nutrient use efficiency. Such systems can very efficiently apply nutrients and water directly to the crop root zone with little waste.

As of 1996, most of the states in the U.S. do not regulate fertilizers or their application to soils. In Arizona, however, where some of the most extensive nitrate contamination has occurred, crop producers must demonstrate compliance with BMPs or face penalties including fines. Such regulation can be expected to increase as the public concern over pollution increases.

14.3 Phosphorus as a Source of Pollution

14.3.1 Phosphorus Fertilizer Use in Crop Production

Approximately 1.68 million metric tons of phosphorus (P) is applied in fertilizers to U.S. cropland yearly (Figure 14-1 on page 212). Phosphorus is usually applied as solid materials, such as triple superphosphate, which is predominantly $Ca(H_2PO_4)_2$, or as various ammonium phosphate compounds (Table 14-1 on page 213). These are soluble compounds that provide a ready source of P for growing plants and microorganisms, which commonly take up P from soil solution as $H_2PO_4^-$ or HPO_4^{2-}. The chemistry of P in soils is quite different from that of N for several reasons. Phosphorus is used neither as a direct source of energy nor an electron acceptor by microorganisms. Furthermore, phosphate anions, unlike nitrate, are bound strongly to soil colloids owing to ligand exchange reactions; therefore, they do not leach or move readily in soils. In addition, the insolubility of most phosphate-containing compounds results in low phosphate concentrations within the soil solution. The predominant reactions that occur following applications of P fertilizers to soils are reactions of P with Al, Fe, or Ca ions in soils to form insoluble precipitates. The insolubility of these compounds has the effect of decreasing the availability of P fertilizers to plants over time and thus necessitates regular P applications to most crops.

14.3.2 Phosphorus Pollution of Waters

Phosphorus fertilizers can pose a pollution hazard under some circumstances. In areas having high water tables under organic soils, for example, P leaching may become significant; in addition, P fertilizers carried via runoff can threaten surface water bodies. In many surface waters, P is a limiting factor to algal and plant growth. Introduction of large amounts of P in leaching or runoff water may remove this constraint and encourage increased plant or algal growth. When these algae or plants die, their decomposition by heterotrophic microbes results in O_2 depletion within the water body. This process, known as eutrophication, can have disastrous consequences for the aquatic ecosystem. Eutrophication may be enhanced by inputs of N into N-limited ecosystems or by introduction of P into P-limited ecosystems. Concentrations of P as low as 10 $\mu g\ L^{-1}$ have been associated with eutrophication.

Well known sites of P pollution include Lake Okechobee and the Everglades in southern Florida, and the Chesapeake Bay on the U.S. East Coast. In Florida P pollution has led to eutrophication and the shift of plant populations from native species with low P requirements, such as sawgrass (*Cladium jamaicense*), to invader species with higher P requirements. A significant portion of the Florida P pollution is attributable to intensive irrigated agriculture on the surrounding organic soils. In these systems, large applications of P fertilizers occur regularly. Furthermore, much of the drainage water from these agricultural areas is deposited in Lake Okechobee or the Everglades. Under natural conditions, P is usually the limiting nutrient in these ecosystems. With the introduction of agricultural P, the nutrient balance has been shifted significantly.

The Chesapeake Bay is an estuary where rivers carrying fresh water mix with seawater. Historically, this has been a rich and varied ecosystem and an important commercial fishing area. The P pollution within the bay probably comes from municipal and agricultural P inputs into rivers that drain into the bay. Unlike the situation in Florida, where intensive irrigation generates P-laden leachate, agricultural P inputs here are carried by surface runoff into the rivers. Within Chesapeake Bay, both N and P have been limiting nutrients, constraining the total biomass production within the entire estuary. Today, however, the N and P inputs have accelerated algal growth, with resultant eutrophication.

14.3.3 Factors Affecting Pollution from Phosphorus Fertilizers

In many areas of the United States, phosphorus pollution from agricultural sources is of no concern because of the low solubility and mobility of P in soils. However, the cases noted above are unique because they involve agricultural areas that abut estuaries (Chesapeake Bay) or heavily fertilized organic soils that overlie shallow groundwater (Everglades). Although estimated P losses are low (about 0.5–2.4 kg P ha^{-1} yearly in southern Florida), they have a disproportionately high impact owing to the sensitivity of the surrounding ecosystems. The major cause for agricultural P pollution in these areas seems to be inadequate timing of fertilizer application, together with excessive surface soil erosion in the case of the Chesapeake Bay area.

The prevention of P pollution also involves BMPs. On the Chesapeake Bay watershed, more efficient fertilization practices in agricultural areas, such as split applications of fertilizers, have resulted in decreased loading of both N and P into the bay. Such split applications of P fertilizers are also being used in the Everglades Agricultural Area to minimize leaching of P into shallow groundwater.

Case Study: The Chesapeake Bay

Concentrations of nutrients entering the Chesapeake Bay have increased for two major reasons. First, land use has changed: it has gone from forests to crop production and urban development. Second, agricultural and urban development have increased nutrient inputs. Because nutrient inputs have generally risen and natural buffer areas have generally declined, greater amounts of nutrients wind up in the bay, accelerating algal growth and eutrophication. Of these nutrients, nitrogen and phosphorus have been identified as the culprits.

In 1983 and 1987, three of the states bordering the Chesapeake Bay (Maryland, Pennsylvania, and Virginia) signed agreements whose objective was to reduce nutrient inputs. The major goal of these agreements was to reduce nutrient inputs into the bay by 40% by the year 2000. The agreements established controls for point and nonpoint sources of pollutants; thus, they banned phosphate detergents and put controls on animal wastes and implemented other agricultural BMPs.

By 1991, P levels had decreased by 20%, largely as result of the ban on such point sources as phosphate detergents. By 1995, however, nitrogen levels had not decreased. This difference is due to sources of these nutrients—most N input is from nonpoint sources, such as animal wastes—and is also due to the greater mobility and reactivity of N in soils compared to P. Furthermore, it has been discovered that air pollution may contribute large amounts of N pollution to the bay. Controlling N levels may be difficult without air pollution controls.

Progress has been made in cleaning up the Chesapeake Bay. Phosphorus levels have been decreased, and N levels have remained constant, even though population in the area has increased. But several critical problems remain. First, the 40% goal for nutrient input reductions was based on model results, which may be incorrect. It is unclear how much nutrient reduction is actually needed to remove the danger to the bay. Second, control of N inputs may be impossible without stringent controls on N fertilizer use and animal production in surrounding areas. Third, N inputs from air pollution remain an unknown, but probably significant, source of N pollution. Continued regional pollution control efforts will be needed if the Chesapeake Bay is to be restored.

Risk Assessment of Agricultural Fertilizers

Risk assessment is the exercise of determining the level of risk involved in environmental pollution, as discussed in Chapter 22, beginning on page 345. Risk assessment can help to determine the priority assigned to cleanup and (or) abatement of the pollution hazard.

In the case of P pollution, the risk is most clearly associated with disturbance of delicate plant and animal communities. Moreover, the risks of P pollution are palpable: we can see them and respond to them. For instance, the P pollution of the Chesapeake Bay area has been the target of aggressive regional action since the 1980s. Similarly, during 1993, many regulatory, environmental, and agricultural groups were able to reach an agreement that in part addresses the problem of agricultural P inputs into the Everglades ecosystem.

The more subtle risks associated with nitrate pollution of groundwater are proving more difficult to assess. Because it involves direct risks to human health, nitrate pollution has attracted considerable media attention in some regions with groundwater contamination. However, the true risks to human health from nitrate ingestion remain largely unknown. It is known that high levels of nitrate intake by newborns cause methemoglobinemia, but this disease is both rare and easily prevented with adequate information. Other reports have suggested possible links between nitrate ingestion in water and various cancers, but no direct link has yet been found. At present, it seems that nitrate contamination of groundwater merits guarded concern but does not apparently pose as high a risk as that posed by some organic compounds. It's worth remembering, however, that nitrate is the most ubiquitous groundwater contaminant.

Cleanup and Abatement of Fertilizer Pollution

Many strategies have been suggested for cleanup of contaminants from groundwater. There have been few attempts to clean up nitrate- or phosphate-contaminated waters. In general this is because of (1) the massive size of contaminated water bodies; (2) the relatively lower risk involved compared to some organic compounds; and (3) the lack of affordable technologies. Nitrate does not lend itself readily to removal from water; therefore pump-and-treat methods would require nitrate removal by distillation, reverse osmosis, or anion-exchange resins, which would be far too expensive. A method of *in situ* removal of nitrate has been proposed that involves enhanced denitrification. In the vadose zone, denitrification is usually limited by inadequate supplies of oxidizable C as an energy source. Addition of C to groundwater under anaerobic conditions should promote the reduction of NO_3^- to N_2. However, these methodologies are unlikely to enjoy widespread use because they involve introduction of another chemical (such as ethanol or sugars) into groundwater, and because the volumes of water involved are too massive to make such an approach practical.

Although nitrate is the most common groundwater contaminant, serious efforts to clean up nitrate-contaminated groundwater are highly unlikely. In some parts of the country, significant amounts of groundwater are contaminated with nitrate, and they will probably remain that way for decades. The only practical answer to the nitrate problem is the prevention of further pollution by nitrate. Because much of the nitrate found in groundwater comes from agriculture, many of the answers to the nitrate problem lie in increased efficiency in crop production and the use of appropriate BMPs.

References and Recommended Reading

Berry J.T. (1992) *Summary Data, 1992*. Tennessee Valley Authority National Fertilizer and Environmental Research Center, Muscle Shoals, Alabama.

Doerge T.A., Roth R.L., and Gardner, B.R. (1991) *Nitrogen Fertilizer Management in Arizona*. Publication 191025. The University of Arizona College of Agriculture, Tucson, Arizona.

Fedkiw J. (1991) *Nitrate Occurrence in U.S. Waters (and Related Questions)*. USDA Working Group on Water Quality, United States Department of Agriculture, Washington, D.C.

Madison R.J. and Burnett J.O. (1985) Overview of the occurrence of nitrate in ground water of the United States. *Water Supply Paper* 2275. U.S. Geological Survey, Washington, D.C.

Spalding R.F. and Exner M.E. (1993) Occurrence of nitrate in groundwater: A review. *Journal of Environmental Quality*. **22**, 392–402.

Stevenson F.J., Editor (1982) *Nitrogen in Agricultural Soils*. Agronomy Monograph No. 22. American Society of Agronomy, Madison, Wisconsin.

Tisdale S.L., Nelson W.L., Beaton J.D., and Havlin J.L. (1993) *Soil Fertility and Fertilizers*. Macmillan, New York.

Problems and Questions

1. Distinguish between biological and nonbiological reactions comprising the nitrogen cycle.
2. Under what conditions is nitrate leaching from crop production likely?
3. In addition to crop production, what are some other potential sources of nitrate to groundwater?
4. Describe some soil and crop management practices that can help in minimizing nitrogen losses.
5. Why has P pollution been a problem in the Chesapeake Bay and the Everglades? Why is P pollution not a problem in most other areas?
6. What is the most ubiquitous groundwater contaminant? Explain why it is so common and discuss the implications of its ubiquity for human health.
7. Why is minimizing future groundwater contamination with nitrate more practical than cleanup of existing contamination?

Once void of its protective vegetative cover, soil is subject to numerous factors which may bring about extensive erosion and with it transport of pollutants sorbed to the soil. Above, a clearing for a housing development near Tucson, Arizona. Photograph: D.F. Post.

15.1 Sediment Pollution

Agricultural technology is one of the real strengths of the United States. Although the population has increased steadily and cropland area has changed little over the past 50 years, food and fiber production has amply met domestic needs and has also provided substantial amounts for export, which is critical to the U.S. trade balance. Fertilizers and pesticides have played a major role in this accomplishment. However, the very success of U.S. agriculture has led to negative effects, including pollutants such as sediment, chemical fertilizers, and pesticides in surface waters (see also Chapters 14 and 17, beginning on pages 211 and 253, respectively). These pollutants may come from point or nonpoint sources. Whereas point sources (which are usually covered by a permit system) are fairly well defined, nonpoint sources (such as mining areas, construction activities, and stormwater runoff from urban areas) are diffuse in nature and discharge pollutants into waters by dispersed pathways (see also Section 16.2, beginning on page 238).

Another major nonpoint source is agricultural runoff, which can produce sediments that are eroded from croplands. For example, the Oklahoma Dust Bowl of the 1930s was largely the result of such an erosion problem—a problem so great that new federal agencies were created to deal with it. Nor is erosion the only problem caused by sediments. Chemicals adsorbed on transported soil particles can be transported to streams and lakes in runoff water.

In this chapter we will concentrate on erosion. We will examine the fundamental processes that transport sediment into water bodies and survey the principal methods by which erosion can be controlled. The focus here is on erosion by water; however it's important to remember that wind erosion under local, arid conditions is also important.

15.2 Extent of Sediments Causing Pollution

The geographic potentials for direct runoff and erosion in the United States are best identified by the

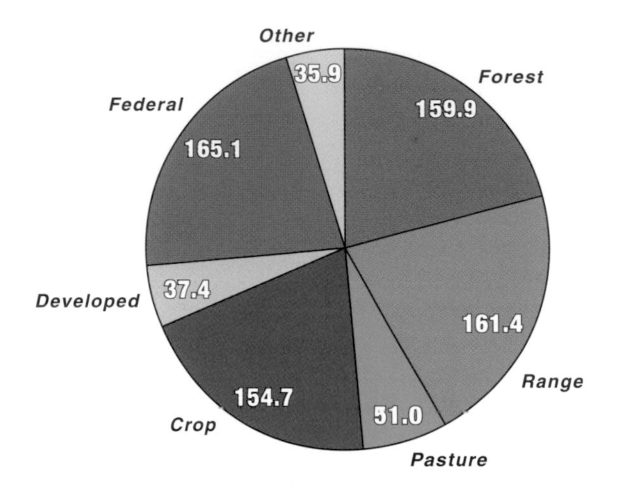

Figure 15-1 U.S. Land use: Nonfederal land 1,483.1 (1992 Data in millions of hectares; excludes Alaska). Source: USDA Soil Conservation Service, 1994.

Land Resource Areas, which are delineated in a classification system devised by the Soil Conservation Service (now called the Natural Resources Conservation Service) of the U.S. Department of Agriculture. These areas are described in detail in Agricultural Handbook No. 296, *Land Resource Regions and Major Land Resource Areas of the United States*. A Land Resource Area is defined as a geographic area characterized by a particular pattern of soil type, topography, climate, water resources, land use, and type of farming.

The type of land resource has a significant bearing on the amount of potential erosion from an area and on the use of agricultural chemicals. The most recent inventory of these resources, the 1992 National Resources Inventory, was completed by the U.S. Department of Agriculture's Soil Conservation Service. It provides updated information on the status, condition, and trends of land, soil, water, and related resources on the nation's nonfederal land. The 1992 inventory covers the 48 contiguous states, Hawaii, Puerto Rico, and the U.S. Virgin Islands. The total area of land monitored was 780 million hectares, about 78% (605 million hectares) of which is nonfederal land, which is mostly privately owned. Data on soil characteristics, land cover, and erosion for more than 800,000 locations have been collected and statistically analyzed. The use of this land is summarized in Figure 15-1. Notice that the land is nearly equally divided between cropland, pasture or rangeland, and forest land.

Most erosion occurs on cropland. Therefore, the U.S. Department of Agriculture has developed a **Land-Capability Classification System**[1] that provides a quick, uniform, useful way to evaluate the potential of land for crop production. In this system, soils are divided into eight classes, I through VIII. Class I–IV soils are those most frequently utilized for cropland.

Class I soils have a low erosion hazard. Suited to a wide range of plants, these soils are nearly level, deep, and well drained; they also have high permeability and water-holding capacity, and are either well supplied with plant nutrients or are highly responsive to fertilizers.

Class II soils are moderately susceptible to wind and water erosion. These soils usually have gentle slopes and exhibit some limitations that may reduce the range of plants they can support. In addition, they often require special soil-conserving cropping systems, soil conservation practices, water-control devices, or tillage methods when they are used for cultivated crops.

Class III soils are more susceptible to water and wind erosion than soils in Class II. Normally found on moderately steep slopes, these soils can be used for cultivated crops, but require highly effective conservation practices to prevent erosion. Moreover, such practices may be difficult to apply and maintain if erosion is to be controlled.

Class IV soils, which are usually found on steep slopes, are highly susceptible to wind and water erosion (see Figure 15-2). Although suitable for cropland, they require careful management and may support only two or three types of crops.

Soils in Classes V–VIII are limited in their use and are generally considered unsuitable for cultivation.

Class V soils have little or no erosion hazard, but have other limitations that are impractical to overcome. Examples are bottomlands subject to frequent

[1] USDA Soil Conservation Service, 1961

Figure 15-2 Poorly drained (wet) cropland with Class III and IV soils in the lowlands. Photograph: D.F. Post.

overflow, stony soils, and ponded soils where drainage is unfeasible.

Class VI soils are usually limited to pasture, range, forest, or wildlife habitat. However, some Class VI soils can be used for common crops with careful management. Some of these soil may also be adapted to special purposes such as sodded orchards, or for cultivation of blueberries and similar crops.

Class VII soils are neither suitable for nor adaptable to cropland.

Class VIII soils are not only unsuited for cropland, but also have limitations so severe that they are restricted primarily to recreation, wildlife habitat, water supply, and esthetic uses.

The erosion hazard of cropland increases sharply from Class I through Class IV soils. Therefore, the larger the cropland area on Class III and IV soils, the greater the erosion hazard. Also, because sediment is a major transport mechanism for the transport of agricultural chemicals, the potential for the loss of such chemicals is much greater for these soils. These soils are therefore less desirable not only because they are more subject to erosion, but also because they are lower in fertility and yield less than Class I and II soils, particularly when no fertilizers are used.

As Figure 15-3 on page 228 shows, the greatest soil limitation restricting the use of land for growing crops is susceptibility to erosion and sedimentation. But only 3% of U.S. soils—those classified as Class I soils—present little or no such susceptibility. Moreover, a large proportion of U.S. cropland requires chemical augmentation of some kind. For example, it is estimated that from one-third to one-half of U.S. agricultural production depends on fertilizer use. This means that if fertilizer use were eliminated, cropland area would have to be greatly expanded to meet current yield levels; and according to the National Resource Inventory, any large increase in cropland would have to come from Class III and IV soils. Without fertilizers, then, the problems of erosion and sedimentation would increase dramatically.

Because sediment loss increases as soil class number increases, one approach to controlling water pollution from cropland is to concentrate crops to the fullest extent possible on Class I and II soils. These soils are naturally more productive and more responsive to fertilizers because of higher water-holding capacities; moreover, they are easier to control with respect to sediment losses. In all likelihood, therefore, a high level of food and fiber production could be achieved with the least impact on the environment by using fertilizers and pesticides on the better

Table 15-1 Conditions indicative of high sediment-yield potential.

Cropland
1. Long slopes farmed without terraces or runoff diversions
2. Rows planted up and down moderate or steep slopes
3. No crop residues on surface after new crop seeding
4. No cover between harvest and establishment of new crop canopy
5. Intensively farmed land adjacent to a stream with no intervening strip of vegetation
6. Runoff from upslope pasture or rangeland flowing across cropland
7. Poor stands or poor quality of vegetation
Other Sources
1. Gullies
2. Residential or commercial construction
3. Highway construction
4. Poorly managed range, idle areas, or wooded areas
5. Unstabilized streambanks
6. Surface mining areas
7. Unstabilized roadbanks
8. Bare areas of noncropland

lands, where their effectiveness is high and their loss is small.

The use of chemicals on more erosive soils presents a substantially greater threat to the environment. However, it is possible to use chemicals safely on poorer soils if a higher level of management is practiced to control sediment and associated chemical losses. Although the extent of the effects of agricultural pollutants on the environment remains uncertain, it appears that sediment, nutrient, and pesticide losses can be controlled at an acceptable level by the selection of proper management systems. The challenge, therefore, is to develop general assessment techniques and institutional mechanisms that

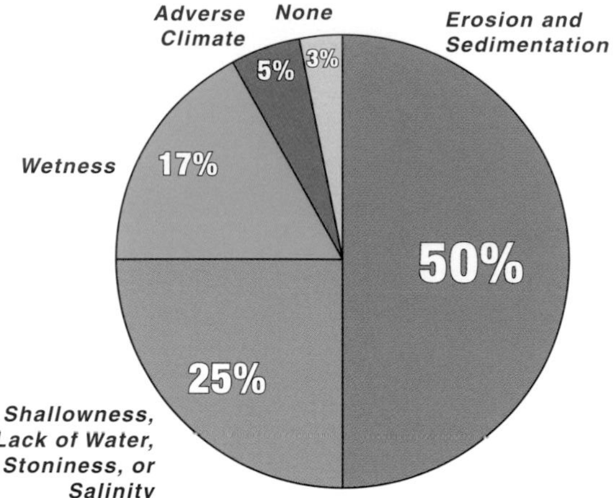

Figure 15-3 Dominant limitations (land capability subclasses) that restrict the suitability of U.S. soils for growing crops expressed as a percentage of total nonfederal land (600.3 million hectares). Note that erosion is the greatest potential problem. Data adapted from USDA Soil Conservation Service, 1992.

can be used for erosion control. It is also important to develop control practices that are site-specific. Because recommending control practices for large areas is extremely difficult—and often impractical—control recommendations for specific sites must be developed by specialists within the area.

Figure 15-4 presents a map that shows the average annual soil erosion rates by water and wind. Although the amount of erosion from U.S. lands has actually decreased in the last ten years, sedimentation remains a major problem. Sediment, which is an end product of soil erosion, is by volume the greatest single pollutant of surface waters; it is also the principal carrier of several chemical pollutants. In some watersheds, sediment from nonagricultural sources exceeds that from cropland; and in regions of relatively low rainfall or sandy soils, wind erosion exceeds water erosion. Table 15-1 describes land conditions where a high sediment yield can be expected.

15.3 Predicting Sediment Losses

Soil erosion is a three-step process: (1) detachment of soil particles from the soil mass; (2) transportation of detached soil particles through the watershed

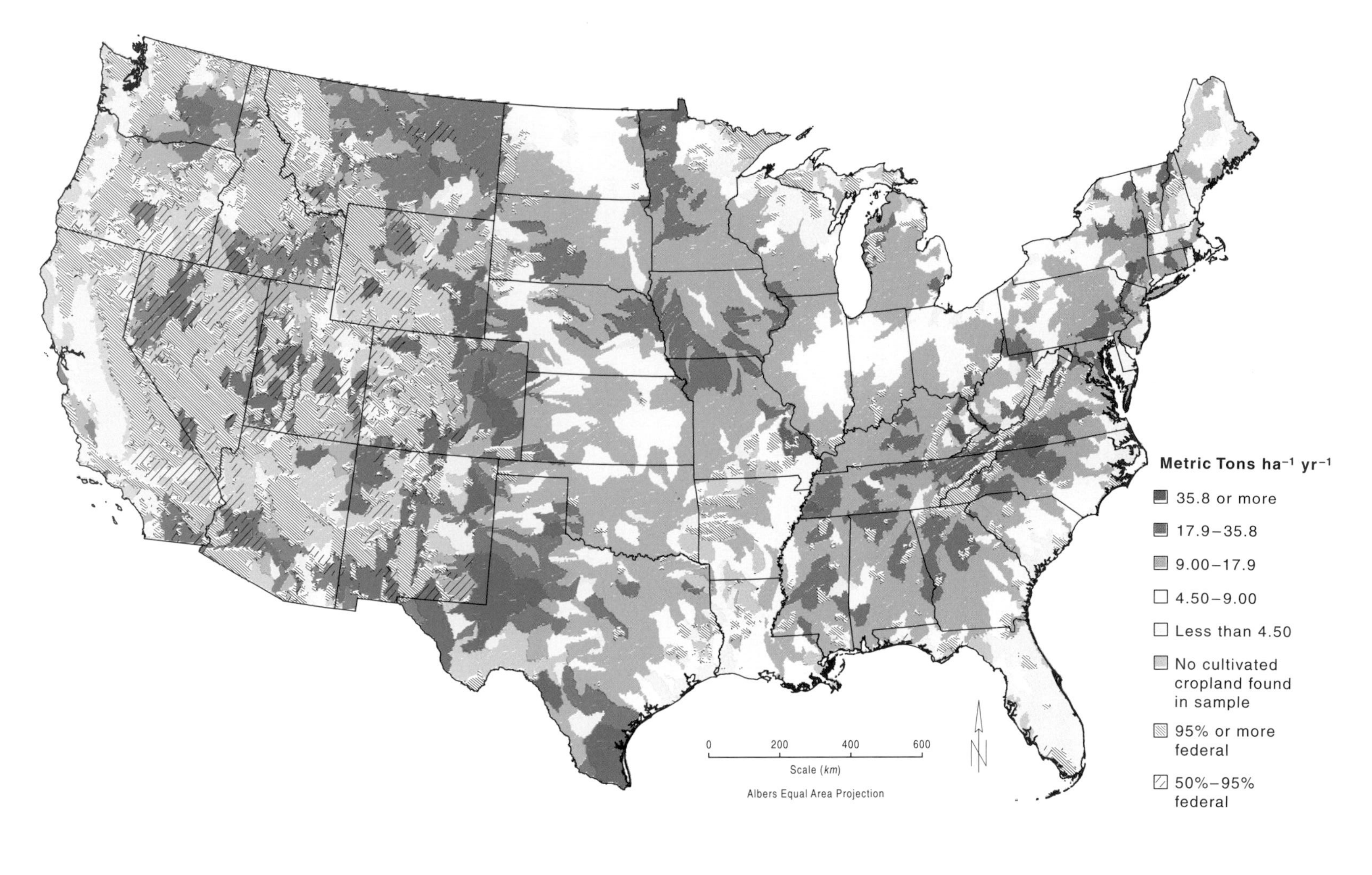

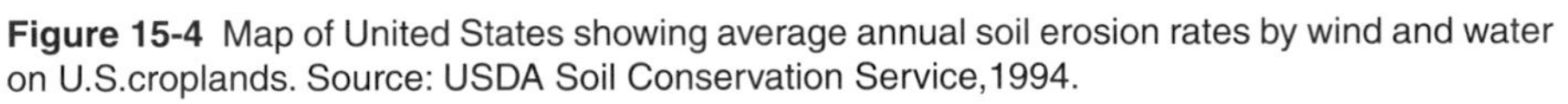

Figure 15-4 Map of United States showing average annual soil erosion rates by wind and water on U.S.croplands. Source: USDA Soil Conservation Service,1994.

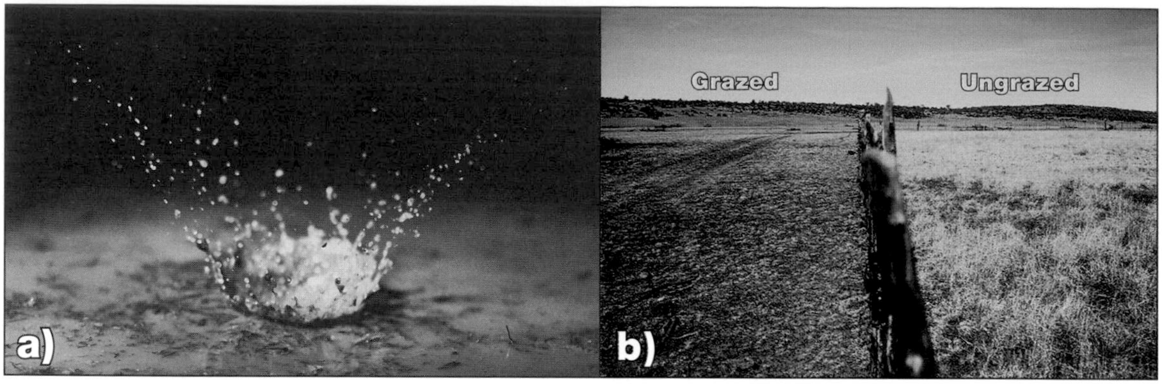

Figure 15-5 (a) A raindrop hitting the soil surface; (b) land management effects on erosion; here plant cover absorbs the raindrop energy and reduces soil erosion. Photographs: (a) Natural Resource Conservation Service, (b) D.F. Post.

system; and (3) sedimentation or deposition of the transported particles.

During the erosion process caused by rainfall, the detachment of particles is the independent variable; thus the impact of a falling raindrop on the soil surface is the most important factor causing detachment of soil particles. Figure 15-5a shows a raindrop hitting a soil surface. As we can see, an unimpeded collision is highly energetic. Any cover or protection on the soil surface that can absorb or dissipate this energy will significantly reduce erosion. Groundcover often provides such protection. In Figure 15-5b we see a "fence-line" contrast, where one side is well-managed rangeland with good grass cover while the other side is overgrazed and unprotected. Clearly, the overgrazed area will be more easily eroded than the well-managed area.

More than 50 years of erosion research by the U.S. Department of Agriculture in cooperation with state agricultural experiment stations has identified the major erosion factors and determined numerical relationships of these factors to soil loss rates. The **Universal Soil Loss Equation** (**USLE**) combines these factors and relationships to predict average annual soil losses from sheet and rill erosion by rainfall and its associated runoff on specific field slopes (see Wischmeier and Smith, 1978). Although the USLE has no geographic bounds, its application requires knowledge of the local values of its individual factors. Therefore, its applicability in some regions is currently limited by lack of research data from which to obtain factor evaluations that are representative of the local climatic, physical and cultural conditions.

The USLE equation is an empirical equation:

$$A = 2.24 R \cdot K \cdot LS \cdot C \cdot P \qquad (15\text{-}1)$$

where

A = the estimated average annual soil loss in metric tons per hectare per year.

R = the **rainfall and runoff erosivity index.** The local value of R can generally be obtained from Figure 15-6. [*Note*: The most erosive rainstorms occur in the southeastern United States, where hurricanes often move inland off the Atlantic Ocean or Gulf of Mexico.]

K = the **soil-erodibility factor**. The average soil loss per unit of R under arbitrarily selected "baseline" conditions (such as continuous fallow), K depends on soil properties. The soil properties that determine the susceptibility of soil to erosion are: (1) the texture of the soil, particularly the silt and very fine sand content; (2) the organic matter content; (3) the structural properties; and (4) the permeability of the soil. Table 15-2 on page 232 lists these soil characteristics and the K factor for different U.S. soils, which can be obtained from soil survey maps or from the

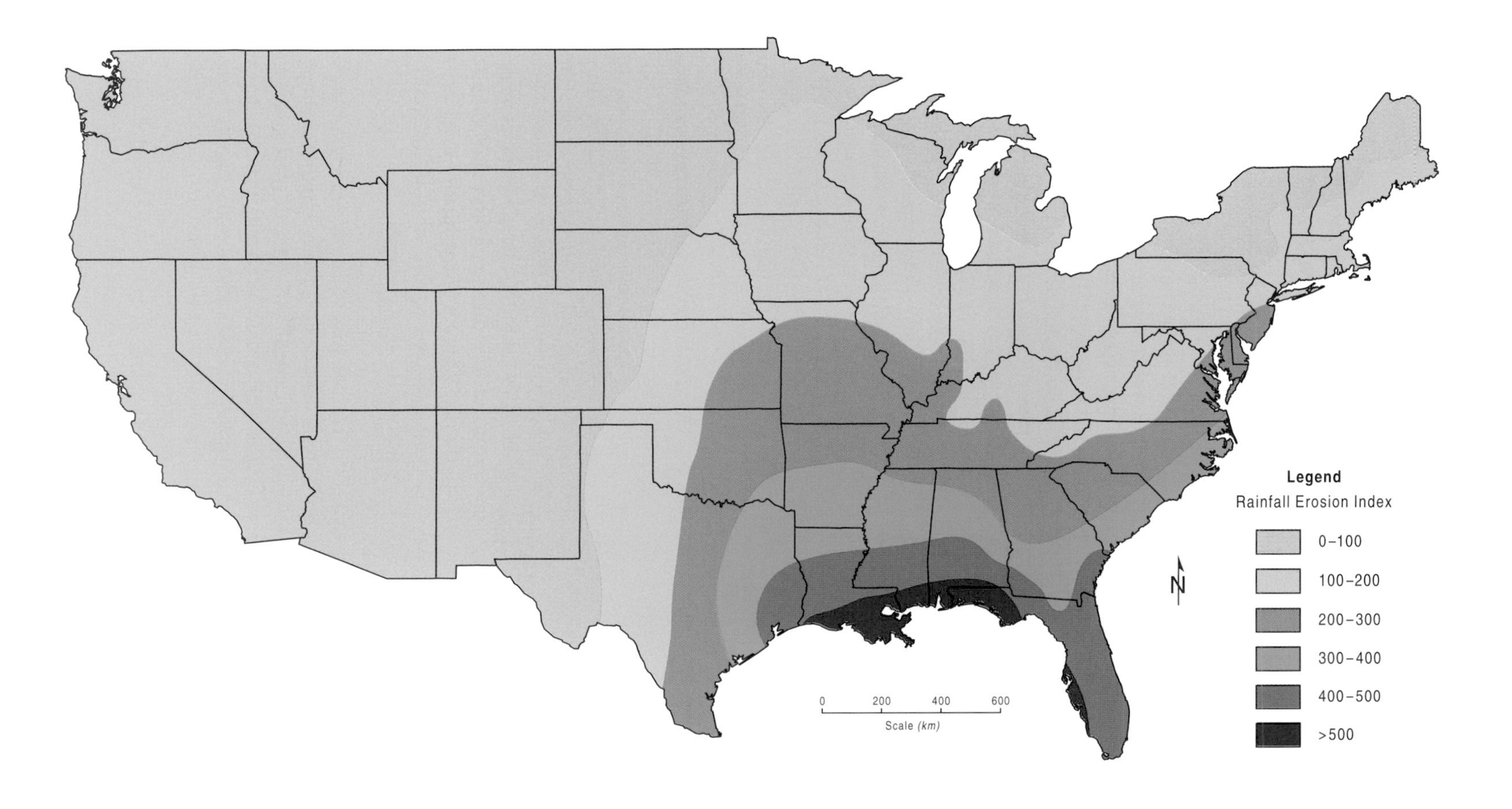

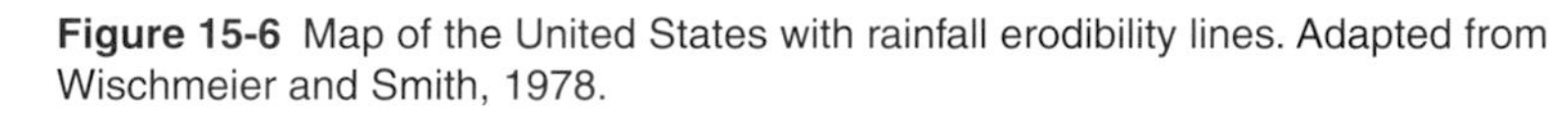

Figure 15-6 Map of the United States with rainfall erodibility lines. Adapted from Wischmeier and Smith, 1978.

Table 15-2 Indications of the general magnitude of the soil-erodibility factor *K* using USDA soil textural classes (see Figure 2-2 on page 12). Source: U.S. EPA, 1975.

Textural Class	K^a		
	<0.5%[b]	2%[b]	4%[b]
Sand	0.05	0.03	0.02
Fine sand	0.16	0.14	0.10
Very fine sand	0.42	0.36	0.28
Loamy sand	0.12	0.10	0.08
Loamy fine sand	0.24	0.20	0.16
Loamy very fine sand	0.44	0.38	0.30
Sandy loam	0.27	0.24	0.19
Fine sandy loam	0.35	0.30	0.24
Very fine sandy loam	0.47	0.41	0.33
Loam	0.38	0.34	0.29
Silt loam	0.48	0.42	0.33
Silt	0.60	0.52	0.43
Sandy clay loam	0.27	0.25	0.21
Clay loam	0.28	0.25	0.21
Silty clay loam	0.37	0.32	0.26
Sandy clay	0.14	0.13	0.12
Silty clay	0.25	0.23	0.19
Clay	0.13–0.29	0.13–0.29	0.13–0.29

[a] This is a generalized approximation of *K* based primarily on soil texture.

[b] Organic matter content.

Natural Resources Conservation Service, of the U.S. Department of Agriculture.

LS = a dimensionless **topographic factor** that represents the combined effects of slope length and steepness. Values of *LS* for uniform slopes are given in Table 15-3.

C = the **cover and management** factor. The *C* factor is the ratio of soil loss from areas with protective cover to the corresponding loss from **continuous fallow** (bare soil with no cover where the rows of plants go up and down the slope). The values of C range from 0.001 for well-managed woodland to 1.0 for tilled, continuous fallow. For a given cropping and management system, *C* varies with rainfall distribution and planting dates. Generalized values for illustrative purposes are given in Table 15-4.

P = the factor for **supporting practices**. The *P* factor is the ratio of the erosion resulting from the practice in question to that which would occur with up-and-down slope cultivation. The values of *P* are listed in Table 15-5. With no support practices, $P = 1.0$.

Factors *R*, *K*, and *LS* are relatively fixed for a given location. Their product is the **erosion-potential index** for the particular combination of rainfall pattern, soil properties, and topographic features. In other words, this product is the average annual soil loss that would occur if no vegetation were present and no erosion-reducing practices were used. When we multiply this potential by the appropriate values of factors *C* and *P*, the resulting product is a number that represents the effects of the given cropping system, cultural management, and supporting control practices. Thus, the complete equation predicts average annual soil loss for specific cropland situations. With the soil loss equation, we can estimate the average annual erosion rates at a particular site. Values of *R*, *K*, *LS*, *C*, and *P* are selected from appropriate tables and the product of all five factors is the soil loss estimate for the cropping and management system represented.

The USLE can be very useful as a planning guide and, until recently, was the most feasible method available for calculating erosion from specific land areas in most of the United States. However, soil losses computed by this equation must be accepted as estimates rather than as absolutes. Derivations of site values of the equation's five factors are based on relationships derived from the erosion research of the past 50 years. These relationships can, in specific situations, be significantly influenced by interactions with other variables.

The soil loss equation and supporting data tables were designed to predict longtime average losses for specific conditions. Specific-year losses may be substantially greater or smaller than the annual averages because of differences in the number, size, and timing of erosive rainstorms and in other weather parameters. A single extreme runoff event shortly after crop seeding on a clean-tilled seedbed could erode as much soil as the annual average for the rotation, but the probability of this is small.

Table 15-3 Values of the erosion equation's topographic factor *LS* for specified combinations of slope length and steepness. Source: U.S. EPA, 1975.

Slope	Slope Length (m)											
(%)[a]	7.60	15.2	22.9	30.5	45.7	61.0	91.4	122[a]	152[a]	183[a]	244[a]	305[a]
1	0.09	0.10	0.12	0.13	0.15	0.16	0.18	0.20	0.21	0.22	0.24	0.26
3	0.19	0.23	0.26	0.29	0.33	0.35	0.40	0.44	0.47	0.49	0.54	0.57
5	0.27	0.38	0.46	0.54	0.66	1.4	0.93	1.1	1.2	1.3	1.5	1.7
8	0.50	0.70	0.86	0.99	1.2	2.6	1.7	2.0	2.2	2.4	2.8	3.1
12	0.90	1.3	1.6	1.8	2.2	5.8	3.1	3.6	4.0	4.4	5.1	5.7
20	2.0	2.9	3.5	4.1	5.0	11.0	7.1	8.2	9.1	10.0	12.0	13.0
30	4.0	5.6	6.9	8.0	9.7	25.0	14.0	16.0	18.0	20.0	23.0	25.0

[a] Values given for slopes longer than 91.4 m or steeper than 18% are extrapolations beyond the range of the research data.

Table 15-4 The *C* factor. Source: Selected data from Wischmeier and Smith, 1978.

Vegetative condition	*C* Value
Cotton after cotton, seedbed period up to 80% soil cover	0.60
Corn mulch, 40% of soil actually covered, no-till cultivation	0.21
Corn mulch, 90% of soil actually covered, no-till cultivation	0.05
Undisturbed forest, 90–100% duff cover, 75–100% area has canopy	0.001–0.0001
Pasture or range, 40% grass cover	0.10
Continuous pasture, 95% grass cover	0.003
Brush 75% plus 40% covered by weeds and broadleaf plants	0.12
Trees 75% plus 20% covered by weeds and broadleaf plants	0.20
Trees 25% with no groundcover	0.42

Table 15-5 The *P* factor. Source: Selected data from Wischmeier and Smith, 1978.

Vegetative Condition		
Slope (%)	Maximum Length of Contour (m)	*P* Value
1–2	122	0.60
3–5	91.4	0.50
6–8	61.0	0.50
9–12	36.6	0.60
13–16	24.4	0.70
21–25	15.2	0.90

As an example for calculating the rate of erosion by water, let's assume that our soil is a silt loam in central Indiana, USA, with 2% organic matter (OM) on a 91.4-m, 5% slope adequately cultivated on the contour with continuous corn (5.0 metric tons per hectare yield) using no-till cultivation, leaving 40% of the soil covered. The needed factors are:

R = 180 (Figure 15-6 on page 231)
K = 0.42 (Table 15-2)
LS = 0.93 (Table 15-3)
C = 0.21 (Table 15-4)
P = 0.5 (Table 15-5)

Using Eq. (15-1) on page 230 and substituting the values, we obtain A = 17 metric tons of soil lost per year. Notice that if continuous pasture with 95% grass cover (C = 0.003 in Table 15-4) were substituted in the equation for planting to continuous corn, the calculated loss would be reduced from 17 to 0.24 metric tons per year.

The USLE was originally proposed in 1965 for estimating sheet and rill erosion sediment losses from cultivated fields in the United States east of the Rocky Mountains. Since then, the equation has been adapted for use in other cultivated areas in the United States, in Europe, and in Africa. It has also been used on rangelands and forestlands. In 1993 a revision of this model was released, called **RUSLE** (Revised USLE). In this revision algorithms have been developed to computerize the calculations; new *R* values have been computed for the western United States; and a subfactor approach for calculating the *C* values, which includes considerations of prior land use and crop canopy, has been adopted. The RUSLE model is now widely used in the United States, particularly for regulatory purposes. USDA Agricultural Research Service scientists are also in the process of developing a new generation of erosion prediction technology. A Water Erosion Prediction Project (WEPP) and Wind Erosion Prediction System (WEPS) are two sophisticated computer models that will further improve our ability to predict, and hence control, soil erosion.

15.4 Practices to Control Erosion

Where the average rainfall is adequate for crop production, some erosion is inevitable, largely because the rainstorms contributing to the average are discrete events that can vary greatly in intensity. Even under the most favorable conditions, rain can fall at rates that exceed the rate at which the soil can absorb it, so that the excess must run off. However, such excesses can usually be controlled by minimizing raindrop impact on the soil surface, thereby reducing runoff velocity and channelization and hence its erosive force.

The goal of Erosion Control Practices (ECP) is to keep erosion rates within tolerances compatible with good water quality, a wholesome environment, and the preservation of the productive capacity of the land. The Environmental Protection Agency (EPA) has published a number of ECPs, many of which are listed in Table 15-6 together with their favorable and unfavorable features. These ECPs include agronomic practices that improve crop residue management, cropping sequences, seeding methods, soil treatments, tillage methods, and timing of field operations.

One of the most important methods of reducing erosion is the practice of farming parallel to the field contours; this practice is known as **contouring**. Contour cultivation produces a ridging of the soil which forms small barriers that slow or stop downhill movement of water. However, this and some of the other agronomic control practices are not effective when the slope angle is steep or the area from which runoff concentrates is excessive. These practices must then be supplemented or replaced by other methods that include terraces, diversions, contour furrows, contour listing, contour strip cropping, waterways, and control structures.

Terraces are level strips of earth constructed on the contour of the land to make steep hillslopes suitable for tillage. **Diversions** are mounds of soil again constructed on hillslopes to direct water away from lands below. **Contour furrows**, **contour listing**, and **contour strip cropping** are all cultivation practices where the farmer plants crops on the same elevation level (contour) to entrap more water and reduce runoff. Listing is a way of tilling the soil that leaves more crop residues on the soil surface to protect the soil from falling rain, and strip cropping means the farmer plants alternate strips of row crops and grasses or legumes to reduce runoff. **Waterways** are structures designed to remove or divert large quantities of water away from lands that might erode.

Under many conditions, it may be necessary to apply various combinations of these practices. Modifications of specific practices within these general types affect their adaptability as well as their effectiveness. These practices are often used together, such as combining contour tillage practices with waterways and control structures to protect the land from erosion.

The technology is available to minimize agricultural pollution; however the economics of implementing these practices on agricultural lands involves long-term investments. The land owner often operates in short time spans, having to meet various financial obligations, and this often compromises the situation. For example, it is expensive to construct terraces and waterways, and often the field size and shape are changed and farmers need to modify their farm equipment to farm these lands. However, yields may be reduced because less row crops such as corn and soybeans (cash crops) can be planted; thus his income is lowered. Conservation practices have long term benefits, but net income is often lowered in the short term.

Table 15-6 Principal types of cropland Erosion Control Practices (ECP). Source: U.S. EPA, 1975.

	Erosion Control Practice	Practice Highlights
ECP 1	No-till, plant in prior-crop residues	Most effective in dormant grass or small grain; highly effective in crop residues; minimizes spring sediment surges and provides year-round control; reduces man, machine, and fuel requirements; delays soil warming and drying; requires more pesticides and nitrogen; limits fertilizer- and pesticide-placement options; some climatic and soil restrictions.
ECP 2	Conservation tillage	Includes a variety of no-plow systems that retain some of the residues on the surface; more widely adaptable but somewhat less effective than ECP 1; advantages and disadvantages generally the same as ECP 1 but to a lesser degree.
ECP 3	Soil-based rotations	Good meadows lose virtually no soil and reduce erosion from succeeding crops; total soil loss greatly reduced, but losses unequally distributed over rotation cycle; aid in control of some diseases and pests; more fertilizer-placement options; less realized income from hay years.
ECP 4	Winter cover crops	Reduce winter erosion where corn stover has been removed and after low-residue crops; provide good base for slot-planting next crop; usually no advantage over heavy cover of chopped stalks or straw; may reduce leaching of nitrate; water use by winter cover may reduce yield of cash crop.
ECP 5	Improved soil fertility	Can substantially reduce erosion hazards as well as increase crop yields.
ECP 6	Timing of field operations	Fall plowing facilitates more timely planting in wet springs, but greatly increases winter and early spring erosion hazards; optimal timing of spring operations can reduce erosion and increase yields.
ECP 7	Contouring	Can reduce average soil loss by 50% on moderate slopes, but less on steep slopes; loses effectiveness if rows break over; must be supported by terraces on long slopes; not compatible with use of large farming equipment on many topographies. Does not affect fertilizer and pesticide rates.
ECP 8	Contour strip cropping	Rowcrop and hay in alternate 15.2- to 30.5-meter strips reduce soil loss to about 50% of that with the same rotation contoured only; fall-seeded grain in lieu of meadow about half as effective; alternating corn and spring grain not effective; area must be suitable for across-slope farming and establishment of rotation meadows; favorable and unfavorable features similar to ECP 3 and ECP 7.
ECP 9	Terraces	Support contouring and agronomic practices by reducing effective slope length and runoff concentration; reduce erosion and conserve soil moisture; facilitate more intensive cropping; conventional gradient terraces often incompatible with use of large equipment, but new designs have alleviated this problem; substantial initial cost and some maintenance costs.
ECP 10	Grassed outlets	Facilitate drainage of graded rows and terrace channels with minimal erosion; involve establishment and maintenance costs and may interfere with use of large implements.
ECP 11	Change in land use	Sometimes the only solution. Well managed permanent grass or woodland effective where other control practices are inadequate; lost acreage can be compensated for by more intensive use of less erodible land.
ECP 12	Other practices	Contour furrows, diversions, subsurface drainage, land forming, closer row spacing, *etc*.

References and Recommended Reading

El-Swaify S.A., Moldenhauer W.C., and Lo A. (1985) *Erosion and Conservation.* Soil and Water Conservation Society of America, Ankeny, Iowa.

Lal R., Editor (1995) *Soil Erosion Research Methods.* Soil and Water Conservation Society of America, Ankeny, Iowa.

National Research Council (1993) *Soil and Water Quality.* National Academy Press, Washington, D.C.

Schwab G.O., Fangmeier D.D., Elliot W.J., and Frevert R.K. (1993) *Soil and Water Conservation Engineering*, 4th Edition. John Wiley and Sons. New York.

Troeh F.R., Hobbs J.A., and Donahue R.L. (1992) *Soil and Water Conservation.* 2nd Edition. Prentice Hall, Englewood Cliffs, New Jersey.

USDA Soil Conservation Service (1961) *Land-Capability Classification*, Agricultural Handbook No. 210. United States Department of Agriculture, Washington, D.C.

USDA Soil Conservation Service (1992) *1992 Summary Report: National Resources Inventory.* United States Department of Agriculture, Washington, D.C.

USDA Soil Conservation Service (1994) *National Resources Inventory*, United States Department of Agriculture, Washington, D.C.

U.S. EPA (1975) *Control of Water Pollution from Cropland.* EPA-600/2-75-026a. U.S. Environmental Protection Agency, Washington, D.C.

Wischmeier W.H. and Smith D.D. (1978) *Predicting Rainfall-Erosion Losses: A Guide to Conservation Planning*. Agriculture Handbook Number 537. United States Department of Agriculture, Washington, D.C.

Problems and Questions

1. Explain how agriculture might cause and/or introduce pollutants into the environment of an area. List the major agricultural pollutants we are concerned about.

2. The U.S. Department of Agriculture classifies land using the Land-Capability Classification System. Describe this system for classifying land for agricultural use.

3. The USLE equation is used to predict average annual soil losses. List and define the seven terms in this equation.

4. Describe six management practices that can be applied to the land to reduce erosion losses from agricultural lands.

Animal wastes can be efficiently managed only through an understanding of the interactions of these wastes with biological factors and transport mechanisms, such as water. Photograph printed by permission; copyright © 1995 by Jeffrey W. Brendecke. All rights reserved.

16.1 Historical Perspective

Animal wastes are pollutants of increasing concern both to the public and regulatory bodies because they have the potential to contaminate both surface and groundwater. Consequently, animal wastes must now be included as part of the agricultural production cycle and figured into the cost of operating a farm or livestock facility. Animal agricultural wastes can be divided into two production types: range and pasture production, and confined or concentrated animal production.

In range and pasture systems, the concentration of wastes is generally much more diffuse or dispersed than it is when large numbers of animals are confined to relatively small areas. Range and pasture systems have two principal measurable effects on surface water quality: (1) increased turbidity through the movement of soil particles into streams, rivers, and lakes; (2) increased fecal coliform counts in areas of heavy animal use. Although we know that grazing systems may adversely affect some measures of water quality, we will focus here on the highly concentrated animal production units and the methods of preventing and controlling pollution from these concentrated units.

In the past, animals were concentrated only intermittently: the period of confinement was a transitory phase followed by a return to pasture, after such management activities as milking or shearing. Today, however, animal production is occurring in increasingly controlled environments owing to the success of efforts to raise productivity and diminish climatic, feeding, and mortality variables. Larger numbers of animals are being raised in **concentrated animal feeding operations** or **CAFOs**—principally, feedlots, dairies, swine operations, poultry houses, and intensive aquaculture.

This shift in production methods has changed the age-old method of reincorporation of animal wastes as manure into the next food production cycle. Following World War II, manure was displaced as the primary fertilizer by fossil-fuel-based fertilizers as farms became increasingly specialized. Specialization has largely divorced animal production from the production of crops: a concentrated animal facility may be located far from crop production, and the same family (or the same corporation) may not pursue the two types of production.

With the breakdown of the traditional cycle of reincorporation of wastes back in to the land, what was once an essential source of nutrients has now become a potential pollutant. Thus the production of large numbers of animals on a small land base has resulted in the stockpiling of wastes, the construction of large waste-storage ponds, and oftentimes, waste applications to land in excess of agronomic crop needs.

16.2 Nonpoint versus Point Source Pollution

In discussing sources of animal wastes, it is important to understand the terms point and nonpoint contaminant sources. The term "nonpoint pollution" is misleading and is often misused. In animal agricultural systems, true nonpoint sources are those in which potential contaminants are not concentrated during production, and do not pass through a single or small number of conduits for disposal. These nonpoint sources include corrals, feedlots, and extensive and intensive pasture systems.

Point sources, are those facilities that concentrate pollutants or contaminants to a significant degree and pass these contaminants through a pipe, ditch, or canal for disposal. The most common point sources are milksheds and barns, dairy and other food-processing plants, intensive indoor swine facilities, anaerobic and aerobic lagoons, and evaporative storage ponds. In addition, certain types of intensive aquaculture may also be point sources of contaminants, with return flows highly nutrient-laden with fish excreta.

According to EPA regulations, however, some concentrated animal feeding operations may be designated as point sources requiring an individual National Pollution Discharge Elimination System (NPDES) permit. In this case, a concentrated animal feeding operation is defined as a lot or facility without vegetation where animals are confined for 45 or more days per year. The number of animals needed to meet this definition as a CAFO depends on several factors; however, the key determinant is whether or not the facility discharges into navigable waters, as determined by the method of discharge. The method of discharge is judged by the **25-year, 24-hour storm event**, which is the required event that a facility must be designed to meet. This design criterion is a storm of 24 hours' duration whose probability of occurrence is once every 25 years. If a facility does not discharge from storms smaller than the 25-year, 24-hour storm event, it may be regulated as if it were a nonpoint source and thus not require a permit. If a facility does discharge from storms smaller than the 25-year, 24-hour storm event, it may be treated as a point source. This "double standard" recognizes the fact that what appear to be agricultural point sources can safely be treated as nonpoint sources because the pollutants being discharged are either not concentrated enough to warrant specific controls at the pipe or are not laden with hazardous materials.

The 25-year, 24-hour storm criterion does not mean that it is certain that once every 25 years a storm of a certain size and duration will occur. Another, possibly clearer, way of looking at this standard is that within 100 years it is likely that four storms that meet the 25-year, 24-hour storm size and length will occur. However, these four storms may occur back-to-back or be spread out over a period longer than 25 years. Therefore, given the hydrologic record, 25 years is the *average* return period for a storm of this size and duration. Figure 16-1 shows rainfall intensity for a 25-year, 24-hour storm event for the whole of the United States. Maps such as this are used in designing waste-handling systems.

The practical application of this design storm is that a facility must be designed so that it does not discharge to surface "navigable waters" in any event that is less than the design storm in size and length. **Navigable waters** have been defined so broadly by the courts that they include essentially all streambeds, even those that are dry most of the year, as well as irrigation canals that eventually discharge into any type of perennial water course.

Nonpoint sources, such as nondischarging concentrated animal feeding operations, require a different approach to prevention and mitigation of pollutants than do point source emissions from a pipe or

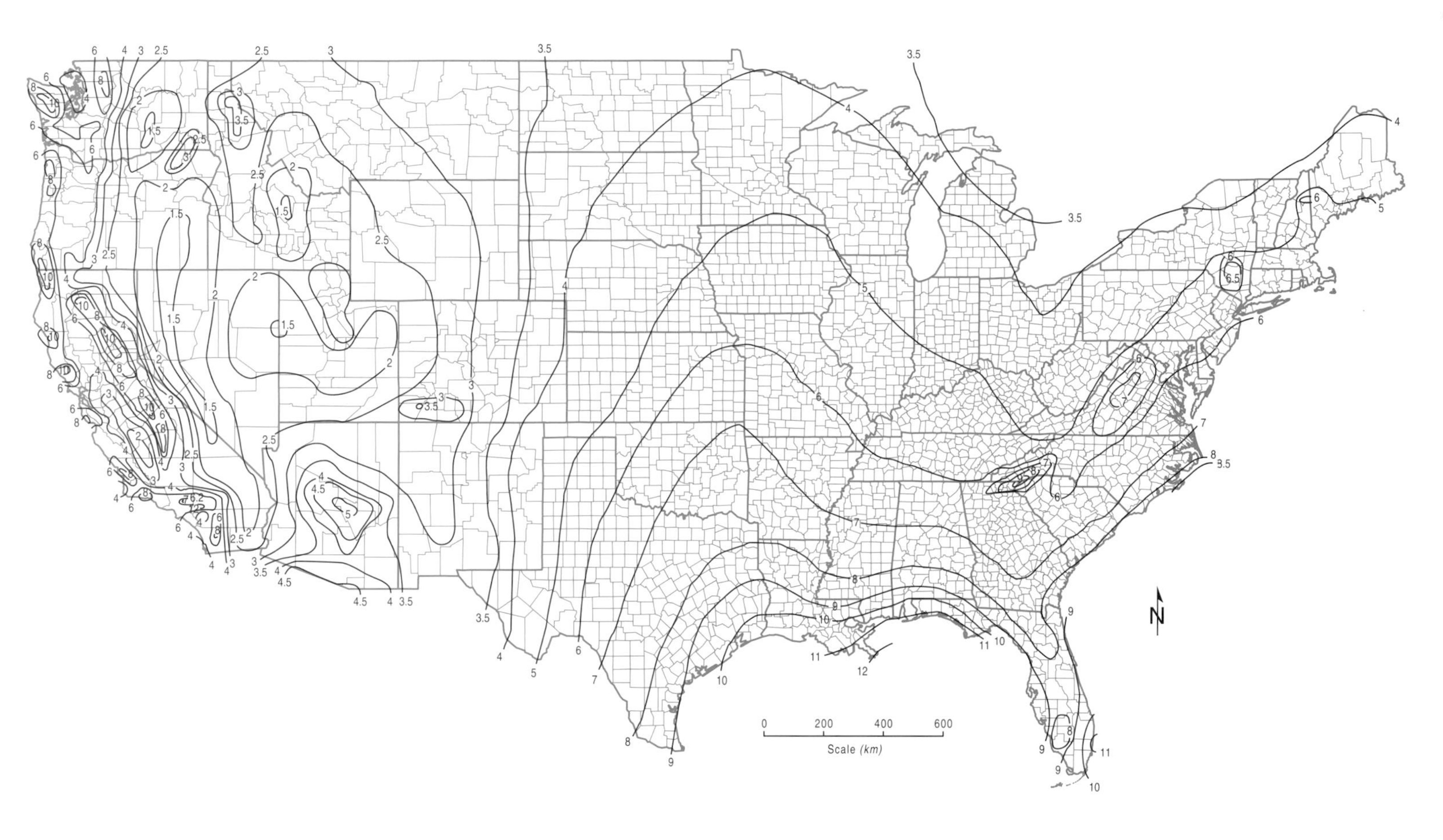

Figure 16-1 25-Year, 24-hour rainfall for the United States (in tenths of an inch, where 1 inch = 2.54 cm). Adapted from USDA Soil Conservation Service, 1992.

conduit. As of 1996, the nonpoint source approach to mitigation employs **Best Management Practices (BMPs)**, as defined by the 1987 Amendments to the Federal Water Pollution Control Act. In contrast, point source methods employ methods termed **Best Available Demonstrated Control Technology** or **Best Available Control Technology**.

This terminology reflects the awareness that simple technical fixes are not appropriate to more diffuse contaminant sources; thus, it is necessary to take an approach that looks at the production system to identify management practices that will reduce or eliminate the potential for contamination. (See also Section 14.2.5, beginning on page 216, for more information on BMPs as applied to crop production.)

16.3 Specific Pollutants

Concentrated animal agriculture produces the following specific pollutants in the wastes resulting from animal metabolic activity:

1. *Nitrate-nitrogen* (NO_3-N) is formed from the ammonification of organic nitrogen contained in plant material that animals consumed. Organic nitrogen compounds in the form of proteins and amino acids are excreted as manure by animals. These compounds are first partially converted by microorganisms into ammonium ions (NH_4^+), which relatively quickly yield nitrite ions (NO_2^-) that react very quickly to afford nitrate ions (NO_3^-). Animal urine is a liquid that contains urea, which can quickly break down as ammonia (NH_3) and carbon dioxide (CO_2). The ammonia can either volatilize as a gas (NH_3) or be converted to NH_4^+ and NO_3^-. Nitrate is subject to leaching into surface or groundwater.

2. *Phosphorus*, as P or as phosphate (PO_4^{3-}), is found in animal manures in concentrations on the order of 0.1–0.4% for both organic and inorganic forms. Organic forms must be mineralized to inorganic forms to be plant-available. Once mineralized, phosphorus is relatively stable and immobile, except when it is carried as part of eroded soil sediment (see Chapter 15, beginning on page 225). When phosphorus from animal agriculture is present in surface water, it is usually the result of direct transmission of liquid and solid manures to the receiving body of water. Phosphorus compounds rarely leach from soils following application, unless the soil particle itself has been translocated.

   Excess P is principally a surface water problem. Phosphorus concentrations as low as 0.0001 mg L^{-1} have been associated with eutrophication. However, excess phosphorus can affect groundwater when the depth to groundwater is very shallow and it is overlain by sandy soils (see also Chapter 14, beginning on page 211).

3. *Fecal coliform bacteria*, measured in counts per 100 mL of water, have been recorded in very high quantities in runoff from concentrated animal production units. These high counts have implications for public health as well as for fish and wildlife. The EPA water-quality criteria include the following standards. For bathing, swimming, and other "full-body contact," fecal coliform counts should not exceed 200 per 100 mL in five samples taken over 30 days, with no samples exceeding 400 at any one time. For consumption of shellfish (which tend to concentrate fecal coliform bacteria), the median concentrations should be no greater than 14, with fewer than 10% of the samples exceeding 43. Drinking water should contain 0 counts of fecal coliform. Fecal coliform bacteria are usually surface-water concerns; they rarely affect the groundwater. Thus, unless wells are very shallow and have not been properly cased, most drinking water is free of these microorganisms.

4. *Pesticides* that are used in concentrated animal facilities can be a concern for both surface and groundwater contamination (see Section 17.4, beginning on page 259). For example, coumaphos, an organophosphate (see Section 17.2.1 on page 255), is commonly used in dips for animals crossing the southern borders of the United States. While relatively immobile, coumaphos is known to be persistent and has been found at some depths when it has not been disposed of

properly. Other pesticides, such as toxaphene, are no longer used in animal dips.

5. *Biochemical oxygen demand* (or BOD) is a measure of the quantity of oxygen (often measured in kilograms of O_2) needed to satisfy biochemical oxidation of organic matter in a waste sample in five days at 20°C. While not a specific pollutant, the BOD is one of the important general indicators of a substance's potential for environmental pollution of surface waters. This measure of pollution capability is important because animal wastes typically contain a high level of BODs, on the order of 0.45–3.6 kg per 454 kg of excreted material (see also Section 13.5, beginning on page 197).

16.4 Air Quality

Animal producers have traditionally employed open lots or pens as the primary facilities for animal containment. However, industry practices are evolving toward greater management intensity, which may require that animals spend a considerably higher percentage of time in closed housing. Producers need to become aware that this change in the ratio of time spent in relatively open lots versus housed time has potential health effects. Unvented areas containing manure gases can be unhealthful and even dangerous. At high manure gas levels, these gases can be immediately dangerous as asphyxiants or toxins. Moreover, the level of manure gases in proportion to normal air does not have to be high before health effects can be noted in both animals and workers.

The four gases of primary concern are as follows:

1. *Carbon dioxide (CO_2):* Present under anaerobic conditions, this gas is nontoxic but becomes an asphyxiant at moderate levels. It is odorless and nondetectable without sampling equipment; however, such symptoms as headaches, rapid breathing, or other nonirritating respiratory distress may indicate high CO_2 levels and dropping oxygen (O_2) levels.

2. *Hydrogen sulfide (H_2S):* This is a toxic, explosive gas that becomes an asphyxiant at relatively low levels. It is present under anaerobic conditions, with levels relatively higher in swine wastes than in cattle or poultry wastes. Its characteristic rotten-egg odor serves as a warning of rising H_2S—a sign that must be taken seriously. When humidity levels are sufficiently high, H_2S can combine with water to produce sulfuric acid (H_2SO_4), which can corrode metals as well as affect respiratory passages and lung tissues. Thus, other signs of H_2S include blackened copper, galvanized steel with a white residue, and obvious respiratory distress.

3. *Ammonia (NH_3):* This gas is toxic and becomes an asphyxiant at relatively low levels. It is easily volatilized and can accumulate in closed structures. Corroded metal structures are a sign of excess NH_3 concentrations. Ammonia also has a characteristic acrid odor. This gas affects nasal and respiratory passages, and can damage lung tissue in both animals and workers.

4. *Methane (CH_4):* This gas is an asphyxiant at very high levels, but it is very flammable over a wide range of concentrations and can explode under certain conditions.

Any closed structures in which manure or urine accumulates should be assumed to generate gases. Thus sump pits, covered lagoons, and receiving areas from flush lines, as well as any housing that features floor slats, should be treated with caution or vented before entering. In addition, any manure that has not been moved for a period of time tends to accumulate gases that will be released when moved or agitated. In general, the accepted solution to the problem of ensuring animal and worker health is adequate venting or forced-air aeration. In addition, workers who must enter a deep sump pit or any other structure that can accumulate heavier-than-air gases should use respirators.

16.5 Fate of Pollutants

16.5.1 Nitrates

As an ionic form of nitrogen, NO_3-N has a high degree of solubility in water and is very mobile, moving with the flow of water. Nitrates have different implications for surface water and groundwater quality. Surface water containing increased NO_3^- levels usually undergoes eutrophication, which results in diminished water quality for drinking uses. Groundwater, however, can be contaminated directly by nitrate-bearing leachate. The presence of nitrates in leachate also diminishes drinking-water quality. For human consumption, levels above 10 mg L^{-1} (as NO_3-N) are considered unacceptable while levels above 100 mg L^{-1} are thought to be unfit for livestock consumption. (For additional discussion of NO_3^-, see Section 14.2, beginning on page 212.)

In surface waters undergoing eutrophication (increased aquatic plant growth), the dissolved O_2 content decreases once the plant growth is in a state of decay, resulting in reduced aquatic animal life. If the source of NO_3^- is diminished or eliminated, the concentration of NO_3^- declines owing to dilution and transformation via heterotrophic denitrification and volatilization as N_2.

Once NO_3^- is leached from the upper soil layers containing higher levels of organic material to the vadose zone containing little organic material, little further denitrification of NO_3^- is likely to take place. Therefore, once NO_3^- has passed the effective plant root zone, it can ultimately be transported to aquifers via saturated or unsaturated flow. The travel time to the groundwater and the degree of contamination are highly variable, and dependent on soil and aquifer characteristics. Coarse-textured soil and a shallow water table contribute to short travel times, thus increasing the likelihood of groundwater contamination. Once NO_3^- reaches the groundwater, it is very difficult to rid the aquifer of the contamination. Figure 16-2 illustrates the potential movement of NO_3^- through a porous soil.

16.5.2 Phosphorus

Phosphorus from animal wastes is relatively immobile and tends to be readily adsorbed to soil particles. Consequently, P does not generally migrate to the groundwater unless the groundwater depth is shallow (less than 3 m) and the soil type is predominantly sandy. But the fate and effect of P from animal wastes in surface water are dramatically different. As an essential plant nutrient, phosphorus added to surface water can dramatically alter the habitat for fish and other organisms by promoting an explosive growth of algae. As is the case with NO_3^-, when algae die and decay, the dissolved oxygen content can diminish drastically. Given sufficient decaying biomass, a stream or lake may be rendered inimical to aerobic organisms. Additionally, P tends to remain and persist in lake sediments and is therefore difficult to remove. (For additional discussion of phosphates as a pollutant, see Section 14.3, beginning on page 219.)

16.5.3 Fecal Coliforms

Fecal coliforms are principally a concern for surface water supplies (see Section 13.4.1, beginning on page 195, and Section 19.3.2, beginning on page 285). While coliforms are present naturally from sources such as wildlife, high concentrations of animals have the potential of vastly increasing the coliform count. Normally, fecal coliforms do not survive for long periods of time in soil or water. Survival periods can be several days or several weeks, depending on the environmental conditions.

In drinking water, disinfection by chlorination can result in the production of trihalomethane (THMs) compounds (limited to 0.10 mg mL^{-1} in drinking water). These compounds are formed in the presence of organic matter that often accompanies fecal coliforms. Trihalomethanes have been implicated as potential carcinogens in laboratory animals and therefore are to be avoided in drinking water (see Section 20.5, beginning on page 315).

16.6 Monitoring and Detection of Pollutants

Water testing is done to determine if surface or groundwater has been contaminated by runoff or leaching from a CAFO. The minimal water analysis should be for NO_3^-, P, fecal coliforms, pH, and total dissolved salts (TDS). Additional analysis may also be needed for volatile organic compounds, metals, and pesticides.

Testing first establishes a water-quality baseline, which serves as a comparative reference. Thus an owner/operator, facility, or regulatory agency can determine when, or if, the water quality changes with time. As both human and animal health can be

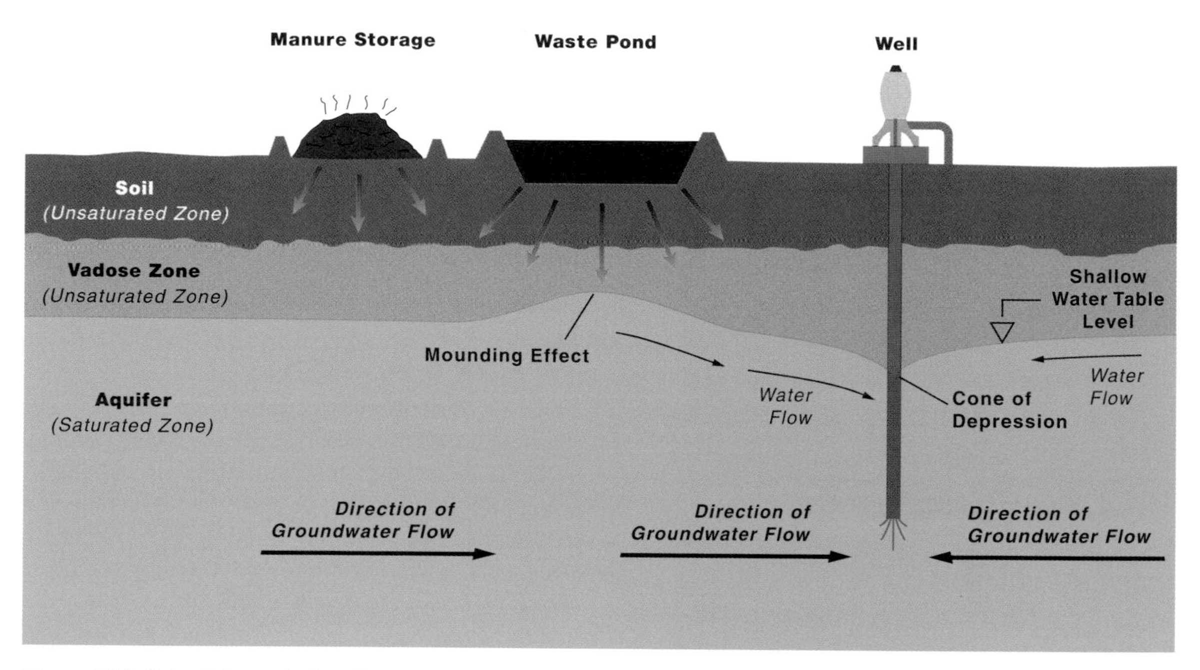

Figure 16-2 Potential percolation of contaminants in porous soils. Adapted by permission from *The Guidebook: Nitrogen Best Management Practices for Concentrated Animal Feeding Operations in Arizona*, by Robert J. Freitas and Thomas A. Doerge; copyright © 1996 by the Arizona Board of Regents. All rights reserved.

affected by poor-quality water, testing is an important quantitative measure to determine the existence and extent of any deterioration in water quality. Well water samples are periodically taken within a facility's boundaries and, if possible, from wells immediately adjacent to the facility. Annual sampling is usually sufficient; however, additional testing should be done if a problem is detected.

16.7 Prevention and Mitigation

Mitigation of contamination, which has been defined in the National Environmental Policy Act (1969), includes the following methods:

1. avoiding the impact altogether by prohibiting a certain action or actions
2. minimizing impacts by limiting the degree of implementation or magnitude of the action
3. rectifying the impact by repairing, rehabilitating, or restoring the affected environment
4. reducing or eliminating the impact over time by preservation operations during the life of the action
5. compensating for the impact by replacing or providing substitute resources or environments.

We often summarize these five methods of mitigation as prevention, minimization, repair, preservation, and compensation.

Best Management Practices (BMPs), which are effective in the mitigation of pollution caused by animal wastes, are the methods, measures, or practices designed to prevent or reduce discharges. They include structural and nonstructural controls and operation and maintenance procedures.

16.7.1 Siting a Facility

Prevention of a problem at the outset is usually more efficient and less expensive than fixing one. For example, locating an animal confinement facility, such as a dairy, feedlot, swine operation, or aquaculture farm, in an area that is hydrogeologically resistant to flooding and leaching is relatively easy and generally inexpensive. When siting such a facility, the preferred locations are those in which

1. the entire facility lies outside of the 100-year flood plain;
2. the soil is relatively high in clay (impervious)—at least in areas where waste ponds will be placed (pen areas, for example, should be of soils no lighter than a sandy loam);
3. the facility is far enough from residences or public buildings to forestall nuisance complaints.

16.7.2 Waste-Management Plan

It is important for a facility owner not only to follow BMPs, but also to be able to document how and where these BMPs have been followed. The documentation and assessment of the waste system should be organized into a single document called the **waste management plan**. The waste-management plan need not be overly complex, but it should contain some basic elements:

1. a drawing of the facility's capital structures showing surrounding elevations and any water courses, dry or not;
2. the location of the 100-year flood plain with respect to the facility;
3. any calculations made of the quantities of runoff/runon water that must be contained from a 25-year, 24 hour storm;
4. any calculations made of solid and liquid wastes generated by the animals that must be managed together with any projected increase in animal numbers;
5. the sizes of the capital structures of the waste-management system (berms, conveyance pipes or canals, solid separators, lagoons or ponds, farmland for nutrient-enriched wastewater applications) and the quantities of wastes they are calculated to handle;
6. the methods that could be used to minimize nitrate movement if the facility should be closed or sold;
7. any soil survey information, well water quality, or other relevant data.

16.8 Waste System Elements

Waste system elements are divided into four general types: collection, conveyance, storage, and reuse. These are illustrated in Figure 16-3.

16.8.1 Collection (Solids)

Removal of relatively dry solid matter, or manure, from animal confinement facilities is a fairly simple, straightforward problem requiring labor and equipment. Manure from animal pens should be removed to a depth of 2–5 cm; this can be done with heavy equipment when pens are empty. According to available evidence, a biological seal is formed below the manure pack; and, if left undisturbed, this seal can diminish leaching potentials.

16.8.2 Conveyance (Liquid and Semiliquid)

A system of piping or culverts conducts wastewater from daily operations—such as milking parlors, holding pens, and, for some facilities, the alleys—to a central point so that the liquid wastes can be stored and treated or applied to cropland. Parlors, pens, and alleys must be designed so that liquid wastes are easily collected and channeled into lined culverts, ditches, or pipes. Increasingly, modern animal facility designs for dairies and swine operations are incorporating a flush system into the waste-removal systems. As flush systems generate greater volumes of wastewater than scrape systems, proper sizing of conveyances becomes more important.

16.8.3 Storm Runoff

The collection and conveyance of storm runoff differs in a number of respects from the routine collection of liquid wastes generated, for example, by a dairy operation. Stormwater runoff is more difficult to accommodate because its frequency, duration, and volume are controlled to a major extent by natural rainfall (Figure 16-4 on page 246). That is, the major determining factor for planning and design purposes is the nature of the rainfall. All drainage systems

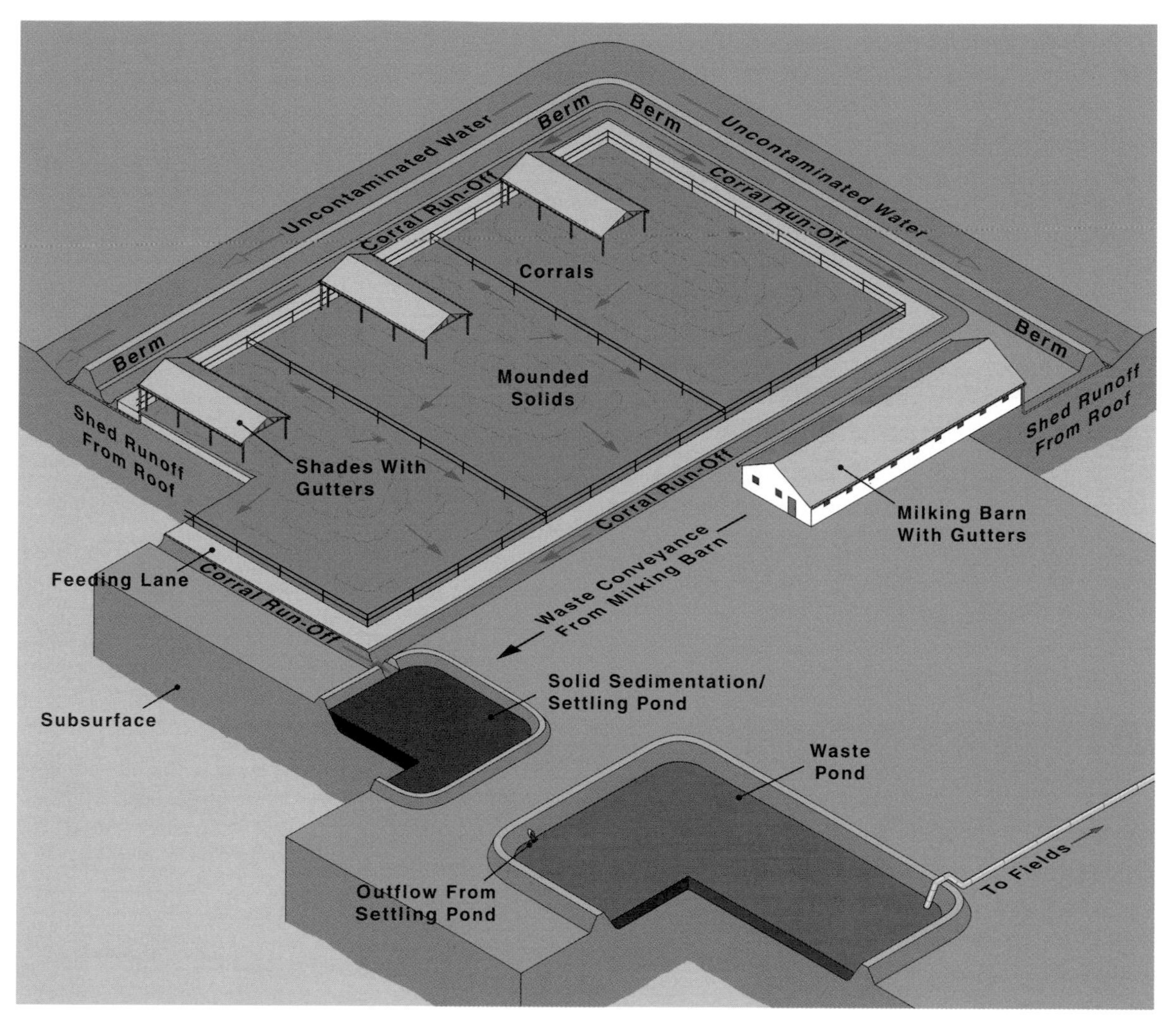

Figure 16-3 General waste flow in a generalized dairy. Adapted by permission from *The Guidebook: Nitrogen Best Management Practices for Concentrated Animal Feeding Operations in Arizona*, by Robert J. Freitas and Thomas A. Doerge; copyright © 1996 by the Arizona Board of Regents. All rights reserved.

should be designed to carry the runoff from a 25-year, 24-hour storm.

16.8.4 Storage

Evaporative ponds or storage ponds are used to contain N-contaminated runoff, which can be a mixture of solid and liquid wastes from a pen. The holding period allows time for microbial action to break down waste materials and permits the volatilization of NH_3, the denitrification of NO_3^-, and the evaporation of liquids to take place.

Solids should be separated from the liquid by screening or settling prior to entry into the storage pond, thereby reducing nutrient loading and minimizing cleanout frequency. Pond design can be either aerobic or anaerobic. An **aerobic pond** is one in which the oxygen level is high enough to support oxygen-requiring microbes, while an **anaerobic pond** (literally "without oxygen") primarily supports microbial degradation that does not depend upon oxygen. Aerobic ponds can also be designed as evaporative ponds in which water is not stored, but

Figure 16-4 Mounding of manure in dairy corrals (as pictured in Figure 16-3 on page 245) facilitates the flow of contaminated run-off water into the appropriate ponds for storage or treatment. In the corral pictured above, a depression in the corral has enabled water to collect in the corral, not only contributing to foot problems in the cows walking in the mud, but also enhancing pollutant leaching and creating an odor problem. Photograph: R.J. Freitas.

rather detained only until it evaporates. Simplified profiles of aerobic and anaerobic ponds are shown in Figure 16-5 on page 247.

16.8.5 Environmental Closure of a Facility

Upon closure, wastes are removed from animal facility pens and storage ponds so that no further pollution potential remains. Manure should be scraped from the pens down to the mineral soil and applied to crop land or otherwise suitably disposed. Any manure stockpiled should be similarly removed. Evaporative or storage ponds should be dried through evaporation, or their liquid contents should be pumped onto crop land.

16.9 Land Application and Disposal

Land application is one of the best methods for disposal of excess manure or manure slurry (see Section 10.4, beginning on page 140, for a discussion of land application as applied to municipal sewage sludge). It works on three premises: (1) that crops will take up excess nutrients contained in manure slurries, (2) that microbial action will result in partial denitrification, and (3) that application rates will be relatively low because excess nutrients are spread over a large area. While supplying some of the essential nutrients for crop production, manure is also a beneficial soil amendment for increasing tilth, aeration, water-holding capacity, organic content, infiltration, and microbial activity. Land application does have some disadvantages, however, including the distribution of weed seeds, the possible accumulation of salts, and the application of excess nutrients that may leach and again become pollutants.

Table 16-1 Nutrient composition of fresh manure (average dairy).

Source	Moisture (%)	Nutrient (kg metric ton^{-1}) Nitrogen (N)	Phosphorus (P_2O_5)	Potassium (K_2O)
Dairy	79	5.60	2.3	6.00
Liquid dairy	91	2.4	0.05	2.3

Before large quantities of manure and/or slurries can be applied, it is necessary to know: (1) the relative nutrient content; (2) the rate of mineralization (decay rate); (3) the salt load; (4) the concentration and type of toxic elements; (5) the proper time and method of application; and (6) the amount and type of weed seed.

The nutrient content of manure can vary greatly, depending upon the age of the livestock in question, the rate of feed consumption, the feed ration, and the manure handling and storage practices prior to land application. Two facts are especially important when deciding upon the application rate for manure: (1) the moisture content of the manure directly affects the nutrient percentage, and (2) manure loses N as NH_3 when it dries. The nutrient composition of fresh dairy manure is shown in Table 16-1.

Application rates for a given N availability are shown in Table 16-2 on page 249. The rate of mineralization of manure is the percentage available for plant uptake each year. This rate is given as a decay series at three levels of nitrogen at the time of application. The amount applied each year is the quantity of manure that should be added to maintain an annual mineralization rate of 220 kg of nitrogen per hectare per year for a given N% and decay rate. Local climatic conditions such as extreme heat or resis-

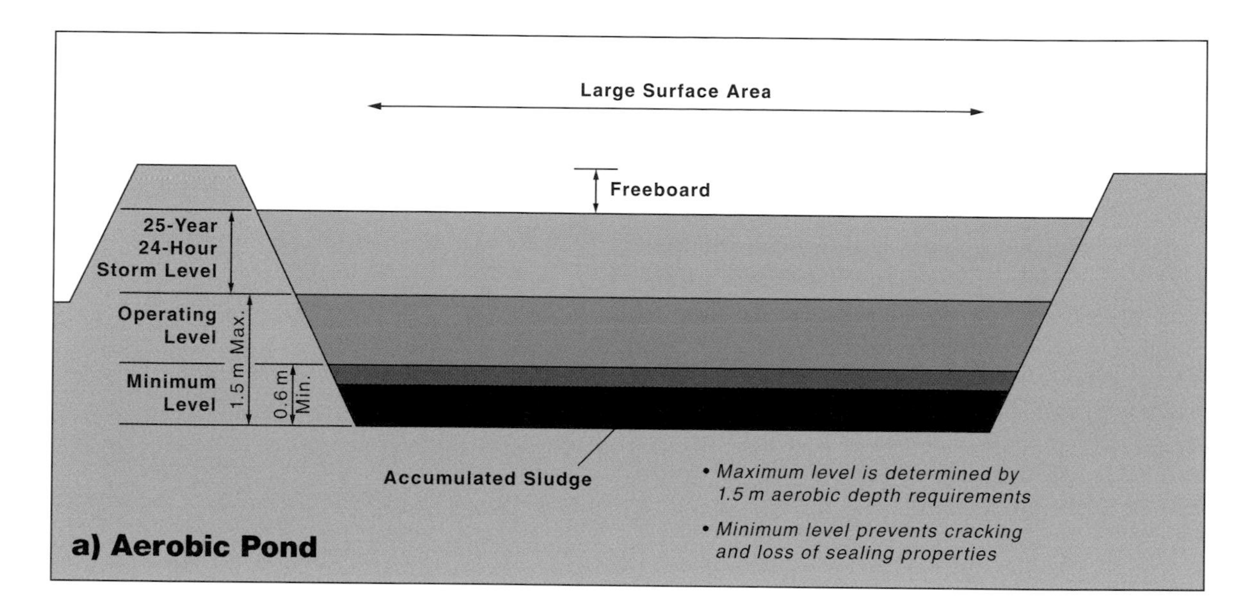

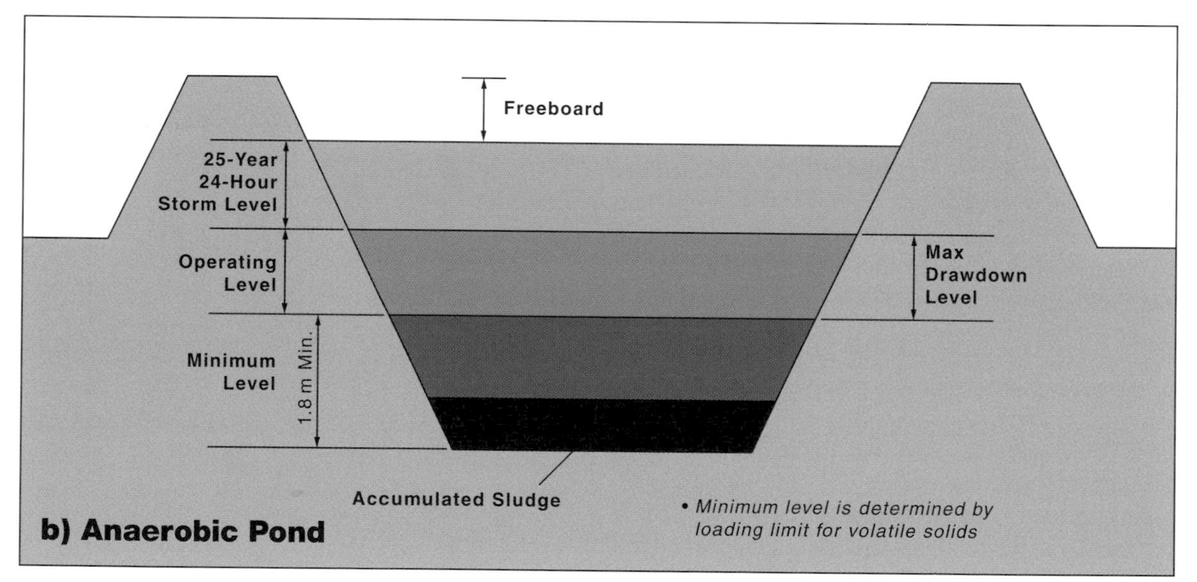

Figure 16-5 Pond profiles: (a) Aerobic waste pond profile, and (b) Anaerobic waste pond profile. Adapted by permission from *The Guidebook: Nitrogen Best Management Practices for Concentrated Animal Feeding Operations in Arizona*, by Robert J. Freitas and Thomas A. Doerge; copyright © 1996 by the Arizona Board of Regents. All rights reserved.

tance may cause the decay rate to change considerably, thereby affecting the application rate as well. Note that the amount of manure that needs to be applied decreases each year; this decrease is due to residual N from the previous application. Note also that the decay rate is maximal during the first year of application, but decreases dramatically in subsequent years.

The most effective means of maintaining the nutrient value of manure depends on the form in which it is applied to cropland. If it is being applied as a slurry, it should be injected below the surface; but if the manure is applied as a solid on the surface, the treated soil should be tilled immediately. This reduces nuisance odors, nutrient losses by volatilization, and the potential for groundwater pollution caused by runoff.

16.10 Microbiology of Animal Wastes Degradation

In nature, animals consume plant material where it is grown and their wastes are distributed over the landscape; that is, primary production, consumption, and decay are linked. In modern livestock production, however, animals consume feed grown on farms that are often great distances from the feedlots and dairies where manure accumulates as a waste. Thus, the link between primary production, consumption, and decay is broken.

Animal wastes, including manure and bedding material, serve as the substrate for microorganisms, especially bacteria, actinomycetes, and fungi (see Chapter 4, beginning on page 31). Decomposition of animal wastes ultimately mineralizes organic forms of C, N, S, and P to inorganic forms. Opposing mineralization is the assimilation or immobilization of these elements into microbial biomass. Mineralization produces energy and oxidized compounds, while assimilation requires energy and produces reduced compounds. Only when mineralization exceeds immobilization do pools of inorganic ions accumulate.

Livestock facilities, their wastes, and waste handling systems vary, but there are common features as shown in Figure 16-3 on page 245. Livestock in feedlots or dairies produce manure, which includes feces and urine. Some carbonaceous bedding, such as straw or sawdust, may also be included. Feedlot wastes are usually handled as solids by earthmoving equipment and applied to agricultural land, composted (see Figure 20-9 on page 313), or just stockpiled. Dairy wastes are often handled by flushing systems and disposed of in lagoons, which may be either aerobic or anaerobic. Aerobic systems may be aerated by bubbling air through pipes, by production of algae, or by passive diffusion from the atmosphere. Although anaerobic and aerobic biochemical pathways are distinct, both aerobic and anaerobic microsites always occur in both systems. Even with forced air, for example, aerobic lagoons are usually anaerobic on the bottom where sludge accumulates.

Regardless of the nature of the animal waste or the waste-handling system, the waste disposal for pollution control should include the following goals:

1. Reduce the total mass and volume of waste, largely by formation of CO_2 from the metabolism of organic compounds.
2. Produce stabilized organic compounds, especially humus, with a balanced C:N ratio and low BOD.
3. Minimize N losses as NO_3^- to surface water and groundwater.
4. Minimize N losses as NH_3 to the atmosphere.
5. Minimize P discharge to surface waters.
6. Minimize gaseous losses of reduced compounds, such as CH_4 and H_2S.
7. Dispose of the final product on land for agriculture.

Consider waste handling at a typical feedlot and assume that the waste consists of feces, urine, and straw bedding. Assume further that the wastes are mechanically scraped and hauled to a storage pile or flushed to a pond, either aerobic or anaerobic. The fecal component contains undigested feed and microbial cells, living and dead, as well as sloughed cells from the animals' gastrointestinal tracts. Some of this material is partially humified. Urea is a major component of urine, while the bedding is mostly cel-

Table 16-2 Average manure decay and application rates to maintain 220 kg N ha^{-1}. Source: Doerge *et al.*, 1991.

	Decay Rate Each Year ($\% \, yr^{-1}$)				Amount Applied Each Year (metric tons ha^{-1})			
Manure Type	1st	2nd	3rd	4th	1st	2nd	3rd	4th
Fresh bovine, 3.5% N	75	15	10	7.5	8.5	8.1	7.9	7.9
Corral (dry), 2.5% N	40	25	6.0	3.0	22.5	15.7	14.8	13.9
Corral (dry), 1.5% N	35	15	10	7.5	42.7	30.8	27.4	25.6
Corral (dry), 1.0% N	20	10	7.5	5.0	112.3	67.4	55.0	53.2

lulose and lignin. Water is also a large and variable component.

From the microorganism's perspective, the waste is a varied feast, consisting of large and small molecules, either resistant or available. Large molecules include polysaccharides, proteins, lipids, nucleic acids, and other organic polymers. These may be free in the mixture, but more likely they are still associated within the structures of plant, microbial, and animal cells. Examples of polysaccharides are cellulose, starch, the peptidoglycan of bacterial cell walls, the chitin of fungal cell walls, the teichoic acid of Gram-positive bacterial cell walls, and the lipopolysaccharide of the Gram-negative bacterial outer membrane. Proteins include enzymes of plant, animal, and microbial origin. Lipids are components of membranes. Nucleic acids derive from genetic material. Small molecules, which include sugars, amino sugars, amino acids, organic acids, alcohols, and nucleotides, are soluble components of the cytoplasm of cells or hydrolysis products of the macromolecules.

Heterotrophic bacteria are the primary decomposers of animal wastes, but chemoautotrophs also have a role. Decomposition is initially a continuation of digestion begun in the animal's gut, so bacterial species excreted from the animal dominate during early periods of decomposition. Decomposition is continued by soil bacteria, actinomycetes, and fungi, which, as a group, are capable of processing all the compounds present in the waste by various pathways. To use a specific substrate, a microorganism may require extracellular hydrolytic enzymes, transport proteins for the products of hydrolysis, and enzymes of the central catabolic pathways inside the cell.

Some carbon is lost as CO_2 and CH_4 while most of the nitrogen is assimilated; thus it is an important feature of the waste stream that the C:N ratio decreases with time. Recall that a goal of waste disposal is to reduce losses of C, N, S, and P in noxious or polluting forms. When these elements are assimilated into microbial biomass, they are temporarily protected from these losses. Of these compounds, N and P are the elements most likely to cause pollution.

Microbial degradation of N compounds such as proteins, nucleic acids, and aminopolysaccharides occurs outside the cell via hydrolysis to amino acids, nucleotides, and amino sugars. Some of these are metabolized for energy, while others are used for biosynthesis. The resulting metabolic N waste is NH_3 or NH_4^+. A very wide range of heterotrophs, including aerobes and anaerobes, are ammonifiers. The appearance of NH_4^+ is an indication of excess N relative to available carbon. If C is plentiful, more biomass will be synthesized and the N will be retained in organic form. Since microbial cells are about 50% C and 5% N on a dry weight basis, net mineralization of N from substrates with C:N ratios greater than about 25:1 cannot occur.

Ammonium ions released by ammonifiers can be assimilated by other microorganisms. A very important fate of NH_4^+ is nitrification, the oxidation to nitrite and then nitrate by chemoautotrophs. The nitrate produced is extremely soluble, which makes it a potential water pollutant (and cause of methemoglobinemia). Nitrate can have various fates. It can be leached to groundwater, assimilated by microorganisms or plants, or denitrified (see also Section 14.2, beginning on page 212).

Lignin complexed with cellulose is a large component of manure. It is derived from plant material in the animal feed or bedding and is very resistant to decomposition. Thus, as more readily degraded substrates are metabolized, the waste actually becomes enriched in lignin. Lignin is a polyphenol—a polymer of complex units containing aromatic rings. Its structure is so complex that a diverse collection of enzymes are necessary to decompose it. Monomers produced during lignin decomposition are oxidized to quinones and assembled into humic substances which are stabilized organic matter. Humus is resistant to further decomposition, has a very low BOD, and can be applied to agricultural land with no risk of pollution. Humus in soil is mineralized at a rate of a few percent a year, and humus mineralization and synthesis of new humus are balanced in nature. Since ammonification of humus is slow, nitrification of NH_4^+ from humus is also slow, thus presenting minimal risk of leaching to groundwater.

Using the principles discussed above, let us summarize an ideal waste disposal system for a confined animal feeding operation.

1. Manure is removed frequently and combined with bedding or some carbonaceous bulking material such as sawdust to produce compost. The high C:N ratio favors immobilization of N.

2. The liquid waste is digested anaerobically in a closed chamber. Methane is produced, collected, and burned for generating heat or electricity. The sludge has a lower BOD than the fresh manure. The ammonia produced is retained as ammonium ions at the pH of digestion. Nitrate is not produced under anaerobic conditions.

3. Sludge from anaerobic digestion is applied to the compost and is composted aerobically. Most of the N remains immobilized in microbial biomass. Small amounts of NH_4^+ and NO_3^- appear as the compost matures.

4. Effluent from anaerobic digestion is aerated and retained long enough to reduce BOD.

5. The mature compost is applied to agricultural land.

6. The effluent is used to irrigate agricultural land, or undergoes further treatment for reuse within the animal facility.

References and Recommended Reading

Blake J., Donald J., and Magette, W., Editors (1992) *National Livestock, Poultry and Aquaculture Waste Management: Proceedings of the National Workshop.* July 1991. American Society of Agricultural Engineers, St. Joseph, Michigan.

Bucklin R., Editor (1994) *Dairy Systems for the 21st Century: Proceedings of the 3rd International Dairy Housing Conference.* February 1994. American Society of Agricultural Engineers. St. Joseph, Michigan.

Doerge T.A., Roth R.L., and Gardner, B.R. (1991) *Nitrogen Fertilizer Management in Arizona.* Publication 191025. The University of Arizona College of Agriculture, Tucson, Arizona.

Follett R.F., Keeney D.R., and Cruse R.M., Editors (1991) *Managing Nitrogen for Groundwater Quality and Farm Profitability: Proceedings of a Symposium, 1988.* Soil Science Society of America, Madison, Wisconsin.

Freitas R.J. and Doerge T.A. (1996) *The Guidebook: Nitrogen Best Management Practices for Concentrated Animal Feeding Operations in Arizona.* Publication 194032. The University of Arizona College of Agriculture, Tucson, Arizona.

Hammer D.A. (1989) *Constructed Wetlands for Wastewater Treatment: Municipal, Industrial and Agricultural.* Lewis Publishers, Chelsea, Michigan.

Merkel J.A. (1981) *Managing Livestock Wastes.* Avi Publishing Company, Westport, Connecticut.

USDA Soil Conservation Service (1992) *Agricultural Waste Management Field Handbook.* April, 1992. United States Department of Agriculture, Washington, D.C.

Problems and Questions

1. Waste systems can be divided into conceptual elements. What are three of those elements?
2. Give three major site characteristics that should be sought in the location of a large animal confinement facility to prevent pollution.
3. Name the three primary potential pollutants of a dairy or feedlot. What are the effects of each of these pollutants on receiving waters and on human health?
4. A large fraction of manure is lignin. What is lignin and why does it resist microbial degradation?
5. Why does composting aid in pollution prevention?

Chapter 17 Pesticides as a Source of Pollution

J.E. Watson

Aerial applications, such as this early morning insecticide spraying of cotton in Arizona, allow pesticide applications within target windows, which are critical to successful, integrated pesticide management at times when the soil cannot support tractor traffic due to wetness. Photograph: J.E. Watson

17.1 Societal Views of Pesticides

The use of pesticides tends to evoke strong emotional responses, both for and against. The truth is that many people regard pesticides as dangerous chemicals whose presence in the environment threatens imminent harm to life on earth, human and otherwise. Others—particularly members of the farming community—regard pesticides as beneficial management tools and tend to react with alarm at the prospect of any pesticide restrictions. Nearly everyone, it seems, takes pesticides *personally.*

Although it is no part of the following discussion to delve into psychology or sociology, it is important to recognize the significant role that emotion plays in the debate about pesticide use, particularly in the area of risk assessment. We need to understand that pesticides, or the lack thereof, scare people largely because they believe that their own well-being will be adversely affected if pesticides are either carelessly applied or not used at all, depending on the viewpoint of those concerned. And we can see that this fear is directly related to the question of control. That is, the level of individual response to a perceived threat is more often correlated with the perceived ability to control the outcome than with any "objective, scientific" evaluation of the true risk.

For example, most of us know—or know of—people who have been seriously injured in accidents, but we willingly continue to operate automobiles. Despite the objective reality presented by news reports and personal experience, most of us choose to take the risk because we *feel* "in control"; thus, on a nonrational level, the feeling of being in control overcomes the feeling of fear. Moreover, on a more rational level, we recognize that the risk of not traveling by automobile outweighs the risk of being involved in a serious accident. In other words, individuals take chances because they *believe* they can influence the outcome and because they *think* the payoff outweighs the risks.

This nonrational/rational approach to risk-taking offers some insight into the emotionally charged debate about pesticides. Although both the benefits and the risks are "real," many members of the community at large don't recognize that the risks associated with pesticides are outweighed by the benefits they offer. Instead, society as a whole tends to respond with its collective gut, preferring to eliminate pesticides rather than run the risk of exposing themselves or loved ones to "dangerous chemicals"—chemicals that others are imposing on them. (Obviously, many gardeners and golf-course owners don't agree, nor do homeowners who saturate nooks and crannies with household pest sprays.) The farming community, however, tends to take the opposite view with equal emotional vigor. Acutely aware of the benefits of pesticides, they often minimize the risks because they feel that legal restrictions will limit their control. Like motorcyclists who balk at helmet use, they respond emotionally to any limitations that may "objectively" benefit them.

Risk assessment with respect to pesticide use is therefore a knotty problem—one that must be untangled from the snarl of emotions that complicate it. In this chapter we will look first at the variety of pesticides available; then we will examine the extent of pesticide use and the methods that can be used to mitigate the problem of pesticide pollution.

17.2 Classes of Pesticides

None of the information contained in this chapter should be considered as recommending use of a particular chemical for a specific pest control problem. All recommendations should be made by certified pest control advisors who are familiar with local conditions. Additionally, many chemicals mentioned in this discussion have been discontinued. Their inclusion in this chapter is for illustrative purposes only.

We generally use the term pesticide colloquially to refer to a broad range of chemicals that kill unwanted life—weeds, insects, rodents, and the like. We apply pesticides to rid ourselves of creatures whose presence adversely affects our persons, our homes, and the components of plant production, storage, or processing. The technical definition, stated in the amended Federal Insecticide, Fungicide and Rodenticide Act (FIFRA, see Section 23.5, beginning on page 368), is that a **pesticide** is any substance or mixture of substances intended for destroying, preventing, or mitigating insects, rodents, nematodes, fungi, weeds, or any other undesirable pests. This definition also includes any substance or mixture of substances intended for use as a plant or insect growth regulator, as well as defoliants (used to cause leaves to drop from plants to facilitate harvest) and desiccants (which dry up unwanted plant tops). Under this definition many chemicals, both newly developed and familiar, may be considered as pesticides and be regulated as such. For example, insect pheromones (sex attractants) may be used to attract certain insect populations, to confuse mating patterns, and thereby control insect populations. Ordinary dishwashing detergent may be used to kill whiteflies or bees. Common table salt (sodium chloride) is used to control weeds in beet fields in humid regions. Are these chemicals pesticides, and if so, how should they be classified?

Classifying pesticides as to their target is one method of categorizing them. Thus, **insecticides** are formulated to control particular insects, **herbicides** are formulated to control weeds, and **fungicides** are formulated to control particular fungi, molds, or mushrooms. Pesticides may also be classified according to their mode of entry into the target pest. **Contact pesticides** enter the target pest upon direct application, while **systemic pesticides** must pass through a host organism before they enter their targets. For example, a contact insecticide, or its residue, kills target insects on direct application, while a systemic insecticide kills insects only after moving through the system of the plant hosting the target insect. Thus, if a particular insect does not feed on the plant, it will not be harmed. Both desiccants and defoliants can be used to remove vegetation. Sodium chlorate and tribufos, for example, are two commonly used defoliants to remove cotton leaves before harvest so that the cotton fiber is not stained with chlorophyll.

Finally, pesticides can be classed by the forms in which they are used. **Fumigants,** for example, are pesticides applied as gases. Fumigants may be used selectively to control drywood termites in houses or to control the pest population in stored products such as fruits, vegetables, and grains. They may also be released over large areas to remove many pests from soil. Metam-sodium and dichloropropene are two commonly used fumigants applied to soil to control nematodes and soil fungi.

17.2.1 Insecticides

Insecticides may be grouped according to their chemical properties and characteristics (see also Section 21.8.3, beginning on page 340, for a discussion of the toxicology of these substances). One broad category of pesticides comprises inorganic substances, which lack the element carbon and therefore cannot be obtained from living things. In general, inorganic pesticides are slow-acting, but have the advantage of having long-term residual activity. Examples of inorganic pesticides include sulfur, which can be used to control thrips which damage citrus, and boric acid, which is often effective in controlling garden pests and is frequently used in cockroach baits and ant poisons. Some of the first pesticides ever used were naturally occurring chemical substances such as sulfur and arsenic. Sometimes, inorganics can be used in their natural state, or they may be chemically modified to make them more effective or safer to use.

Other pesticides are broadly classed as organic substances. Organic insecticides differ from inorganic insecticides not because they are "natural," but because they contain the element carbon. This means that organic pesticides may be obtained from living things or derived from natural compounds. They may also be synthesized "from scratch." That is, the chemical structures of synthetic organic insecticides may either be modified from the original biological sources, as in the case of synthetic pyrethroids derived from the natural compound pyrethrum, or they may be novel, as in the case of organophosphates, which were developed during World War II by German chemists who were looking for a substitute for nicotine.

Chlorinated hydrocarbons (also called organochlorines), which are composed of carbon, hydrogen, and chlorine, were the first group of synthetic insecticides to be widely used. One such chlorinated hydrocarbon is 1,1,1-trichloro-2,2-bis(*p*-chlorophenyl)ethane, which is popularly known as DDT. Although this compound was used extensively throughout the world for many years, it was banned for use in the United States in 1971 for a number of reasons, primarily because of its deleterious effects on bird reproduction. (Arizona, the first U.S. state to limit its use, restricted it because of problems with DDT-contaminated milk in dairies.) Modern chlorinated hydrocarbons include lindane, methoxychlor, and dicofol. Methoxychlor has a very low toxicity to humans and warm blooded animals. Even before the banning of DDT it was used as a replacement for DDT where the DDT application could constitute a hazard to warm-blooded animals. Most chlorinated hydrocarbon pesticides, including DDT, have a low toxicity to humans.

Organophosphates contain the element phosphorus linked to oxygen to form the phosphate group. Developed to replace some of the chlorinated hydrocarbons, organophosphates include such compounds as diazinon, chlorpyrifos, and acephate. Organophosphorus compounds are anticholinesterase chemicals which damage or destroy cholinesterase, the enzyme required for nerve function in the animal body. Careful handling is required to prevent serious health problems for the person applying the chemical, as the human toxicity is usually high.

Carbamates contain carbon atoms linked to nitrogen and oxygen. First developed in the 1950s, carbamates include such insecticides as carbaryl, propoxur, and bendicarb. Like the organophosphorus compounds, the carbamates inhibit cholinesterase.

Microbial insecticides are microorganisms—or the toxins they manufacture—that can be used to control insects. For example, *Bacillus thuringiensis* is a bacterium that produces an endotoxin during fermentation. Different serotypes (*e.g.*, *israelensis,* Serotype H-14; *kurstaki,* Serotype H-3a3b; *morrisoni,* Serotype 8a8b) produce various endotoxins which control assorted insects. For example, *Bacillus thuringiensis berliner* var. *israelensis*, Serotype H-14 is the most potent subspecies of *Bacillus thuringiensis* against mosquito and black fly larvae. *Bacillus thuringiensis* var. *kurstaki*, Serotype H-3a3b is a caterpillar larvicide for use with most lepidopterous larvae with high gut pH, that is, it is known to disrupt the gut of caterpillars. When these bacteria are sprayed on the plants the caterpillar is feeding on, the caterpillar consumes the bacteria and the toxin as well as the plant.

Alternatively, the toxin produced by the bacteria can be extracted and applied directly. In some cases, the bacterial gene that controls the production of a toxin can be inserted into the plant itself. For example, some cotton varieties have been genetically altered to produce the toxin from *B. thuringiensis* themselves without the bacteria. Abamectin, derived from the soil microorganism *Streptomyces avermitilis*, is another microbial insecticide that is used as a miticide. Mites are tiny organisms closely related to ticks. Some mites, such as the chigger, are parasitic on higher animals. (Take a walk through the rangelands of the southwest and you will likely become a host for some chiggers.) A large family of mites, known as the spider mites, derive their name from the habit of spinning a web on undersides of leaves where they feed. Spider mites are often pests of agricultural crops. The avermectins derived from *Streptomyces avermitilis* are used to control these mites. Avermectins are also used in fire ant bait.

Many plants contain materials that deter feeding by insects, thus acting as **botanical insecticides**. Some of these natural compounds have been extracted and used as the basis for developing new insecticides. **Pyrethroids**, for example, were developed by chemists interested in modifying the basic chemical structure of the botanical insecticide **pyrethrum**. Pyrethrum is extracted from the plant *Chrysanthemum cinerariaefolium*, with the flowers being the source of the pyrethrins. Pyrethrum has a low order of toxicity to warm-blooded animals, being about 30 times less toxic than the insecticide methyl parathion. Pyrethroids, which are also known for their ability to kill insects quickly, include such common insecticides as permethrin and fenvalerate. These pyrethroids, and others, are synthetic compounds, whose structure is based on the chemistry of pyrethrum. [*Note:* Just because chemicals are derived from plants does not imply they are less toxic to humans or other animals. Rotenone is derived from a plant, but it is highly toxic to fish. Nicotine is also plant-derived, but its toxicity rivals that of the synthetic methyl parathion, which is about 20 times more toxic to warm-blooded animals than pyrethrum.] Allethrin was the first synthetic pyrethroid, introduced in 1949. It has a toxicity similar to pyrethrum. Fenvalerate, on the other hand is somewhat more toxic than pyrethrum.

Some organic pesticides don't kill anything directly. **Insect growth regulators,** for example, are considered pesticides because they stimulate or disrupt growth and/or development in insects. Among these regulators are **juvenile hormone analogs**. These are synthetic or derived compounds that mimic the natural insect hormone known as juvenile hormone, which is involved in controlling molting and other processes in insects. An example of a juvenile hormone analog is methoprene. Methoprene products are available for the control of larval mosquitoes in water, sciarid flies in mushroom production facilities, and pharaoh's ants in buildings. Fenoxycarb is also a growth regulator used for fire ant control. After consuming fenoxycarb, the queen fire ant can lay her eggs as usual, but the eggs never develop into worker ants. Hydroprene is the active ingredient in a commercial product which inhibits normal molting processes, causing the mortality or sterility of the insect at or before the molt. It is often used to kill pests of stored-products, including cockroaches.

17.2.2 Rodenticides

A rodenticide is a material used primarily for the control of rodents (rats, mice, *etc.*) and related animals. A variety of chemical compounds are or have been used, including coumarins, indandiones, organochlorines, organophosphates, pyriminilureas, and botanicals.

17.2.3 Herbicides

Like insecticides, herbicides may be inorganic or organic, derived or synthesized (see Section 21.8.4, beginning on page 341, for more information on the toxicology of these substances). The earliest herbicides were inorganic materials such as salt (NaCl) or ashes. It is reported, for example, that after their final battle with Carthage in 146 BC, the Romans spread salt over the site of the ruined city, thereby destroying the crops that remained and preventing replanting. Other salts are also good herbicides, such as copper sulfate, which is toxic to algae, and sodium chlorate, which is used as a desiccant.

Petroleum oils (which contain only carbon and hydrogen) were the first organic herbicides. In the late nineteenth and early twentieth century, hydrocarbon materials such as used motor oil, kerosene, and diesel fuel were used to keep areas free of weeds. These petroleum products are no longer recommended: they can contaminate surface or groundwater and they are fire hazards.

Carbamate compounds (which contain oxygen as well as carbon and hydrogen) are organic substances that have herbicidal, as well as pesticidal and fungicidal, properties. Carbamates are commonly used as pre-emergence herbicides, although some have post-emergence activity. Examples of carbamate herbicides are chlorpropham, which is a plant growth regulator that kills several weed species after they emerge, and asulam, used in a number of situations to control various weed species after they emerge, including johnsongrass, crabgrass, foxtail, and goosegrass.

Triazines, which are made up of carbon and nitrogen atoms that form a six-sided ring, are commonly available as herbicides. Well known examples of a triazine herbicide include atrazine, prometryn, and cyanazine. These compounds tend to be more water soluble than the carbamate herbicides or insecticides, or the chlorinated hydrocarbon insecticides. Therefore, they are more likely to contaminate groundwater or surface water. A variety of weed species are controlled by these compounds.

Phenoxy herbicides, which are aromatic (ring) compounds that contain oxygen, were developed in the 1940s. These compounds may also be substituted aromatics, such as chlorphenoxy herbicides, which contain the element chlorine. Phenoxy herbicides mimic auxins, which are natural plant hormones used as growth regulators. An example is 2,4-dichlorophenoxy acetic acid (2,4-D), which is used for post-emergent control of Canada thistle, dandelion, annual mustard, ragweed, and others. It acts as a selective hormone-type herbicide. Many broadleaf crops, such as cotton and grapevines, are extremely sensitive to 2,4-D.

Amide herbicides, which contain the element nitrogen, are generally simple molecules that break down readily and do not persist in the soil. Some amide herbicides inhibit root elongation in seedlings, whereas others interrupt photosynthesis. Examples include propanil (whose structure is shown in Figure 7-4 on page 80) and cypromid.

Other nitrogen-containing herbicides are also common. Like the amides, **dinitroaniline** and other substituted anilines contain the element nitrogen, but the nitrogen atoms are linked to oxygen. These herbicides interfere with enzymes produced by the plant and inhibit root or shoot growth. An example is trifluralin. **Substituted ureas**, which also contain nitrogen, block photosynthesis as well. Examples of substituted ureas include diuron and fenuron.

17.2.4 Plant-Growth Regulators, Fungicides, and Disinfectants

Plant growth regulators are also used to control weeds. These include auxins, gibberellins, and cytokinins, as well as ethylene generators and growth retardants, all of which affect either plant growth or fruit ripening.

Plants are often susceptible to the same disease agents that strike animals; these disease agents include fungi, rickettsia, and mycoplasma-like organisms, as well as viruses and bacteria. Moreover, such diseases are often difficult to control in plants because it is hard to kill the disease agent without harming the host. Thus the use of fungicides and disinfectants is an important part of achieving healthy crops. Some fungicides and disinfectants cure a problem after it has appeared, whereas others are effective only as preventatives.

Elements such as sulfur or copper are often used as fungicides. For instance, copper is an enzyme inhibitor, while sulfur is a metabolism inhibitor. Salts of these elements, such as copper sulfate and zinc sulfate, are examples of inorganic fungicides that have been used. As is the case for insecticides and herbicides, a number of different types of organic compounds can also act as fungicides. These include dithiocarbamates (*e.g.*, thiram, maneb), thiazoles (*e.g.*, terrazole), triazines (*e.g.*, anilazine), substituted aromatics (*e.g.*, PCNB, chlorothalonil), dicarboximides (*e.g.*, captan, folpet, captafol), benzimidazoles (*e.g.*, benomyl, thiophanate), dinitrophenols (*e.g.*, dinocap), quinones (*e.g.*, dichlone), and organotins (*e.g.*, fentin hydroxide).

17.3 Extent of Pesticide Use

Pesticides are all "toxins"—that is, they kill or otherwise adversely affect their target organisms. However, not all pesticides are pollutants. Even the most dangerous (to humans and other desirable life forms) chemical is not a pollutant until it shows up where it's not wanted, say in groundwater, accumulated in the food chain, or as a residue on food. In attempts to prevent pesticides from becoming pollutants, it is first necessary to find out how much and where pesticides are being, and have been, used. And, someone has to keep track of this information.

17.3.1 Reporting Criteria

Before a commercial pesticide is sold or distributed by intra- or interstate commerce in the United States, it must be registered by the U.S. Environmental Protection Agency (EPA). The EPA has compiled substantial lists of pesticide ingredients whose application must be reported. (The EPA estimates that the average cost for a registrant to comply with The EPA's data requirements for registration varies between $2.4 to $4 million per ingredient.) The EPA is also authorized, by the Federal Food, Drug, and Cosmetic Act (FFDCA), to establish tolerances for pesticide residues in raw and processed foods. The Food and Drug Administration (FDA) of the Department of Health and Human Services monitors and enforces the established tolerances. When a pesticide is registered for use on a food or feed crop, a tolerance level, or an exemption from such a requirement, must be established. The purpose of establishing such tolerance levels is to ensure that U.S. consumers are not exposed to unsafe levels of pesticide residues. Any commodities with residue levels in excess of the tolerance level are subject to immediate seizure. Generally, however, only a small percentage of the produce imported into the U.S. is actually sampled for compliance.

In addition, many individual states in the United States have established other regulatory agencies to control pesticide applications to protect wildlife and water supplies. For example, Arizona has compiled a list of chemicals—the Groundwater Protection List—whose use must be reported. Similar requirements exist for the sales of these pesticides, so that significant under-reporting of applications cannot occur without alerting the regulatory agency.

How, then, is the decision made to include chemicals on federal or state lists for regulation? One important criterion is a chemical's potential for migrating to groundwater or surface water. Additional criteria include acute toxicity (to humans) and carcinogenicity, as well as whether the chemical has been found in groundwater or surface water in other locations.

However, neither the federal nor the state lists are definitive. For example, these lists usually only consider the active ingredient of commercial pesticide products, not the inert ingredients. The component that actually controls the target pest is called the **active ingredient**, while the components that are not active against the target pests are the **inert ingredients**. Inert ingredients usually include solvents, surfactants, and carriers or substrates. The trouble is that not all inert ingredients are innocuous. Although some of these ingredients may be subject to control or regulation because of environmental or health concerns, others may not be regulated, even though they may cause undesirable secondary effects. (Such an undesirable secondary effect is produced by a spray for wasp control; while the active ingredient kills wasps, the "inert" solvent kills a sensitive landscape plant.)

17.3.2 Extent of Pesticide Use

About 45,000 pesticide products are marketed in the U.S., amounting to 0.55 billion kilograms of pesticides per year, valued at over 6 billion dollars. Most of these products are organic chemicals; the use of inorganic pesticides, such as compounds containing arsenic, has declined over time. The U.S. EPA estimates that, in 1986, about 70% of these products were used in agricultural production and 7% in home and garden settings, while the remainder was used in forestry, industry, and government programs. One reason for the marked contrast between the amount applied through agriculture and that applied in other settings is obvious—agriculture covers more land surface than do residences or industries. These per-

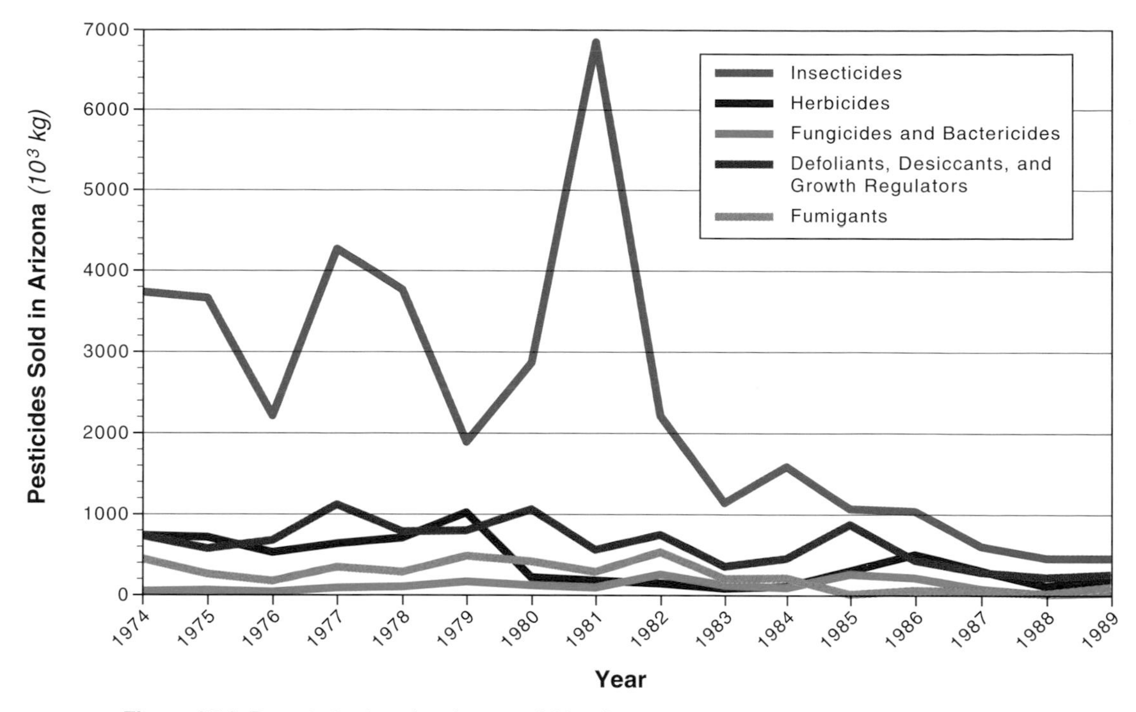

Figure 17-1 Reported sales of various pesticides (in 1,000 kg of technical material) in Arizona from 1976 to 1989. Data derived from Sherman *et al.*, 1993.

centages are extremely tenuous, however, because of the lack of records.

Unfortunately, the extent of pesticide use—and the capacity for pesticides to become pollutants—is neither well known nor adequately documented. In the United States, for example, individual states have begun to track such information only since the mid 1970s, particularly with respect to specific crops. This lack of data is further exacerbated by the fact that registration procedures are variable and subject to change. Moreover, although the use of agricultural pesticides is relatively well reported, applications by homeowners are generally not included at all. To make matters worse, many pesticides could be used without reporting them if their active ingredients did not appear on the EPA's special list of controlled chemicals, or on similar lists, such as Arizona's Groundwater Protection List.

Figure 17-1, for example, presents estimates of the volume of sales of pesticides in Arizona alone beginning in 1974. The numbers are only estimates because sales and applications of materials were not reported in the earlier years. Notice that the annual sales substantially vary; such variability is frequently due to changes in the amount of sulfur (insecticide) and sodium chlorate (defoliant/desiccant) sold.

17.4 Fate of Pesticides

Depending upon their physicochemical properties, patterns of use, and local conditions, some pesticides may leach through the crop root zone and eventually contaminate groundwater at certain locations. The two most important properties of a chemical that determine whether a pesticide represents a threat to groundwater are its persistence and mobility in soil. During the registration of new pesticides, computer programs are used to estimate the potential for groundwater to be contaminated by the specific use of a particular chemical at various locations in the United States. Several states, including Arizona and California, consider the capacity of a compound to

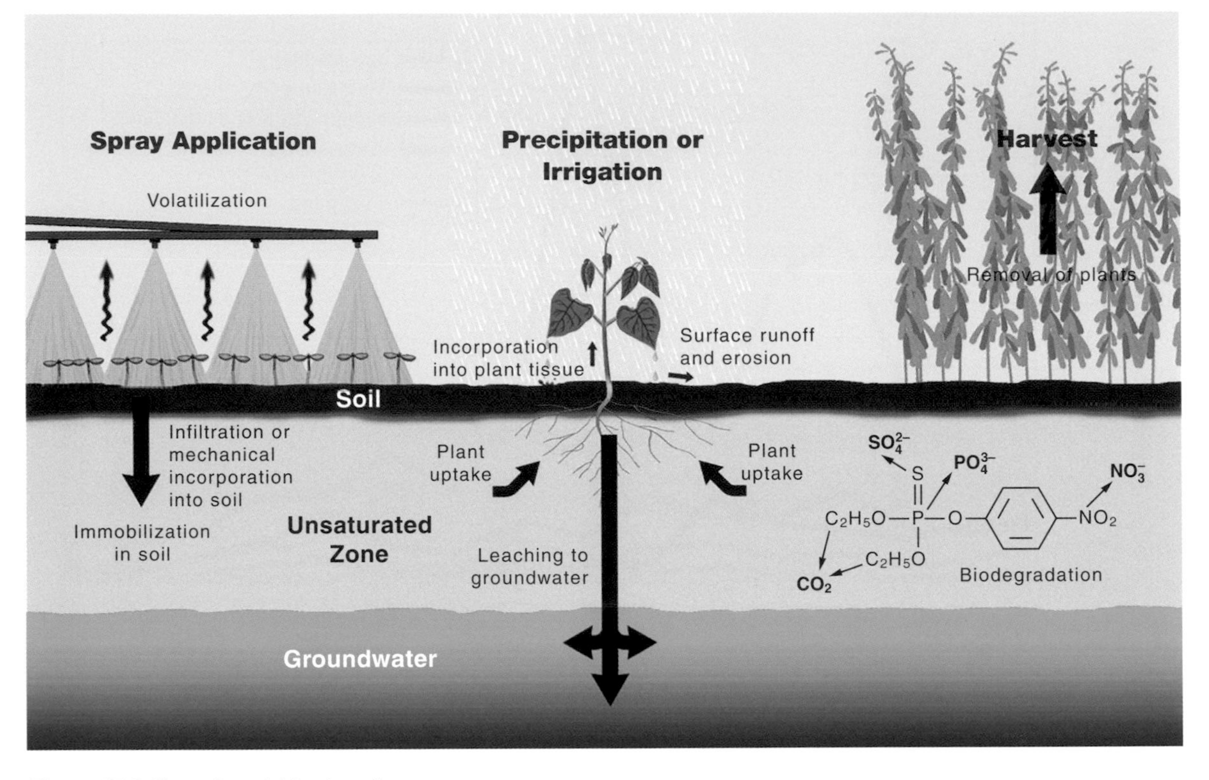

Figure 17-2 Fate of pesticides in soil.

leach through the soil into groundwater as a criterion for inclusion in their lists of controlled chemicals.

The **persistence—**that is, the staying power—of a pesticide depends on whether it is subject to chemical or biological degradation. Of these factors, biological degradation, or biodegradation, is the most important (see Chapter 7, beginning on page 77, and Section 11.4.1, beginning on page 160). While the ease of biodegradation depends on several factors (as discussed in Section 7.3, beginning on page 80), in general, the more similar a chemical compound is to a natural microbial metabolite, the more likely it will be easily degraded.

The other major factor affecting the potential groundwater contamination hazard of pesticides is the **mobility** of the pesticide. Factors affecting mobility are soil bulk density, soil volumetric water content, soil organic carbon content, the amount of excess irrigation and rainfall, and the degree of attraction the pesticide has for the soil organic matter or soil mineral surfaces (as discussed in Chapters 5 and 6, beginning on pages 45 and 63, respectively). As pointed out in those chapters, the sorption mechanism controlling nonpolar organic chemicals differs greatly from the mechanisms controlling sorption of inorganic chemicals.

Many pesticides used in agricultural settings have such a low persistence that they are not considered threats to groundwater quality. One example is acephate, which is considered mobile in most soils, but which has a half-life ($T_{1/2}$, see Eq. 6-17 on page 74) of less than 3 days in aerobic soils. Apparently, most of its degradation products are considered immobile, even in sandy soils. Other chemicals, such as trifluralin, are so tightly adsorbed to soil particles that even excessive irrigation cannot move them below the soil root zone before they degrade. In contrast, such chemicals as triazine herbicides are not so strongly adsorbed to soil and can leach below the root zone before they degrade if excessive irrigation water is applied to soils.

Two fungicides that used to be common, EDB (ethylene dibromide) and DBCP (dibromochloropropane), have been found in several groundwater wells. In one case (in Arizona), the contaminated groundwater lies beneath land areas that were previously citrus orchards, in which these compounds were used together with irrigation. It is not known whether this contamination was due to a point source (*e.g.*, backflow of pesticide into a well after injection into an irrigation system or rinsing of spray equipment at the wellhead) or to a nonpoint source (*i.e.*, leaching of chemical below the root zone of the crop). However, these two compounds are both persistent and mobile enough to have leached below the crop root zone under typical agricultural management practices of the 1970s.

Among the pesticides that have a low potential for groundwater contamination are some of the chlorinated hydrocarbons, which strongly adsorb to soil (*e.g.*, DDT and toxaphene). However, these compounds are no longer available for use in the United States because they can move to surface water when the soil to which they are adsorbed moves to lakes and streams during erosive storm events (see Section 15.2, beginning on page 225). These compounds do not degrade easily; consequently, they tend to accumulate in the food chain, creating long-term environmental problems. A newly developed herbicide class of pyridines has low soil mobility, providing protection for groundwater.

After a pesticide is applied to a field, it may meet a variety of fates, as shown in Figure 17-2. Some may be lost to the atmosphere through volatilization, carried away to surface waters by runoff and erosion, or photodegraded by sunlight. Pesticides that have entered into soil may be taken up by plants (and subsequently removed), biodegraded into other chemical forms, or leached downward with water below the crop root zone. The amount of any particular chemical that ends up volatilized, leached, degraded, or in surface runoff depends upon site conditions, weather conditions, management practices, soil properties, and pesticide properties (as described in several previous chapters). Generally, the insecticides presently used in agricultural operations have short degradation half-lives. Consequently, the length of time they are available for environmental contamination is relatively short.

The retention or adsorption of a compound in the soil may be measured empirically using its sorption coefficient (see Section 6.4, beginning on page 71). However, since most pesticides are organic compounds, the **organic carbon sorption coefficient** (K_{oc}) is often used to define the **retention** of the compound in the soil:

$$K_{oc} = \frac{K}{C_{oc}} \qquad (17\text{-}1)$$

where K is the sorption coefficient of the compound for a particular soil, and C_{oc} is the organic carbon content of the soil (in grams of organic carbon per gram of soil). Although this term is not appropriate for inorganic pesticides or for highly polar organic pesticides (which are ionized in soil solution), it is very useful for most organic pesticides. Because a pesticide's adsorption is related to the organic carbon content of the soil, we can estimate adsorption for a variety of soils of known organic carbon content even without making site-specific measurements of retention by first converting K_{oc} to K using Eq. (17-1) and the value for C_{oc} for the soil of interest and then applying Eq. (6-3) on page 70.

As noted in Section 6.4 (see Eq. 6-5 on page 71), the soil bulk density and volumetric water content are also important factors in estimating the retention of a pesticide in soil. As a general rule, however, we can say that for a given soil, the larger the pesticide's measured organic carbon sorption coefficient, the more slowly the pesticide moves downward through the soil and the less likely it is to contaminate groundwater. By the same token, the larger a pesticide's organic carbon sorption coefficient, the more likely it is to contaminate surface water; that is, it is more subject to soil erosion, since it will be retained near the soil surface for a greater period of time. Obviously, then, a pesticide's half-life is also an important consideration in estimating the potential for contamination.

The use of sorption coefficients and half-lives permits us to obtain only approximations for any general situation, as the specific measurements that make up such "constants" are subject to particular environmental conditions that may or may not be applicable in a selected field application. Thus, users of these data should explicitly recognize the potential errors implicit in them. Table 17-1 contains a list of

Table 17-1 Chemical properties of selected pesticides. Data: Watson and Baker, 1990

Common Name	Organic Carbon Sorption Coefficient (K_{oc}) (mL g^{-1})	Half-life ($T_{1/2}$) (days)
1,3-D	26	10
Alachlor	190	7
Aldicarb	12	28
Atrazine	163	48
Benefin	11,000	30
Bromacil	72	106
Captan	33	3
Carbofuran	29	37
Chlordane	38,000	3,500
Chlorpyrifos	6,070	63
Cyanazine	168	108
Cypermethrin	10,000	30
DBCP	70	180
Diazinon	580	30
Dicamba	2	14
Dicofol	>990,000	60
Dimethoate	8	7
Dinoseb	120	30
Disulfoton	1,603	5
Diuron	383	328
Endrin	8,100	4,300
EPTC	280	30
Fenamiphos	171	10
Fenvalerate	100,000	50
Glyphosate	10,000	30
Heptachlor	24,000	2,000
Linuron	863	75
Malathion	1,797	1
Methomyl	28	8
Methyl Parathion	5,102	4
Mevinphos	1	3
Napropamide	900	70
Oxamyl	9	6
Pendimethalin	24,300	60
Phorate	320	14
Picloram	26	138
Prometryn	194	60
Pronamide	990	30
Sethoxydin	50	5
Simazine	138	75
Terbacil	46	50
Trifluralin	14,000	70
Vinclozolin	98	20

pesticides, their persistence as defined by their degradation half-lives, and an estimate of their retention potential in soil. Many of the pesticides listed in this chapter do not have data available for them. This is particularly true for those pesticides which are no longer available for sale.

In evaluating the contamination potential of a particular pesticide, it is essential to consider its sorption coefficient and half-life jointly. For example, a pesticide with a small organic carbon sorption coefficient (*e.g.*, less than 100 mL g^{-1}) and a long half-life (*e.g.*, more than 100 days) poses a considerable threat to groundwater through leaching, particularly in soils having low organic matter. On the other hand, a pesticide with a large organic carbon sorption coefficient (*e.g.*, more than 1000 mL g^{-1}) and a long half-life is more likely to remain on or near the surface of soils with moderate levels of organic carbon content, thereby increasing its chances of being carried to a lake or stream in runoff water. For pesticides with short half-lives (less than about 30 days), the possibility of surface or groundwater pollution depends primarily on whether heavy rains or irrigations occur soon after application. Without water to transport them, pesticides with short half-lives remain in the biologically active root zone of soil and may degrade rapidly. In terms of water-quality protection, pesticides with intermediate sorption coefficients and short half-lives may be considered the "safest." Although they are not readily leached, they move into the soil with water, thereby reducing their potential for loss from erosion, and they degrade fairly rapidly, thereby reducing the chance for losses below the root zone. Figure 17-3 provides a schematic representation of the depth of movement of a strongly sorbed (glyphosate), a moderately sorbed (atrazine), and a weakly sorbed (aldicarb) chemical. It was assumed that the rainfall and irrigation amounts exceeded the crop water use by twice the amount of water contained in the root zone at an optimum water content which moved the chemicals downward.

As shown in the figure, glyphosate would be concentrated in the root zone to a depth of about 25 cm, atrazine would be concentrated near the bottom of the root zone (about 125 cm), and aldicarb would be concentrated at a depth of about 250 cm. A slightly

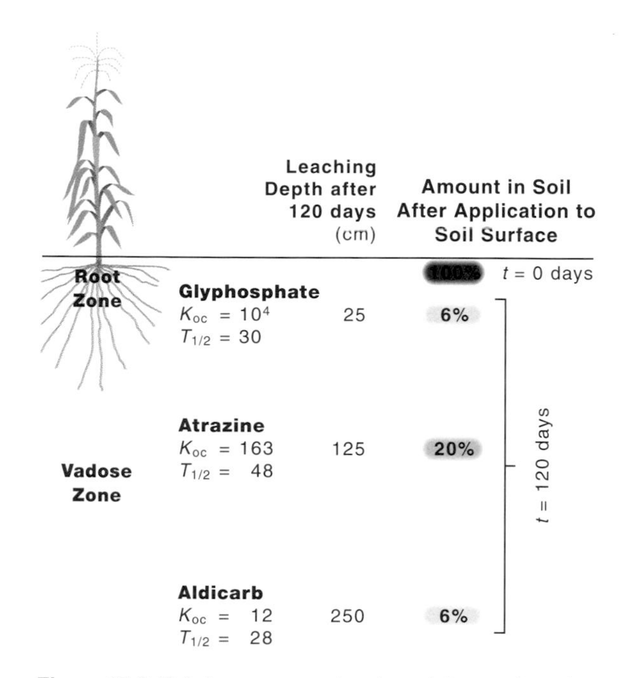

Figure 17-3 Relative movement and persistence of pesticides in soil. The transport of a particular pesticide is strongly influenced by its organic carbon sorption coefficient. Here the degree of sorption increases in the order aldicarb < atrazine < glyphosate. The amount of pesticide remaining in the soil is illustrated by the intensity of color 120 days after the application.

higher percentage of the applied atrazine would exist in the system, compared with the other two pesticides, because it has a slightly larger half-life. For a growing season of about 120 days, about 6% of the applied aldicarb and glyphosate would remain, while about 20% of the atrazine would remain. This example does not account for numerous differences in management practices that would influence the persistence and soil distribution of these pesticides.

17.5 Mitigation of Pesticide Pollution

When pesticides have a great potential for off-target losses to surface or groundwater, users need to apply management practices that reduce the likelihood of such losses. Consider irrigated areas, for example. In this case, most of the water applied to a field after a pesticide application comes from irrigation rather than rainfall. Thus the *method* of pesticide application, the *timing* of the application, and the *irrigation control* are all factors that must be taken into account.

These factors are particularly important for herbicides, some of which have a moderate potential for leaching losses to groundwater. By maintaining high irrigation efficiencies following herbicide application, the applied irrigation water, and hence the herbicide, will be maintained within the root zone of the crop, where it is available for uptake by weeds. In addition, applying the herbicide to the side of the row beds—rather than across the entire furrow and bed—will maintain more herbicide in the top portion of the soil surface. Also tilling the soil reduces or eliminates large macropores in the soil surface, thereby increasing the sorption of the applied herbicide to the soil particles and reducing the potential for losses.

The actual concentration of a pesticide that makes its way into the water is also of concern. The EPA has issued guidelines for **Health Advisory Levels (HALs)** for some pesticides, which include a human-health safety factor of several orders of magnitude. Water containing pesticides in concentrations at or below the HAL is designated as acceptable for drinking every day for a lifetime. However, such safety factors do not address the psychological problem that most people don't want *anything* in their water supply. (For example, even fluoride additions for cavity control often raise great local battles for water providers.) Nor do they take account of the fact that wildlife may be exposed to pesticides lost to streams. Toxicity to humans and toxicity to other animals tend to differ significantly. But HALs do provide a defensible rationale for establishing priorities for cleanup and regulation.

The concept of **Integrated Pest Management** (IPM) is frequently employed to address the problem of mitigating the negative effects of environmental pollution by improper use of pesticides. The principle behind Integrated Pest Management is the desire to reduce pesticide use to the minimum required for maintaining the quality of food that consumers will purchase while simultaneously maximizing human health and environmental quality. One example of an IPM approach is the use of **Heat Units** to track the growth of a cotton crop and pests and to plan insect control during times that coincide with critical insect growth stages. Heat units are a measure of the total amount of time that a particular pest or crop has to develop. They are related to both the length of time for development and the temperature during development. By studying the results of previous research, scientists have been able to predict the periods, or windows, when particular pest populations will peak depending upon the accumulation of heat units during the year. Thus, they have developed planting windows and pesticide application windows that growers can use to manage plant pests. Tracking the planting windows allows growers to establish a crop before a particular pest becomes a problem, while tracking the pesticide application windows allows optimum pest control with a significant reduction in the number of applications needed for control. The result is more efficient pest eradication, a reduced load of potential pollutants, and more effective maintenance of desirable insects in cropped fields.

Generally, point-source pesticide pollution arises from inadequate controls at mixing and loading areas, particularly when these areas are near wells. Failure to prevent back-siphoning to water supplies when pesticides are being mixed or applied in the irrigation water (chemigation) significantly increases the possibility of contaminating the water supply. Spills are also a frequent point-source of contamination by pesticides. Corrective action for such situations, which is fairly straightforward, includes the construction and maintenance of impermeable (*e.g.*, concrete) pads for mixing and loading areas, the installation of appropriate antisiphon devices on water delivery lines, and the updating and frequent review of a response plan should a spill occur. Pollution can also be caused by failure to follow label instructions during an application. When application instructions require that certain distances be maintained from surface water or prohibit use when groundwater depths are shallow, pollution may result if the instructions are not followed. However, such failures are seldom the documented causes of observed pollution.

Once soil, surface water, or groundwater becomes contaminated with a pesticide, remediation is sometimes deemed necessary. Remediation (which is discussed in greater detail in Chapter 11, beginning on page 151) generally involves removal of the offending chemical from the site and/or enhancement of its degradation process. Both of these methods can be difficult, expensive, and time-consuming. If, for example, the pollutant must be removed off-site, the costs for transportation and/or pumping may greatly exceed any management and maintenance costs that would have incurred preventing the original contamination. Much research is under way to discover efficient methods to biodegrade selected chemicals. One approach is **phytoremediation** of organic and nutrient contaminants involving the use of vegetation for the *in situ* treatment of contaminated soils. In phytoremediation, the contaminated soil or even the soil above a shallow, contaminated water table is planted with varieties of plants that absorb the contaminating pesticide. Once harvested at the appropriate growth stage, the concentration of contaminant either in the soil or groundwater is reduced. Phytoremediation appears to be best suited for sites with shallow contamination (within 5 m of the soil surface), moderately hydrophobic pollutants, and/or excess nutrients.

Note: The author acknowledges the generous support of Ms. Roberta Gibson, Research Specialist, and Dr. David Langston, Extension Entomologist, through their contributions to each of the sections on classes of pesticides.

References and Recommended Reading

Alford H.G. and Ferguson M.P. (1982) *Pesticides in Soil and Groundwater: Proceedings of a Conference at University of California.* Davis Special Publication No. 3300. Agricultural Sciences Publications, University of California, Berkeley, California.

Sine, C. (1993) *Farm Chemicals Handbook.* Meister Publishing Company, Willoughby, Ohio.

Parsons D.W. and Witt J.M. (1988) *Pesticides in Groundwater in the United States of America: A Report of 1988 Survey of State Lead Agencies.* Department of Agricultural Chemistry, Oregon State University, Corvallis, Oregon.

Sherman W., Erwin W., Short D., DeWalt D, and Hayes T. (1993) *1993 Arizona Agricultural Statistics.* Arizona Agricultural Statistics Service, Phoenix, Arizona.

Smith E.G., Knutson R.D., Taylor C.R., and Penson J.B. (1990) *Impacts of Chemical Use Reduction on Crop Yields and Costs.* Agricultural and Food Policy Center, Department of Agricultural Economics, Texas A & M University, College Station, Texas.

van Es H.M. and Trautmann N.M. (1990) *Pesticide Management for Water Quality.* Extension Series No. 1. Distribution Center, Cornell University, Ithaca, New York.

Watson J. and Baker P. (1990) *Pesticide Transport through Soils.* Bulletin No. 9030. Arizona Cooperative Extension, College of Agriculture, The University of Arizona, Tucson.

Williams W.M., Holden P.W., Parsons D.W., and Lorber M.W. (1988) *Pesticides in Groundwater Data Base: 1988 Interim Report.* U.S. Environmental Protection Agency, Office of Pesticide Programs, Washington, D.C.

Problems and Questions

1. Explain to your 10-year old city cousin why some natural compounds are "bad for us" while some synthetic chemicals are "good for us."

2. List some of the criteria used to determine whether a particular pesticide should be included in the EPA's list of controlled chemicals. Explain why some organic pesticides such as DDT have been banned for use in the United States. Explain your answer to your 10-year old country cousin.

3. What is a pesticide and how are different pesticides classified? What is the difference between a contact pesticide and a systemic pesticide? Explain why a growth regulator can be considered a "pesticide."
4. What is an active ingredient? An inert ingredient? Why are some ingredients considered inert even though they may be toxic?
5. Explain why it is so difficult to make accurate estimates of pesticide usage. Suggest some ways in which you would improve the situation.
6. What are the two most important properties of a chemical which determine whether its inclusion in a pesticide represents a threat to groundwater?
7. What is the organic carbon sorption coefficient and how is it used? Why isn't this coefficient useful when the contaminant is a highly polar organic compound?
8. If a compound has a high organic sorption coefficient, what does that tell you about its potential to leach into groundwater? Surface water?
9. What is Integrated Pest Management? Explain why it is important in mitigating pesticide pollution.
10. What three factors can be taken into account to minimize pesticide losses to groundwater or surface water?

Treatment of industrial effluents often results in aqueous wastes. which themselves must be disposed of. In the above photograph, saline waste from a flue gas desulfurization plant has lead to the formation of a barren playa capable of supporting only sparse, salt-tolerant vegetation. Photograph: J.F. Artiola.

18.1 Industrial Wastes

In the United States, the nineteenth century was an era of rapid growth during which many new and developing industries were systematically processing raw materials into finished goods. Throughout this century—and well into the twentieth—these industries concentrated on the quality of the final product, while discarding the residues produced during increasingly complicated manufacturing processes. Like the British industrial revolution (which began a century earlier), the U.S. industrial revolution was a period of unparalleled productivity. But it was also a period of unrestrained "by-productivity," during which unprecedented quantities of raw industrial wastes were discharged into the environment.

Mining was one of the earliest contributors to the industrial revolution, starting with coal and metal ores. When transformed, these minerals became both the fuel and the building blocks of western society's industrialization. With the discovery of petroleum and natural gas deposits, chemical processing industries quickly followed. From the early 1900s to the late 1960s, chemical industries refined, produced, and developed hundreds of thousands of petroleum-derived and synthetic chemicals used in the production of goods.

Although carbon-based plastic materials, together with such organic chemicals as pesticides and solvents, have dominated industrial production since the early 1950s, metal-based goods remain fundamental to modern industry. Numerous modern goods—from cars to paints—require the use of such common metals as iron, aluminum, and copper. And, in addition to these three metals, other less common metals and metalloids, such as lead, cadmium, nickel, mercury, arsenic, and selenium, are essential for the manufacture of these and other goods. Metallic elements are therefore commonly found in industrial wastes, where they have complex and ill-understood effects on the environment.

Although it is known that uncontrolled and concentrated releases of metals into the environment present both short- and long-term hazards to human health, less is known about the ultimate fate of these elements once they are released into the environment (see Section 21.8.1, beginning on page 338). For example, while most elements are quickly washed out of the atmosphere soon after their release, they may remain in the water phase for long periods of time. It is worth pointing out that most elements are biologically active (*i.e.*, are taken up by living organisms) and actively participate in natural cycles when they are in the water phase. Nonetheless, most elements eventually partition (accumulate) into solid phases such as aquatic sediments and soils. As these elements become part of solid phases, the transformations previously mentioned are reduced dramatically. However, most elements can become soluble again, as environmental changes take place, becoming bioactive once more.

Industries that mine and process ores, drill for oil and gas, or burn coal also generate large volumes of salt-containing wastes. For these industries the predominant chemical species include sodium, calcium, sulfate, chloride, and carbonate ions, which are also very abundant in the natural environment. Because these wastes are not intrinsically hazardous or acutely toxic, they do not pose an immediate risk to health and the environment. Nonetheless, the volumes of these wastes that are generated each year are massive enough to be of concern.

In this chapter we will focus on the production of metallic and salt-based wastes as sources of pollution.

Industrial wastes are classified as *hazardous* or *nonhazardous* by the **Code of Federal Regulations** (CFR 40, Part 261). Most of the materials classified as nonhazardous are regulated under such specific industry categories as mining and oilfield wastes. Of all the wastes generated in the United States, more than 95% are classified as nonhazardous municipal, industrial, mining, oil- and gasfield waste products. Municipal wastes are discussed in Chapters 10 and 20, beginning on pages 135 and 301, respectively. Although industrial, mining, oil- and gasfield wastes are regulated separately, in this chapter we will label all of them as "industrial wastes" and discuss them together.

In recent years, industrial wastes classified as hazardous have received considerable attention owing to their obvious potential for producing a deleterious impact on human health and the environment. Consequently, government scrutiny has focused on new regulations for storage and disposal of these wastes. Nonetheless, less regulated, nonhazardous wastes constitute the bulk of the waste generated by industry. Moreover these wastes often contain the same potentially polluting components found in hazardous wastes, but at much lower concentrations.

18.2 Major Forms of Industrial Wastes

Regardless of how wastes are technically categorized by regulatory agencies, their pollutant-releasing capacities depend on their physical characteristics, which determine their mode of transport into the environment. The four major waste types based on physical characteristics—combustible wastes, solid wastes, sludge and slurry wastes, and wastewaters—are presented in the schematic in Figure 18-1, which shows how each of these types can eventually release pollutants into the atmospheric, terrestrial, and aquatic environments. **Combustible wastes** yield by-products that can be released directly into the atmosphere as gases or particulate pollutants when they are not properly filtered. **Solid wastes** can release pollutants into the atmosphere via dust or particulate transport. Or, when these wastes come into contact with water, their soluble constituents can be leached out into the soil surface or below. **Sludge and slurry wastes** can release pollutants into the soil and groundwater from both their solid and liquid phases. ***Wastewaters***, owing to their liquid state, are always potential sources of pollution if discharged directly into the aquatic or terrestrial environment without proper treatment. In general, we can see that the pollutants whose impact on the environment is most severe are those usually found in liquid or gas phases. By their fluid nature, these pollutants can be transported large distances and thus affect large segments of the environment. (The various physical methods by which pollutants are translocated within and among environments have been discussed in Chapters 5 and 6, beginning on pages 45 and 63.)

Except for a few isotopic forms, none of the 40 elements that are economically important to modern industrial society can be made synthetically.

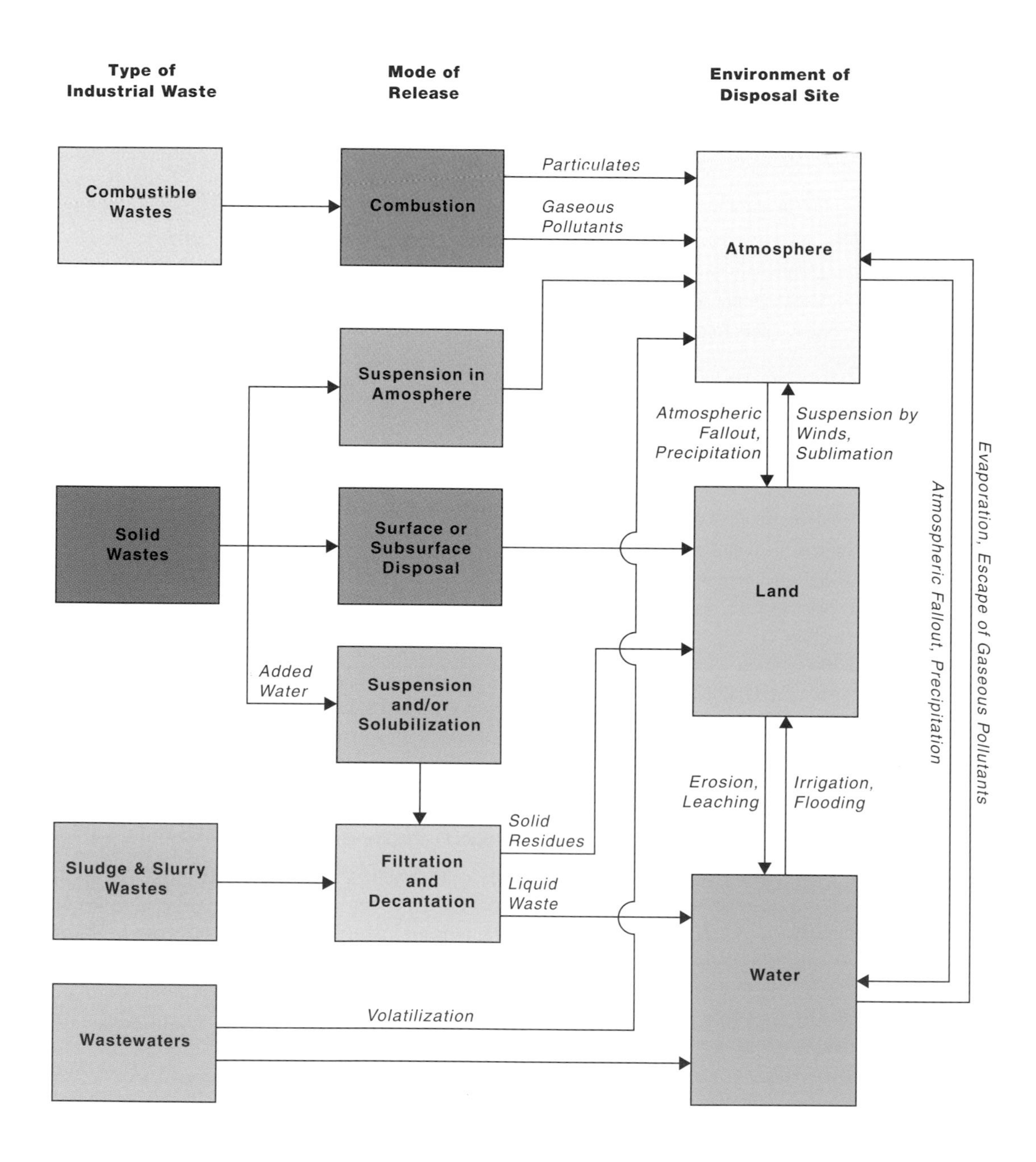

Figure 18-1 Sources and modes of releases of pollutants into the environment.

Table 18-1 Elements of economic importance and the forms in which they are commonly found in wastes.

Element	Common Form in Wastes
Aluminum	Al^{3+} oxides, Al metal
Antimony	Sb^{3+}, Sb^{5+} (oxides, halides)
Arsenic	As^{n+} oxides (in anionic form)
Barium	$BaSO_4$
Boron	B^{3+} oxides (in anionic form)
Bromine	Br^- ion
Cadmium	Cd^{2+} and Cd^{2+} halides and oxides
Calcium	$CaCO_3$, $CaSO_4$, and Ca^{2+} oxides
Carbon	CO_3^{2-}, organic compounds
Chlorine	Cl^- ion
Chromium	Cr^{n+} oxides, Cr metal
Cobalt	Co^{2+} (amines, oxides)
Copper	Cu^{2+} oxides, Cu metal
Fluorine	F^- ion
Gold	Au^{3+} (halides, oxides, cyanides)
Iron	Fe^{n+} oxides, Fe metal
Lead	Pb^{2+} oxides and carbonates, Pb metal
Magnesium	Mg^{2+} as $MgCO_3$, $MgSO_4$, Mg oxides, asbestos
Mercury	Hg^{2+} oxides and halides, organomercury complexes
Molybdenum	Mo^{n+} oxides (in anionic form)
Nickel	Ni^{2+} ion, Ni metal, amines
Nitrogen	NO_3^- and NH_4^+ ions, organic compounds
Phosphorus	PO_4^{3-} ion
Platinum	Pt^{2+} (halides, cyanides, organic compounds)
Potassium	K^+ ion, numerous complexes with oxygen and halides
Selenium	Se^{n+} oxides (in anionic form)
Silicon	Si^{4+} oxides, asbestos
Silver	AgCl, Ag metal
Sodium	Na^+ ion
Sulfur	SO_4^{2-} ion, organic compounds
Titanium	Ti^{4+} oxides
Tungsten	W^{6+} oxides
Uranium	Numerous complexes with oxygen and halides
Vanadium	V^{n+} oxides
Zinc	Zn^{2+} oxides, Zn metal

Thus, these elements have to be mined, extracted from their natural state, and subsequently purified. Because of such processes, as well as other processes used in manufacturing, these elements—including many precious and strategic metals such as gold (Au), platinum (Pt), cobalt (Co), antimony (Sb), and tungsten (W)—can be found in significant amounts in some industrial wastes. A few elements, such as mercury (Hg) and arsenic (As), can be found in gaseous, liquid, or solid wastes. Yet others, such as lead (Pb) and chromium (Cr), can be discharged into the atmosphere as particulate matter associated with other elements such as sulfur (S) and carbon (C).

However, most of the elements listed in Table 18-1, including metals, metalloids, and salts, can be dissolved in the liquids (usually the aqueous phase) found in wastewaters, sludges, and solid industrial wastes. That is, with few exceptions, most industrial liquid, slurry, and sludge wastes have a water phase that can range from 99% to less than 10% by weight. Therefore, the process of dewatering can be used to reduce much of the mass of these wastes. Moreover, many forms of inorganic and organic pollutants are also water-soluble to various degrees, and are therefore found in the water phase of these wastes. These aqueous phases, once separated from solids, must be processed as wastewaters prior to discharge into the open environment.

18.2.1 Wastewaters

Many industries, such as lumbermills, papermills, and food-processing plants, use large volumes of water as an integral part of a process or as a means to clean, and transport away, residues. Other industries that use large quantities of water either intrinsically or as a cleaning and transport agent include metal-finishing plants and many chemical industries that manufacture such diverse products as plastics, pharmaceuticals, fibers, detergents, and paints. Therefore, industrial processes are often major point sources of pollution because they generate large volumes of wastewaters that must either be treated or discharged directly into the environment. For example, metal-finishing plants can generate large volumes of wastewaters that contain significant amounts of metals: concentrations of up to 1,000 mg L^{-1} of Cr, Ni, Cu, Zn, Fe, and Cd metals may be found in raw, unprocessed wastewaters. On the other hand, food-processing plants and papermills usually generate wastewaters that are low in metals but high in organic constituents and salts containing Na^+, Ca^{2+}, Cl^-, SO_4^{2-}, and CO_3^{2-} ions.

Stormwaters originating in industrial sites constitute a significant nonpoint source of pollution from industry. Runoff from industrial sites can be contaminated when rainfall comes into contact with storage piles; fuel, chemical, and mining spills; dusts and sediments; or even smokestack emissions. Stormwaters from industrial sites are currently regulated, and the concentration of pollutants released therefrom have been reduced. However, their cumulative impacts on the surrounding environment are often evident. Such long-term effects are particularly obvious in the case of industrial activities that are traditionally located near waterways: the historical record of pollution emissions can be read in the sediment strata found there.

18.2.2 Solid Wastes and Sludges

In virtually all industrial processes that produce solid wastes, the waste products are either present in water as colloidal suspensions or mixed with water as slurries or sludges. Thus, many industrial sludges may contain metals, metalloids, and other elements in a suspended or mixed liquid/solid phase that makes their disposal complicated. The metal plating and metal finishing industries, for example, produce large quantities of sludges that are typically acidic ($pH < 2$) and contain more than 1% (10,000 mg kg^{-1}) of one or more of such metals as Cr, Ni, Cu, Zn, Fe, and Cd. Because these wastes have both acute and chronic toxic characteristics, they must be neutralized and/or oxidized and physically stabilized before disposal (see Section 11.4.2 on page 167).

Chemical manufacturing industries also generate large quantities of solid wastes, but these wastes tend to be high in organic constituents such as oils, resins, and biological residues, which do not readily dissolve in or mix with water. Solid wastes generated in industrial wastewater treatment plants, for example, usually contain low concentrations of pollutants, depending on the treatment processes to which they have been subjected. If the residual solid wastes have an insoluble organic matrix and low metal concentrations, landfilling, incineration, and even land treatment offer safe alternatives for final disposal (see Chapter 10, beginning on page 135).

18.2.3 Industrial Wastes with High Metal Content

18.2.3.1 Mining

Strip and underground mining of metal ores produces vast quantities of residues or mine spoils that contain significant concentrations of metals. These mine spoils are usually disposed of in terraces, ponds, and natural depressions. Thus, these residues, which are usually composed of primary minerals, can alter the environment physically and chemically. Strip mining for copper, for example, produces large quantities of spoils that often contain concentrations of 100–10,000 mg kg^{-1} of such metals as cadmium and lead. Similarly, iron pyrites (FeS_2), which are often associated with copper, silver, and lead ores, can have a devastating impact on the aquatic environment because their oxidation releases sulfuric acid into the environment:

$$FeS_2 + 4O_2 + 3.5H_2O \longrightarrow Fe(OH)_3 + 2SO_4^{2-} + 4H^+ \quad (18\text{-}1)$$

In addition mining operations that treat or leach ores and/or store acid chemicals for the extraction of metals can generate large volumes of acidic metal-containing wastewaters as well as leachates containing metal-cyanide complexes.

18.2.3.2 Air Emissions

Metal smelting and refining processes generate wastes that may contain ten or more of the most hazardous metals and metalloids listed in Table 18-1, such as Pb, Zn, Ni, Cu, Cd, Cr, Hg, Se, As, and Co. These elements may be found in the ores used, or they may be added as mixed metals into the melts to produce metal alloys. Thus, metal-containing smelter wastes have to be treated and disposed of as hazardous wastes. Smelting and refining require very high temperatures to reduce the metal ores (such as pyrite and bauxite for iron and aluminum production) into pure metal and to refine metals and alloys.

For example, iron melts at 1536°C, copper melts at 1083°C, and aluminum melts at 660°C. At these temperatures, many other metals and metal compounds volatilize; for example, the boiling points of mercury, cadmium, zinc, and arsenic are 357°C, 765°C, 906°C, and 613°C, respectively. Therefore, smelter and metal refining stacks that do not have gas scrubbers can release significant amounts of relatively volatile toxic metals into the atmosphere.

On the other hand, incineration is often used to destroy toxic organic wastes (see Section 10.6.1, beginning on page 146). Low-temperature (up to 850°C) and high-temperature (~1200°C) incineration oxidizes most carbon-based wastes into carbon dioxide and water vapor. Incineration efficiencies are closely regulated and must exceed 99.99% destruction of all organic compounds. However, metals and salts in wastes cannot be destroyed and usually end up in the ash residues. Thus, incineration residues tend to be high in metals and salts. Ultimately, these metal- and salt-containing residues must be disposed of as any other industrial waste that contains metals. These wastes are usually disposed of in a landfill after physical stabilization (solidification, see Section 18.3.4, beginning on page 274). However, low-efficiency or improperly operated incinerators or incinerators that burn wastes high in volatile metals (*e.g.*, Hg, Cd, Zn, As) can discharge significant amounts of metals into the atmosphere. In addition, other metals may also be released into the air as particulate matter during the process of incineration if the incinerator gas is not filtered.

Atmospheric contaminants from industry can also come from many other sources besides smokestacks; these sources include low-temperature gaseous emissions and particulates such as aerosols and dusts originating from industrial processes and on-site activities. Airborne contaminants move and are deposited over great distances and large areas of land (see Section 12.4, beginning on page 177, for more information on pollutant dispersal in the atmosphere). Deposition rates in urbanized areas can range from 40 to 400 tons km^{-2} per year or higher, depending on the location. These airborne materials can consist of relatively benign chemicals such as phosphorus or sulfur, or they may contain toxic materials such as heavy metals or complex organic pesticides.

18.2.4 Industrial Wastes with High Salt Content

18.2.4.1 Oil Drilling and Refining

The process of drilling for crude oil requires powerful drill rigs that use large quantities of drilling fluids. These fluids contain high density weighing agents such as barium sulfate (barite). Other drilling fluids are composed of sodium chloride solutions, which are used to force crude oil up to the surface. Once spent, these fluids must be disposed of. Prior to 1985, these spent fluids were stored in ponds near the drill sites and often simply bulldozed over when the oil well was completed. Consequently, many older oilfields have large tracts of land contaminated with spent drilling wastes. These wastes are not considered hazardous because they do not contain significant amounts of metals. Although free barium is very toxic, the mineral barite ($BaSO_4$) is quite inert in the environment (see Table 18-2). On the other hand, NaCl is very soluble in water and can increase the salinity of surface waters, rendering them impotable.

Oil refineries generate large volumes of wastewaters and oily sludges that can contain significant amounts of metals and salts. In general, however, most of the metals associated with refinery waste come from crude oil refining processes that utilize metals. For example, the production and storage of leaded gasoline generates sludges that may contain more than 10,000 mg kg^{-1} of lead. Other refinery sludges may contain residues of metals that are used as catalysts in oil refining processes, such as Ni, Mo, and Co.

18.2.5 Coal-Burning Electric Power Plants

Electric power plants produce millions of tons of fly ash and flue gas desulfurization wastes every year. Because these residues are not considered hazardous, they may be stored either in ponds or landfills; in the case of fly ash, they may also be used as fill material. **Fly ash** is recovered from precipitators that scrub out silt-size particulate matter from the flue gases generated from coal combustion. These particles generally arise from the incombustible silt land clay found in coal deposits. Upon exposure to high temperatures, silt and clay (which consists mostly of

Table 18-2 Water solubilities of minerals found in waste.

Mineral Names and Chemical Compositions	Water Solubility (mg L^{-1})
Calcite ($CaCO_3$)	14
Gypsum ($CaSO_4 \cdot 2H_2O$)	2,400
Salt (NaCl)	370,000
Barite ($BaSO_4$)	2

silica and alumina) combine to yield amorphous Si/Al-based spheres onto which other elements may condense. Typically, fly ash spheres also include Ca, Na, Fe, Mg, K, and Ti with small amounts of other elements sorbed onto them, such as As, B, Ba, Cd, Cr, Cu, F, Mo, Ni, Pb, S, Zn, and others. The concentrations of these elements in fly ash vary widely, depending on the source of the coal. A typical empirical composition of fly ash is

$$Si_{10}\,Al_5\,Ca_{0.5}\,Na_{0.5}\,Fe_{0.4}\,Mg_{0.2}\,K_{0.2}\,Ti_{0.1}\,S_{0.5}$$

with trace amounts of more than 15 other elements. The removal (scrubbing) of sulfur dioxide (SO_2) gas from flue gases produces large quantities of **flue gas desulfurization wastes**, which consist largely of calcium carbonates, sulfates, and sulfites. These wastes may also contain trace quantities of some of the elements in fly ash, but the concentration of these elements depends on the source of the coal and the type of scrubbing systems used. Because flue gas desulfurization products are usually more than 70% water, these wastes are disposed of in drying ponds and are often treated along with power-plant wastewaters. This waste mixing may add significant amounts of soluble salts (*e.g.*, NaCl) that increase the salinity of sludges (see Table 18-2).

18.2.6 Industrial Wastes High in Organic Chemicals

Most industrial wastes contain varying amounts of organic chemicals. With few exceptions, carbon-based chemicals, reagents, solvents, feedstocks, and raw materials are extensively used in most phases of industrial processing. Exceptions to this rule may include mine tailings and metal plating wastes. Wastes high in organic chemicals include those originating from oil refineries, as well as petrochemical, chemical, pharmaceutical, and food-processing industries.

18.3 Treatment and Disposal of Industrial Wastes

Technologies and practices used for the treatment and disposal of metal- and salt-containing wastes vary widely. Methods that separate metals from the other waste constituents are driven by the need to reduce waste disposal costs and ultimately by the costs associated with liability. Organic wastes that contain low residual concentrations of metals and salts often can be degraded by using thermal or biological destruction processes that completely transform the waste into carbon dioxide and water (see Chapter 7, beginning on page 77). On the other hand, wastes that contain significant amounts of metals and salts always leave indestructible residues that may or may not be recycled economically. Therefore, these wastes that cannot be eliminated have to be disposed of in a manner that minimizes their impact on the environment.

18.3.1 Gas and Particulate Emissions

Gases and dust particulates that are generated in the thermal destruction of wastes and smelting can be prevented from escaping into the air using one or more of the following processes: electrostatic precipitators, baghouse and cyclone separators, and wet scrubbers. This technology is expensive and difficult to operate efficiently. However, without this equipment, smelters and incinerators would discharge large quantities of toxic metals into the atmosphere, thereby contaminating large tracts of land. In the United States, the Clean Air Act requires the control of hazardous emissions by industries. Originally passed in 1970, by 1990 this act had been extended to cover 189 industrial chemicals, requiring the installation of air pollution control equipment on all major industrial sources. These regulations also include requirements for the removal of metal-containing particulate matter (PM-10) from air emissions.

18.3.2 Chemical Precipitation

Wastewater and slurried sludges can be treated with chemical agents to precipitate out and remove metals from the rest of the waste components. For example, solutions containing alkaline materials, such as calcium carbonate, sodium hydroxide, aluminum oxide, or sodium sulfide, are commonly used to precipitate metals from waste streams. These chemicals help form insoluble metal hydroxides, carbonates, and sulfides. Similarly, aluminum and iron oxides sorb metals such as cadmium and metalloids such as arsenic, thus removing them from solution.

General precipitation reactions can be represented by the following where M = Cd, Zn, Cu, or Pb.

Carbonate precipitation:

$$M^{2+}(aq) + CaCO_3(aq) \longrightarrow MCO_3(aq) + M(OH)_2(s)\downarrow \quad (18\text{-}2)$$

Sulfide precipitation:

$$M^{2+}(aq) + Na_2S(aq) \xrightarrow{pH > 8} 2Na^{+}(aq) + MS(s)\downarrow \quad (18\text{-}3)$$

18.3.3 Flocculation, Coagulation

Particulate pollutants in wastewaters can be made to settle out quickly by using chemicals known as flocculants and coagulants. Flocculants, which react with dissolved chemicals, facilitate the formation of aggregates or clumps, which can be decanted or filtered out of solution. On the other hand, coagulants destabilize colloids, thereby permitting suspended particles to form aggregates that can settle out of solution. These chemicals can be useful in removing all kinds of pollutants from wastewaters, including metals and organic constituents. Flocculants and coagulants include iron and copper sulfates and chlorides, as well as complex synthetic organic polymers.

18.3.4 Solidification or Stabilization

Sludges containing metals must be chemically and physically stabilized prior to final disposal. Since metals are more soluble in water at low pH, acidic wastes must be neutralized with such basic compounds as lime ($CaCO_3$) to form metal complexes that have very low water solubilities. The process of waste solidification usually involves trapping or encapsulating the waste into a physically stable matrix. For example, when wet cement is mixed in with sludges, it forms a stable block after a few days of curing (drying). This solidification method encapsulates wastes in a matrix that is relatively low in porosity and cannot easily be deformed or cracked under typical landfill overburden pressures. Consequently, percolating water does not readily infiltrate into the matrix and metals or salts are less likely to leach out (see also Section 10.6.2, beginning on page 147).

18.3.5 Oxidation

Carbon-based waste streams can be oxidized to detoxify and completely destroy organic pollutants. There are two major types of treatment processes: thermal and chemical. The overall goals are the same in both processes. Thermal oxidation reactions can be described as follows:

$$RC_2NOCl + O_2 + \text{heat} \longrightarrow CO_2 + H_2O + N_2 + NO_x + SO_x + HCl + \text{intense heat} \quad (18\text{-}4)$$

where RC_2NOCl is a representative organic waste compound. Thermal oxidation processes include incineration using conventional-fuel-driven burners. Solid or liquid organic wastes having low water content but high heat values are good candidates for thermal oxidation because they burn hotly enough to sustain the energy these processes require. The major disadvantages of this process include high costs (usually more than $200 a barrel), limited reliability, and a negative public perception of their safety. In addition, large emissions of pollutants can result when the systems are not operated and maintained properly. Incinerators operate most efficiently when designs are tailored to specific waste stream characteristics. Thus, incinerators are not suitable for the efficient oxidations of mixed waste streams (see also Section 10.6.1 on page 146).

Chemical oxidation generally proceeds via the following reaction pathway, here using the representative organic compound RC_2NOCl:

$$RC_2NOCl + (O_2, Cl_2, \textit{or } O_3) \longrightarrow R'CO + CO_2 + H_2O + N_x + Cl^- + \text{intense heat} \quad (18\text{-}5)$$

where R′ is the remaining part of the organic molecule and R′CO is the detoxified product. For example, chemical oxidation is widely used in the destruction of cyanide-containing wastewaters. This process involves chlorination, which can be summarized as follows:

$$CN^-(aq) + Cl_2(g) + NaOH(aq) \xrightarrow{pH > 8\text{–}9} N_2(g) + Na^+(aq) + Cl^-(aq) + HCO_3^-(aq) + H_2O \quad (18\text{-}6)$$

Many forms and combinations of chemical oxidation processes utilize chlorine (Cl_2), chlorine dioxide (ClO_2), ozone (O_3), or ultraviolet (UV) radiation to eliminate low concentrations of organics from industrial wastewaters. The chemicals oxidized using these processes include acids, alcohols, and aldehydes (such as oxalic acids and phenols), as well as more stable chlorinated pesticides, such as DDT, and some petroleum-derived solvents, *e.g.*, benzene or the xylenes.

18.3.6 Landfilling

Wastes that contain significant concentrations of metals that cannot be recycled or recovered economically are usually good candidates for landfill disposal. However, sludges high in metals must be neutralized and solidified prior to landfill disposal, and only landfills approved for the disposal of solidified hazardous wastes may be used for the disposal of these wastes. Similarly, wastes high in salts may also be buried in special landfills or used as fill materials for road and dam construction. Although such wastes are not considered toxic, they are very soluble in water; thus it is important to minimize water contact to prevent the leaching of salts into the environment (see also Section 10.6.2, beginning on page 147).

18.3.7 Stockpiling, Tailing, and Muds

Mining activities produce vast quantities of tailings, which are usually stockpiled in the form of terraces on or near the ore-processing mills. The environmental impacts of mining activities, which vary widely from site to site, are usually associated with runoff or percolation of waters contaminated with sediments. They are also related to pH, metal solubility, salt concentrations, and the quantities of windblown particulates that are contaminated with metals.

Oil- and gasfields produce large quantities of well cuttings (muds). These well cuttings, which contain barite, salts, and crude oil, are usually stockpiled. The potential for offsite releases of pollutants from these sites is normally associated with water and wastewater releases that contain varying concentrations of these chemicals.

18.4 Treatment and Reuse of Industrial Wastes

Because metal- and salt-containing wastes cannot be completely destroyed, they can be a hazard to humans and the environment when improperly disposed of. At the same time, these wastes have potential economic value—a value that is becoming more evident as their natural sources diminish. Precious and strategic metals, such as gold, platinum, cobalt, antimony, and tungsten, are routinely mined out of wastes that contain significant concentrations of these elements. However, wastes that contain less valuable metals, such as aluminum and iron, are far less likely to be treated for the removal of these elements. Even more improbable is the extraction of salts with low solubility, such as the sulfates and carbonates of magnesium and calcium, which are found in most industrial wastes (see also Section 10.8 on page 148).

18.4.1 Metal Recovery

Such economically valuable elements as precious metals can be recovered from waste streams by using complex chemical reactions, including chemical separation or precipitation (compare Section 7.5.2 on page 90). Silver, for example, can be recovered from photographic wastes by acidifying the liquid wastes and separating the Ag sludge that precipitates out. Then the supernatant can be neutralized and disposed of safely, while the Ag sludge is sent to a smelter for purification.

Metals can also be made to react selectively with synthetic organic chemicals known as **chelates** (literally, "claws"). In such reactions, metal atoms are "grabbed" by the chelate structure while other, undesirable atoms slip away. Once reacted, the metal of interest is either precipitated out of the waste solution or is removed with a sorbent. The recovered metal sludge can be smeltered and melted into pure solid metal. Metal-containing residues resulting from the smelting and refining of ores can be further refined by using a combination of chemical and physical separation techniques, then refined again using high-temperature furnaces. These processes are too numerous to describe individually and are generally tailored to specific waste sources.

18.4.2 Energy Recovery

The majority of wastes containing organic carbon forms have large quantities of stored energy. Thus, wastes containing high concentrations of reduced carbon (usually organic carbon), such as organic liquids, woody materials, oils, resins, or asphalts, can be used in incinerators and electrical generators as energy sources. For example, kerosene and natural gas have about 4.4×10^7 to 4.9×10^7 J kg^{-1} of energy. Many solvents such as hexane, xylenes, and paraffins, which are found in oils and alcohols, have similar stored energies (4.2×10^7 to 5.8×10^7 J kg^{-1}). Therefore, many waste mixtures of organic chemicals have energy values approaching those of commercial fuels. However, the limitations of this technology, when applied to wastes, are set by very strict air emission standards promulgated under the Resource Conservation and Recovery Act (RCRA) of 1976 (see Table 23-1 on page 367).

18.4.3 Solvent Recovery

Industrial wastes high in solvents are being successfully recycled using recovery systems that include distillation techniques and chemical and physical fractionation processes. For example, spent solvents used to clean and paint metal parts can be redistilled as a means to separate the heavy impurities from the volatile solvents. Some examples of solvents that can then be reused via solvent recovery stills include chloroform, acetone, xylene, hexane, and methylene chloride.

18.4.4 Waste Reuse

Some mine tailings can be used as fills for earthworks. However, economic issues related to transportation costs usually limit their reuse. Consequently, mine tailings are most often left in place. New regulations will require that physical stabilization and revegetation practices be applied to these sites.

Oilfield wastes that are high in salts may be leached with water to remove the soluble salts. The salt-free muds are then dried and stockpiled for use as fill material. These treated solid oilfield wastes, which have been shown to be safe in earthworks, release insignificant amounts of residual salts into the surrounding environment.

Coal-burning wastes such as fly ash and flue gas desulfurization sludges have been used successfully as soil amendments. Fly ash has also been used as a fill material for earthworks (see also Section 10.8 on page 148). These materials can be good sources of gypsum and calcite, which can neutralize acidity in soils and replenish macronutrients such as Ca, Mg, and S, and even trace elements, such as Zn, Cu, Fe, and Mn. Fly ash additions to agricultural soils have also been shown to improve the soil porosity and overall structure.

References and Recommended Reading

Bunce N. (1991) *Environmental Chemistry.* Wuerz Publishing Ltd., Winnipeg, Canada.

Cope C.B., Fuller W.H., and Willets S.L. (1983) *The Scientific Management of Hazardous Wastes.* Cambridge University Press, Cambridge.

Dohlido J.R., and Best G.A. (1993) *Chemistry of Water and Water Pollution.* Ellis Horwood, New York.

Fuller W.H. and Warrick A.W. (1985) *Soils in Waste Treatment and Utilization,* Volume II. CRC Press, Boca Raton, Florida.

Hesketh H.E., Cross F.L., and Tessitore J.L. (1990) *Incineration for Site Cleanup and Destruction of Hazardous Wastes.* Technomic Publishing, Lancaster, Pennsylvania.

Jackman A.P. and Powell R.L. (1991) *Hazardous Waste Treatment Technologies: Biological Treatment, Wet Air Oxidation, Chemical Fixation, Chemical Oxidation.* Noyes Data Corporation, Park Ridge, New Jersey.

LaGrega M.D., Buckingham P.L., and Evans J.C. (1994) *Hazardous Waste Management*. McGraw-Hill, New York.

U.S. EPA (1980) *Hazardous Waste Land Treatment.* United States Environmental Protection Agency, Office of Water and Waste Management, Washington, D.C.

Wang L.K. and Wang M.H.S. (1992) *Handbook of Industrial Waste Treatment*. Marcel Dekker, Inc., New York.

Problems and Questions

1. Using Table 18-1 on page 270, select the 10 most prevalent elements found in industrial wastes. Give examples of the chemical forms of these elements. Example: Oxygen (O_2) is found in the aqueous phase of wastes. [*Hint*: What is the bulk composition of most solid and liquid wastes?]

2. What is the most important difference between organic (carbon-based) and inorganic wastes?

3. What types and forms of pollutants are commonly released into the atmosphere as a result of industrial activities? Give four examples.

4. One hundred liters of an industrial wastewater contains the following materials:

   (a) 4.5 kg of table salt

   (b) 5 kg of gypsum

   (c) 1 kg of iron rust [($Fe(OH)_3$ (solubility $\cong$ 0.1 mg L^{-1})]

   (d) 100 g of the pesticide DDT (solubility of approximately 1 mg L^{-1})

   (e) Using the data from Table 18-2 on page 273, what amounts of each of these four materials are dissolved in the wastewater and what amounts are in solid forms? [*Hint*: Assume that 1 L of water weighs 1 kg.]

5. Which elements and/or chemicals would be found in solution in the wastewater of Problem 4? List four, in order of importance.

6. Would you consider incinerating the wastewater in Problem 4? Explain your answer.

A cyst of Giardia lamblia, *the protozoan organism responsible for the gastrointestinal disease giardiasis, is seen in the lower right corner amidst debris from sewage using immunofluorescent antibody staining (magnification = 1900×). Photograph: D.C. Johnson.*

19.1 Water-Related Microbial Disease

London's Dr. John Snow (1813–58) was one of the first to make a connection between certain infectious diseases and drinking water contaminated with sewage. In his famous study of London's Broad Street pump, published in 1854, he noted that people afflicted with cholera were clustered in a single area around the Broad Street pump, which he isolated as the source of the infection. When, at his insistence, city officials removed the handle of the pump, Broad Street residents were forced to obtain their water elsewhere. Subsequently, the cholera epidemic in that area subsided. However successful the effect, Snow's explanation of the cause was not generally accepted because disease-causing germs had not yet been discovered at the time.

In the United States, the concept of **waterborne disease** was equally poorly understood. During the Civil War (1860–65), encamped soldiers often disposed of their waste upriver, but drew drinking water from downriver. This practice resulted in widespread dysentery. In fact, dysentery, together with its sister disease typhoid, was the leading cause of death among the soldiers of all armies until the twentieth century. It was not until the end of the nineteenth century that this state of affairs began to change. By that time, the germ theory was generally accepted, and steps were taken to properly treat wastes and protect drinking water supplies.

In 1890 more than 30 people out of every 100,000 in the United States died of typhoid. But by 1907 water filtration was becoming common in most U.S. cities, and in 1914 chlorination was introduced. Because of these new practices, the national typhoid death rate in the United States between 1900 and 1928 dropped from 36 to 5 cases per 100,000 people. The lower death toll was largely the result of a reduced number of outbreaks of waterborne diseases. In Cincinnati, for instance, the yearly typhoid rate of 379 per 100,000 people in the years 1905–1907 decreased to 60 per 100,000 people between 1908 and 1910 after the inception of sedimentation and filtration treatment. The introduction of chlorination after 1910 decreased this rate even further.

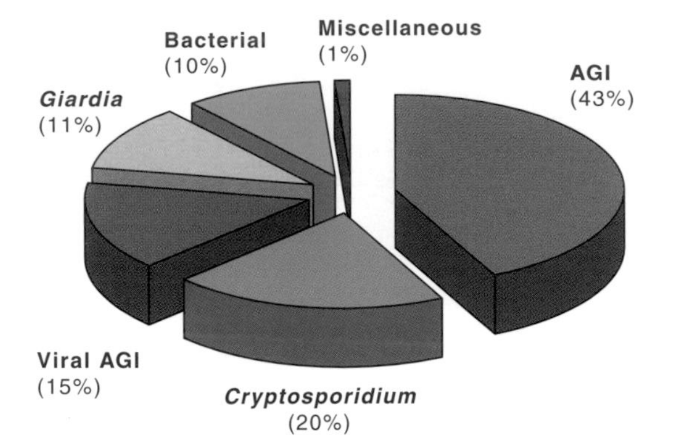

Figure 19-1 Etiological agents associated with cases of waterborne diseases. AGI = acute gastrointestinal illness of unknown etiology.

Worldwide, however, the lack of safe drinking water and proper waste treatment remains a major factor in human illness and mortality. Consider the following statistics, for example: an estimated 25,000 people died every day in 1980 from the consumption of contaminated water; approximately one in four hospital beds in the world is still occupied by someone who has become ill from drinking polluted water; and nearly five billion waterborne infections occur annually in Africa, Asia, and Latin America. Moreover, diarrhea caused by microorganisms found in water is still one of the leading causes of mortality in developing countries. Children in developing countries experience 10–12 episodes of diarrhea per year. For the most part, these statistics apply to developing countries, where neither the quality nor the quantity of drinking water is equal to that available in the developed countries of the industrialized world. The waterborne disease agents that cause typhoid, cholera, and infectious hepatitis are now relatively uncommon, or far less prevalent, in the United States than in developing countries.

Although many diseases have been eliminated or controlled in the developed countries, microorganisms continue to be the major cause of waterborne illness today. Most outbreaks of such diseases are attributable to the use of untreated water, inadequate or faulty treatment (*i.e.*, no filtration or disinfection), or contamination after treatment. In addition, some pathogens, such as *Cryptosporidium*, are very resistant to removal by conventional drinking water treatment and disinfection. Moreover, an increasing proportion of waterborne disease outbreaks are associated with nonbacterial microorganisms such as enteric viruses and protozoan parasites, because of their successful resistance to water treatment processes (see Figure 19-1).

The true incidence of waterborne disease in the United States is not known because neither investigation nor reporting of waterborne disease outbreaks is required. Investigations are difficult because waterborne disease is not easily recognized in large communities, and epidemiological studies are costly to conduct. Nevertheless, between 12 and 20 waterborne disease outbreaks per year have been documented in the United States, and the true incidence may be 10 to 100 times greater. It also appears that conventionally treated drinking water—water that meets all current EPA Standards—may contain as-yet unidentified microorganisms that are responsible for as much as 35% of the incidence of gastroenteritis in Canada. While gastroenteritis is usually not life-threatening in developed countries, it has a significant economic impact: it is estimated that the United States loses up to $20 billion per year to "stomach flu."

19.2 Classes of Diseases and Types of Pathogens

Disease-causing organisms, or pathogens, that are related to water can be classified into four groups (as shown in Table 19-1).

Waterborne diseases are those transmitted through the ingestion of contaminated water that serves as the passive carrier of the infectious or chemical agent. The classic waterborne diseases, cholera and typhoid fever, which frequently ravaged densely populated areas throughout human history, have been effectively controlled by the protection of water sources and by treatment of contaminated water supplies. In fact, the control of these classic diseases gave water supply treatment its reputation, and played an important role in the reduction of infectious diseases. Other diseases caused by bacteria or by viruses, protozoa, and helminths may also be

Table 19-1 Classification of water-related illnesses associated with microorganisms. Modified from White and Brandley (1972).

Class	Cause	Example
Waterborne	Pathogens that originate in fecal material and are transmitted by ingestion.	Cholera, typhoid
Water-washed	Organisms that originate in feces and are transmitted through contact because of inadequate sanitation or hygiene.	Trachoma
Water-based	Organisms that originate in the water or spend part of their life cycle in aquatic animals and come in direct contact with humans in water or by inhalation.	Schistosomiasis
Water-related	Microorganisms with life cycles associated with insects that live or breed in water.	Yellow fever

transmitted by contaminated drinking water. However, it is important to remember that waterborne diseases are transmitted through the fecal-oral route, from human to human or animal to human, so that drinking water is only one of several possible sources of infection.

Water-washed diseases are those closely related to poor hygiene and improper sanitation. In this case, the availability of a sufficient quantity of water is generally considered more important than the quality of the water. The lack of water for washing and bathing contributes to diseases that affect the eye and skin, including infectious conjunctivitis and trachoma, as well as to diarrheal illnesses, which are a major cause of infant mortality and morbidity in the developing countries. The diarrheal diseases may be directly transmitted through person-to-person contact, or indirectly through contact with contaminated foods and utensils used by persons whose hands are fecally contaminated. When enough water is available for hand washing, the incidence of diarrheal diseases has been shown to decrease, as has the prevalence of enteric pathogens such as *Shigella.*

Water-based diseases are caused by pathogens that either spend all (or essential parts) of their lives in water or depend upon aquatic organisms for the completion of their life cycles. Examples of such organisms are the parasitic helminth *Schistosoma* and the bacterium *Legionella*, which cause schistosomiasis and Legionnaires' disease, respectively.

The three major schistosome species that develop to maturity in humans are *Schistosoma japonicum*, *S. haematobium*, and *S. mansoni*. Each has a unique snail host and a different geographic distribution. It is estimated that more than 200 million people in Asia, Africa, South America, and the Caribbean are currently infected with one, or perhaps two, of these species of schistosome. Although schistosomiasis is not indigenous to North America, schistosomiasis dermatis has been documented in the United States, immigrants to the United States have been found to be infected with schistosomiasis, and some 300,000 persons in Puerto Rico are probably infected. The economic effects of schistosomiasis have been estimated at some $642 million annually—a figure that includes only the resource loss attributable to reduced productivity, not the cost of public health programs, medical care, or compensation of illness.

Legionella pneumophila, the cause of **Legionnaires' disease**, was first described in 1976 in Philadelphia, Pennsylvania. This bacterium is ubiquitous in aquatic environments. Capable of growth at temperatures above 40°C, it can proliferate in cooling towers, hot water heaters, and water fountains. If growth occurs at high temperatures, these bacteria become capable of causing pneumonia in humans if they are inhaled as droplets or in an aerosol.

Water-related diseases, such as yellow fever, dengue, filariasis, malaria, onchocerciasis, and sleeping sickness, are transmitted by insects that breed in water (like the mosquitoes that carry malaria) or live near water (like the flies that transmit the filarial infection onchocerciasis). Such insects are known as **vectors**.

19.3 Types of Pathogenic Organisms

Pathogenic organisms identified as capable of causing illness when present in water include such microorganisms as viruses, bacteria, protozoan parasites, and blue-green algae, as well as some macroorganisms—the helminths, or worms, which can grow to considerable size. Some of the characteristics of these organisms are listed in Table 19-2.

The major waterborne and water-related pathogens and the diseases they cause are shown in Table 19-3.

- **Viruses** are organisms that usually consist solely of nucleic acid (which contains the genetic information) surrounded by a protective protein coat or **capsid**. The nucleic acid may be either ribonucleic acid (RNA) or deoxyribonucleic acid (DNA). They are always obligate parasites; as such, they cannot grow outside of the host organism (*i.e.*, bacteria, plants, or animals), but they do not need food for survival. Thus, they are potentially capable of surviving for long periods of time in the environment. Viruses that infect bacteria are called **bacteriophages**, and those bacteriophages that infect intestinal, or coliform, bacteria are known as **coliphages** (see Section 19.3.2.4 on page 287).
- **Bacteria** are prokaryotic single-celled organisms surrounded by a membrane and cell wall. Bacteria that grow in the human intestinal or gastrointestinal (GI) tract are referred to as **enteric bacteria**. Enteric bacterial pathogens usually cannot survive for prolonged periods of time in the environment.
- **Protozoa** are single-celled animals. Protozoan parasites that live in the GI tract are capable of producing environmentally resistant cysts or oocysts. These cysts or oocysts have very thick walls, which make them very resistant to disinfection.
- **Helminths** (literally "worms") are multicellular animals that parasitize humans. They include roundworms, hookworms, tapeworms, and flukes. These organisms usually have both an intermediate and a final host. Once these parasites enter their final human host, they lay eggs that are excreted in the feces of infected persons and spread by wastewater, soil, or food. These eggs are very resistant to environmental stresses and to disinfection.
- **Blue-green algae**, or **cyanobacteria**, are prokaryotic organisms that do not contain an organized nucleus—unlike the green algae. Cyanobacteria, which may occur as unicellular, colonial, or filamentous organisms, are responsible for algal blooms in lakes and other aquatic environments. Some species produce toxins that may kill domestic animals or cause illness in humans.

19.3.1 Viruses

More than 140 different types of viruses are known to infect the human intestinal tract, from which they are subsequently excreted in feces. Viruses that infect and multiply in the intestines are referred to as **enteric viruses**. Some enteric viruses are capable of replicating in other organs such as the liver and the heart, as well as in the eye, skin, and nerve tissue. For example, hepatitis A virus infects the liver, causing hepatitis. Enteric viruses are generally very host-specific; therefore, human enteric viruses cause disease only in humans and sometimes in other primates. During infection, large numbers of virus particles, up to 10^8–10^{12}, per gram may be excreted in feces, whence they are borne to sewer systems.

Table 19-2 Characteristics of waterborne and water-based pathogens.

Organism	Size (μm)	Shape	Environmentally Resistant Stage
Viruses	0.01–0.1	variable	virion
Bacteria	0.1–10	rod, spherical, spiral, comma	spores or dormant cells
Protozoa	1–100	variable	cysts, oocysts
Helminths	$1–10^9$	variable	eggs
Blue-green algae	1–100	coccoid, filamentous	cysts

Table 19-3 Waterborne and water-based human pathogens.

Group	Pathogen	Disease or Condition
Viruses	Enteroviruses (polio, echo, coxsackie)	Meningitis, paralysis, rash, fever, myocarditis, respiratory disease, diarrhea
	Hepatitis A and E	Hepatitis
	Norwalk virus	Diarrhea
	Rotavirus	Diarrhea
	Astrovirus	Diarrhea
	Calicivirus	Diarrhea
	Adenovirus	Diarrhea, eye infections, respiratory disease
	Reovirus	Respiratory, enteric
Bacteria	*Salmonella*	Typhoid, diarrhea
	Shigella	Diarrhea
	Campylobacter	Diarrhea
	Vibrio cholerae	Diarrhea
	Yersinia enterocolitica	Diarrhea
	Escherichia coli (certain strains)	Diarrhea
	Legionella	Pneumonia, other respiratory infections
Protozoa	*Naegleria*	Meningoencephalitis
	Entamoeba histolytica	Amoebic dysentery
	Giardia lamblia	Diarrhea
	Cryptosporidium	Diarrhea
Blue-green algae	*Microcystis*	Diarrhea,
	Anabaena	possible production of
	Aphanitomenon	carcinogens
Helminths	*Ascaris lumbricoides*	Ascariasis
	Trichuris trichiora	Trichuriasis-whipworm
	Necuter americanus	Hookworm
	Taenia saginata	Beef tapeworm
	Schistosoma mansoni	Schistosomiasis (complications affecting the liver, bladder, and large intestines)

Enteroviruses, which were the first enteric viruses ever isolated from sewage and water, have been the most extensively studied viruses. The more common enteroviruses include the polioviruses (3 types), coxsackieviruses (30 types), and the echoviruses (34 types). Although these pathogens are capable of causing a wide range of serious illness, most infections are mild. Usually only 50% of the people infected actually develop clinical illness. However, coxsackieviruses can cause a number of life-threatening illnesses, including heart disease, meningitis, and paralysis; they may also play a role in insulin-dependent diabetes.

Infectious viral hepatitis is caused by **hepatitis A** virus (HAV) and **hepatitis E** virus (HEV). These types of viral hepatitis are spread by fecally contaminated water and food, whereas other types of viral hepatitis, such as hepatitis B virus (HBV), are spread by exposure to contaminated blood. Hepatitis A and E virus infections are very common in the developing world, where as much as 98% of the population may exhibit antibodies against HAV. HAV is not only associated with waterborne outbreaks, but is also commonly associated with foodborne outbreaks, especially shellfish. HEV has been associated with large waterborne outbreaks in Asia

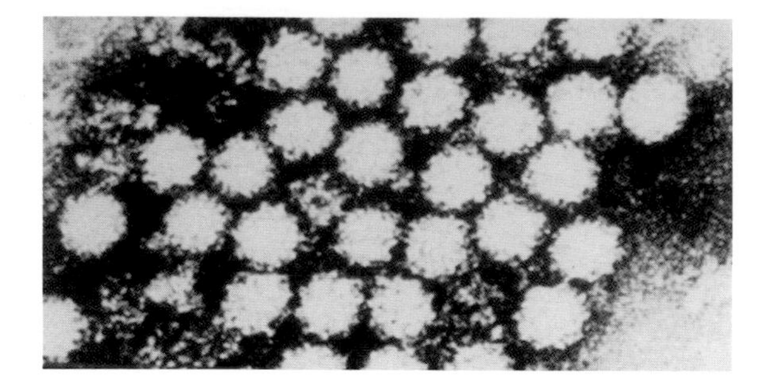

Figure 19-2 Norwalk virus: a leading cause of waterborne gastroenteritis (magnification = 175,000×). Photograph courtesy of C.P. Gerba.

and Africa, but no outbreaks have been documented in developed countries. HAV is one of the enteric viruses that is very resistant to inactivation by heat.

Rotaviruses (5 types) have been identified as the major cause of infantile gastroenteritis, that is, acute gastroenteritis in children under two years of age. This condition is the leading cause of mortality in children and is responsible for millions of childhood deaths per year in Africa, Asia, and Latin America. These viruses are also responsible for outbreaks of gastroenteritis among adult populations, particularly among the elderly, and can cause "traveler's diarrhea" as well. Several waterborne outbreaks have been associated with rotaviruses.

The **Norwalk virus**, first discovered in 1968 after an outbreak of gastroenteritis in Norwalk, Ohio, causes an illness characterized by vomiting and diarrhea that lasts a few days. This virus is the agent most commonly identified during water- and foodborne outbreaks of viral gastroenteritis in the United States (see Figure 19-2). Although it has not yet been grown in the laboratory, the Norwalk virus appears to be related to the calicivirus group. In addition, a large number of enteric viruses similar in size to the Norwalk virus also seem to cause gastroenteritis; these viruses are often referred to as Norwalk-like viruses.

The ingestion of just a few viruses is enough to cause infection (see Section 22.4, beginning on page 357). But because enteric viruses usually occur in relatively low numbers in the environment, large volumes of environmental samples must usually be collected before the presence of these viruses can be

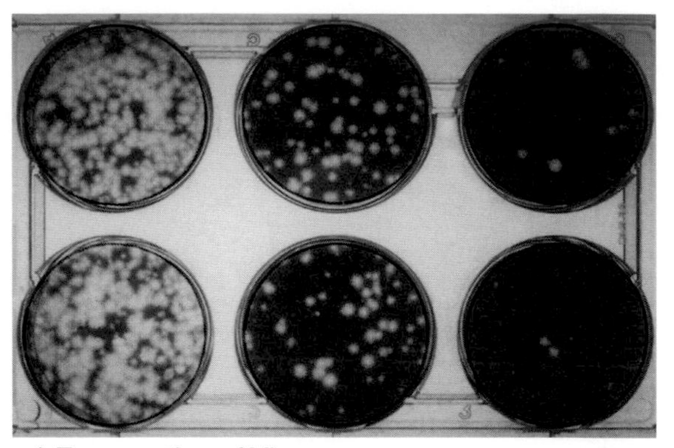

a) Enumeration of Viruses

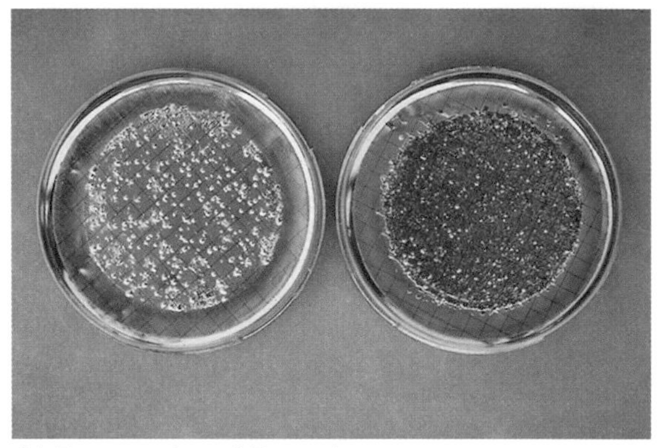

b) Enumeration of Bacteria

Figure 19-3 Quantitative assays for viruses and bacteria: (a) Viruses can be enumerated by infection of buffalo green monkey kidney cells (BGMK cells). The infections lead to lysis of the host animal cells resulting in a clearing (plaque) on the medium. This yields a measure of the infectious virus in terms of plaque forming units (PFUs). (b) Fecal coliforms can be enumerated by growth on a selective medium, such as the above pictured mEndo agar. The resulting bacterial colonies are counted as colony forming units (CFUs). Note that in both (a) and (b), more than one original virus or bacterium may have participated in the formation of macroscopic entities. Photographs courtesy of (a) C.P. Gerba and (b) I.L. Pepper.

detected. For example, from 10 to 1,000 L of water must be collected in order to assay these pathogens in surface and drinking water. This volume must first be reduced in order to concentrate the pathogen population. The water sample is thus passed through microporous filters to which the viruses adsorb; then the adsorbed viruses are eluted from the filter. This process is followed by further concentration, down to a few milliliters of sample, leaving a highly concentrated pathogen population. Next, the con-

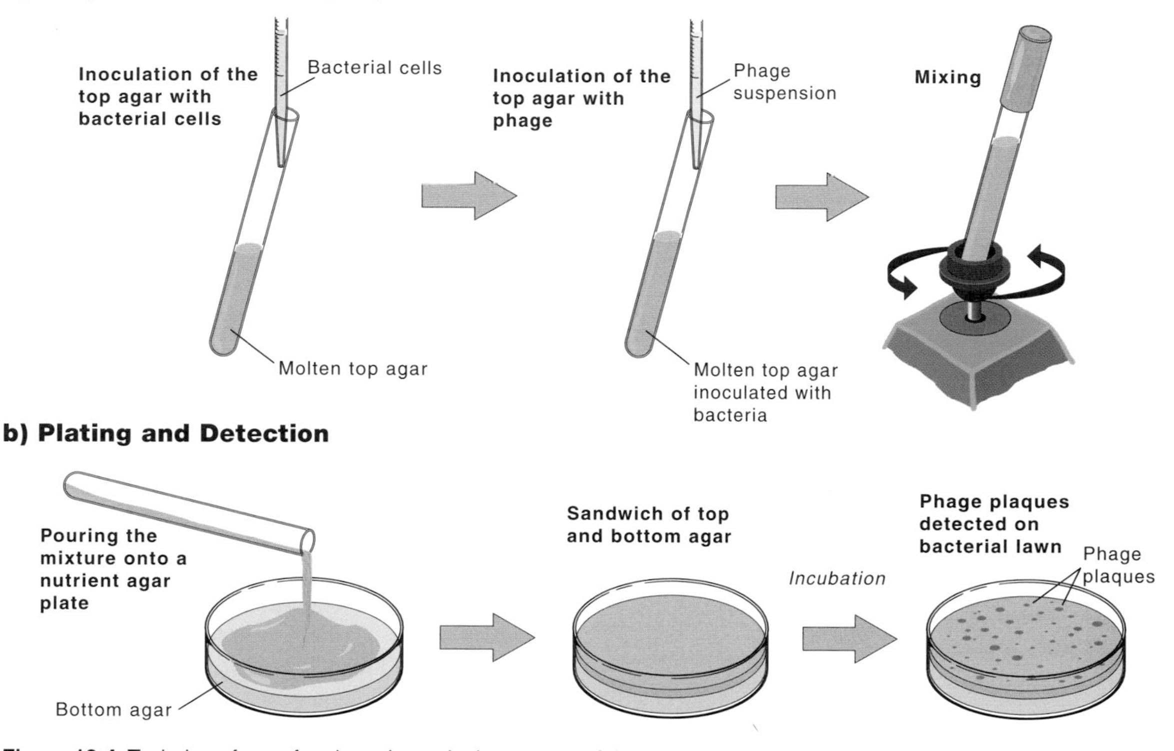

Figure 19-4 Technique for performing a bacteriophage assay. Adapted by permission from Figure 11-2 on page 79 of *Environmental Microbiology: A Laboratory Manual*, by I.L. Pepper, C.P. Gerba, and J.W. Brendecke; copyright © 1995 by Academic Press, Inc. All rights reserved.

centrate is assayed by using either cell culture or newer molecular techniques. Cell culture techniques involving animal cells are effective (see Figure 19-3a), but they may require several weeks for results; thus, bacteriophages may sometimes be used as timely and cost-effective surrogates (Figure 19-4). For example, coliphages are commonly used as models to study virus fate during water and wastewater treatment and in natural waters.

19.3.2 Bacteria

19.3.2.1 Enteric Bacteria

The existence of some enteric bacterial pathogens has been known for more than a hundred years. At the beginning of the twentieth century, modern conventional drinking water treatment involving filtration and disinfection was shown to be highly effective in the control of such enteric bacterial diseases as typhoid fever and cholera. Today, outbreaks of bacterial waterborne disease in the United States are relatively rare: they tend to occur only when the water treatment process breaks down, when water is contaminated after treatment, or when nondisinfected drinking water is consumed. The major bacteria of concern are members of the genus *Salmonella*, *Shigella*, *Campylobacter*, *Yersina*, *Escherichia*, and *Vibrio*.

Salmonella is a very large group of bacteria comprising more than 2,000 known serotypes. All these serotypes are pathogenic to humans and can cause a range of symptoms from mild gastroenteritis to severe illness or even death. *Salmonella* are capable of infecting a large variety of both cold- and warm-blooded animals. Typhoid fever, caused by *S. typhi*, and paratyphoid fever, caused by *S. typhoid*, are both enteric fevers that occur only in humans and primates. In the United States, salmonellosis is primarily due to foodborne transmission since the bacteria infect beef and poultry and are capable of growing in foods. The pathogen produces a toxin that causes fever, nausea, and diarrhea and may be fatal if not properly treated.

Shigella spp. infect only human beings, causing gastroenteritis and fever. They do not appear to survive long in the environment, but outbreaks from drinking and swimming in untreated water continue to occur in the United States. ***Campylobacter*** and ***Yersina*** *spp.* occur in fecally contaminated water and food and are believed to originate primarily from animal feces. *Campylobacter,* which infects poultry and contaminates foods, is often implicated as a source of foodborne outbreaks; it is also associated with the consumption of untreated drinking water in the United States. ***Escherichia coli*** is found in the gastrointestinal tract of all warm-blooded animals and is usually considered a harmless organism. However, several strains are capable of causing gastroenteritis; these are referred to as **enterotoxigenic** (**ETEC**), **enteropathogenic** (**EPEC**), or **enterohemorrhagic** (**EHEC**) strains of *E. coli*. Enterotoxigenic *E. coli* causes gastroenteritis with profuse watery diarrhea accompanied by nausea, abdominal cramps, and vomiting. This bacterium is another common cause of travelers' diarrhea. EPEC strains are similar to ETEC isolates but contain toxins similar to those found in the shigellae. Enterohemorrhagic *E. coli* almost always belong to the single serological type **0157:H7**. This strain generates a potent group of toxins that produce bloody diarrhea, which can be fatal in infants and the elderly. This organism can contaminate both food and water.

The genus ***Vibrio*** comprises a large number of species, but only a few of these species infect human beings. One such is *V. cholerae,* which causes cholera exclusively in humans. Cholera can result in profuse diarrhea with rapid loss of fluid and electrolytes. Fatalities exceed 60% for untreated cases, but death can be averted by replacement of fluids. Cholera outbreaks were unknown in the Western Hemisphere in this century until 1990, when an outbreak that began in Peru spread through South and Central America. The only cases that occur in the United States are either imported or result from consumption of improperly cooked crabs or shrimp harvested from Gulf of Mexico coastal waters. *Vibrio cholera* is a native marine microorganism that occurs in low concentrations in warm coastal waters.

Usually, the survival rate of enteric bacterial pathogens in the environment is just a few days, which is less than the survival rates of enteric viruses and protozoan parasites. They are also easily inactivated by disinfectants commonly used in drinking water treatment. Analysis of environmental samples for enteric bacteria is not often performed because they are difficult to isolate. Instead, indicator bacteria (see Section 19.3.2.4, beginning on page 287) are used to indicate their possible presence.

19.3.2.2 Legionella

The pathogen *Legionella pneumophila* was unknown until 1976, when 34 people died after an outbreak at the annual convention of the Pennsylvania Department of the American Legion in Philadelphia. **Legionellosis**, the acute infection resulting from *L. pneumophila,* is currently associated with two different diseases: Pontiac fever and Legionnaires' disease. Since 1976, numerous deaths from Legionnaires' disease have been reported. Pontiac fever is a milder type of legionellosis. Both these diseases are *noncommunicable,* that is, not transmitted person-to-person. The Centers for Disease Control estimates that between 50,000 and 100,000 cases of legionellosis occur annually in the United States, an unknown number of which are due to contaminated drinking water.

Scientists, however, point out the error of referring to *L. pneumophila* as a classical contaminant. Although this organism occupies an ecological niche (just as do hundreds of other microorganisms in the water environment), no outbreak of legionellosis has yet been directly associated with a natural waterway such as a lake, stream, or pond. The only scientifically documented habitats for *Legionella pneumophila* are damp or moist environments. Evidently, it takes human activity—and certain systems like cooling towers, plumbing components, or even dentist water lines—to harbor or grow the organisms. Therefore, while *L. pneumophila* may be common to natural water, they can proliferate only when taken into distribution systems where water is allowed to stagnate and temperatures are favorable.

Legionella pneumophila can grow to a level that can cause disease in areas that restrict water flow and cause buildup of organic matter. Moreover, the optimum temperature for the growth of *L. pneumophila* is 37°C. Thus, *L. pneumophila* has been discovered in the hot water tanks of hospitals, hotels, factories, golf course clubhouses, and homes. Ironically, some hospitals and hotels keep their water-heater temperatures low to save money and to avert lawsuits, thereby rendering themselves vulnerable to *Legionella* growth. And once established, *Legionella* tends to be persistent. One survey of a hospital water system showed that *L. pneumophila* can exist for long periods under such conditions, collecting in showerheads and faucets in the system. It is believed that showerheads and faucets can emit aerosols composed of very small particles that harbor *L. pneumophila*. Such aerosols, owing simply to their small size, can reach the lower respiratory tract of humans.

A link between the presence of *L. pneumophila* in the water system and Legionnaires' disease in susceptible hospital patients has been established by the medical community. It is this abundance of susceptible people, together with the nature of the water system, that has resulted in outbreaks in hospitals. The great majority of people who have contracted Legionnaires' disease were immunosuppressed or -compromised because of illness, old age, heavy alcohol consumption, or heavy smoking. Although some healthy people have come down with Legionnaires' disease, outbreaks that included healthy individuals have usually resulted in the milder Pontiac fever. But the fact that *L. pneumophila* exists in a water system does not necessarily mean disease is inevitable. *Legionella* bacteria have been detected in systems where no disease or only a few random cases were found. Therefore, the condition or susceptibility of the host or patient is considered to be the single most important factor in whether the infection develops.

Legionella has the ability to survive conventional water treatment. It appears to be considerably more resistant to chlorination than coliform bacteria, and can survive for extended periods in water with low chlorine levels. In addition, it can gain access to municipal water systems through broken or corroded piping, water-main work, and cross connections.

19.3.2.3 Opportunistic Bacterial Pathogens

This group includes heterotrophic Gram-negative bacteria belonging to the following genera: *Pseudomonas*, *Aeromonas*, *Klebsiella*, *Flavobacterium*, *Enterobacter*, *Citrobacter*, *Serratia*, *Acinetobacter*, *Proteus*, and *Providencia*. Segments of the population particularly susceptible to opportunistic pathogens are the newborn, the elderly, and the sick. These organisms have been reported in high numbers in hospital drinking water, where they may attach to water distribution pipes or grow in treated drinking water. However, their public health significance with regard to the population at large is not well understood. Other opportunistic pathogens are the nontubercular mycobacteria, which cause pulmonary and other diseases. The most frequently isolated nontubercular mycobacteria belong to the species *Mycobacterium avium-intracellular*. Potable water, particularly that found in hospital water supplies, can support the growth of these bacteria, which may be linked to infections of hospital patients.

19.3.2.4 Indicator Bacteria

The routine examination of water for the presence of intestinal pathogens is currently a tedious, difficult, and time-consuming task. Thus, scientists customarily tackle such examinations by looking first for certain indicator bacteria whose presence indicates the possibility that other bacteria may also be present. Developed at the turn of the last century, the indicator concept depends upon the fact that certain nonpathogenic bacteria occur in the feces of all warm-blooded animals. These bacteria can easily be isolated and quantified by simple bacteriological methods. Detecting these bacteria in water means that fecal contamination has occurred and suggests that enteric pathogens may also be present.

For example, **coliform bacteria**, which normally occur in the intestines of all warm-blooded animals, are excreted in great numbers in feces. In polluted water, coliform bacteria are found in densities roughly proportional to the degree of fecal pollution. Because coliform bacteria are generally hardier than disease-causing bacteria, their absence from water is an indication that the water is bacteriologically safe for human consumption. Conversely, the presence of the coliform group of bacteria is indicative that other kinds of microorganisms capable of causing disease also may be present, and that the water is unsafe to drink.

The coliform group, which includes *Escherichia, Citrobacter, Enterobacter,* and *Klebsiella* species, is relatively easy to detect: specifically, this group includes all aerobic and facultatively anaerobic, Gram-negative, non-spore forming, rod-shaped bacteria that produce gas upon lactose fermentation in prescribed culture media within 48 hours at 35°C. In short, they're hard to miss.

Scientists commonly use three methods to identify total coliforms in water. These are the **most probable number (MPN)**, the **membrane filter (MF)**, and the **presence-absence (P-A)** tests.

19.3.2.5 The Most Probable Number (MPN) Test

The MPN test allows scientists to confirm the presence of coliforms in a sample and to estimate their numbers. This test consists of three steps: a presumptive test, a confirmed test, and a completed test. In the **presumptive test** (Figure 19-5a), lauryl sulfate tryptose lactose broth is added to a set of test tubes containing different dilutions of the water to be tested. Usually, three to five test tubes are prepared per dilution. These test tubes are incubated at 35°C for 24 to 48 hours, then examined for the presence of coliforms, which is indicated by gas and acid production. Once the positive tubes have been identified and recorded, it is possible to estimate the total number of coliforms in the original sample by using an MPN table that gives numbers of coliforms per 100 mL. In the **confirming test** (Figure 19-5b), the presence of coliforms is verified by inoculating such selective bacteriological agars as Levine's Eosin Methylene Blue (EMB) agar or Endo agar with a small amount of culture from the positive tubes. Lactose-fermenting bacteria are indicated on the media by the production of a colonies with a green sheen or colonies with a dark center. In some cases a **completed test** (not shown in Figure 19-5) is performed in which colonies from the agar are inoculated back into lauryl sulfate tryptose lactose broth to demonstrate the production of acid and gas.

19.3.2.6 The Membrane Filter (MF) Test

The MF test also allows scientists to confirm the presence and estimate the number of coliforms in a sample, but it is easier to perform than the MPN test because it requires fewer test tubes and less handling (see Figure 19-6 on page 290). In this technique, a measured amount of water (usually 100 mL for drinking water) is passed through a membrane filter (pore size 0.45 μm) that traps bacteria on its surface. This membrane is then placed on a thin absorbent pad that has been saturated with a specific medium designed to permit growth and differentiation of the organisms being sought. For example, if total coliform organisms are sought, a modified Endo medium is used. For coliform bacteria, the filter is incubated at 35°C for 18 to 24 hours. The success of the method depends on using effective differential or selective media that can facilitate identification of the bacterial colonies growing on the membrane filter surface (see Figure 19-6g). To determine the number of coliform bacteria in a water sample, the colonies having a green sheen are enumerated.

19.3.2.7 The Presence-Absence (P-A) Test

Presence-absence tests are not quantitative tests—rather they answer the simple question of whether the target organism is present in a sample or not. The use of a single tube of lauryl sulfate tryptose lactose broth as used in the MPN test, but without dilutions, would be used as a P-A test. In recent years, enzymatic assays have been developed that allow for the detection of both total coliform bacteria and *E. coli* in water and wastewater at the same time. These assays can be a simple P-A test or an MPN assay. One commercial P-A test commonly used is the Colilert® test, also called the ONPG-MUG (for *O*-nitrophenyl-*β*-D-galactopyranoside 4-methylumbelliferyl-*β*-D-glucuronide) test. The test is performed by adding the sample to a single bottle (P-A test) or MPN tubes that contain(s) powdered ingredients consisting of salts and specific enzyme substrates that serve as the only carbon source for the organisms (see Figure 19-7a on page 291). The enzyme substrate used for detecting total coliforms is ONPG, and that used for detecting of *E. coli* is MUG. After 24 hours of incubation, samples positive for total coliforms turn

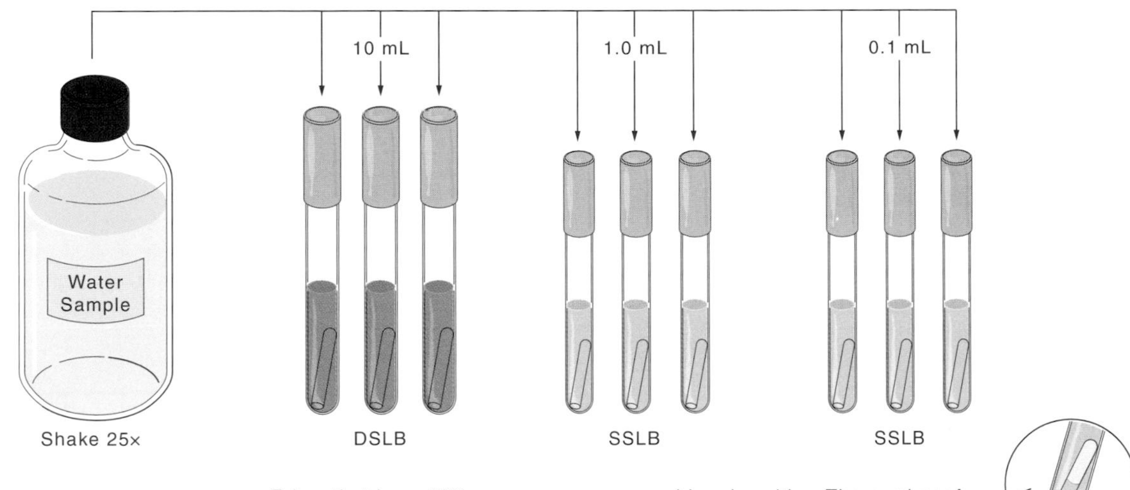

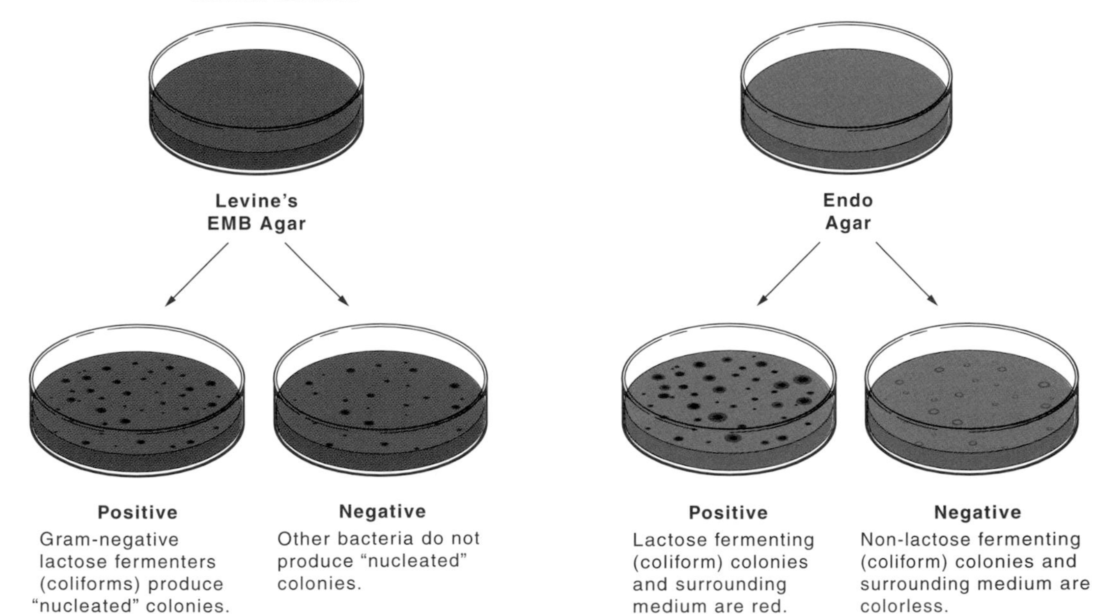

Figure 19-5 Procedure for performing an MPN test for coliforms on water samples: (a) presumptive test and (b) confirmed test. Adapted by permission from Figure 9-1 on page 68 of *Environmental Microbiology: A Laboratory Manual*, by I.L. Pepper, C.P. Gerba, and J.W. Brendecke; copyright © 1995 by Academic Press, Inc. All rights reserved.

a)

Using sterile forceps, place a sterile blotter pad in the bottoms of 3 special petri plates for the mEndo broth-MF.

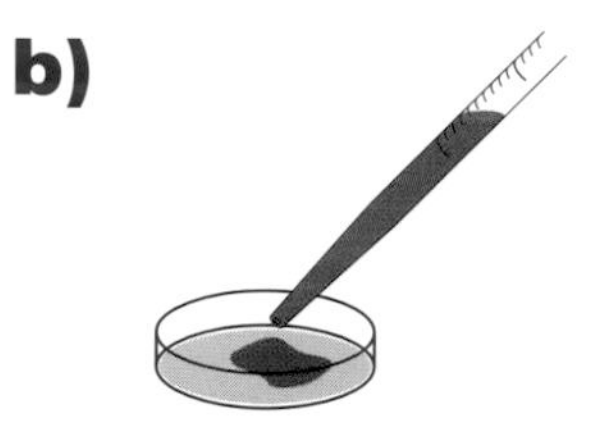

Pipette 2 mL of mEndo broth-MF onto each pad and replace covers. Additionally, prepare 3 mFC agar plates.

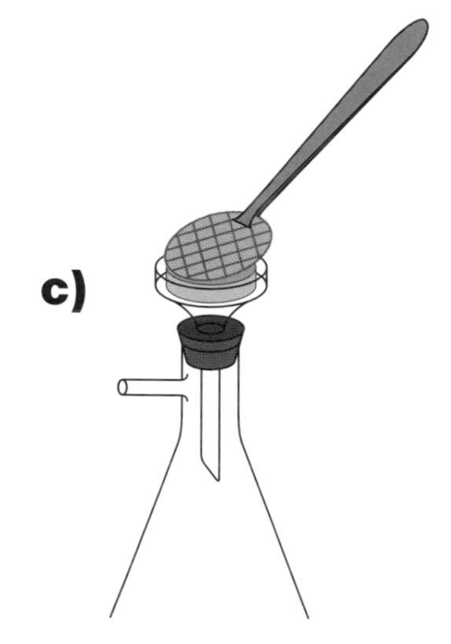

Assemble the filter funnel on the flask. Place a sterile membrane filter using sterile forceps with the grid side up. Center the filter.

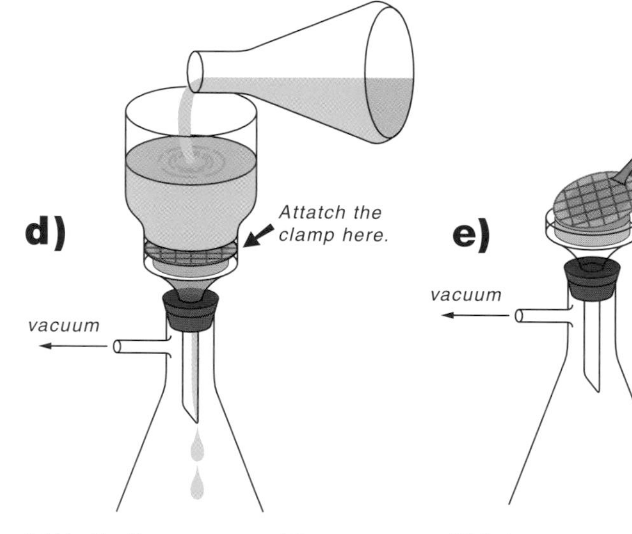

Add buffer if necessary and then add the prescribed volume of sample. Filter under gentle vacuum.

With the vacuum still applied, remove the filter with sterile forceps.

Place the filter on the appropriate medium prepared in steps (a) and (b).

g)

After incubation, count the colonies to determine the concentration of organisms in the original water sample.

Figure 19-6 The membrane filtration method for determining the coliform count in a water sample.

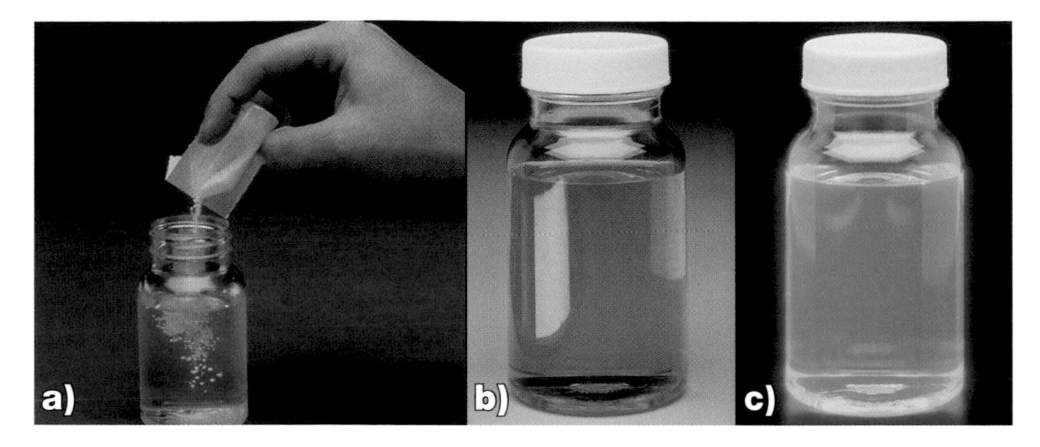

Figure 19-7 Detection of indicator bacteria with Colilert® reagent. (a) Addition of salts and enzyme substrates to water sample; (b) yellow color indicating the presence of coliform bacteria; (c) fluorescence under longwave ultraviolet light indicating the presence of *E. coli*. Photographs used with permission of IDEXX Laboratories, Inc., Westbrook, Maine.

yellow (Figure 19-7b), whereas *E. coli*-positive samples fluoresce under longwave UV illumination (Figure 19-7c).

Although the total coliform group has served as the main indicator of water pollution for many years, many of the organisms in this group are not limited to fecal sources. Thus methods have been developed to restrict the enumeration to those coliforms that are more clearly of fecal origin—that is, the **fecal coliforms** (see Figure 19-3b on page 284). These organisms, which include the genera *Escherichia* and *Klebsiella,* are differentiated in the laboratory by their ability to ferment lactose with the production of acid and gas at 44.5°C within 24 hours. In general, then, this test indicates fecal coliforms; it does not, however, distinguish between human and animal contamination.

Although coliform and fecal coliform bacteria have been successfully used to assess the sanitary quality of drinking water, they have not been shown to be useful indicators of the presence of enteric viruses and protozoa. While outbreaks of enteric bacterial waterborne disease have been virtually eliminated in the United States, fairly recent outbreaks of waterborne disease have occurred in which coliform bacteria were not found. The fact is that enteric viral and protozoan pathogens are more resistant to inactivation by disinfectants and filtration than are the fecal indicator bacteria. For this reason, other potential indicators have been investigated. These include bacteriophages (*i.e.*, bacterial viruses) of the coliform bacteria (known as coliphages), fecal streptococcus, enterococci, or *Clostridium perfringens*. The criteria for an ideal indicator organism are shown in Table 19-4 on page 293. None of these potential indicators has yet been proven ideal.

19.3.1 Protozoa

19.3.1.1 Giardia

Anton van Leeuwenhoek (1632–1723), the inventor of the microscope, was the first person to identify the protozoan *Giardia* in 1681. However, *Giardia lamblia,* the specific microorganism responsible for giardiasis, was unknown in the United States until 1965, when the first case of giardiasis was reported in Aspen, Colorado. Giardiasis is a particularly nasty disease whose acute symptoms include gas, flatulence, explosive watery foul diarrhea, vomiting, and weight loss. In most people these symptoms last from one to four weeks, but have been known to last as little as three or four days or as long as several months. The incubation period ranges from one to three weeks.

Giardia lamblia occurs in the environment—usually water—as a cyst, which can survive in cold water for months and is fairly resistant to chlorine disinfection. When ingested, the organism can enter its reproductive stage and become infectious. Almost half of all waterborne disease outbreaks in community water systems have been attributed to *Giardia lamblia*, making it the most commonly identified pathogen in the United States. Moreover, the number of reported cases probably represents only a fraction of the actual number that occur.

Outbreaks of giardiasis have occurred throughout the United States, but most commonly in mountainous areas in the New England, Rocky Mountain, and Pacific Northwestern states, where high-quality water is often expected. That is, most of these areas are mistakenly thought to be void of sewage or microbial contamination, and communities there tend to use smaller, less comprehensive treatment plants. The majority of these outbreaks are the result of consumption or contact with surface water that is either untreated or treated solely with chlorine. In fact, outbreaks in swimming pools used by small children have also been identified.

Because some communities use chlorination as the sole method for treating surface waters, they can be vulnerable to giardiasis outbreaks which are resistant to chlorination. For example, one outbreak in Rome, New York, in 1974, resulted in 4,800 cases of illness. Another such outbreak occurred in August 1983 in Red Lodge, Montana, where about 780 people were infected.

Researchers have identified three factors contributing to the greatest risk of *Giardia* infection: drinking untreated water, contact with contaminated surface water, and having children in a day-care center.

Current estimates indicate that 60% of the *Giardia* infections in the United States originate from drinking water or contact with surface waters. However, the specific origin of waterborne *Giardia* in the vast majority of outbreaks remains unknown. One recent study concluded that, contrary to previous theories, *Giardia* cysts may be constantly present at low concentrations, even in isolated and pristine watersheds. Moreover, wild animals have been implicated as the cause of giardiasis outbreaks. For example, beavers have been blamed as the source that originally transferred the disease to humans. Another study showed that, although giardiasis outbreaks occur most often in surface water systems, *Giardia* cysts may also be present in groundwater supplies. But no definitive conclusions have yet been reached. For instance, groundwater sources are usually contaminated by surface water such as springs and wells. And in outbreaks in Washington and New Hampshire, it was found that human waste or sanitary violations of individual sewage disposal systems probably infected the beaver first; that is, the beavers actually acquired *Giardia* from humans and only acted as reservoirs for the growth of the disease.

Whatever the original source, we do know that increased numbers of *Giardia* outbreaks occur in the summer months, especially among visitors of recreational areas. Either contamination increases at this time, or the supply is always contaminated, and the extent of the outbreak is determined by the larger number of people using the supply.

19.3.1.2 *Cryptosporidium*

Cryptosporidium, an enteric protozoan first described in 1907, has been recognized as a cause of waterborne enteric disease in humans since 1980. Within five years of its recognition as a human pathogen, the first disease outbreak associated with *Cryptosporidium* was described in the United States. Since then, many outbreaks have been reported, and several studies have documented that *Cryptosporidium parvum* is widespread in U.S. surface waters. Moreover, this species is responsible for infection both in human beings and domesticated animals. For example, this species infects cattle, which, in turn, serve as a major source of the organism in surface waters.

The prevalence of *Cryptosporidium* infection is largely attributable to its life cycle. The organism produces an environmentally stable oocyst that is released into the environment in the feces of infected individuals. The spherical oocysts of *Cryptosporidium parvum* range in size from 3 to 6 μm in diameter. After ingestion, the oocyst undergoes excystation, releasing sporozoites, which then initiate the intracellular infection within the epithelial cells of the gastrointestinal tract. Once in the GI tract, *Cryptosporidium* in humans causes **cryptosporidiosis**, characterized by profuse watery diarrhea, which can result in fluid losses averaging three or more liters a day. Other symptoms of cryptosporidiosis may include abdominal pain, nausea, vomiting, and fever. These symptoms, which usually set in about three to six days after exposure, can be severe enough to cause death.

Studies have indicated that *Cryptosporidium* oocysts occur in 55 to 87% of the world's surface waters. Thus, *Cryptosporidium* oocysts are generally more common in surface water than are the cysts of *Giardia*. But like *Giardia* cysts, *Cryptosporidium* oocysts are very stable in the environment, especially at low temperatures, and may survive for many weeks. Levels are lowest in pristine waters and protected watersheds, where human activity is minimal and domestic animals scarce.

Occurrence of *Cryptosporidium* outbreaks associated with conventionally treated drinking water suggests that the organism is unusually resistant to removal by this process. So far, the oocysts of *Cryptosporidium* have proved to be the most resistant of any known enteric pathogens to inactivation by common water disinfectants. Concentrations of chlorine commonly used in drinking water treatment (1–2 mg L^{-1}) are not enough to kill the organism. Consequently, water filtration is the primary technique used to protect water supplies from contamination by this organism.

In one study, however, *Cryptosporidium* oocysts were identified in 27% of the filtered drinking water supplies in the United States. In addition, several large outbreaks of waterborne disease in the United States and Europe attest to the fact that conventional drinking water treatments involving filtration and disinfection may not be sufficient to prevent disease outbreaks when large concentrations of this organism occur in surface waters. For example, one U.S. outbreak in Carrollton, Georgia, involved illness in 13,000 people—fully one-fifth of the county's total population. In this case, oocysts were identified in the drinking water and in the stream from which the conventional drinking water plant drew its water. Another outbreak of *Cryptosporidium*, which occurred in Milwaukee, Wisconsin, was the largest outbreak of waterborne disease ever documented in the United States (see *Case Study: Cryptosporidiosis in Milwaukee* on page 294). In other cases, *Cryptosporidium* infection has been transmitted by contact with, or swimming in, contaminated water.

Table 19-4 Criteria for an ideal indicator organism.

- The organism should be useful for all types of water.
- The organism should be present whenever enteric pathogens are present.
- The organism should have a reasonably longer survival time than the hardiest enteric pathogen.
- The organism should not grow in water.
- The testing method should be easy to perform.
- The density of the indicator organism should have some direct relationship to the degree of fecal pollution.

Much remains to be learned about the survival and occurrence of this organism, and the ability of water treatment processes to remove it.

As is the case with other enteric organisms, only low numbers of *Giardia* cysts and *Cryptosporidium* oocysts need to be ingested to cause infection. Thus, large volumes of water are sampled (from 100 to 1,000 L) for analysis. The organisms are entrapped on filters with a pore size smaller than the diameter of the cysts and oocysts. After extraction from the filter, they can be further concentrated and detected by observation with the use of a microscope. Antibodies tagged with a fluorescent compound are used to aid in the identification of the organisms (see Figure 19-8 on page 295 and the photograph at the top of page 279).

19.3.2 Helminths

In addition to the unicellular protozoa, some multicellular animals—the helminths—are capable of parasitizing humans. These include the Nematoda (roundworms) and the Platyhelminthes, which are divided into two subgroups: the Cestoda, or tapeworms, and the Trematoda, or flukes. In these parasitic helminths, the microscopic ova, or eggs, constitute the infectious stage. Excreted in the feces of infected persons and spread by wastewater, soil, or food, these ova are very resistant to environmental

Case Study: Cryptosporidiosis in Milwaukee

Early in the spring of 1993, heavy rains flooded the rich agricultural plains of Wisconsin. These rains produced an abnormal run-off into a river that drains into Lake Michigan, from which the city of Milwaukee obtains its drinking water. The city's water treatment plant seemed able to handle the extra load: it had never failed before, and all existing water quality standards for drinking water were properly met. Nevertheless, by April 1, thousands of Milwaukee residents came down with acute watery diarrhea, often accompanied by abdominal cramping, nausea, vomiting, and fever. In a short period of time, more than 400,000 people developed gastroenteritis, and more than 100—mostly elderly and infirm individuals—ultimately died, despite the best efforts of modern medical care. Finally, after much testing, it was discovered that *Cryptosporidium* oocysts were present in the finished drinking water after treatment. These findings pointed to the water supply as the likely source of infection; and on the evening of April 7, the city put out an urgent advisory for residents to boil their water. This measure effectively ended the outbreak. All told, direct costs and loss of life are believed to have exceeded $150 million dollars.

The Milwaukee episode was the largest waterborne outbreak of disease ever documented in the United States. But what happened? How could such a massive outbreak occur in a modern U.S. city in the 1990s? And how could so many people die? Apparently, high concentrations of suspended matter and oocysts in the raw water resulted in failure of the water treatment process—a failure in which *Cryptosporidium* oocysts passed right through the filtration system in one of the city's water treatment plants, thereby affecting a large segment of the population. And among this general population were many whose systems could not withstand the resulting illness. In immunocompetent people, cryptosporidiosis is a self-limiting illness; it's very uncomfortable, but it goes away of its own accord. However, in the immunocompromised, cryptosporidiosis can be unrelenting and fatal.

stresses and to disinfection. The most important parasitic helminths are listed in Table 19-5.

Ascaris, a large intestinal roundworm, is a major cause of nematode infections in humans. This disease can be acquired through ingestion of just a few infective eggs. Since one female *Ascaris* can produce approximately 200,000 eggs per day and each infected person can excrete a large quantity of eggs, this nematode is very common. Worldwide estimates indicate that between 800 million and one billion people are infected, with most infections being in the tropics or subtropics. In the United States, infections tend to occur in the Gulf Coast. The life cycle of this parasite includes a phase in which the larvae migrate through the lungs and cause pneumonitis (known as Loeffler's syndrome). Although the eggs are dense, and hence readily removed by sedimentation in wastewater treatment plants, they are quite resistant to chlorine action. Moreover, they can survive for long periods of time in sewage sludge after land application unless removed by sludge treatment (see Section 20.3, beginning on page 302).

Although ***Taenia saginata*** (beef tapeworm) and ***Taenia solium*** (pig tapeworm) are now relatively rare in the United States, they can still be found in developing countries around the world. These parasites develop in an intermediate animal host, where

Table 19-5 Major parasitic helminths.

Organism	Disease (Main Site Affected)
Nematodes (roundworms)	
Ascaris lumbricoides	Ascariasis: intestinal obstruction in children (small intestine)
Trichuris trichiura	Whipworm (trichuriasis): (intestine)
Hookworms	
Necator americanus	Hookworm disease (intestine)
Ancylostoma duodenale	Hookworm disease (intestine)
Cestodes (tapeworms)	
Taenia saginata	Beef tapeworm: results in abdominal discomfort, hunger pains, chronic indigestion
Taenia solium	Pork tapeworm (intestine)
Trematodes (flukes)	
Schistosoma mansoni	Schistosomiasis (liver [cirrhosis], bladder, and large intestine)

they reach a larval stage called **cysticercus.** Then these larvae are passed, via meat products, to humans, who serve as final hosts. For instance, cattle that ingest the infective ova while grazing serve as intermediate hosts for *Taenia saginata*, while pigs are the intermediate hosts for *Taenia solium*. The cysticerci invade the muscle, eye, and brain tissue of the intermediate host and can cause severe enteric disturbances, such as abdominal pains and weight loss in their final hosts.

Toxocara canis is another nematodal parasite that infects humans. Found primarily in dogs (*T. cati* is found in cats), this roundworm mainly infects children who have ingested dirt. In addition to causing intestinal disturbances, the larvae of this parasite can migrate into the eyes, causing severe ocular damage, sometimes resulting in loss of the eye.

19.3.3 Blue-Green Algae

Blue-green algae, or **cyanobacteria**, occur commonly in all natural waters, where they play an important role in the natural cycling of nutrients in the environment and the food chain. However, a few species of blue-green algae, such as *Microcystis*, *Aphanitomenon*, and *Anabaena*, produce toxins capable of causing illness in humans and animals. These toxins can cause gastroenteritis, neurological disorders, and possibly cancer. In this case, illness is caused by the ingestion of the toxin produced by the organisms, rather than ingestion of the organism itself, as is the case with helminths. Numerous cases of livestock, pet, and wildlife poisonings by the ingestion of water blooms of cyanobacteria have been reported, and evidence has been mounting that humans are also affected. Heavy blooms of cyanobacteria can occur in surface waters when sufficient nutrients are available, resulting in contaminated water supplies. In the United States, a large outbreak of gastroenteritis in 1975 in Sewickly, Pennsylvania, was attributed to the presence of high levels of algal toxins in the conventionally treated water.

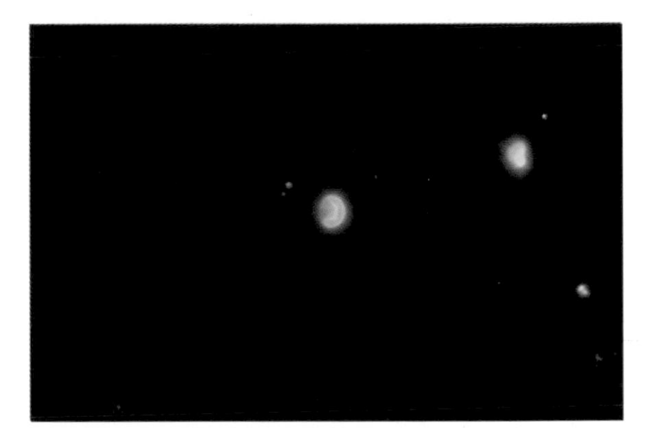

Figure 19-8 *Cryptosporidium* oocysts (green) observed under UV light after staining with immunofluorescent antibodies (magnification = 800×). Photograph: D.C. Johnson.

19.4 Sources of Pathogens in the Environment

Waterborne enteric pathogens are excreted, often in large numbers, in the feces of infected animals and humans, whether or not the infected individual exhibits the symptoms of clinical illness. In some cases, infected individuals may excrete pathogens without ever developing symptoms; in other cases, infected individuals may excrete pathogens for many months—long after clinical signs of the illness have passed. Such asymptomatic infected individuals are known as carriers, and they may constitute a potential source of infection for the community (see Figure 19-9). Owing to such sources, pathogens are almost always present in the sewage of any community. However, the actual concentration of pathogens in community sewage depends on many factors: the incidence of enteric disease (*i.e.*, the number of individuals with the disease in a population), the number of carriers in the community, the time of year, sanitary conditions, and per capita water consumption.

The peak incidence of many enteric infections is seasonal in temperate climates. Thus, the highest incidence of enterovirus infection is during the late summer and early fall, while rotavirus infections tend to peak in the early winter, and *Cryptosporidium* infections peak in the early spring and fall. The reason for the seasonality of enteric infections is not completely understood, but several factors may play a role. It may be associated with the survival of different agents in the environment during the different seasons: *Giardia,* for example, can survive winter temperatures very well. Alternatively, excretion differences among animal reservoirs may be involved, as is the case with *Cryptosporidium.* Or it may well be that greater exposure to contaminated water, as in swimming, is the explanation for increased incidence in the summer months.

Figure 19-9 Bathers can be a common source of pathogens: bather density in Honolulu, Hawaii, has sometimes been correlated with fecal coliform counts. Photograph: I.L. Pepper.

Table 19-6 Incidence and concentration of enteric viruses and protozoa in feces in the United States.

Pathogen	Incidence (%)	Concentration in Stool (per gram)
Enterovirus	10–40	10^3–10^8
Hepatitis A	0.1	10^8
Rotavirus	10–29	10^{10}–10^{12}
Giardia	3.8	10^6
	18–54[a]	10^6
Cryptosporidium	0.6–20	10^6–10^7
	27–50[a]	10^6–10^7

[a] Children in day-care centers.

Table 19-7 Estimated levels of enteric organisms in sewage and polluted surface water in the United States. Source: U.S. EPA, 1988.

	Concentration (per 100 mL)	
Organism	Raw Sewage	Polluted Stream Water
Coliforms	10^9	10^5
Enteric viruses	10^2	1–10
Giardia	10	0.1–1
Cryptosporidium	10–10^3	0.1–10^2

Certain populations and subpopulations are also more susceptible to infection. For example, enteric infection is more common in children because they usually lack previous protective immunity. Thus the incidence of enteric virus and protozoa infections in day-care centers, where young children are in close proximity, is usually much higher than that in the general community (see Table 19-6). A greater incidence of enteric infections is also evident in lower socioeconomic groups, particularly where lower standards of sanitary conditions prevail. Concentrations of enteric pathogens are much greater in sewage in the developing world than the industrialized world. For example, the average concentration of enteric viruses in sewage in the United States has been estimated at 10^3 L^{-1} (Table 19-7), while concentrations as high as 10^5 L^{-1} have been observed in Africa and Asia.

19.4.1 Sludge

During municipal sewage treatment, **biosolids**—or sludge—are produced. Biosolids are a by-product of physical (primary treatment), biological (activated sludge), and physicochemical (precipitation of suspended solids by chemicals) treatment processes. (The process of sludge production is described in detail in Section 20.3, beginning on page 302.) Although treatment by anaerobic or aerobic digestion and/or dewatering reduces the numerical population of disease agents in these biosolids, significant numbers of the pathogens present in raw sewage often remain in sewage biosolids. On a volume basis, the concentration of pathogens in sewage biosolids can be fairly high because of settling (of the large organisms, especially helminths) and adsorption (especially viruses). Moreover, most microbial species found in raw sewage are concentrated in sludge during primary sedimentation. And although enteric viruses are too low in mass to settle alone, they are also concentrated in sludge because of their strong binding affinity to particulates.

The densities of pathogenic and indicator organisms in **primary sludge** shown in Table 19-8 represent typical, average values detected by various investigators. Note that the indicator organisms are normally present in fairly constant amounts. But bear in mind that different sludges may contain significantly greater or fewer numbers of any organism, depending upon the kind of sewage from which the sludge was derived. Similarly, the quantities of pathogenic species are especially variable because these figures depend on which kind are present in a specific community at a particular time. Finally, note that concentrations determined in any study are dependent on assays for each microbial species; thus, these concentrations are only as accurate as the assays themselves, which may be compromised by such factors as inefficient recovery of pathogens from environmental samples.

Table 19-8 Densities of microbial pathogens and indicators in primary sludges. Modified from Straub *et al.*, 1993.

Type	Organism	Density (number per gram of dry weight)
Virus	Various enteric viruses	10^2–10^4
	Bacteriophages	10^5
Bacteria	Total coliforms	10^8–10^9
	Fecal coliforms	10^7–10^8
	Fecal streptococci	10^6–10^7
	Salmonella sp.	10^2–10^3
	Clostridium sp.	10^6
	Mycobacterium tuberculosis	10^6
Protozoa	*Giardia* sp.	10^2–10^3
Helminths	*Ascaris* sp.	10^2–10^3
	Trichuris vulpis	10^2
	Toxocara sp.	10^1–10^2

Table 19-9 Densities of pathogenic and indicator microbial species in secondary sludge biosolids. Modified from Straub *et al.*, 1993.

Type	Organism	Density (number per gram dry weight)
Virus	Various enteric viruses	3×10^2
Bacteria	Total coliforms	7×10^8
	Fecal coliforms	8×10^6
	Fecal streptococci	2×10^2
	Salmonella sp.	9×10^2
Protozoa	*Giardia* sp.	10^2–10^3
Helminths	*Ascaris* sp.	1×10^3
	Trichuris vulpis	$<10^2$
	Toxocara sp.	3×10^2

Secondary sludges are produced following the biological treatment of wastewater. Microbial populations in sludges following these treatments depend on the initial concentrations in the wastewater, die-off or growth during treatments, and the association of these organisms with sludge. Some treatment processes, such as the activated sludge process, may limit or destroy certain enteric microbial species. Viral and bacterial pathogens, for example, are reduced in concentration by activated sludge treatment (see Section 20.3.2.2, beginning on page 304). Even so, the ranges of pathogen concentration in secondary sludges obtained from this and most other secondary treatments are usually not significantly different from those of primary sludges, as shown in Table 19-9.

19.4.2 Solid Waste

Municipal solid waste may contain a variety of pathogens, a source of which is often disposable diapers—it has been found that as many as 10% of the

Table 19-10 Environmental factors affecting enteric pathogen survival in natural waters.

Factor	Remarks
Temperature	Probably the most important factor; longer survival at lower temperatures, freezing kills bacteria and protozoan parasites, but prolongs virus survival.
Moisture	Low moisture content in soil can reduce bacterial populations.
Light	UV in sunlight is harmful.
pH	Most are stable at pH values of natural waters. Enteric bacteria are less stable at pH >9 and <6.
Salts	Some viruses are protected against heat inactivation by the presence of certain cations.
Organic matter	The presence of sewage usually results in longer survival.
Suspended solids or sediments	Association with solids prolongs survival of enteric bacteria and viruses
Biological factors	Native microflora is usually antagonistic

fecally soiled disposable diapers entering landfills contain enteroviruses. Another primary source of pathogens is sewage biosolids, where co-disposal is practiced. Pathogens may also be present in domestic pet waste (*e.g.*, cat litter) and food wastes. Municipal solid wastes from households have been found to average 7.7×10^8 coliforms and 4.7×10^8 fecal coliforms per gram. *Salmonella* have also been detected in domestic solid waste. In unlined landfills, such pathogens may be present in the leachate beneath landfills (see Figure 10-2 on page 137).

19.5 Fate and Transport of Pathogens in the Environment

There are many potential routes for the transmission of excreted enteric pathogens. The ability of an enteric pathogen to be transmitted by any of these routes depends largely on its resistance to environmental factors, which control its survival, and its capacity to be carried by water as it moves through the environment. Some routes can be considered "natural" routes for the transmission of waterborne disease, but others—such as the use of domestic wastewater for groundwater recharge, large-scale aquaculture projects, or land disposal of disposable diapers—are actually new routes created by modern humankind.

Table 19-11 Reported survival times of enteric viruses in various environments.

Environment	Survival Time (days)
Sea or estuary water	2–130
River water	2–188
Tapwater	5–168
Oysters	6–90
Marine sediment	8–436

Human and animal excreta are sources of pathogens. Humans become infected by pathogens through consumption of contaminated foods, such as shellfish from contaminated waters or crops irrigated with wastewater; from drinking contaminated water; and through exposure to contaminated surface waters as may occur during bathing or at recreational sites. Furthermore, those individuals infected by the above processes become sources of infection through their excrement, thereby completing the cycle.

In general, viral and protozoan pathogens survive longer in the environment than enteric bacterial pathogens. How long a pathogen survives in a particular environment depends on a number of complex factors, which are listed in Table 19-10. Of all the factors temperature is probably the most important. Temperature is a well-defined factor with a consistently predictable effect on enteric pathogen survival in the environment. Usually, the lower the temperature, the longer the survival time. But freezing temperatures generally result in the death of enteric bacteria and protozoan parasites. Viruses, however, can remain infectious for months or years at freezing temperatures. Moisture—or lack thereof, can cause decreased survival of bacteria. The UV light from the sun is a major factor in the inactivation of indicator bacteria in surface waters; thus, die-off in marine waters can be predicted by amount of exposure to daylight. Viruses are much more resistant to inactivation by UV light.

Many laboratory studies have shown that the microflorae of natural waters and sewage are antagonistic to the survival of enteric pathogens. It has been shown, for example, that enteric pathogens survive longer in sterile water than in water from lakes, rivers, and oceans. Bacteria in natural waters can feed upon indicator bacteria such as fecal coliforms. The adsorption of enteric bacteria to suspended matter (clays, organic debris, and the like) and fresh or marine sediments has been shown to prolong their survival time. The range of survival time of viral pathogens is shown in Table 19-11.

References and Recommended Reading

Bennett J.V., Homberg S.D., Rogers M.F., and Solomon S.L. (1987) Infectious and parasitic diseases. *American Journal of Preventative Medicine.* **55**, 102–114.

Bitton G. (1994) *Wastewater Microbiology.* Wiley-Liss, New York.

Craun G.F. (1986) *Waterborne Diseases in the United States.* CRC Press, Boca Raton, Florida.

Jeffrey, H.C. and Leach R.M. (1972) *Atlas of Medical Helminthology and Protozoology.* Churchill Livingstone, Edinburgh.

LeChevallier M.W., Norton W.D., and Lee R.G. (1991) *Giardia* and *Cryptosporidium* spp. in filtered drinking water supplies. *Applied and Environmental Microbiology.* **57**, 2617–2626.

Lin S.D. (1985) *Giardia lamblia* and water supply. *Journal of the American Water Works Association.* **77**, 40–47.

McFeters G.A. (1990) *Drinking Water Microbiology.* Springer-Verlag, New York.

Pepper I.L., Gerba C.P., and Brendecke J.W. (1995) *Environmental Microbiology: A Laboratory Manual.* Academic Press, San Diego, California.

Straub T.M., Pepper I.L., and Gerba C.P. (1993) Hazards of pathogenic microorganisms in land-disposed sewage sludge. *Reviews of Environmental Contamination and Toxicology.* **132**, 55–91.

U.S. EPA (1988). *Comparative Health Effects Assessment of Drinking Water.* United States Environmental Protection Agency, Washington, D.C.

White G.F. and Bradley D.J. (1972) *Drawers of Water.* University of Chicago Press, Chicago.

Problems and Questions

1. What are pathogens? What is an enteric pathogen?
2. What is the difference between a waterborne and a water-based pathogen?
3. Which group of enteric pathogens survives the longest in the environment and why?
4. Describe some of the methods that can be used to detect indicator bacteria in water.
5. What are some of the criteria for indicator bacteria?
6. Protozoan parasites are the leading cause of waterborne disease outbreaks in the United States when an agent can be identified. Why?
7. What are some of the factors that control the survival of enteric pathogens in the environment?
8. Why is waterborne disease still a problem in the United States?

9. What are some of the niches in which *Legionella* bacteria can grow to high numbers?
10. What are the factors that determine the concentration of enteric pathogens in sewage?
11. By what other routes can enteric pathogens be transmitted besides drinking water?

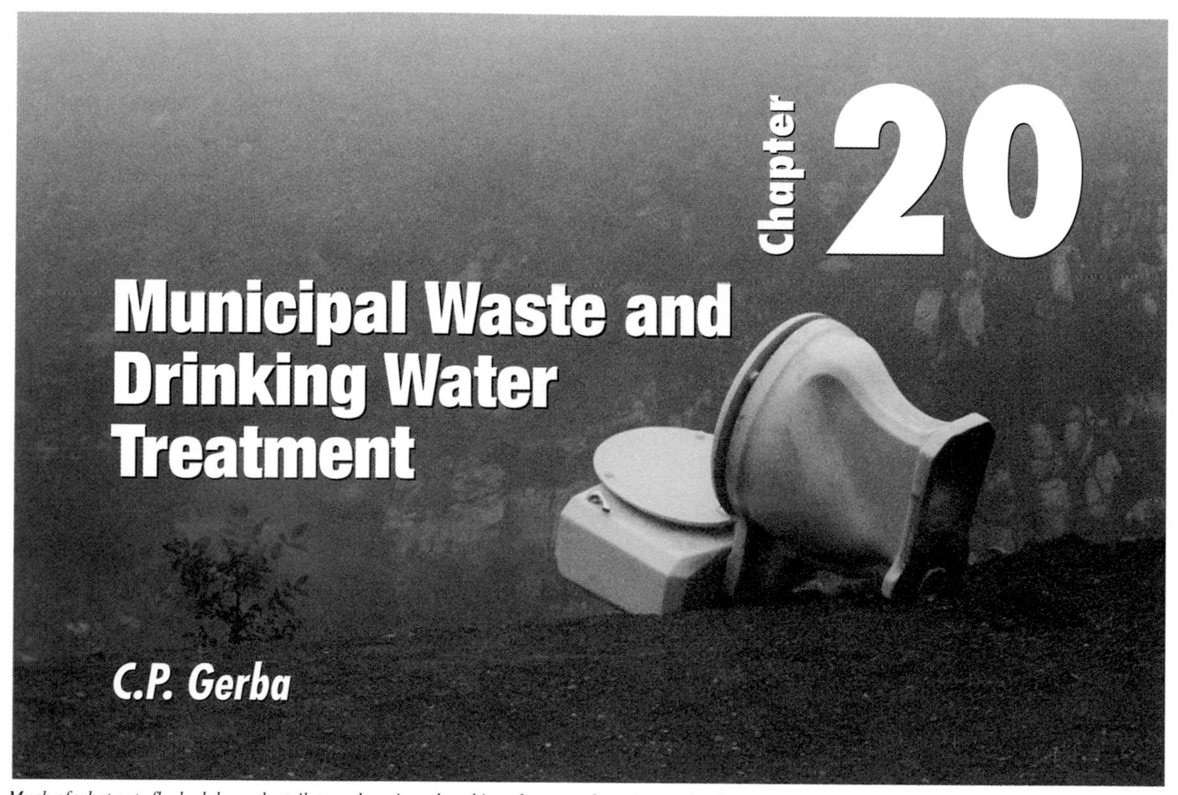

Much of what gets flushed down the toilet can be reintroduced into the natural nutrient cycles discussed in previous chapters. However, along with the valuable nutrients come numerous hazards, including pathogens, necessitating adequate treatment of the sewage. Photograph printed by permission; copyright © 1995 by Alyson B. Brendecke. All rights reserved.

20.1 Water and Civilization

Civilization—literally, "citification"—is built on water. Historically, as small communities became cities, their need for safe, dependable water supplies and reliable waste-disposal systems became critical. More than two thousand years ago, for example, the Romans met this need by building a sophisticated system of aqueducts and sewers, some of which are still in use. Modern cities, however, require vast quantities of potable water and generate much larger amounts of wastes than did ancient Rome. In addition, we are aware of the indissoluble link between disease and sewage, as well as the relationships between adequate waste treatment and a wholesome environment (see Figure 20-1 on page 303 for a pictorial description of the numerous routes through which municipal wastes can contaminate groundwater). Thus we face a major challenge in the search for effective, economical solutions to water supply and waste disposal problems that will protect not only human health but also the quality of the environment.

20.2 The Nature of Wastewater (Sewage)

The *cloaca maxima*, the "biggest sewer" in Rome, had enough capacity to serve a city of one million people. This sewer, and others like it, simply collected wastes and discharged them into the nearest lake, river, or ocean. This expedient made cities more habitable, but its success depended on transferring the pollution problem from one place to another. And what worked reasonably well for the Romans doesn't work well today. Current population densities are too high to permit a simple dependence on transference. Thus, modern-day sewage is treated before it is discharged into the environment. In the latter part of the nineteenth century, the design of sewer systems allowed collection with treatment to lessen the impact on natural waters. Today, more than 15,000 wastewater treatment plants treat approximately 150 billion liters of wastewater per day in the United States alone. In addition, septic tanks, which were also introduced at the end of the nine-

Table 20-1 Typical composition of untreated domestic wastewater. Modified from Metcalf and Eddy, 1991.

	Concentration (mg L^{-1})		
Contaminants	Low	Moderate	High
Solids, total	350	720	1200
Dissolved, total	250	500	850
Volatile	105	200	325
Suspended solids	100	220	350
Volatile	80	165	275
Settleable solids	5	10	20
Biochemical oxygen demand[a], (mg L^{-1})	110	220	400
Total organic carbon	80	160	290
Chemical oxygen demand	250	500	1000
Nitrogen (total as N)	20	40	85
Organic	8	15	35
Free ammonia	12	25	50
Nitrites	0	0	0
Nitrates	0	0	0
Phosphorus (total as P)	4	8	15
Organic	1	3	5
Inorganic	3	5	10

[a] 5-day, 20°C (BOD, 20°C)

Table 20-2 Types and numbers of microorganisms typically found in untreated domestic wastewater. Modified from Metcalf and Eddy, 1991.

Organism	Concentration (per mL)
Total coliform	10^5–10^6
Fecal coliform	10^4–10^5
Fecal streptococci	10^3–10^4
Enterococci	10^2–10^3
Shigella	Present
Salmonella	10^0–10^2
Clostridium perfringens	10^1–10^3
Giardia cysts	10^{-1}–10^2
Cryptosporidium oocysts	10^{-1}–10^1
Helminth ova	10^{-2}–10^1
Enteric virus	10^1–10^2

teenth century, serve approximately 25% of the U.S. population, largely in rural areas.

Domestic wastewater is a combination of human feces, urine, and "graywater." Graywater results from washing, bathing, and meal preparation. Water from various industries and businesses may also enter the system. It is, however, the amount of organic matter in domestic wastes that determines the degree of biological treatment required. Three tests are used to assess the amount of organic matter: total organic carbon (TOC), biochemical oxygen demand (BOD), and chemical oxygen demand (COD) (the latter two of which are discussed in Sections 13.4 and 13.5, beginning on page 195).

The major pollutants in domestic wastewater are biodegradable organics, recalcitrant organics, toxic metals, suspended matter, and pathogenic microorganisms. Some of the components of domestic sewage are listed in Table 20-1, together with some of their physicochemical characteristics and concentrations. Several types of indicator bacteria and pathogens are also present in wastewater in various concentrations, as shown in Table 20-2.

20.3 Wastewater (Sewage) Treatment

20.3.1 Septic Tanks

Until the middle of the twentieth century in the United States, many rural families and quite a few residents of towns and small cities depended on pit toilets or "outhouses" for waste disposal. These pit toilets, however, often allowed untreated wastes to seep into the groundwater, allowing pathogens to contaminate drinking water supplies. This risk to public health led to the development of septic tanks and properly constructed drainfields. Primarily, septic tanks serve as repositories where solids are separated from incoming wastewater and biological digestion of the waste organic matter can take place under anaerobic conditions. In a typical septic tank system (see Figure 20-2 on page 305), the wastewater and sewage enters a tank made of concrete, metal, or fiberglass. There, grease and oils rise to the top as scum, and solids settle to the bottom. The wastewater and sewage then undergo anaerobic bacterial decomposition, resulting in the production of a sludge. The wastewater usually remains in the septic tank for just 24–72 hours, after which it is channeled out to a drainfield. This drainfield, or leachfield, is composed of small perforated pipes that are embedded in gravel below the surface of the soil. Periodically, the residual septage in the septic tank is pumped out into a tank truck and taken to a treatment plant for disposal.

Although the concentration of contaminants in septic tank septage is typically much greater than

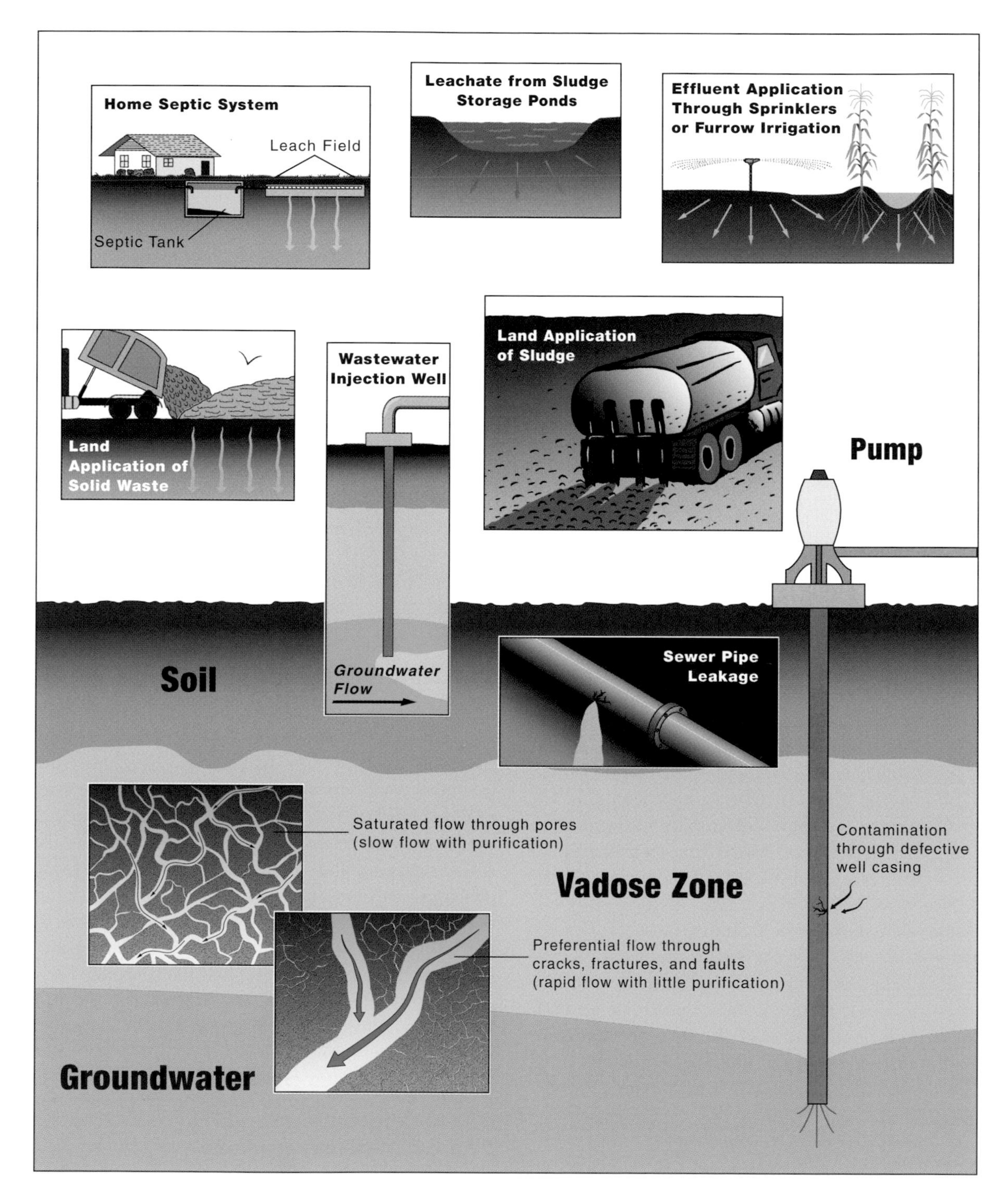

Figure 20-1 Pathways by which municipal wastes can contaminate groundwater.

Table 20-3 Typical characteristics of septage. Modified from Metcalf and Eddy, 1991.

	Concentration (mg L^{-1})	
Constituent	Range	Typical
Total solids	5,000–100,000	40,000
Suspended solids	4,000–100,000	15,000
Volatile suspended solids	1,200–14,000	7,000
5-day, 20°C BOD	2,000–30,000	6,000
Chemical oxygen demand	5,000–80,000	30,000
Total Kjeldahl nitrogen (as N)	100–1,600	700
Ammonia, NH_3, as N	100–800	400
Total phosphorus as P	50–800	250
Heavy metals[a]	100–1,000	300

[a] Primarily iron (Fe), zinc (Zn), and aluminum (Al).

that found in domestic wastewater (see Table 20-3), septic tanks can be an effective method of waste disposal where land is available and population densities are not too high. Thus, they are widely used in rural and suburban areas. But, as suburban population densities increase, groundwater and surface water pollution may arise, indicating the need to shift to a commercial municipal sewage system. (In fact, private septic systems are sometimes banned in many suburban areas.) Moreover, septic tanks are not appropriate for every area of the country. They do not work well, for example, in cold, rainy climates where the drainfield may be too wet for proper evaporation, nor in areas where the water table is shallow. High densities of septic tanks can also be responsible for nitrate contamination of groundwater. And finally, most of the waterborne disease outbreaks associated with groundwater in the United States are thought to result from contamination by septic tanks.

20.3.2 The Overall Process of Wastewater Treatment

Modern sewage treatment began at the turn of the nineteenth century when attempts were made to develop processes that would reduce the amount of degradable organic matter in sewage before discharge into natural bodies of water. The presence of organic matter in discharge stimulates rapid microbial growth, resulting in microbial consumption of available oxygen in water. The resulting anaerobic conditions can kill fish and other forms of aquatic life while producing characteristically unpleasant odors. Designed to reduce the amount of degradable organic matter by conducting these processes under controlled conditions, complete sewage treatment comprises three major steps, as shown in Figure 20-3 on page 307.

- **Primary treatment** is a physical process that involves the separation of large debris, followed by sedimentation.
- **Secondary treatment** is a biological process that is carried out by microorganisms.
- **Tertiary treatment** is usually a physicochemical process that removes turbidity caused by the presence of nutrients (*e.g.*, nitrogen), dissolved organic matter, metals, or pathogens.

20.3.2.1 Primary Treatment

Primary treatment is the first step in municipal sewage treatment. It physically separates large solids from the waste stream. As raw sewage enters the treatment plant, it passes through a metal grating that removes large debris, such as branches, tires, and the like. A moving screen then filters out smaller items such as diapers and bottles, after which brief residence in a grit tank allows sand and gravel to settle out. The waste stream is then pumped into the primary settling tank (also known as a sedimentation tank or clarifier) where about half the suspended organic solids settle to the bottom as **sludge** or **biosolids**. The resulting sludge is referred to as **primary sludge**. Microbial pathogens are not effectively removed in the primary process.

20.3.2.2 Secondary Treatment

Secondary treatment consists of biological degradation, in which the remaining suspended solids are decomposed and the number of pathogens reduced. In this stage, the effluent from primary treatment may

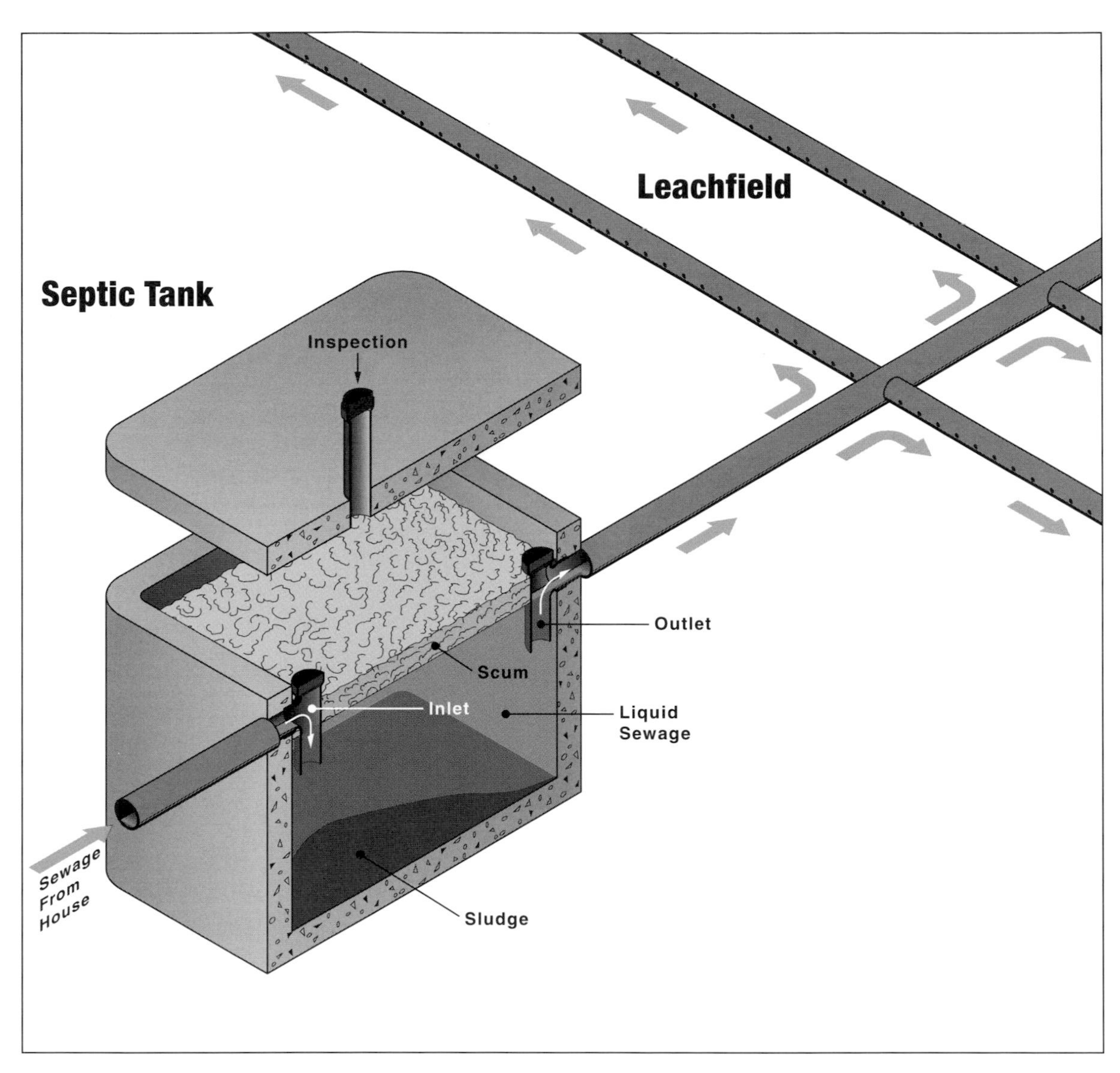

Figure 20-2 Septic tank system.

be pumped into a trickling filter bed, an aeration tank, or a sewage lagoon. A disinfection step is generally included at the end of the treatment.

The **trickling filter bed** is simply a bed of stones or corrugated plastic sheets through which water drips (see Figure 20-4 on page 308). The effluent is pumped through a system of perforated pipes or through a sweeping overhead sprayer onto this bed, where bacteria and other microorganisms reside. These microorganisms intercept the organic material as it trickles past and decompose it aerobically.

Aeration-tank digestion is also known as the activated sludge process. In the United States, wastewater is most commonly treated by this process (this is the process illustrated in Figure 20-3 on page 307). Effluent from primary treatment is pumped into a tank and mixed with a bacteria-rich slurry, known as **activated sludge**. Air or pure oxygen pumped through the mixture encourages bacterial growth and decomposition of the organic material. It then goes to a secondary settling tank, where water is siphoned off the top of the tank and sludge is removed from the bottom. Some of the sludge is used as an inoculum for incoming primary effluent. The remainder of the sludge, known as secondary sludge, is removed. The concentration of pathogens is reduced in the activated sludge process by antagonistic microorganisms as well as adsorption to or incorporation into the secondary sludge.

Sewage lagoons are often referred to as **oxidation** or **stabilization ponds**, which are the oldest of the wastewater treatment systems. Usually, no more than a hectare in area and just a few meters deep, oxidation ponds serve as natural "stewpots" where wastewater is detained while organic matter is degraded. This period of time, ranging from 1 to 4 weeks (and sometimes longer), is necessary to complete the decomposition of organic matter. Light, heat, and settling of the solids can also effectively reduce the number of pathogens present in the wastewater.

There are three categories of oxidation ponds, which are often used in series: aerobic ponds, aerated ponds, and anaerobic ponds.

- **Aerobic ponds** (Figure 16-5a on page 247), which are naturally mixed, must be shallow because they depend upon the penetration of light to stimulate algal growth, which promotes subsequent oxygen generation. The detention time of wastewater is generally 3 to 5 days.
- **Aerated ponds**, which are mechanically aerated, may be 1–2 meters deep with a detention time of less than 10 days. In general, treatment depends on the amount of aeration time and temperature, as well as the type of wastewater. For example, at 20°C, an aeration period of 5 days results in an 85% BOD removal.
- **Anaerobic ponds** (Figure 16-5b on page 247) may be 1 to 10 meters deep and require a relatively long detention time of 20–50 days. These ponds, which do not require expensive mechanical aeration, generate small amounts of sludge. Often, anaerobic ponds serve as a pretreatment step for high-BOD organic wastes rich in protein and fat (*e.g.*, meat wastes) with a heavy concentration of suspended solids.

Because sewage lagoons require a minimum of technology and are relatively low in cost, they are most common in developing countries and in small communities in the United States, where land is available at reasonable prices. However, biodegradable organic matter and turbidity are not as effectively reduced as they are in activated sludge treatment.

The effluent from all these secondary treatment processes usually receives a disinfection treatment to eliminate residual bacterial pathogens. Chlorination is the most common form of disinfection, but irradiation with ultraviolet light is becoming more common. However, enteric viruses and protozoan pathogens often survive disinfection, and chlorination can produce by-products that are toxic to aquatic animals including fish. Thus, wastewater should be dechlorinated before discharge into sensitive bodies of water, or ultraviolet light disinfection should be used exclusively.

20.3.2.3 Tertiary Treatment

Tertiary treatment involves a series of additional steps after secondary treatment to further reduce organics, turbidity, nitrogen, phosphorus, metals, and pathogens. Most processes involve some type of physicochemical treatment such as coagulation, filtration, activated carbon adsorption of organics, and additional disinfection. Tertiary treatment of wastewater is practiced for additional protection of wildlife after discharge into rivers or lakes. Even more commonly, it is performed when the wastewater is to be reused for irrigation (*e.g.*, food crops, golf courses), for recreational purposes (*e.g.*, lakes, estuaries), or for drinking water.

20.3.3 Sludge Processing

The sludge resulting from the various stages of sewage processing (both the primary sludge and the activated sludge generated by secondary processes) also requires treatment, in this case to stabilize its organic matter and reduce its water content. Treating the organic matter prevents the formation of odors and decreases the number of pathogens, such as enteric parasites and viruses, which are often concentrated in the sludge. Reducing the water content

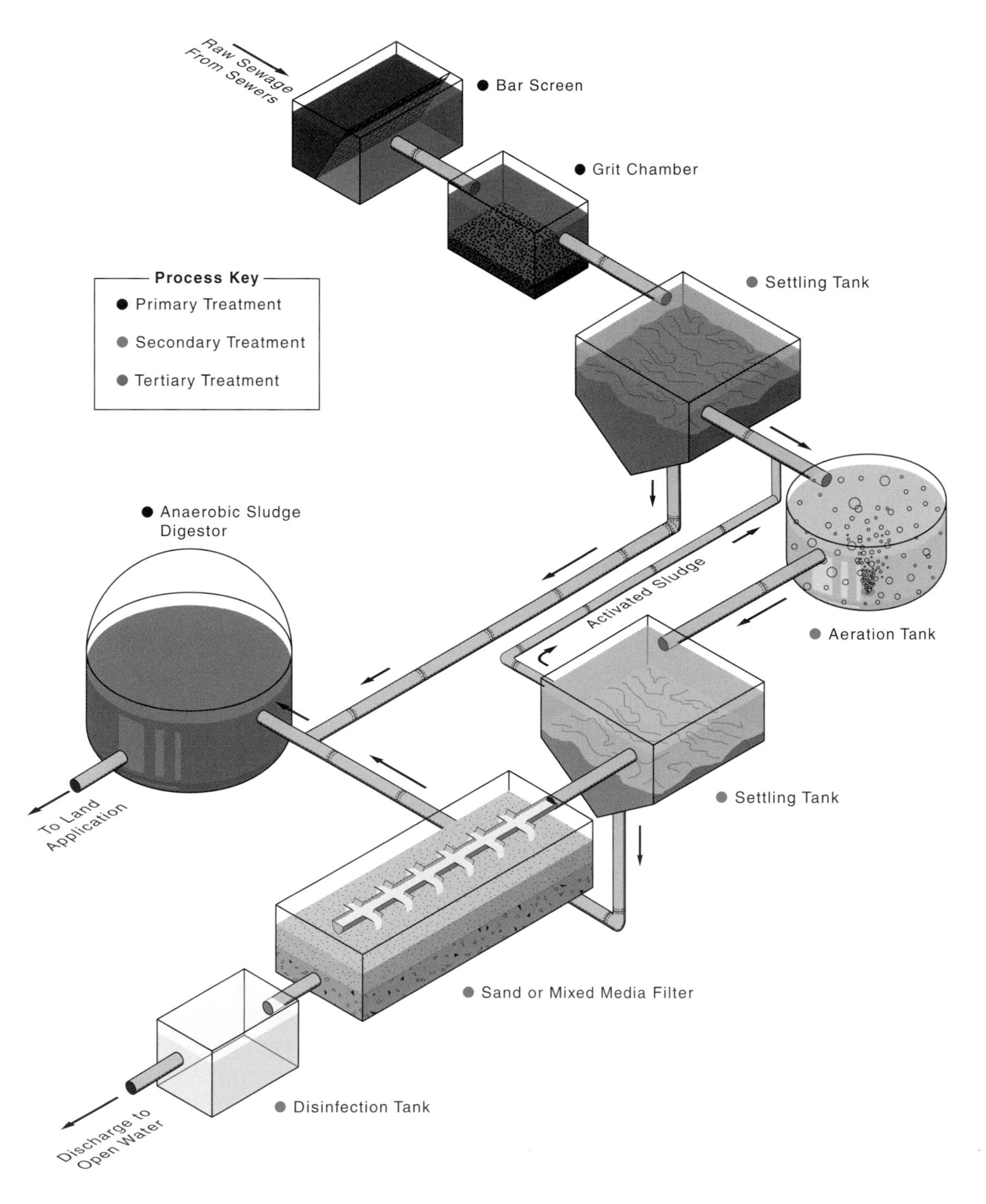

Figure 20-3 Schematic of the treatment processes typical of modern wastewater treatment plants utilizing aeration tank digestion. Note that as wastewater treatment varies with the type of inputs as well as local climactic and regulatory factors, the types of treatment processes used vary widely.

Figure 20-4 A trickling filter bed, such as this unit in Puerto Rico, can be a simple method of secondary treatment. Here, rocks provide a matrix supporting the growth of a microbial biofilm that actively degrades the organic material in the wastewater under aerobic conditions. Photograph: C.P. Gerba.

reduces the weight of the sludge, making it more economical to transport it to its final disposal site.

Sludge treatment typically involves several steps, as shown in Figure 20-5. **Thickening** reduces the volume of the sludge and can be accomplished by allowing the solids to settle in a tank or by centrifugation. **Digestion** is a microbial process that results in the stabilization of the organic matter and some destruction of pathogens because of the high temperatures generated. Sludge can be digested both anaerobically and aerobically. Anaerobic digestion takes place over a period of 2–3 weeks in large covered tanks at the sewage treatment facility. This process has the advantage of producing methane gas, which can be recovered as an energy source. Aerobic digestion also stabilizes sludge. In this process, air or oxygen is passed through the sludge in open tanks. The advantages of aerobic digestion are low capital costs, easy operation, and production of odorless, stabilized sludge. However, it also produces greater

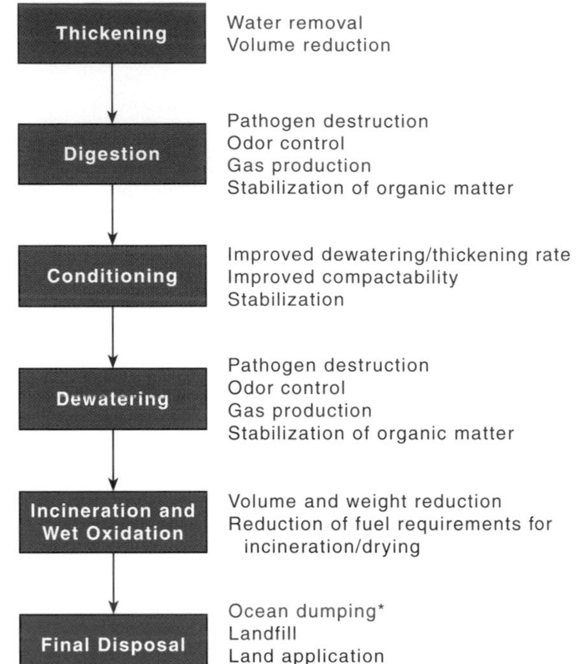

Figure 20-5 Sludge treatment processes and their function. *No longer allowed in the United States. Adapted from U.S. EPA, 1974.

Table 20-4 General characteristics of the three methods used for land application of sewage effluent.

	Application Method		
Factor	Low-Rate Irrigation	Overland Flow	High-Rate Infiltration
Main objectives	Reuse of nutrients and water Wastewater treatment	Wastewater treatment	Wastewater treatment Groundwater recharge
Soil permeability	Moderate (sandy to clay soils)	Slow (clay soils)	Rapid (sandy soils)
Need for vegetation	Required	Required	Optional
Loading rate	1.3–10 cm $week^{-1}$	5–14 cm $week^{-1}$	>50 cm
Application technique	Spray, surface	Usually spray	Usually spray
Land required for flow of 10^6 L day^{-1}	8–66 hectares	5–16 hectares	0.25–7 hectares
Needed depth to groundwater	About 2 meters	Undetermined	About 5 meters
BOD and suspended solid removal	90–99%	90–99%	90–99%
N removal	85–90%	70–90%	0–80%
P removal	80–90%	50–60%	75–90%

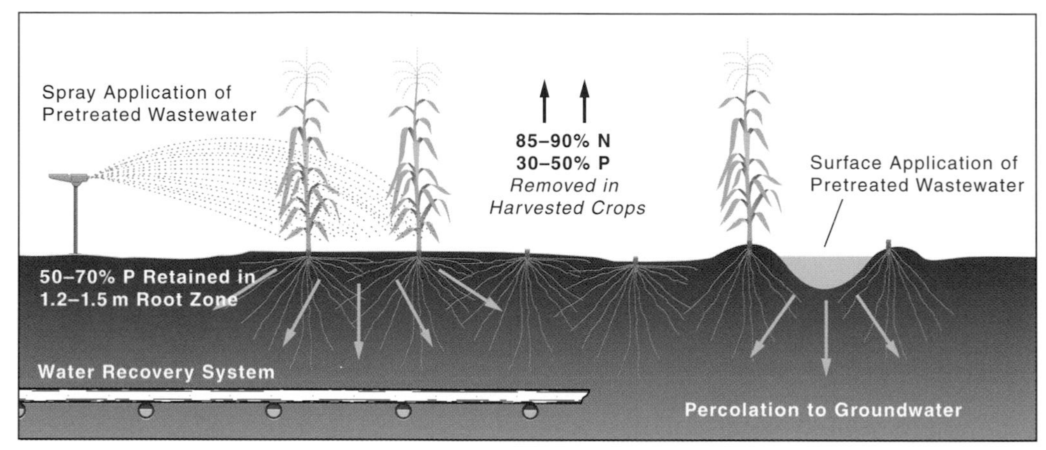

a) Low-Rate Irrigation

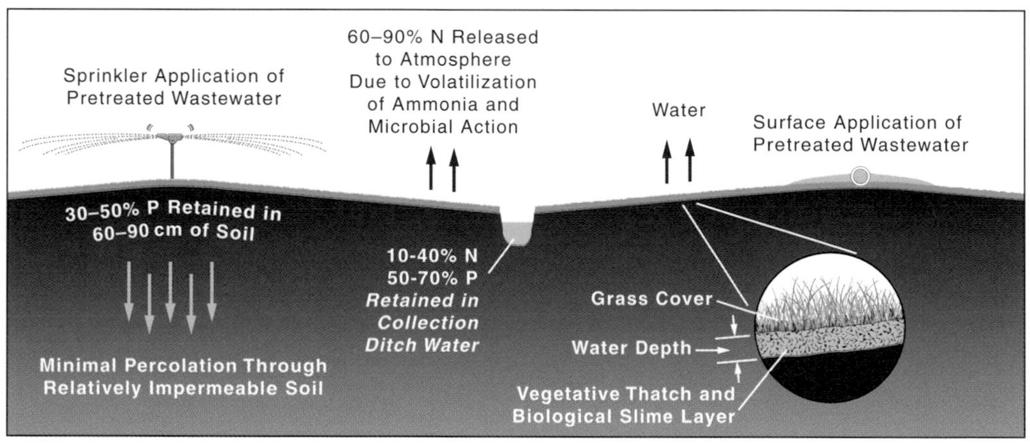

b) Overland Flow

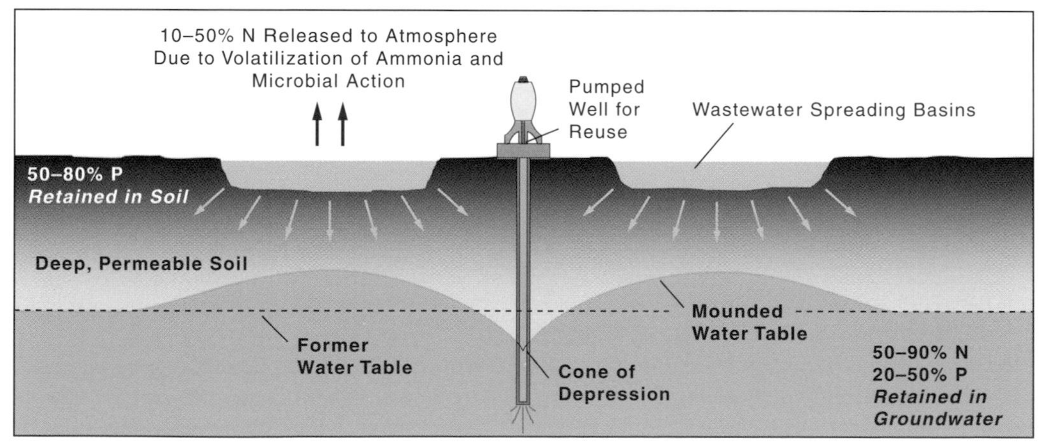

c) High-Rate Infiltration

Figure 20-6 Three basic methods of land application of wastewater.

amounts of waste sludge that has to be disposed of. Sludges may be disposed of after they are digested, but they are usually further treated to reduce the volume of water. This is accomplished by a processes called **conditioning** in which chemicals such as alum, ferric chloride, or lime are added to aggregate suspended particles. This is followed by **dewatering**, which can be accomplished by a number of methods, such as air drying in spreading basins, centrifugation, or vacuum filtration.

Coagulation with alum, lime, or polyelectrolytes (see Section 20.5, beginning on page 315) is effective in removing suspended matter and phosphorus from sludges. It is also very effective in removing viruses. The precipitate that forms is removed and the residual effluent is usually passed through sand or mixed-media filters. Mixed-media filters are composed of granular coal, garnet, and sand which together enhance filtration performance. Disinfection is usually more effective in killing viral pathogens after coagulation and filtration because they remove interfering solids and soluble organic matter.

20.3.4 Land Application of Wastewater

While treated domestic wastewater is usually discharged into bodies of water, it may also be disposed of via land application—sometimes for crop irrigation and sometimes as a means of additional treatment or disposal (see Section 10.4, beginning on page 140). The three basic methods used in the application of sewage effluents to land include low-rate irrigation, overland flow, and high-rate infiltration (see Figure 20-6 on page 309), whose leading characteristics are listed in Table 20-4 on page 308. The choice of a given method depends on the conditions prevailing at the site under consideration (loading rates, method of irrigation, crops, and expected treatment).

With **low-rate irrigation** (Figure 20-6a on page 309), sewage effluents are applied by sprinkling or by surface application at a rate of 1.5 to 10 cm per week. Two-thirds of the water is taken up by crops or lost by evaporation while the remainder percolates through the soil matrix. The system must be designed to maximize denitrification in order to avoid pollution of groundwater by nitrates. Phosphorus is immobilized within the soil matrix by fixation or precipitation. The irrigation method is used primarily by small communities and requires large areas, generally on the order of 5–6 hectares per 1000 people.

In the **overland flow** method (Figure 20-6b), wastewater effluents are allowed to flow for a distance of 50–100 m along a 2–8% vegetated slope and are collected in a ditch. The loading rate of wastewater ranges from 5 to 14 cm a week. Only about 10% of the water percolates through the soil as compared to 60% that runs off into the ditch. The remainder is lost as evapotranspiration. This system requires clay soils with low permeability and infiltration.

The primary objective of **high-rate infiltration** (Figure 20-6c)—also referred to as **soil aquifer treatment** (**SAT**) or **rapid infiltration extraction** (**RIX**)—is the treatment of wastewater at loading rates exceeding 50 cm per week. The treated water, most of which has percolated through coarse-textured soil, is used for groundwater recharge and may be recovered for irrigation. This system requires less land than irrigation or overland-flow methods. Drying periods are often necessary to aerate the soil system and avoid problems due to clogging.

The selection of a site for land application is based on many factors, including soil types, drainability and depth, distance to groundwater, groundwater movement, slope, underground formations, and degree of isolation of the site from the public. As far as virus removal is concerned, the soil type and depth and distance to groundwater are important factors to be taken into account. Several meters of moderately fine-textured, continuous soil layer are necessary for virus removal, particularly if groundwater is directly beneath the soil.

20.3.5 Wetlands and Aquaculture Systems

Wetlands, which are typically less than one meter in depth, are areas that support aquatic vegetation and foster the growth of emergent plants such as cattails, bulrushes, reeds, sedges, and trees. They also provide important wetland habitat for many animal species. Recently, wetland areas have been receiving

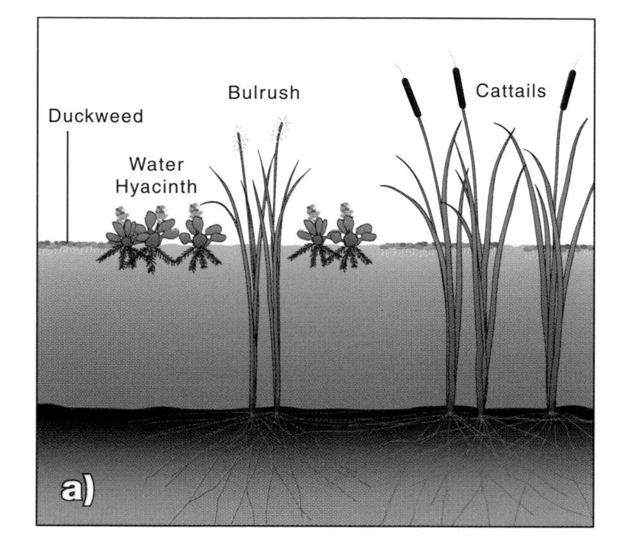

Figure 20-7 (a) Common aquatic plants used in constructed wetlands. (b) An artificial wetlands system in San Diego, California, utilizing water hyacinths. Photograph: C.P. Gerba.

increasing attention as a means of additional treatment for secondary effluents. The vegetation provides surfaces for the attachment of bacteria and aids in the filtration and removal of such wastewater contaminants as biological oxygen and excess carbon. Although both natural and constructed wetlands have been used for wastewater treatment, recent work has focused on constructed wetlands because of regulatory requirements. Two types of constructed wetland systems are in general use: (1) **free water surface** (FWS) systems and (2) **subsurface flow systems** (SFS). An FWS wetland is similar to a natural marsh since the water surface is exposed to the atmosphere. Floating and submerged plants, such as those shown in Figure 20-7a, may be present. SFS systems consist of channels or trenches with relatively impermeable bottoms filled with sand or rock media to support emergent vegetation.

During wetland treatment, the wastewater is usable. It can, for instance, be used to grow aquatic plants such as water hyacinths (Figure 20-7b) and/or to raise fish for human consumption. The growth of such aquatic plants provides not only additional treatment for the water but a food source for fish and other animals as well. Such aquaculture systems, however, tend to require a great deal of land area; moreover, the health risk associated with the production of aquatic animals in this manner for human consumption must be better defined.

20.3.6 An Example of a Modern Sewage Treatment Plant

Roger Road Wastewater Treatment Plant, serving Pima County and the Tucson metropolitan area, exemplifies many of the principles of sewage treatment covered in Section 20.3, beginning on page 302. An aerial view of the facility appears in Figure 20-8 on page 312.

Raw sewage enters the plant at the headworks, which houses the bar screen and grit chamber (compare Figure 20-3 on page 307). Once out of the headworks, the raw sewage flows into the primary settling tanks, which perform an initial separation of the solids from the wastewater. The primary sewage flows on to the biotowers, a means of aerobic treatment which trickles the effluent over a plastic latticework exhibiting a high surface-to-volume ratio and supporting an active biofilm—the microorganisms in the biofilm degrade organic materials in the wastewater. (The biotowers take the place of the more simplistic trickling filters shown in Figure 20-4 on page 308, which are used in some communities.) In the cooler winter months when levels of microbial activity are reduced, additional treatment of the effluent occurs in the adjoining aeration basins under forced aeration. The aerobically treated effluent proceeds to the final or secondary settling tanks, where further suspended solids are removed from the effluent. The secondary sewage is then chlorinated before discharge into a dry stream bed. However, part of the wastewater receives tertiary treatment,

Figure 20-8 Roger Road Wastewater Treatment Plant, Tucson, Arizona. Photograph: C.P. Gerba.

passes through mixed-media filters to reduce turbidity, and is further disinfected before being used to irrigate area golf courses or further treated by rapid filtration extraction (RIX) (Figure 20-6c on page 309) through application into dry, constructed basins.

The sludges collected from the primary and final settling tanks are combined and further processed in the anaerobic digesters. The anaerobically treated sludge is piped to a neighboring sewage treatment plant where the combined output from both plants is centrifuged. The treated sludge (now 8–9% solids) then resides in a holding pond awaiting land application to local agricultural fields by a contracted company (see Figure 10-6 on page 141). Currently, all of Tucson's municipal sludge is land applied.

20.4 Solid Waste

All major cities in the United States generate two types of solid waste that must either be disposed of or treated and reused: municipal solid waste (MSW) and sludge (biosolids). Municipal solid waste originates in households as garbage or trash, in commercial establishments, or at construction/demolition sites. Sludge (biosolids) comes from municipal sewage treatment plants, either as activated sludge or as nonhazardous industrial waste. Disposal and treatment methods for MSW and sludges are discussed in Chapter 10, beginning on page 135.

One additional way to deal with solid waste is **composting**, in which the organic component of solid waste is biologically decomposed under controlled aerobic conditions. Heat produced during this decomposition destroys human pathogens, including many that survive other treatment methods. The result is a product that can be safely handled, easily stored, and readily applied to the land without adversely affecting the environment. In other words, composting is an engineered biological system.

Composting systems are generally divided into three categories: windrow, static pile, and in-vessel. In the **windrow** approach, the MSW mixtures are

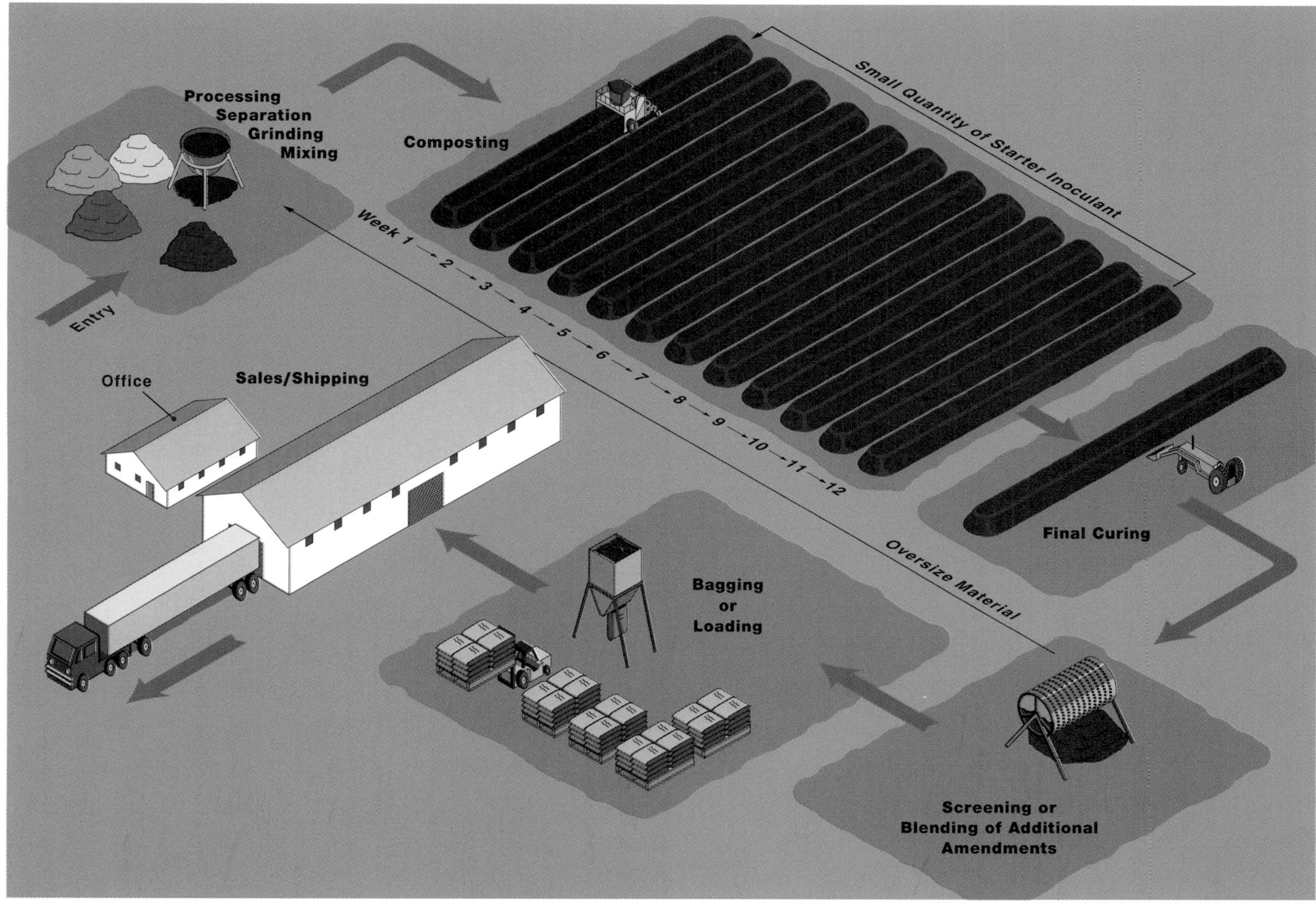

Figure 20-9 An overview of a compost facility using the windrow process. Such a system is appropriate for both municipal and agricultural waste. See Section 16.10, beginning on page 248. Reprinted by permission from *The Guidebook: Nitrogen Best Management Practices for Concentrated Animal Feeding Operations in Arizona*, by Robert J. Freitas and Thomas A. Doerge; copyright © 1996 by the Arizona Board of Regents. All rights reserved.

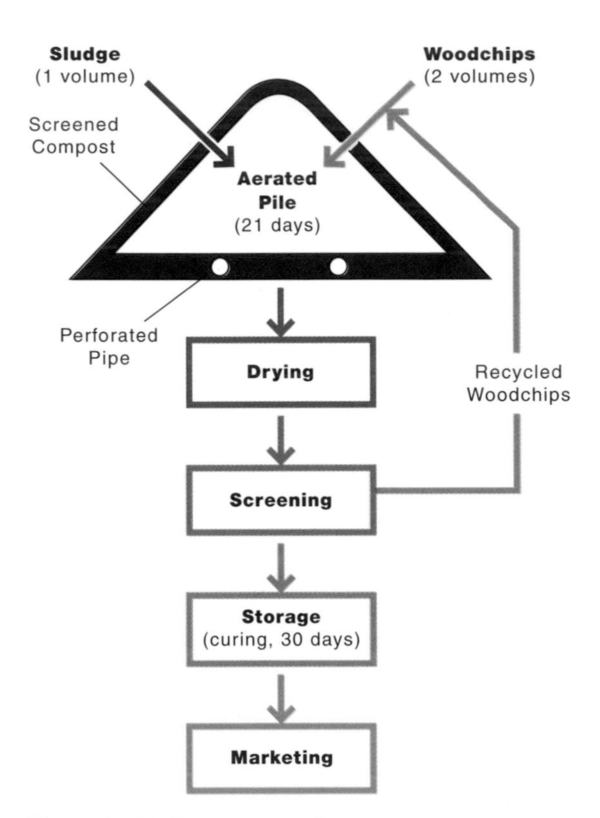

Figure 20-10 Static aerated pile composting method. Adapted by permission from *Introduction to Environmental Virology* by G. Bitton; copyright © 1980 by Wiley-Interscience. All rights reserved.

composted in long rows (called windrows) and aerated by convective air movement and diffusion. In this case, the mixtures are turned periodically by mechanical means to expose the organic matter to ambient oxygen. An overview of a compost facility using the windrow approach is shown in Figure 20-9 on page 313. In the **static pile** (or forced-aeration) approach, piles of MSW mixture are aerated by using a forced-aeration system, which is installed under the piles to maintain a minimum oxygen level throughout the compost mass (see Figure 20-10). **In-vessel** composting (also known as mechanical or enclosed-reactor composting) takes place in partially or completely enclosed containers in which environmental conditions can be controlled. In-vessel systems may incorporate the features of windrow and/or static pile methods of composting.

Table 20-5 lists the advantages of each composting system. At present, an estimated 90% of the operational facilities in the United States use static-pile composting; the remainder use windrow methods. As shown in Table 20-5 these two methods both yield stable products, and they are relatively low in cost. However, the more costly in-vessel systems are currently being evaluated in Europe and the United States for large-scale MSW treatment that can meet stringent regulations with respect to pathogens and metals.

Compost—the end product of the composting process—is a stable, humus-like substance with valuable properties as a soil conditioner. It also contains several macro- and micronutrients favorable to plant growth, but not enough nitrogen to be considered a fertilizer. Although it is usually pathogen-free, the product is not completely stabilized; that is, the organic matter is not 100% degraded. Rather, compost is sufficiently stabilized to reduce the potential for odor generation, thereby allowing the product to be stored and marketed.

For the composting process, large debris and easily recycled materials, such as metals (*e.g.*, aluminum cans, batteries), plastics, and the like, must first be removed. Then the remaining solids are ground and mixed. During the actual composting, the mix is aerated and biological processes decompose the organic matter, thus generating temperatures (above 55°C) high enough to destroy any pathogenic microorganisms that may be present. The oxygen required to fuel the biological processes can be supplied in two ways:

- By mechanically turning the mixture so that the compost is periodically exposed to oxygen in the atmosphere. (In the windrow approach, for example, convective air movement and diffusion move oxygen; thus, turning is necessary to increase porosity, which facilitates air movement, and to distribute anaerobic areas to the aerobic zone.)

- By using a blower to force or draw air through the mix, as in the static-pile approach.

Table 20-5 Advantages of three composting systems. Modified from U.S. EPA, 1985.

Windrow Systems	• Rapid drying of the compost because moisture is released as the piles are turned over. • Drier compost material, which results in easier separation of bulking agent from the compost during screening and relatively high rates of recovery for bulking materials. • High volume of material can be utilized. • Good product stabilization. • Relatively low capital investment: materials required are a pad for the piles, a windrow machine, and generally a front-end loader.
Static Pile Systems	• Low capital costs. The capital equipment required consists of a paved surface, some front-end loaders, a screen, relatively inexpensive blowers, and a water trap. • A high degree of pathogen destruction. The insulation over the pile and uniform aeration throughout the pile help maintain pile temperatures that destroy pathogens.
In-Vessel Systems	• Better odor control than windrow composting. The pile is kept aerobic at all times. Also with blowers in the suction mode, the odors can be treated as a point source. • Good product stabilization. Oxygen and temperature can be maintained at optimal levels. • Space efficiency. • Better process control than outdoor operations. • Protection from adverse climatic conditions. • Good odor control should be possible. • Potential heat recovery depending on system design.

Composting usually takes about 3–4 weeks, after which the material is frequently cured for about 30 days. During this phase, further decomposition, stabilization, and degassing take place. Some systems have an additional drying stage, which can vary from a few days to several months.

20.5 Drinking Water Treatment

Rivers, streams, lakes, and underground aquifers are all potential sources of potable water. In the United States, all water obtained from surface sources must be filtered and disinfected to protect against the threat of microbiological contaminants (see Chapter 19, beginning on page 279). [*Note*: Such treatment of surface waters also improves esthetic values such as taste, color, and odors.] In addition, groundwater under the direct influence of surface waters (*e.g.*, nearby rivers) must be treated as if it were a surface supply. In many cases, however, groundwater needs either no treatment or only disinfection before use as drinking water—this is because soil itself acts as a filter to remove pathogenic microorganisms, decreasing their chances of contaminating drinking supplies.

Modern water treatment processes provide barriers, or lines of defense, between the consumer and waterborne disease. These barriers, when implemented as a succession of treatment processes, are known collectively as a **treatment process train** (Figure 20-11 on page 316). The simplest treatment process train, known as "**chlorination**," consists of a single treatment process, disinfection by chlorination (Figure 20-11a). The treatment process train known as "**filtration**" entails chlorination followed by filtration through sand or coal, which removes particulate matter from the water and reduces turbidity (Figure 20-11b). At the next level of treatment, "**in-line filtration**," a coagulant is added prior to filtration (Figure 20-11c). Coagulation alters the physical and chemical state of dissolved and suspended solids and facilitates their removal by filtration. More conservative water treatment plants add a flocculation (stirring) step before filtration, which enhances the agglomeration of particles and further improves the removal efficiency in a treatment process train called "**direct filtration**" (Figure 20-11d). In direct filtration, disinfection is enhanced by adding chlorine (or an alternative disinfectant, such as chlorine dioxide or ozone) both at the beginning and end of the process train.

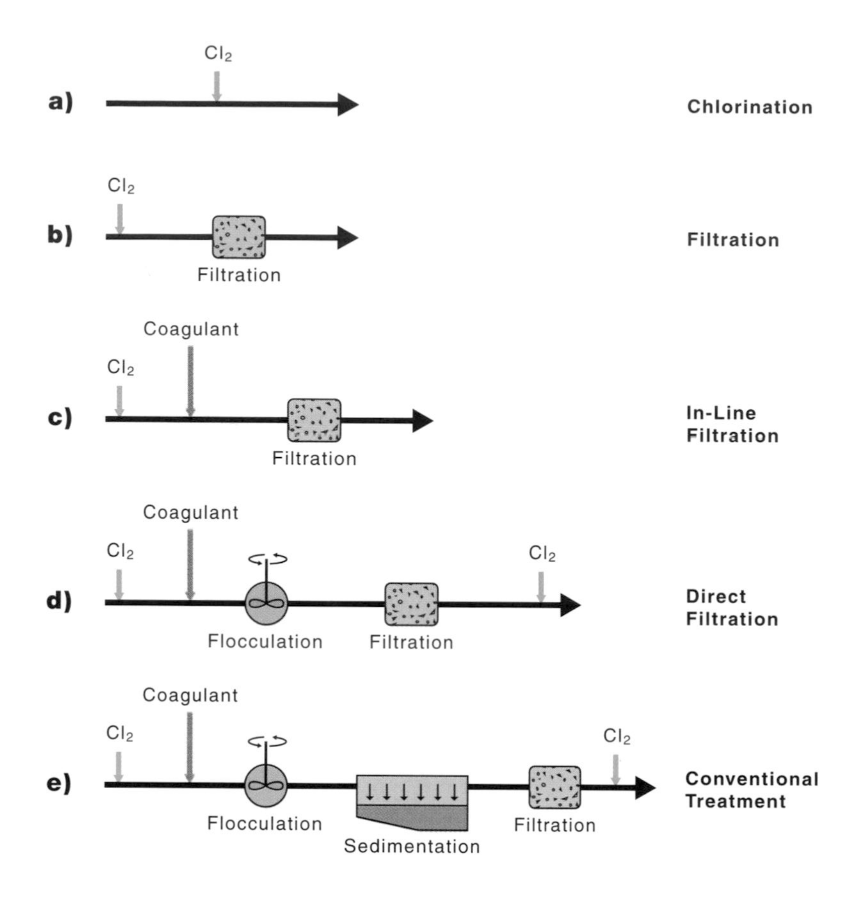

Figure 20-11 Typical water treatment process trains. Adapted from U.S. EPA, 1988.

The most common treatment process train for surface water supplies, known as "conventional treatment" consists of disinfection, coagulation, flocculation, sedimentation, filtration, and disinfection (Figure 20-11e). **Coagulation** involves the addition of chemicals to alter the physical state of dissolved and suspended solids, thereby facilitating their removal by sedimentation and filtration. The most common primary coagulants are hydrolyzing metal salts, most notably alum [$Al_2(SO_4)_3 \cdot 14H_2O$], ferric sulfate [$Fe(SO_4)_3$], and ferric chloride ($FeCl_3$). Additional chemicals that may be added to enhance coagulation include charged organic molecules called polyelectrolytes; these include high-molecular-weight polyacrylamides, dimethyldiallylammonium chloride, polyamines, and starch. These chemicals ensure the aggregation of the suspended solids during the next treatment step—flocculation. Sometimes, polyelectrolytes (usually polyacrylamides) are added after flocculation and sedimentation as an aid to the filtration step.

Coagulation can also remove dissolved organic and inorganic compounds. Hydrolyzing metal salts added to the water may react with the organic matter to form a precipitate, or they may form aluminum hydroxide or ferric hydroxide floc particles on which the organic molecules adsorb. The organic substances are then removed by sedimentation and filtration, or filtration alone if direct filtration or in-line filtration is used. Adsorption and precipitation also remove inorganic substances.

Flocculation is a purely physical process in which the treated water is gently stirred to increase interparticle collisions, thus promoting the formation of large particles. After adequate flocculation, most of the aggregates will settle out during the 1 to 2 hours of sedimentation. Microorganisms are entrapped or adsorbed to the suspended particles and removed during sedimentation.

Sedimentation is another purely physical process, involving the gravitational settling of suspended particles that are denser than water. The resulting effluent is then subjected to **rapid filtration** to separate out solids that are still suspended in the water. Rapid filters typically consist of 50–75 cm of sand and/or anthracite having a diameter between 0.5 and 1.0 mm. Particles are removed as water is filtered through the media at rates of 4–24 L $minute^{-1}$ $(10\ dm)^{-2}$. Rapid filtration is commonly used in the United States. Another method—**slow sand filtration**—is also used. Seen primarily in the United Kingdom and Europe, this method operates at low filtration rates without the use of coagulation.

Taken together, coagulation, flocculation, sedimentation, and filtration effectively remove many contaminants, as shown in Table 20-6. Equally important, they reduce turbidity, yielding water of good clarity and hence enhanced disinfection efficiency. If not removed by such methods, particles may harbor microorganisms and make final disinfection more difficult. Filtration is an especially important barrier in the removal of the protozoan parasites *Giardia lamblia* and *Cryptosporidium*. The cysts and oocysts of these organisms are very resistant to inactivation by disinfectants, so disinfection alone cannot be relied upon to prevent waterborne illness. However, because of their smaller size, viruses and bacteria can pass through the filtration process. Thus, disinfection remains the ultimate barrier to these microorganisms.

Table 20-7 Methods of disinfection. Source: Craun, 1993.

Drinking water systems serving a population >10,000	
• Chlorination	70%
• Chloramination	25%
• Chlorine dioxide	5%
• Ozonation	1%
Drinking water systems serving a population <10,000	
• Chlorination	>95%

Generally, disinfection is accomplished through the addition of an oxidant. Chlorine is, by far, the most common disinfectant used to treat drinking water; but other oxidants, such as chloramines, chlorine dioxide, and even ozone, are also used (see Table 20-7). However, each of these disinfectants can also produce **disinfection by-products**, which may be carcinogenic or otherwise deleterious. A summary of the types of disinfectant by-products is shown in Table 20-8 on page 318. The most widely recognized chlorination by-products include chloroform, bromodichloromethane, dibromochloromethane, and bromoform. These compounds are collectively known as the **trihalomethanes** (THM), and the term **total trihalomethane** (TTHM) refers to their combined concentrations. These compounds are formed by the reaction of chlorine with organic matter—largely humic acids—naturally present in the water. Chloramine is known to produce many of the same types of compounds, as well as cyanogen chloride.

Table 20-6 Coagulation, sedimentation, filtration: typical removal efficiencies and effluent quality. Source: U.S. EPA, 1988.

	Coagulation and Sedimentation (% removal)	Filtration (% removal)	Filtered Water Concentrations
Total coliforms	74–97	50–98	<1 in 100 mL (after disinfection)
Fecal coliforms	76–83	50–98	<1 in 100 mL (after disinfection)
Enteric viruses	88–95	10–98	*NDA*[a]
Giardia lamblia	58–99	97–99.9	*NDA*[a]
Giardia muris	*NDA*[a]	*NDA*[a]	*NDA*[a]
Turbidity	40–96	*NDA*[a]	<1 NTU[b]

[a] *NDA* = No data available.

[b] NTU = nephelometric turbidity unit.

Table 20-8 Classes of disinfectant by-products. Source: Craun, 1993.

Disinfectant	Inorganic By-Products	Organic By-Products	
		Halogenated	Nonhalogenated
Chlorine	Chlorate	Trihalomethanes Haloacetates Haloacetonitriles Haloaldehydes Haloketones Halofuranones Chloropicrin	Aldehydes Carboxylic acids
Chloramine		Cyanogen chloride Others generally thought to be the same as chlorine, but lower in concentration	
Chlorine dioxide	Chlorite Chlorate	Not well characterized	
Ozone	Bromate Hydrogen peroxide	Bromomethanes Bromoacetates Bromoaldehydes Bromoketones Iodinated analogs	Aldehydes Carboxylic acids

The by-products of ozone have not been as well studied, but in waters containing small amounts of bromide, the inorganic by-product bromate is produced.

Many of these by-products are classified as possible carcinogens. Bromate, for example, has been found to be carcinogenic in rats and mice. However, the results of numerous epidemiological studies of populations consuming chlorinated drinking water in the United States show that the risks of cancer appear low (Craun, 1993). Maximum contaminant levels (MCL) have been recommended for some of these by-products and further regulation is expected in the future. Meanwhile, it is fair to say that the risks of illness and death posed by waterborne microorganisms far outweigh the risk from low levels of potentially toxic chemicals produced during water treatment (Craun, 1993).

References and Recommended Reading

Bitton G. (1980) *Introduction to Environmental Virology.* Wiley-Interscience, New York.

Bitton G. (1994) *Wastewater Microbiology.* Wiley-Liss, New York.

Craun G.F. (1993) *Safety of Water Disinfection: Balancing Chemical and Microbial Risks.* ILS1 Press, Washington, D.C.

Freitas R.J. and Doerge T.A. (1996) *The Guidebook: Nitrogen Best Management Practices for Concentrated Animal Feeding Operations in Arizona.* The University of Arizona Press, Tucson, Arizona.

Metcalf and Eddy, Inc. (1991) *Wastewater Engineering. Treatment, Disposal, and Reuse* (Revised by G. Tchobanaglous and F.L. Burton), 3rd Edition. McGraw-Hill, New York.

Montgomery J.M. (1985) *Water Treatment Principles and Design.* John Wiley and Sons, New York.

Reed S.C., Crites R.W., and Middlebrooks E.J. (1995) *Natural Systems for Waste Management and Treatment*, 2nd Edition. McGraw-Hill, New York.

U.S. EPA (1974) *Process Design Manual for Sludge Treatment and Disposal.* EPA 625/1-74-006. United States Environmental Protection Agency Technology Transfer, Washington, D.C.

U.S. EPA (1985) *Composting of Municipal Wastewater Sludges*. EPA 625/4-85-014. United States Environmental Protection Agency, Washington, D.C.

U.S. EPA (1988) *Comparative Health Effects Assessment of Drinking Water Treatment Technologies.* Office of Drinking Water, United States Environmental Protection Agency, Washington, D.C.

Problems and Questions

1. What are the three major steps in modern wastewater treatment?
2. Why is it important to reduce the amount of biodegradable organic matter and nutrients during sewage treatment?
3. When would tertiary treatment of wastewater be necessary?
4. What are some types of tertiary treatment?
5. What are some methods of sludge (biosolid) reuse?
6. What is the major component of biosolids?
7. What are two major contaminant classes in biosolids?
8. Describe the major steps in the conventional treatment of drinking water.
9. What are disinfection by-products? What kinds of risks do they pose and how can these risks be evaluated?

Part 4

Risk Assessment and Risk Management

On Reverse *An unlined lagoon holds metal-containing sulfuric acid waste from an electroplating industry. This lagoon was neutralized with lime, drained, lined with a clay liner, and reclaimed* in situ *using organic amendments. Environmental regulations and regulatory agencies are designed not only to prevent such environmental and health hazards as pictured here from happening in the first place, but also serve to allocate scarce financial resources to the most hazardous sites and to determine the degree to which remediation must occur.*

Photograph *J.F. Artiola*

Principles of toxicology are valuable for determining the toxicity of trace levels of pollutants, such as those which may be present in treated sewage effluent being used to recharge groundwater. Photograph: I.L. Pepper.

21.1 History of Modern Toxicology in the United States

In **toxicology** we study both the adverse effects of chemicals on health and the conditions under which those effects occur. A natural outgrowth of biology and chemistry, this study began to assume a well-defined shape in just the past four to five decades. Newer still, environmental toxicology is concerned with the effects of chemical contaminants on various ecological systems, both large and small. Here, we focus on some of the basic principles of toxicology as related to environmental contaminants.

The first federal law for regulating potentially toxic substances was the Pure Food and Drug Act, passed by Congress in 1906. Much of the impetus for this law came from the work of Harvey Wiley, who was the chief chemist of the Department of Agriculture under Theodore Roosevelt. Wiley and his "Poison Squad" had a very personal interest in their work—he and his team of chemists not infrequently dosed themselves with suspect chemicals to test for their deleterious effects (Rodricks, 1992).

The systematic study of toxic effects in laboratory animals (other than chemists) began in the 1920s, spurred by concerns about the unwanted side effects of food additives, drugs, and pesticides (DDT and related pesticides became available in this era). The 1930s saw issues raised about occupational cancers and other chronic diseases resulting from chemical exposures. These concerns and issues prompted increased legislative activity culminating in the modern version of the Food, Drug, and Cosmetic Act. This law, enacted by Congress in 1938, was passed in response to a tragic episode in which more than 100 people died from acute kidney failure after ingesting contaminated sulfanilamide—the antibiotic had been improperly prepared in a diethylene glycol solution (Rodricks, 1992).

But the real growth of toxicology largely paralleled that of the chemical industry, especially after World War II. The thousands of new compounds produced by chemical manufacturers created a need for information about their possible harmful effects. And this growth received a significant stimulus from public opinion. Sporadically during the 1940s and 1950s, the public was presented with a series of seemingly unconnected announcements about poisonous pesticides in their foods, food additives of dubious safety, chemical disasters in the workplace, and air pollution episodes that claimed thousands of victims in urban centers throughout the world. Then, in 1962, marine biologist Rachel Carson (1907–1964) drew together these various environmental horror stories in her book, *Silent Spring*. Menaced by the presence of synthetic chemical killers in the environment, the public responded with predictable outrage, which, among other things, fostered renewed interest in the science of toxicology. It also helped pave the way for the introduction of several major federal environmental laws in the late 1960s and early 1970s, and for the creation of the Environmental Protection Agency (EPA) in 1970.

The relative newness of the science of toxicology is reflected in the fact that, even today, we have little solid information about the toxicity of a large number of chemicals. Of 6,000,000 known chemicals, about 50,000 are in common use, and detailed chronic toxicity tests have been performed on only a few hundred of these. Even for those that have been tested, many questions remain about the interpretation of the results obtained, including serious reservations about the applicability of laboratory test results to human populations in everyday situations. In many cases we lack a basic understanding of how toxicants act.

21.2 Toxic versus Nontoxic

The term *safe* commonly means "without risk." But this common definition has no meaning in scientific study. Scientists cannot ascertain conditions under which a given chemical exposure is absolutely without risk of any type. On the other hand, they can describe conditions under which risks are so low that they have no practical consequence to a specific population. In technical terms, the safety of chemical substances—whether in food, drinking water, air, or the workplace—has typically been defined as a condition of exposure under which there is a "practical certainty" that no harm will result to exposed individuals. In terms of mortality, this is usually accepted as a risk of 1:1,000,000 chance of dying during a lifetime (see Section 22.1, beginning on page 345).

Another fundamental concept is the classification of chemical substances as either *safe* or *unsafe* (or as *toxic* and *nontoxic*). This type of classification can be highly problematic. All substances, even those that we consume in high amounts every day, can be made to produce a toxic response under some conditions of exposure. In this sense, all substances can be "toxic." Thus, safety involves not simply the degree of toxicity of a substance, but rather the degree of risk under given conditions. In other words, we ask "What is the probability that the toxic properties of a chemical will be expressed under actual or anticipated conditions of human or animal exposure?" The science of risk assessment attempts to link toxicological information on adverse effects to the probability of toxic effects during likely exposure scenarios (see Section 22.1, beginning on page 345).

21.3 Exposure and Dose

Humans and other organisms can be exposed to substances in different environmental media—air, water, soil, or food—or they may have direct contact with a sample of the substance. The exposure concentration is the amount of a substance present in the medium with which an organism has contact. The dose is the amount of the chemical that is received by the target (organ). The exposure concentration may differ from the dose owing to biochemical transformations in living organisms.

Suppose, for example, a substance is present in drinking water. The amount of this substance in the water is the exposure concentration. For many environmental substances, this amount ranges from less than 1 μg to greater than 1 mg, and is usually reported as milligrams or micrograms of the substance present in 1 liter of water (*i.e.*, in mg L^{-1} or μg L^{-1}).[1]

An individual's intake—or *dose*—of this substance depends on the amount present in a given vol-

[1] These two units are sometimes expressed as the more ambiguous units of parts per million (ppm) or parts per billion (ppb), respectively. See also Footnote 3 on page 340.

ume of water and on the amount consumed in a given period of time. Given the concentration of the substance in water (say in mg L^{-1}) and the human consumption of water per unit of time, it is possible to estimate the total amount of the substance an individual will consume through use of contaminated water. For instance, adults are assumed to consume 2 L of water each day through all uses (see Table 22-3 on page 350). Thus, if a substance is present at 10 mg L^{-1} in water, the average daily individual intake of the substance is

$$\frac{10\,\text{mg}}{\text{L}} \cdot \frac{2\,\text{L}}{\text{day}} = \frac{20\,\text{mg}}{\text{day}}$$

Toxicity measures must also take body size differences into account, usually by dividing daily intake by the weight of the individual. That is, the toxicity of a substance is usually dependent upon concentration per unit of body weight. Thus, for a man of average weight (usually assumed to be 70 kg), the daily dose of this substance is

$$\frac{20\,\text{mg}}{\text{day}} \cdot \frac{1}{70\,\text{kg}} = \frac{0.28\,\text{mg}}{\text{kg day}}$$

For a person of lower weight, such as a female or child, the **daily dose** at the same intake rate would be larger. For example, a 50-kg woman ingesting this substance would receive a dose of

$$\frac{20\,\text{mg}}{\text{day}} \cdot \frac{1}{50\,\text{kg}} = \frac{0.40\,\text{mg}}{\text{kg day}}$$

Using the same equation, a child of 10 kg would receive a dose of 2.0 mg kg^{-1} day^{-1}. However, children drink less water each day than do adults (say, 1 L), so a child's dose would be

$$\frac{10\,\text{mg}}{\text{L}} \cdot \frac{1\,\text{L}}{\text{day}} \cdot \frac{1}{10\,\text{kg}} = \frac{1.0\,\text{mg}}{\text{kg day}}$$

In general, the smaller the body size, the greater the dose (in mg kg^{-1} day^{-1}) received from drinking water. This is also true of experimental animals. Usually rats or mice will receive a much higher dose of drinking water contaminants than humans because of their much smaller body size.

Because each medium (air, soil, water) of exposure must be treated separately, some calculations are more complex than those of dose per liter of water. Many calculations may simply be additive. For example, a human may be simultaneously exposed to the same substance through several media (*e.g.*, through inhalation, ingestion, dermal contact). Thus, if an individual can both ingest and inhale (say in the shower) some volatile compound in tapwater, the **total dose** received by that individual is the sum of doses received through each individual route. In some cases, however, it is inappropriate to add doses in this fashion because the toxic effects of a substance may depend on the route of exposure. For example, inhaled chromium is carcinogenic to the lung, but it appears that ingested chromium is not. In general, though, as long as a substance acts at an internal body site (*i.e.*, acts systematically rather than at the particular point of initial contact), it is usually an acceptable procedure to add doses received from the various routes.

Absorption, or **absorbed dose**, is another factor that requires special attention when considering dose and exposure. When a substance is ingested in food or drinking water, it enters the gastrointestinal tract. When it is present in air (as a gas, aerosol, particle, dust, fume, *etc.*), it enters the upper airways and lungs. A substance may also come into contact with the skin and other body surfaces as a gas, liquid, or solid. Some substances may cause toxic injury at the point of initial contact (skin, gastrointestinal tract, upper airways, lungs, eyes). Indeed, at high concentrations, most substances do cause at least irritation at these points of contact. However, for many substances, toxicity occurs after they have been absorbed—that is, after they pass through certain barriers (*e.g.*, the wall of the gastrointestinal tract or the skin itself), enter blood or lymph, and gain access to the various organs or systems of the body. Some chemicals may be distributed in the body in various ways and then excreted. However, some chemical types—usually lipid-soluble substances such as the pesticide dichlorodiphenyltrichloroethane (DDT)—can be stored for long periods of time, usually in body fat (see Section 21.8.3, beginning on page 340).

Substances vary widely in extent of absorption. The fraction of a dose that passes through the wall of the gastrointestinal tract may be very small (1 to 10% for some metals) or it may be substantial (close to 100% for certain types of organic molecules). Absorption rates also depend on the medium in which a chemical is present: a substance present in water might be absorbed differently than the same substance present in, say, a fatty diet. Absorption rates also vary among animal species and among individuals within a species. Ideally, an estimation of a **systemic dose** should consider absorption rates. Unfortunately, data on absorption are limited for most substances, especially in humans, so absorption is not always included in dose estimation. In some cases, dose estimates may be crudely adjusted on the basis of the molecular characteristics of a particular substance and/or general principles of absorption. In many cases, however, absorption is simply considered to be complete by default.

The technique of **extrapolation**, or drawing inferences, from experimentally observed results can also be a major factor in predicting the likelihood of toxicity, say from one route of exposure to other routes, or from one organism to another. Experiments for studying toxicity usually involve intentional administration of substances to subjects (usually mice or rats) through ingested food or inhaled air or through direct application to skin; or they may include other routes of administration, such as injection under the skin (subcutaneous), into the blood (usually intravenous), or into body cavities (intraperitoneal). Such toxicity studies in experimental animals are of greatest value when experimental exposures mimic the mode of human exposure. Thus, if both animals and humans are exposed to a contaminant via drinking water, it is generally assumed that the data in animals can be applied directly to man. But when experimental routes differ from human routes (*e.g.*, **animal dose** via injection; human exposure via drinking water), a correction or safety factor must be used to apply such data to human exposures (see Section 22.2.3, beginning on page 351).

21.4 Evaluation of Toxicity

Information on the toxic properties of chemical substances is obtained through plant, bacterial, and animal studies, controlled epidemiological investigations of exposed human populations or microcosms, and clinical studies or case reports of exposed humans or ecosystem studies (*e.g.*, oil spills). Other information bearing on toxicity derives from experimental studies in systems other than whole animals (*e.g.*, isolated organics in cells or subcellular components) and from analysis of the molecular structures of the substances of interest. These last two sources of information are generally less certain as indicators of toxic potential.

Many types of toxicity studies can be conducted to identify the nature of health damage produced by a substance and the range of doses over which such damage is produced. Each of the many different types of toxicological studies has a different purpose. The usual starting point for such investigations is a study of the **acute (single-dose) toxicity** of a chemical in experimental animals, plants, or bacteria. Acute toxicity studies are used to calculate doses that will not be lethal to any organism and can be used in toxicity studies of longer duration. Moreover, such studies provide an estimate of the compound's comparative toxicity and may indicate the target organ system (*e.g.*, kidney, lung, or heart) affected in an animal. Once the acute toxicity is known, organisms may be exposed repeatedly or continuously for several weeks or months in **subchronic toxicity** studies, or for close to their full lifetimes in **chronic toxicity** studies.

When toxicologists examine the lethal properties of a substance, they estimate its $\mathbf{LD_{50}}$, which is the lethal dose for 50% of an exposed population. A group of well-known substances and their LD_{50} values are listed in Table 21-1. LD_{50} studies reveal one of the basic principles of toxicology:

- *Not all individuals exposed to the same dose of a substance will respond in the same way.*

Thus, the same dose of a substance that leads to the death of some experimental individuals will impair some organisms while not affecting other organisms at all.

The premise underlying animal toxicity studies is the long-standing assumption that effects in humans can be inferred from effects observed in animals. This principle of extrapolating animal data to humans has been widely accepted in scientific and regulatory communities. For instance, all of the chemicals that have been demonstrated to be carcinogenic in humans are carcinogenic in some, although not all, animal species typically used for toxicological studies. In addition, the acutely toxic doses of many chemicals are assumed to be similar in humans and a variety of experimental animals. The foundation of this inference is the evolutionary relationships between animal species. That is, at least among mammals, the basic anatomical, physiological, and biochemical parameters are expected to be much the same across species.

On the whole, the general principle of making such **interspecies inferences** is well founded. But some exceptions have been noted; for example, guinea pigs are much more sensitive to dioxin (2,3,7, 8-tetrachlorodibenzo-*p*-dioxin) than are other laboratory animals. Many of these exceptions arise from differences in the ways various species handle exposure to a chemical and to differences in **pharmacokinetics**, which includes the rates at which a specific chemical is distributed among tissues, the manner in which it is excreted, and the types of metabolic changes it undergoes. Because of these potential differences, it is essential to evaluate all interspecies differences carefully when inferring human toxicity from animal toxicologic studies.

In the particular case of long-term animal studies conducted to assess the carcinogenic potential of a compound, certain general observations increase the overall strength of the evidence that the compound is carcinogenic. Thus, for example, an increase in the number of tissue sites affected by the agent is a strong indicator of carcinogenicity, as is an increase in the number of animal species, strains, and sexes showing a carcinogenic response. Other observations that affect the strength of the evidence may involve a high level of statistical significance of the increase of tumor incidence in treated versus control animals, as well as clear-cut dose-response relationships in the data evaluated, such as dose-related shortening of the time-to-tumor occurrence or time-to-death with tumor and a dose-related increase in the proportion of tumors that are malignant.

21.4.1 Manifestations of Toxicity

Toxic effects can take various forms. A toxic effect can be immediate, as in strychnine poisoning, or delayed, as in lung cancer. Indeed, cancer typically affects an individual many years after continuous or intermittent exposure to a carcinogen. An effect can be local (*i.e.*, at the site of application) or systemic (*i.e.*, carried by the blood or lymph to different parts of the body). There are several important factors to consider when examining toxic effects, some of which are dosage-related.

Table 21-1 Approximate oral LD_{50} in a species of rat for a group of well-known chemicals. Source: U.S. EPA, 1989.

Chemical	LD_{50} (mg kg^{-1})
Sucrose (table sugar)	29,700
Ethyl alcohol	14,000
Sodium chloride (common salt)	3,000
Vitamin A	2,000
Vanillin	1,580
Aspirin	1,000
Chloroform	800
Copper sulfate	300
Caffeine	192
Phenobarbital, sodium salt	162
DDT	113
Sodium nitrite	85
Nicotine	53
Aflatoxin B1	7
Sodium cyanide	6.4
Strychnine	2.5

1. The *severity* of injury can increase as the dose increases and vice versa. Some organic chemicals, for example, are known to affect the liver. High doses of such a chemical (*e.g.*, carbon tetrachloride) will kill liver cells—perhaps killing so many cells that the whole liver is destroyed so that most or all of the experimental animals die. As the dose is lowered, fewer cells are killed, but the liver exhibits other forms of injury that indicate an impairment in cell function and/or structure. At still lower doses, no cell death may occur, and only slight changes are observed in cell function or structure. Finally, a dose may be so low that no effect is observed, or the biochemical alterations that are present have no known adverse effects on the health of the animal. One of the goals of toxicity studies is to determine this dose level, known as the **no-observed-adverse-effect level (NOAEL)** (see Section 22.2.3, beginning on page 351).

2. The *incidence* of an effect, but not its severity, may increase with increasing dosage. In such cases, as the dose increases, the fraction of experimental organisms experiencing diverse effects (*i.e.*, disease or injury) increases. At sufficiently high doses, all experimental subjects will experience the effect. Thus, increasing the dose increases the probability (*i.e.*, the risk) that an abnormality will develop in an exposed population.

3. Both the severity and the incidence of a toxic effect may increase as the level of exposure increases. The increase in severity is a result of increased damage at higher doses, while the increase in incidence is a result of differences in individual sensitivity. In addition, the site at which a substance acts (*e.g.*, liver, kidney) may change as the dosage changes. Many toxic effects, including cancer, fall in this category. Generally, as the duration of exposure increases, the critical NOAEL dose decreases; in some cases, new effects not seen with exposures of short duration appear after long-term exposure.

The *seriousness* of a toxic effect must also be considered. Certain types of toxic damage, such as asbestosis caused by inhalation of asbestos fibers, are clearly adverse and are a definite threat to health. However, the health significance of other types of effects observed during toxicity studies may be ambiguous. For example, at a given dose a chemical may produce a slight increase in body temperature. If no other effects are observed at this dose, researchers cannot be sure that a true adverse response has occurred. Determining whether such slight changes are significant to health is one of the critical issues in assessing safety.

Toxic effects also vary in degree of reversibility. In some cases, an adverse health effect will disappear almost immediately following cessation of exposure. At the other extreme, some exposures will result in a permanent injury—for example, a severe birth defect arising from exposure to a substance that irreversibly damaged the fetus at a critical moment of its development. Furthermore, some tissues, such as the liver, can repair themselves relatively quickly, while others such as nerve cells have no ability to repair themselves. Most toxic responses fall somewhere between these extremes.

21.4.2 Toxicity Testing

Any organism can be used to assess the toxicity of a substance. The choice of test organism depends on several factors, including budget, time, and the organism's occurrence in a given environment. The simplest and least costly tests are performed with unicellular animals or plants and may last only a few hours or days. In a water environment, for example, *Daphnia* or algae may be used in testing for potential aquatic pollutants: short-term tests may look at the death or immobilization of swimming *Daphnia*, while longer term tests may look at the growth of the organisms (increase in biomass) or numbers of offspring. Animals higher in the food chain are also important in aquatic toxicity tests, and experiments involving fish, amphibians, and other macroinvertebrates are familiar standbys (Table 21-2). For terrestrial toxicity tests, higher plant, rodent, or bird toxicity tests can be used. Strains of genetically characterized rats and mice, for example, are often used in such studies. Avian toxicity tests have also been developed; for instance, birds are frequently used in evaluating the effects of pesticides on nontarget species.

In environmental toxicology, researchers often perform toxicity tests in artificially contained communities to assess the environmental impacts of toxic substances after release into the environment. These artificial communities, which serve as laboratory models of natural ecosystems, are referred to as **microcosms**. While many microcosms are elaborate systems that effectively mimic whole ecosystems, some microcosms may be nothing more than a set of glass jars containing soil or water with sediment at the bottom. But even simple glass jars allow researchers to examine the effect of substances on multispecies, such as algae, bacteria, and microinvertebrates.

Toxicity experiments vary widely in design and protocols. Some tests and research-oriented investigations are conducted using prescribed study designs, as is the case with carcinogenicity assays in fish. In connection with premarket testing requirements for certain classes of chemicals, however, regulatory and public agencies have developed relatively few standardized tests for various types of toxicity.

Rats and mice are the most commonly used laboratory animals for toxicity testing. These rodents are inexpensive, and they can be handled relatively easily; moreover, the genetic background and disease susceptibility of these species are well established. In addition, the full life span of these small rodents is complete in two to three years; thus, the effects of lifetime exposure to a substance can be measured relatively quickly. Other rodents such as hamsters and guinea pigs are also common laboratory subjects, as are rabbits, dogs, and primates. Usually, the choice of experimental animal depends on the system being studied. Reproductive studies, for example, often use primates such as monkeys or baboons because their reproductive systems are similar to that of humans. Similarly, rabbits are often used for testing dermal toxicity because their shaved skin is more sensitive than that of other animals.

Animals are usually exposed by a route that is as close as possible to the route by which humans will be exposed. In some cases, however, it may be necessary to use other routes or conditions of dosing to achieve the desired experimental dose. For example, some substances are administered by stomach tube (gavage) because they are too volatile or unpalatable to be placed in the animals' feed at the high levels needed for toxicity studies.

A toxicity experiment is of limited value unless researchers find a dose of sufficient magnitude to cause some type of adverse effect within the duration of the experiment. If no effects are seen at any dose administered, the toxic properties of the substance cannot be characterized; thus experiments may be repeated at higher doses or for longer times until distinct adverse effects are observed. The most distinctive adverse effect is, of course, death. Therefore, researchers frequently begin their experiments by determining the LD_{50} since the endpoint of this experiment (death) is easily measured. Next, researchers usually look at the effects of lower doses administered over longer periods to find the range of doses over which adverse effects occur and to identify the NOAEL for these effects.

Table 21-2 Organisms commonly used in toxicity testing.

Type of Organism	Organism
Invertebrates	*Daphnia magna*
	Crayfish
	Mayflies
	Midges
	Plandria
Aquatic Vertebrates	Rainbow Trout
	Goldfish
	Fathead minnow
	Catfish
Algae	*Chlarnydomonas reinhard* (green algae)
	Microcystis aueriginosa (blue-green algae)
Mammals	Rats
	Mice
Avian Species	Bobwhite
	Ring-necked pheasant

Studies may be characterized according to the **duration of exposure**. Acute toxicity studies involve a single dose, or exposures of very short duration (*e.g.*, 8 hours of inhalation). Chronic studies involve exposures for nearly the full lifetime of the experimental animals, while subchronic studies vary in duration between these two extremes. Although many different dose levels are needed to develop a well-characterized dose-response relationship, practical considerations usually limit the number to two or three, especially in chronic studies. Experiments involving a single dose are frequently reported, but these leave great uncertainty about the full range of doses over which effects are expected.

21.4.3 Toxicity Tests for Carcinogenicity

One of the most complex and important of the specialized tests is the carcinogenesis bioassay. This type of experiment is used to test the hypothesis of carcinogenicity—that is, the capacity of a substance to produce malignant tumors.

Usually, a test substance is administered over most of the adult life of a laboratory animal; then the animal is observed for formation of tumors. In this kind of testing, researchers generally administer high doses of the chemical to be tested—specifically, the **maximum tolerated dose** (MTD), which is the maximum dose that an animal can tolerate for a major portion of its lifetime without significant impairment of growth or observable toxic effect other than carcinogenicity. The MTD and one-half of that, or MTD_{50}, are the usual doses used in a National Cancer Institute (NCI) carcinogenicity bioassay so that the animals survive in relatively good health over their normal lifetime. The main reason for using the MTD as the highest dose in a bioassay is that these very high doses help to overcome the statistical insensitivity inherent in small-scale experimental studies.

Owing largely to cost considerations, experiments are carried out with relatively small groups of animals—typically, 50 or 60 animals of each species and sex at each dose level, including the control group. At the end of such an experiment, the incidence of cancer (including tumor incidence in control animals) is tabulated and plotted as a function of dose. Then the data are analyzed to determine whether any observed differences in tumor incidence (fraction of animals having a tumor of a certain type) are due to exposure to the substance under study or to random variations. In an experiment of this size, assuming none of the control animals develop tumors, the lowest incidence of cancer that is detectable with statistical reliability is in the range of 5%, which is equivalent to 3 out of 60 animals developing tumors. If control animals develop tumors (as they frequently do), the lowest range of statistical sensitivity is even higher. A cancer incidence of 5% is very high; but ordinary experimental studies are not capable of detecting lower rates, and most are even less sensitive.

21.4.4 Epidemiological Studies

Information on adverse health effects in human populations is obtained from four major sources: (1) summaries of self-reported symptoms in exposed persons; (2) case reports prepared by medical personnel; (3) correlation studies, in which differences in disease rates in human populations are associated with differences in environmental conditions; and (4) epidemiological studies. The first three of these sources are characterized as descriptive epidemiology, while the fourth category—**epidemiological studies**—is generally reserved for studies that compare the health status of a group of persons who have been exposed to a suspected agent with that of a nonexposed control group. [*Note*: Such studies cannot identify cause-and-effect relationships between exposure to a substance and particular diseases or conditions; however, they can draw attention to previously unsuspected problems and generate hypotheses that can be tested further.]

Most epidemiological studies are either case-control studies or cohort studies. **Case-control studies** first identify a group of individuals who have a specific disease, then attempt to ascertain commonalities in exposures that the group may have experienced. For example, the carcinogenic properties of diethylstilbestrol (DES), a drug once used to prevent miscarriages, were brought to light through studies of women afflicted with certain types of vaginal and cervical cancer. **Cohort studies**, on the other hand, begin by examining the health status of individuals known to have had a common exposure. These studies then attempt to determine whether any specific condition is associated with that exposure by comparing the exposed group's health with that of an appropriately matched control population. For example, a cohort study of lab workers exposed to benzene revealed an excessively high incidence of leukemia, thereby providing strong evidence in support of a benzene leukemogenesis hypothesis. Generally, epidemiologists have used individuals who belong to an identifiable group, such as those in certain occupational settings or patients treated with certain drugs, to conduct such studies—hence the name "cohort."

Convincing results from epidemiological investigations can be enormously beneficial because the data provide information about humans under actual conditions of exposure to a specific agent. Therefore, results from well-designed, properly controlled studies are usually given more weight than results from animal studies. Although no study can provide complete assurance that a chemical is harmless, negative data from epidemiological studies of sufficient size can assist in establishing the maximum level of risk due to exposure to the agent.

Obtaining and interpreting epidemiological results, however, can be quite difficult. Appropriately matched control groups are difficult to identify, because the factors that lead to the exposure of the study group (*e.g.*, occupation or residence) are often inextricably linked to the factors that affect health status (*e.g.*, lifestyle and socioeconomic status). Thus, controlling for related risk factors (*i.e.*, cigarette smoking) that have strong effects on health is difficult. Moreover, the statistical detection power of epidemiological studies depends on the use of very large populations (see Section 22.2.1, beginning on page 348), and data may be hard to come by or incomplete. Few types of health effects—other than death—are recorded systematically in human populations, and even the information on cause of death is limited in reliability. For example, infertility, miscarriages, and mental illness are not, as a rule, systematically recorded by public health agencies, while death is often attributed to "heart failure," whatever its proximate cause.

In addition, accurate data on the degree of exposure to potentially hazardous substances are only rarely available, especially when exposures have taken place years or decades earlier. Establishing dose-response relationships is therefore frequently impossible. Nor can current data, however carefully obtained, immediately help researchers who are investigating slowly developing diseases such as cancer; rather, epidemiologists must wait many years to determine the absence of an effect. Meanwhile, exposure to suspect agents could continue during these extended periods of time, thereby increasing risk further.

For these and other reasons, interpretations of epidemiological studies are sometimes subject to extreme uncertainties. Independent confirmatory evidence is usually necessary, including supporting results from a second epidemiological study or supporting data from experimental studies in animals. Such confirmatory evidence is particularly necessary in the case of negative findings, which must be interpreted with great caution (EPA, 1989).

For example, suppose we have a drinking water contaminant that is known to cause cancer in 1 out of every 100 people exposed to 10 mg L^{-1}. Further suppose that the average time required for cancer to develop from 10 mg L^{-1} of exposure is 30 years (not uncommon for a carcinogen). After our townspeople have been exposed to the drinking water contaminant for 15 years, we conduct a study. For this study, we collect death certificates of 20 people exposed to the contaminant, but we have little information on actual exposure. We know that some of the deceased were exposed when the contaminant was first introduced into the water supply, and that others were exposed several years later. When we turn to the health records, we find that they are incomplete. Finally, the results of our study reveal that 20 cancer deaths is not an excessive number when compared to an appropriate control group. Is it then correct for us to conclude that our known carcinogen is not carcinogenic?

21.4.5 Short-Term Tests for Toxicity

The lifetime animal study is the primary method used for detecting the carcinogenic properties and general toxicity of a substance. Short-term tests for toxicity, however, are used to measure effects that appear to be correlated with specific toxic effects. For example, those for carcinogenicity include assays for gene mutations in bacteria, yeast, fungi, insects, and mammalian cells; mammalian cell transformation assays; assays for DNA damage and repair, and *in vitro* (outside the animal) or *in vivo* assays (within the animal) for chromosomal mutations in animal cells. There are also a number of short-term toxicity assays that are based on inhibiting the functions of necessary enzymes in organisms, such as ATPases, phosphatases, and dehydrogenase. Phosphatase measurements, for example, can be used to assess the activity of specific substances, such as the toxicity of heavy metals in soils. In addition, short-term bioassays can use enzymes or microorganisms to assess general toxicity of environmental samples (Bitton, 1994).

Several tests involving whole animals are also available. These tests, which are usually of intermediate duration, include the induction of skin and lung tumors in female mice, breast cancer in certain species of female rats, and anatomical changes in the livers of rodents.

Many carcinogenic (cancer-causing), mutagenic (mutation-causing), and teratogenic (defect-causing) agents act in the same way: they cause changes in DNA that eventually affect cell development. Because of this relationship, initial screening for such substances can often be accomplished quickly by testing their capacity to cause mutations in a particular strain of bacteria—*Salmonella typhimurium*. This unique strain of bacteria requires the essential amino acid histidine to grow, so it can grow on histidine-free media only if it has first mutated. Thus, if we observe these bacteria growing on histidine-free media after exposure to a test chemical, we can safely assume that the chemical caused the bacteria to mutate. The test chemical is therefore likely to be a carcinogen, mutagen, or teratogen. This short-term test is called the **Ames test** after its developer, Dr. Bruce Ames of the University of California at Berkeley.

Another short-term test is a bioassay based on the light output of the bioluminescent bacterium, *Photobacterium phosphorecum*. This bioassay has been used to assess the general toxicity of wastewater effluents, industrial wastes, sediment extracts, and hazardous waste leachates. As toxic substances diminish the viability of the bacterium, bioluminescent activity decreases and the light output can be quantitatively measured by an instrument.

21.4.6 Threshold Effects

Commonly accepted theory suggests that most biological effects of noncarcinogenic chemical substances occur only after a certain concentration or level is achieved. This level, known as the threshold dose, is approximated by the NOAEL. Another widely accepted premise, at least in the setting of public health standards, is that the human population is likely to have a much more variable response to toxic agents than do the small groups of well-controlled, genetically homogeneous animals ordinarily used in experiments. The NOAEL is itself subject to some uncertainty owing to variabilities in the data from which it was obtained. For these reasons, public health agencies divide experimental NOAELs by large uncertainty factors, known as safety factors, when examining substances that display threshold effects (see Section 22.2.3, beginning on page 351). The magnitude of these safety factors varies according to the following: the nature and quality of the data from which the NOAEL is derived; the seriousness of the toxic effects; the type of protection being sought (*e.g.*, protection against acute, subchronic, or chronic exposures); and the nature of the population to be protected (*i.e.*, the general population versus identifiable subpopulations expected to exhibit a narrower range of susceptibilities). Safety factors of 10, 100, 1000, and 10,000 have been used in various circumstances.

At present, only agents displaying carcinogenic properties are treated as if they display no thresholds (see Section 22.2.3, beginning on page 351). Thus, the dose-response curve for carcinogens in the human populations achieves zero risk only at zero dose; as the dose increases above zero, the risk immediately becomes finite and thereafter increases as a function of dose. Risk in this case is the probability of producing cancer, and at very low doses the risk can be extremely small.

21.5 Responses to Toxic Substances

In general, an organism's response to a toxic chemical depends on the dose administered. However, once a toxicant enters the body, the interplay of four processes—absorption, distribution, excretion, and metabolism—determines the actual effect of a toxic chemical on the target organ, which is the organ that can be damaged by that particular chemical. Carbon tetrachloride, for example affects, the liver and kidneys, while benzene affects the blood-cell forming system of the body. Figure 21-1 summarizes routes of absorption, distribution, and excretion.

21.5.1 Absorption

Absorption of toxicants across body membranes and into the bloodstream can occur in the gastrointestinal (GI) tract, in the lungs, and through the skin. Contaminants present in drinking water, for example, enter the body primarily through the GI tract. Once they enter the GI tract, most chemicals must be absorbed to exert their toxic effect. Owing largely to differences in solubility, some compounds are absorbed more readily than others. Lipid-soluble, nonionized organic compounds, such as DDT and polychlorinated biphenyls (PCBs), are more readily absorbed by diffusion in the GI tract than are lipid-insoluble, ionized compounds such as lead and cadmium salts. The GI tract also employs specialized active transport systems for compounds, such as sugars, amino acids, pyrimidines, calcium, and sodium. Although these active transport systems do not generally play a major role in absorption of toxicants, they can contribute to their absorption in some cases; lead, for example, can be absorbed via the calcium transport system.

The behavior of toxicants in the GI tract also depends on the action of digestive fluids. These digestive fluids can be beneficial or harmful. For example, snake venom, a protein that is quite toxic when injected, is nontoxic when administered orally because stomach enzymes attack the protein structure, breaking it down into amino acids. However, these enzymes can also contribute to the conversion of nitrates in the GI tract to carcinogenic compounds known as nitrosamines.

Age is also an important factor affecting the intestine's ability to act as a barrier to certain toxicants. The GI tract of newborns, for instance, has a higher pH and a higher number of *E. coli* bacteria than that in adults. These conditions promote the conversion of nitrate, a common drinking water pollutant from agricultural run-off, into the more toxic chemical nitrite. The resulting nitrite then interferes with the blood's ability to carry oxygen, causing methemoglobinemia or "blue-baby syndrome" (see also Section 14.2, beginning on page 212). Lead is also absorbed more readily in newborns than in adults. [*Note*: Even though a chemical has been absorbed through the GI tract, it can still be excreted or metabolized by the intestine or liver before it reaches the systemic circulation.]

The lungs are anatomically designed to absorb and excrete chemicals, as is shown by their continuous absorption of oxygen and excretion of carbon dioxide. The alveoli have a large surface area (50–100 m^2) and are well supplied with blood, and the blood is very close (10 μm) to the air space within the alveoli. These characteristics make the lungs particularly good vehicles for the absorption of toxicants. Toxicants may have to pass through as few as two cells to travel from the air into the bloodstream.

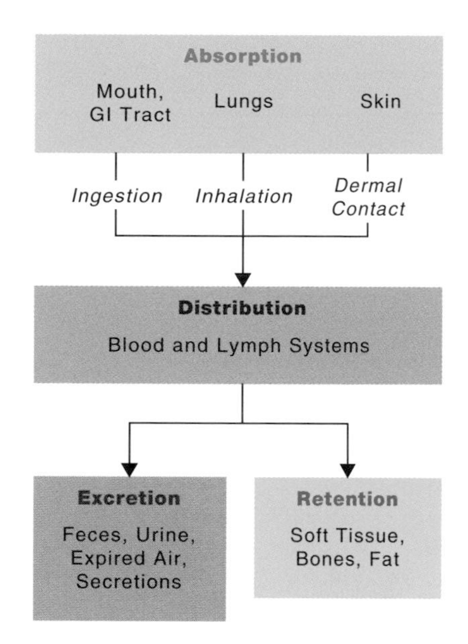

Figure 21-1 Key routes of chemical absorption, distribution, and excretion. Adapted from U.S. EPA, 1989.

The skin is relatively impermeable to toxicants. However, some toxicants, such as carbon tetrachloride, can be absorbed through the skin in sufficient quantities to cause liver injury. In addition, a few chemicals, such as dimethyl sulfoxide (DMSO), have been shown to penetrate the skin fairly readily. Absorption through the skin is possible through the hair follicles, through the cells of the sweat glands and sebaceous glands, and through cuts or abrasions, which increase the rate and degree of absorption. The sole means of absorption through the skin appears to be passive diffusion.

21.5.2 Distribution

Distribution of a toxicant to various organs depends on the ease with which it crosses cell membranes, its affinity for various tissues, and the blood flow through the particular organ. A toxicant's site of concentration is not necessarily the target organ of toxicity. For example, lead can be stored harmlessly in

Table 21-3 Cancer incidence per year in the U.S. by site and sex excluding nonmelanoma skin cancer and carcinoma *in situ*. Adapted from Kolluru, 1994.

Male		Female	
Cancer Type	Incidence	Cancer Type	Incidence
Prostate	122,000	Breast	175,000
Lung	101,000	Colon and rectum	78,500
Colon and rectum	79,000	Lung	60,000
Bladder	37,000	Uterus	46,000
Lymphoma	23,800	Lymphoma	20,800
Oral	20,600	Ovary	20,700
Melanoma of the skin	17,000	Melanoma of the skin	15,000
Kidney	15,800	Pancreas	14,500
Leukemia	15,800	Bladder	13,200
Stomach	14,500	Leukemia	12,200
Pancreas	13,700	Oral	10,200
Larynx	10,000	Kidney	9,500
Other	74,800	Other	79,400
All sites	545,000	All sites	555,000

bone, and many lipid-soluble toxicants (such as the chlorinated hydrocarbon insecticides) are stored in fat, where they cause relatively little harm. However, a stored contaminant can be released back into the bloodstream under various conditions. Thus, fat-stored chlorinated pesticides can be released during starvation, dieting, or illness when fat is consumed.

A number of anatomical barriers in the body are thought to prevent or hinder the entrance of certain toxicants into organs. However, these barriers are not impenetrable walls. The so-called blood/brain barrier, for example, does not prevent toxicants from entering the **central nervous system (CNS)**; rather, the physiological conditions at the blood/brain interface make it more difficult for some toxicants to leave the blood and enter the CNS. In general, lipid-soluble toxicants can cross the blood/brain barrier, but some water-soluble toxicants cannot. Then there's the "placental barrier," which is even less of a barrier: the fact is, any chemical absorbed into the mother's bloodstream can and will cross her placenta and enter the bloodstream of the fetus to some degree.

21.5.3 Excretion

Chemicals can be excreted from the body in several ways. The kidney removes toxicants from the blood in the same way that the end products of metabolism are eliminated, that is, by glomerular filtration, passive tubular diffusion, and active secretion. Glomerular filtration is simply a filtration process in which compounds below a certain molecular weight (and hence bulk) pass through pores in a part of the kidney known as the glomeruli. All compounds whose molecular weight is less than 60,000 can filter through the glomeruli unless they are bound to plasma proteins (only a few do so). The molecular weight of most toxicants in drinking water is between 100 and 500; thus, these compounds easily pass through the glomeruli. The toxicants then pass through collecting ducts and tubules through which water-soluble toxicants may be excreted with urine. However, lipid-soluble toxicants can defeat the excretion process at this point by moving (via passive diffusion) through the tubule wall and back into the bloodstream.

The liver eliminates toxicants through the bile, which passes into the intestine through the gall bladder and bile duct, and finally exits from the body via the feces. As in the kidney, the transport mechanisms used are passive diffusion and carrier-mediated transport. Toxicants that have been excreted into the intestine through the bile can be reabsorbed (especially if they are lipid-soluble) into the bloodstream while in the intestine.

Toxicants are also excreted through several other routes, including the lungs, GI tract, cerebrospinal fluid, milk, sweat, and saliva. Milk, for example, has a relatively high concentration of fat (3.5%); thus, lipid-soluble compounds such as DDT and PCBs can concentrate in milk. And because milk is slightly acidic (with a pH of 6.5), basic compounds may concentrate in it as well. In this way, toxicants may be passed from mother to child or from cows to humans.

An important concept in excretion is a toxicant's half-life $T_{1/2}$ (see Eq. 6-17 on page 74), which is the time it takes for one-half of the chemical to be eliminated from the body. Thus, if a chemical has a half-

Table 21-4 Cancer deaths per year in the U.S. out of a total of 514,000 by site and sex excluding nonmelanoma skin cancer and carcinoma *in situ*. Adapted from Kolluru, 1994.

Male		Female	
Cancer Type	Incidence	Cancer Type	Incidence
Lung	92,000	Lung	51,000
Prostate	32,000	Breast	44,500
Colon and rectum	30,000	Colon and rectum	30,500
Pancreas	12,000	Pancreas	13,200
Lymphoma	10,600	Ovary	12,500
Leukemia	9,800	Uterus	10,000
Stomach	8,100	Lymphoma	9,700
Esophagus	7,300	Leukemia	8,300
Bladder	6,400	Liver	5,800
Kidney	6,300	Brain	5,300
Liver	6,300	Stomach	5,300
Brain	6,200	Multiple myeloma	4,500
Other	45,000	Other	41,400
All sites	272,000	All sites	242,000

life of 1 day, 50% of it will remain with the body one day after absorption, 25% will remain after two days, 12.5% after three days, and so on. The concept of the half-life is important since it indicates how long a compound will remain within the body. Generally a compound is considered eliminated after a period of five half-lifes.

21.5.4 Metabolism

Because lipid-soluble compounds can cross cell membranes to be reabsorbed in the kidney and intestine, they are subject to metabolic processes, which are the biochemical reactions whereby cells transform food into energy and living tissue. In many cases, the body metabolizes these toxicants into water-soluble compounds, which can be excreted easily. However, in some instances, metabolism of a chemical creates a more toxic chemical or does not change the chemical's toxicity.

Two types of reactions occur in **metabolism**: (1) relatively simple reactions involving oxidation, reduction, and hydrolysis; and (2) more complex reactions involving conjugation and synthesis. All of these reactions occur primarily in the liver. Oxidation is the mechanism of metabolism for many compounds. When considering metabolism, researchers also look at species, strain, and gender differences. Age is also an important factor in both humans and laboratory animals: both the very young and very old are more susceptible to certain chemicals.

21.6 Carcinogens, Mutagens, and Teratogens

Carcinogens are agents that cause cancer, which is the uncontrolled growth of cells. Each of us is made up of approximately 100 trillion cells, and any one of these cells can be transformed to a malignant (cancerous) cell by a variety of agents, which may be chemical (*e.g.*, disinfection by-products), biological (*e.g.*, cancer-causing viruses), or physical (*e.g.*, ultraviolet light, gamma irradiation) in origin. Approximately a hundred different types of cancer, which can be found in every organ and system of the body, have been identified. Table 21-3 shows the incidence of cancer at various sites in the human body.

As of the mid 1990s in the United States, cancer is second only to heart disease as a cause of death. Table 21-4 shows an estimate of yearly cancer deaths in the United States, broken down by site and sex. As we can see, more than 500,000 people die from it each year, and lung cancer is by far the leading killer. In fact, lung cancer has increased more than 200 percent during the last 35 years. Cancer ultimately kills one out of about every four Americans. However, aside from the increased incidence of lung cancer, the incidence of all other forms of cancer has collectively declined by about 13% over the past 30 years.

Carcinogens trigger uncontrolled cell growth in many different ways. In general, we can think of carcinogens as initiators or promoters, depending upon the stage of carcinogenesis in which they are active.

Initiation—the first stage of carcinogenesis, or conversion of a normal cell to a cancer cell—is a rapid, essentially irreversible change caused by the interaction of a carcinogen with cell DNA. This step in the development of cancer involves an **initiator**, a type of carcinogen that structurally modifies a gene that normally controls cell growth. A gene is a specific segment of a DNA molecule, so these altered growth-regulating genes are called **oncogenes** (from *onco*, meaning tumor, and gene). For a cell to begin to grow uncontrollably, at least two different growth-regulating genes must be altered. Such changes prime the cell for subsequent neoplastic development (from *neo*, meaning new, and *plastic*, referring to something formed, literally, new growth).

Some chemicals are initiators in their own right. For instance, formaldehyde, a widely used chemical in industrial glues, is thought to be an initiator. We know that vapors of formaldehyde trigger the development of malignant tumors in the respiratory tracts of rats, even though this compound has not been shown to cause cancer in humans. In other cases, however, an initiator can be a metabolic by-product. Thus, benzo(a)pyrene, a natural product of the incomplete combustion of organic materials (including tobacco), is not itself an initiator, but the body metabolizes it to a related chemical, benzopyrene-7,8-diol-9,10-epoxide, which is an initiator.

The **promotion** process triggers the progressive multiplication of abnormal cells known as neoplastic development. **Promoters**, then, are carcinogens that activate the oncogenes, which would otherwise remain dormant. These carcinogens may act in several ways. For example, normal cells appear to prevent the activation of an oncogene in an adjacent initiated cell. A substance may act as a promoter by killing the normal cells that surround an initiated cell. Alternatively, promoters may activate oncogenes by inhibiting the action of **suppressor genes**, which prevent oncogenes from initiating uncontrolled cell growth. If suppressor genes are inactivated, oncogenes can then spur tumor formation.

Many compounds can act as promoters, some of them seemingly innocuous. For instance, dietary factors such as salts and fats apparently act as promoters by killing normal cells. Ingestion of such promoters is not immediately harmful, but lifelong exposure significantly increases the risk of cancer. Thus, people whose diets are high in salts or fats are more likely to develop stomach cancer and colon cancer, respectively. Fortunately, removing promoters from the area of an initiated cell that has not yet completed the promotion stage prevents the formation of a cancer cell. Therefore, if we reduce the salt and fat we consume, we can greatly reduces the risk of developing cancer.

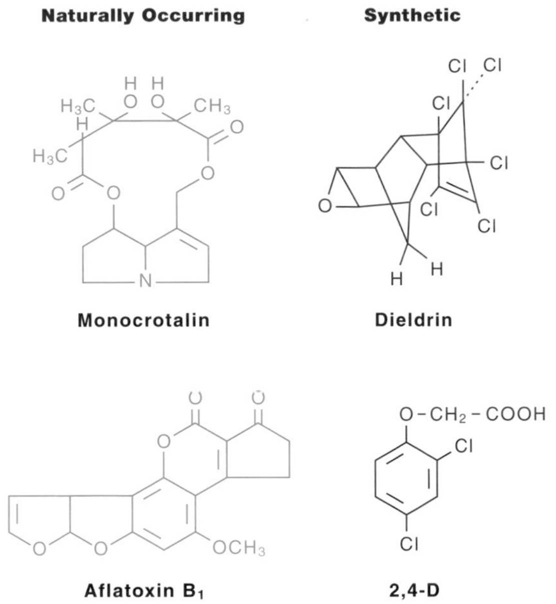

Naturally Occurring **Synthetic**

Monocrotalin

Dieldrin

Aflatoxin B_1

2,4-D

Figure 21-2 Structures of natural and synthetic toxic chemicals found in the environment.

Carcinogens and toxins differ in one very important respect: the incidence of cancer (number of cases per million population) is dose-dependent, but the severity of the response (cancer) is independent of dose. This means that we would expect more cases of cancer to develop as a population is exposed to higher levels of a carcinogen, just as we would expect more cases of poisoning in a population exposed to higher levels of a toxin. But while the severity of toxic response is also dose-dependent, the dose of the carcinogenic agent has little to do with the severity of the disease once an individual has contracted cancer. This distinction between toxins and carcinogens explains why exposure regulations for substances classified as carcinogens are much more stringent than for toxins.

Like carcinogens, mutagens and teratogens affect DNA. **Mutagens** cause **mutations**, which are inheritable changes in the DNA sequences of chromosomes. Mutations involve a random change in the natural functioning of chromosomes or their component genes; such changes rarely benefit the organism's offspring. **Teratogens** affect the DNA in a developing fetus, often causing gross abnormalities or severe deformities such as the shortening or absence of arms or legs.

Perhaps the most famous (or infamous) teratogen is thalidomide, a sedative that was taken by thousands of pregnant women during the early 1960s. Sometimes, however, the deleterious effect of a teratogen does not appear until many years after the mother has been exposed. This was the case for DES (diethylstilbestrol), a drug that was prescribed for pregnant women in the United States for more than thirty years. Developed to prevent miscarriages, DES has been implicated in cervical and vaginal abnormalities. Nor are drugs the only teratogens. For example, the rubella virus, which causes a mild viral infection (German measles), is a teratogen during the first trimester of pregnancy. This virus can cross the transplacental barrier to produce cardiac defects and deafness in the offspring.

21.7 Chemical Toxicity: General Considerations

The toxic effects of a substance on a particular individual depend on both the chemical and the individual. However, the variability in the toxic potential of different compounds greatly exceeds the variability in toxic response from individual to individual. That is, if we expose a particular individual to a whole series of different chemicals, we would see that some substances cause toxic effects in minute amounts while others must be present in huge quantities. And this range is enormous: the toxicity of one chemical can be millions or billions of times greater than that of another chemical. Thus, it can take millions or billions of times as much of one chemical to cause the same effect as another. The range of human variability is not nearly so great. If a particular chemical causes an effect in one individual when a particular amount is administered, it is not likely that an amount a billion times less will cause a toxic effect in another individual. The exact range of human variability is not well established, but it is probable that it is closer to a tenfold than a billionfold.

In considering toxicity, we cannot make a distinction between human-made (synthetic) and naturally occurring chemicals—that is, everything is chemical in composition. It is not the source of the chemical that is important, but its characteristics. In Figure 21-2, for example, we see the structures of some natural and synthetic compounds, each of which is considered a toxin at certain dosages. However, most chemicals, synthetic or natural, are not very toxic.

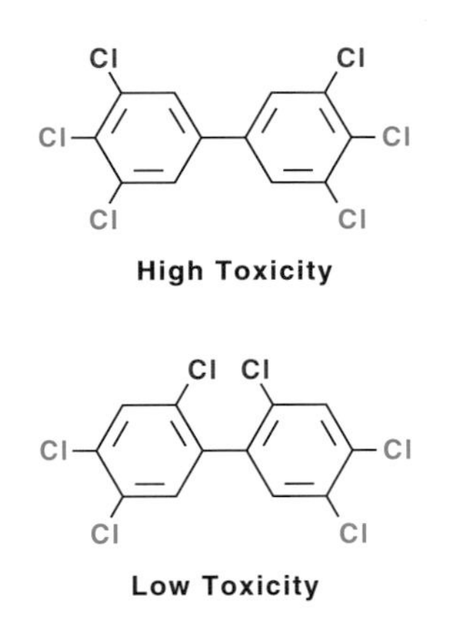

Figure 21-3 Example of structurally similar chemicals with different toxic potency.

The molecular shape or structure of a chemical is one of the most important characteristics to consider in determining its toxicity. Current theory suggests that a living organism "recognizes," and hence reacts to, most chemicals that enter the body by their shape. These body-recognition responses can be very sensitive to subtle differences in shape or conformation. Two molecules, for example, might be very similar in structure, but exhibit slightly different configurations or three-dimensional isomers, one of which one may induce a toxic response in the body while the other will not (see Figure 21-3). In fact, toxicity is frequently the result of recognition gone wrong. That is, at a cellular or chemical level, the body "prefers" the toxin over its structurally similar counterpart.

A second important characteristic in determining the degree of toxicity of a chemical is its solubility in different solvents. In particular, compounds are divided into those that are soluble in water or water-like (polar) solvents and those that are soluble in fat (oil) or fat(oil)-like (nonpolar) solvents. This difference is very important in determining how easily a chemical can enter the body, how it is distributed inside the body, and how easily it can be excreted. Animals are most efficient at excreting polar compounds, so an ionic compound like sodium chloride, which is very soluble in water, is easily excreted. But a nonpolar compound like the pesticide DDT, which dissolves in fats, is not so easily eliminated. In some cases, the body is capable of converting nonpolar compounds into polar variants, thus facilitating their removal from the body. Unfortunately, this is not true of DDT, so it can remain in the body for long periods of time.

21.8 Chemical Toxicity: Selected Substances

21.8.1 Heavy Metals

Heavy metals, such as lead, mercury, cadmium, chromium, arsenic, selenium, and the like, comprise a major category of inorganic pollutants. Heavy-metal contamination often originates from manufacturing and the use of various synthetic products (*e.g.*, pesticides, paints, batteries; see Section 18.2, beginning on page 268), but these chemicals may also occur naturally. Heavy metals, many of which are toxic to plants as well as animals, tend to be mobile in the food chain, which means they can be bioconcentrated in animals—including humans, who are at the top of the food chain. The toxicity of some heavy metals such as lead and mercury has been known for centuries, but was not fully understood until recent times. Because of their widespread use and/or occurrence in nature, some heavy metals are of particular concern.

Lead borders on ubiquitous. It can be found in drinking water, where it comes from several sources. The most significant of these sources is lead solder and piping in water distribution systems, particularly when contacted by corrosive water. (While the use of lead solder and piping in repairs and construction of water systems has been banned, many existing water distribution systems still contain lead materials.) Lead is also found in food, tetraethyl lead in gasoline (which ends up in the air, soil, and water), lead-based paint, and improperly glazed earthenware. In addition, it can come from industrial sources, such as smelters and lead-acid battery manufacturing. Since the toxic effects of lead depend on total exposure, environmental assessment must take into account all of these sources.

Compared with fat-soluble substances, lead is relatively poorly absorbed, as lead compounds are water soluble. In adults, only about 10% of the lead ingested through the GI tract is absorbed into the bloodstream. Lead is initially distributed to the kidney and liver and then redistributed, mostly to bone (about 95%). Moreover, lead does not readily enter the central nervous system in adults because the blood/brain barrier can keep it out. However, in children, the blood/brain barrier is not yet fully developed, so exposure in children can affect their mental development.

Acute lead poisoning is rare; however, chronic lead poisoning is not uncommon. In adults, chronic lead poisoning sometimes results in the painful gastrointestinal symptoms known as lead colic. Because of the pain, lead colic often compels exposed persons to seek medical help, whereupon an accurate diagnosis of lead exposure can prevent the development of more serious problems. Lead can also affect the neuromuscular system, decreasing muscle tone in the wrists and feet. Exposure to lead also affects the body's blood-forming system. Lead can interfere with the synthesis of heme (part of the oxygen-carrying compound hemoglobin), thereby causing anemia; it can also damage red blood cells in a condition known as basophilic stippling. The most serious effects of lead is the brain-degenerative condition called lead encephalopathy. This condition can occur in adults. For example, historians now speculate that Caligula, the most insane of the Roman emperors, suffered from lead encephalopathy caused by eating food from lead-containing pewter dishes. Today, however, this condition is more common in children and can be quite serious; approximately 25% of children with lead encephalopathies die, and about 40% of the survivors experience neurologic aftereffects. Finally, lead has been shown to affect the kidneys of laboratory animals, causing impairment of function and cancer.

Cadmium, like lead, has many sources. It is a by-product of lead and zinc mining, it is used as a pigment, and it is found in corrosion-resistant coatings and nickel–cadmium batteries. It is also released when fossil fuels are burned. Cadmium can enter drinking water when corrosive water contacts certain types of water piping; or it can enter the food chain through application of sewage sludge to the land, where it is taken up by plants and stored in leaves and seeds. Cadmium's adverse effect on health was originally made public by an incident in Japan, where rice paddies were contaminated with cadmium-rich drainage from zinc mines. Rice grown on the paddies concentrated the cadmium and those eating it suffered such characteristic symptoms as easily broken bones and extreme joint pain; thus, cadmium poisoning is known as *itai-tai* ("ouch-ouch") disease.

Since cadmium is water-soluble, only 1–5% of a given dose is absorbed in the GI tract (although 10–40% can be absorbed through the lung), and cadmium distributes to the kidney and liver. Acute cadmium poisoning causes GI disturbances. Chronic cadmium poisoning most severely affects the kidney. Animal studies have shown cadmium to be carcinogenic, and some researchers have suggested that it may increase the incidence of prostate cancer in elderly men.

Arsenic occurs naturally in bedrock and soil and is a waste product from smelting operations as well as the manufacture of products, such as pesticides and herbicides. Airborne particles of arsenic may travel considerable distances and penetrate deeply into the lungs. Arsenic is also readily taken up by food plants, the degree of uptake being dependent on soil pH. Arsenic is considered an essential dietary element, although in very small amounts. There are four main types of arsenic: organoarsenics, pentavalent arsenic, trivalent arsenic, and arsine gas.

Arsenic can be excreted relatively quickly and it has a half-life of two days. Although the effects of acute arsenic poisoning are seen primarily in mystery fiction, chronic arsenic poisoning manifests in a wide variety of chronic toxic effects. Many of these effects stem from the capacity of arsenic to increase the permeability of capillaries in various locations in the body. This capacity has been exploited for centuries, from ancient times well into the modern era. The empress Theodora, for example, took arsenic to improve her appearance; the arsenic broke the capillaries in her cheeks, creating a much-desired "milk and roses" complexion. This increased permeability is, however, harmful; it allows plasma to leak into the tissues, sometimes leading to severe diarrhea and kidney injury. Arsenic also damages both the central nervous system, inflaming peripheral nerves and causing brain injuries, and the liver, infiltrating fatty deposits and causing tissue necrosis. Moreover, the EPA has classified arsenic as a Type A human carcinogen (*i.e.*, a human carcinogen based on epidemiological studies), with skin and lung cancer as the two principal types of cancer arising from arsenic exposure.

21.8.2 Inorganic Radionuclides

Certain unstable elements spontaneously decay into different atomic configurations, in the process releasing radiation consisting of alpha particles, beta particles, or gamma rays. These particles and rays can damage living tissue and/or cause cancer to develop, with the degree of damage depending on the type of radiation and means of exposure (*i.e.*, inhalation, ingestion, or external radiation). As an element undergoes radioactive decay, it progresses through a series of atomic configurations, known as isotopes. Isotopes are simply atoms identified by atomic weight, a number indicating the number of neutral and charged particles in the nucleus. The uranium in pitchblende, for example, is a mixture of three isotopes which have atomic weights of 234, 235, and 238.

Radionuclides are atoms (nuclei) that are undergoing spontaneous decay, emitting radiation as they disintegrate to form isotopes of lower atomic weight. The radioactivity of an isotope is expressed in units called picocuries (pCi),[2] which represent the isotope's number of disintegrations per second; 1 pCi is equal to 3.7×10^{-2} disintegrations per second. [*Note*: Radionuclides continue to decay until they achieve a stable configuration, which is often the configuration of another element altogether.]

Radioactive decay is a natural process. In fact, everyone is exposed to some background radiation

[2] Another unit commonly used to express radioactivity is the becquerel (Bq), which is defined as disintegrations per second (s^{-1}). 1 Bq = 3.7×10^2 pCi.

both from cosmic rays and from radioactive soil and rock. The three naturally occurring series of isotopes arise from the decay of the isotopes uranium-238, uranium-235, and thorium-232. Thus, the naturally occurring radionuclides found in drinking water—the uranium-238 series (whose decay isotopes include uranium-234, radium-226, and radon-222) and the thorium series (whose decay products include radium-228)—are of the greatest concern. Synthetic radioactive isotopes, such as strontium-90, also pose health risks, but such isotopes generally occur in lower concentrations in the environment than the naturally occurring radionuclides. However, site-specific contamination, such as nuclear waste disposal sites and nuclear power plant accidents (*e.g.*, Chernobyl), may release concentrations of anthropogenic radionuclides that threaten human and environmental health.

The concern over radionuclides focuses largely on their potential to cause cancer. Radium-226, which has a half-life of 1622 years, is perhaps the single most important radioactive isotope found in drinking water. It is deposited in bone and can cause bone cancer. One of the decay products of radium-226 is radon gas, that is, the isotope radon-222, which has a half-life of 3.85 days. Inhalation of the short-lived decay products of radon-222 can cause lung cancer; however, less is known about the risks of ingested radon. This isotope has recently become the subject of great public concern because it has been discovered in homes and other buildings. Most of the total amount of radon-222 that enters homes comes through the soil, but it can also enter by degassing from a dissolved state in drinking and washing water. This degassing occurs when water is heated and/or aerated, as it is during clothes and dish washing, showering, and bathing.

21.8.3 Insecticides

Insecticides can be divided into organochlorine, organophosphorus, and carbamate compounds, as well as botanical insecticides (see also Section 17.2.1, beginning on page 255). Within each group, the pesticides have similar characteristics.

Organochlorine Insecticides: This category, which was developed in the 1930s and 1940s, includes the chlorinated ethanes, chlorinated cyclodienes, and other chlorinated compounds. Dichlorodiphenyltrichloroethane (DDT) is the most famous of the chlorinated insecticides. First synthesized in 1943, it was used extensively (worldwide) in agriculture from the end of World War II until 1972, when it was banned in the United States. This highly lipid-soluble compound is stored in fat—in fact, the fat of most U.S residents contains DDT concentrations of 5–7 mg kg^{-1}.[3] DDT is very persistent in the environment and is **biomagnified** in the food chain. That is, smaller organisms absorb the compound, then they are eaten by larger organisms, and the progression continues until DDT attains a relatively high concentration in macrovertebrates, such as fish, which are then eaten by humans and other large animals.

In general, DDT is toxic to humans and most other higher animal life only in extremely high doses. But, because of its low toxicity, it was applied in much greater quantities than were necessary. Then, in the 1960s, the effects of these massive applications became noticeable. For example, certain birds (for example, the peregrine falcon) began to produce overly fragile egg shells that broke before hatching, thereby threatening their survival as a species. Fish, too, are extremely vulnerable to DDT, and die-offs have occurred following heavy rains when the pesticide was washed into streams and rivers.

The chlorinated cyclodienes comprise an important subgroup of organochlorines. Developed in the late 1940s, these chemical pesticides include aldrin, dieldrin, endrin, heptachlor, and chlordane. These cyclodienes are similar to, but more toxic than DDT and have caused many human fatalities (mostly among farm workers who handle the pesticide directly). Their toxicity is much greater for humans because they are more efficiently absorbed transdermally (through the skin). Like DDT, they are lipid-soluble and stored in fat and hence are biomagnified in the environment. They have also caused cancer in laboratory animals. Thus, their registration for use on agricultural crops was suspended in the mid-1970s.

[3] This unit is sometimes expressed as the more ambiguous unit of parts per million (ppm). See also Footnote 1 on page 324.

The other chlorinated hydrocarbons, such as lindane, toxaphene, mirex, and kepone, are also similar to DDT. In general, then, we can say that all organochlorine insecticides cause some CNS stimulation, increase cancer incidence in laboratory animals (lindane, less so), and persist in the environment to some degree.

Organophosphorus Insecticides: Organophosphorus compounds do not persist in the environment and have an extremely low potential to produce cancer; thus, insecticides based on these compounds have largely replaced the chlorinated hydrocarbon insecticides. However, these phosphorus-containing compounds have a much higher acute toxicity in humans than the organochlorines and are the most frequent cause of human insecticide poisoning. Fortunately, both laboratory tests and antidotes are available for acute organophosphorus poisoning.

A typical organophosphorus insecticide is parathion, which must be metabolized to the compound paraxon to exert its toxic effect. This toxic effect stems from the compound's ability to inhibit the enzyme cholinesterase, a crucial chemical for the regulation of the nerve transmitter acetylcholine. Thus, acute effects of poisoning with organophosphorus insecticides include fibrillation of muscles, low heart rate, paralysis of respiratory muscles, confusion, convulsions, and eventually death. Other organophosphorus pesticides, such as malathion, are less toxic in acute doses than parathion.

In general, the organophosphorus insecticides that are currently widely available do not cause delayed neurotoxicity. However, some previously available pesticides have been found to cause delayed neurological problems, such as weakness, lack of muscle coordination, and sensory disturbances.

Carbamates: The carbamate insecticides, which include carbaryl and aldicarb, have toxicities very similar to those of the organophosphorus insecticides. Like the organophosphorus pesticides, these widely used chemicals also act by inhibiting cholinesterase, but the toxic effects of carbamates may be more easily reversed than those of the organophosphorus compounds. In addition, current evidence does not seem to suggest carcinogenicity as a toxic effect of the carbamates. Although most of these chemicals are not persistent in the environment, aldicarb may be the exception. Used on potato crops in Long Island, New York, aldicarb has contaminated the groundwater aquifers there. It has been estimated that levels of 6 $\mu g\ L^{-1}$ may persist up to 20 years.

21.8.4 Herbicides

In the United States, herbicides—chemicals that kill plants—are used in greater quantities than insecticides. Herbicides are added directly to the soil; thus they can, in some cases, readily enter the groundwater. Chlorophenoxy compounds include 2,4-dichlorophenoxyacetic acid (2,4-D), 2,4,5-trichlorophenoxyacetic acid (2,4,5-T), and 2,4,5-trichlorophenoxypropionic acid (2,4,5-TP or silvex). These herbicides act as growth hormones, forcing plant growth to outstrip the ability to provide nutrients. Although these compounds can have toxic effects on the liver, kidney, and central nervous system, clinical reports of poisoning are rare. The compounds, which have a half-life of about 24 hours, are rapidly excreted into the urine in humans.

One of the best known, and most controversial, of the chlorophenoxy compounds is 2,4,5-T, which was combined with 2,4-D to create the defoliant Agent Orange used in the Vietnam War. However, during the industrial synthesis of 2,4,5-T and 2,4,5-TP, a hazardous by-product can also be inadvertently formed: tetrachlorodioxin (TCDD) or **dioxin** for short. Dioxin, which can also be produced during the combustion of certain substances, is the most toxic manufactured chemical known. In sufficient doses, dioxin is a potent teratogen and carcinogen in laboratory animals, and causes liver injury and general tissue wasting. At lower doses, it causes a form of acne called chloracne, which concentrates between the eyes and hairline.

Clinical reports of acute dioxin poisoning are rare, and thus far chloracne seems to be the worst effect seen in humans. Dramatic interspecies differences exist for the effects of dioxin; the LD_{50} for guinea pigs is about 1/10,000 of the LD_{50} for hamsters. Fortunately, current evidence indicates that the human reaction to dioxin tends to resemble that of the hamster rather than that of the guinea pig. However, di-

Table 21-5 Health effects of some chlorinated halogenated hydrocarbons. (+) = harmful effect, (–) = no effect, (+–) = a less significant effect. Adapted from U.S. EPA, 1989.

	CNS Depression	Sensitization of Heart	Liver Injury	Kidney Injury	Cancer
Methanes					
Carbon tetrachloride	+	+	++++	++	+
Chloroform	+	+	+++	+++	+
Dichloromethane (methylene chloride)	+	–	+–	–	+
Ethanes					
1,1-Dichloroethane	+	+	+	*NDA*[a]	*NDA*[a]
2,2-Dichloroethane	+	*NDA*[a]	+	–	+
1,1,1-Trichloroethane	+	+	+–	–	–
1,1,2-Trichloroethane	+	*NDA*[a]	++	–	+
1,1,2,2-Tetrachlorethane	+	*NDA*[a]	++	++	*NDA*[a]
Hexachloroethane	+	*NDA*[a]	*NDA*[a]	+	+
Ethylenes					
Chloroethylene (vinyl chloride)	+	*NDA*[a]	++	–	+++
1,1-Dichloroethylene (vinylidine chloride)	+	*NDA*[a]	+++	–	+
1,2-*trans*-Dichloroethylene	+	*NDA*[a]	++	*NDA*[a]	*NDA*[a]
Trichloroethylene	+	+	+	+–	+
Tetrachloroethylene (perchloroethylene)	+	–	+–	+–	+

[a] *NDA* = No data available.

oxin contamination must be considered a serious environmental problem. In Times Beach, Missouri, for example, dioxin contamination forced the abandonment of 800 homes in 1984.

21.8.5 Halogenated Hydrocarbons

Halogenated hydrocarbons are common because they are widely used as effective, yet relatively nonflammable solvents, unlike kerosene or gasoline. Halogenated hydrocarbons are also formed during the chlorination of drinking water when chlorine combines with organic material in the water. One of the oldest and simplest of these compounds is carbon tetrachloride (CCl_4), which was extensively used as a solvent and dry-cleaning agent; it is also so nonflammable that it was used in fire extinguishers. (The use of carbon tetrachloride as a dry-cleaning agent was banned after it was shown to cause liver damage.)

The halogenated hydrocarbons (trichloroethene, halothone, chloroform, *etc.*) tend to be similar to carbon tetrachloride in their health effects. In high doses carbon tetrachloride causes CNS depression—so much so that it was once used as an anesthetic. It can also sensitize the heart muscle to catecholamines (hormones such as epinephrine) and thus can cause heart attacks. And it can cause kidney injury, liver injury, and cancer in laboratory animals. High blood alcohol levels can act as a potentiator for carbon tetrachloride's damaging effects on internal organics. It owes its toxicity to the fact that it is converted in the liver to carbon trichloride ($\cdot CCl_3$), which is a free radical capable of inducing peroxidation of lipid double bonds and poisoning protein-synthesizing enzymes.

Table 21-5 summarizes the health effects of various halogenated hydrocarbons. Plus signs (+) indicate a harmful effect, minus signs (–) indicate a lack of effect, and both a plus and a minus indicates a less significant effect. Among the methanes is chloroform, a trihalomethane formed during drinking water chlorination. Chloroethylene (vinyl chloride) receives three plus signs under cancer because it has been established as a human carcinogen, while the others have been established as probable human carcinogens based on results of animal studies. Also noteworthy are trichloroethylene and tetrachloroethylene (perchloroethylene). These chemicals are very common contaminants of drinking water; although they are listed as probable carcinogens, they cause cancer in laboratory animals only at very high doses. Also, it has been observed that when perchloroethylene and tetrachloroethylene degrade naturally in groundwater, vinyl chloride is formed as a degradation product.

References and Recommended Reading

Bitton G. (1994) *Wastewater Microbiology.* Wiley-Liss, New York.

Chiras D.D. (1985). *Environmental Science.* Benjamin/Cummings, Menlo Park, California.

Cockerham L.G. and Shane B.S., Editors (1994) *Basic Environmental Toxicology.* CRC Press, Boca Raton, Florida.

Hodgson E. and Levi P.E. (1987) *A Textbook of Modern Toxicology.* Appleton and Lange, Norwalk, Connecticut.

Kolluru R.V. and Kamrin M.A. (1993) *Environmental Strategies Handbook.* McGraw-Hill, New York.

Landis W.G., and Yu M.H. (1995) *Introduction to Environmental Toxicology.* Lewis Publishers, Boca Raton, Florida.

Philp R.B. (1995) *Environmental Hazards and Human Health.* Lewis Publishers, Boca Raton, Florida.

Rodricks J.V. (1992) *Calculated Risks: The Toxicity and Human Health Risks of Chemicals in Our Environment.* Cambridge University Press, Cambridge, United Kingdom.

U.S. EPA (1989) *Risk Assessment, Management and Communication of Drinking Water Contamination.* EPA/625/4-89/024. U.S. Environmental Protection Agency, Washington, D.C.

Problems and Questions

1. What event in the year 1962 had an impact in creating an interest in the environment? Why?
2. Why are small animals used in laboratory tests involving toxic materials?
3. How is safety of chemical substances defined with regard to exposure?
4. What is exposure concentration? Exposure dose? How are the two related?
5. Adults are assumed to consume 2 L of water daily. If a substance is present at 10 mg L^{-1}, give the average individual intake of the substance.
6. Explain the similarities and differences between carcinogens, mutagens, and teratogens.
7. What is LD_{50}? How is it used?
8. What are the advantages of short-term toxicity testing?
9. What role do initiators, promoters, and suppressor genes play in cancer formation?
10. Why are some compounds highly toxic when injected, but innocuous when ingested?
11. Can one predict if a new chemical is a carcinogen by its chemical structure? How would you do this?
12. What are some sources of heavy metals in the environment?

13. Give an example of a chemical that is made more toxic through biotransformation in the body.
14. Define toxicity.
15. What are some of the limitations in using epidemiological studies to assess the toxicity or carcinogenicity of a chemical? Do negative findings indicate that the substance is not a carcinogen? Why or why not?
16. What are some bioassays that can be used to assess the toxicity of effluents from wastewater treatment plants?

Risk assessments can be used to assess risks to recreational bathers from pathogenic microorganisms in the water. Photograph: K.D. Crabtree.

22.1 The Concept of Risk Assessment

Risk, which is common to all life, is an inherent property of everyday human existence. It is therefore a key factor in all decision making. Risk assessment or analysis, however, means different things to different people: Wall Street analysts assess financial risks and insurance companies calculate actuarial risks, while regulatory agencies estimate the risks of fatalities from nuclear plant accidents, the incidence of cancer from industrial emissions, and habitat loss associated with increases in human populations. All these seemingly disparate activities have in common the concept of a measurable phenomenon called risk that can be expressed in terms of probability. Thus, we can define **risk assessment** as the process of estimating both the probability that an event will occur and the probable magnitude of its adverse effects—economic, health/safety-related, or ecological—over a specified time period. For example, we might determine the probability that a chemical reactor will fail and the probable effect of its sudden release of contents on the immediate area in terms of injuries and property loss over a period of days. In addition, we might estimate the probable incidence of cancer in the community where the chemical was spilled over a period of years. Or, in yet another type of risk assessment, we might calculate the health risks associated with the presence of pathogens in drinking water or pesticides in food.

There are, of course, several varieties of risk assessment. Risk assessment as a formal discipline emerged in the 1940s and 1950s, paralleling the rise of the nuclear industry. Safety-hazard analyses have been used since at least the 1950s in the nuclear, petroleum-refining, and chemical-processing industries, as well as in aerospace. Health-risk assessments, however, had their beginnings in 1986 with the publication of the *Guidelines for Carcinogenic Risk Assessment* by the Environmental Protection Agency (EPA).

In this chapter, we are concerned with two types of risk assessment:

- **Health-based risks**. For these risks, the focus is on general human health, mainly outside the workplace. Health-based risks typically involve

Table 22-1 Examples of some commonplace risks in the United States. Based on data in Wilson and Crouch, 1987; Gerba and Rose, 1992.

Risk	Lifetime Risk of Mortality
Cancer from cigarette smoking (one pack per day)	1:4
Death in a motor vehicle accident	2:100
Homicide	1:100
Home accident deaths	1:100
Cancer from exposure to radon in homes	3:1000
Exposure to the pesticide afla-toxin in peanut butter	6:10,000
Diarrhea from rotavirus	1:10,000
Exposure to typical EPA maximum chemical contaminant levels	1:10,000–1:10,000,000

high-probability, low-consequence, chronic exposures with long latency periods and delayed effects that make cause-and-effect relationships difficult to establish. [*Note*: This category also includes microbial risks, which usually have acute short-term effects. However, the consequences of microbial infection can persist throughout an individual's lifetime.]

- **Ecological risks**. For these risks, the focus is on the myriad interactions among populations, communities, and ecosystems (including food chains) at both the micro and the macro level. Ecological risks typically involve both short-term catastrophes, such as oil spills, and long-term exposures to hazardous substances.

But whatever its focus, the **risk assessment process** consists of four basic steps:

- **Hazard identification**—Defining the hazard and nature of the harm; for example, identifying a chemical contaminant, say lead or carbon tetrachloride, and documenting its toxic effects on humans.
- **Exposure assessment**—Determining the concentration of a contaminating agent in the environment and estimating its rate of intake in target organisms; for example, finding the concentration of aflatoxin in peanut butter and determining the dose an "average" person would receive.
- **Dose-response assessment**—Quantitating the adverse effects arising from exposure to a hazardous agent based on the degree of exposure. This assessment is usually expressed mathematically as a plot showing the response in living organisms to increasing doses of the agent.
- **Risk characterization**—Estimating the potential impact of a hazard based on the severity of its effects and the amount of exposure.

Once the risks are characterized, various regulatory options are evaluated in a process called **risk management**, which includes consideration of social, political, and economic issues as well as the engineering problems inherent in a proposed solution. One important component of risk management is **risk communication**, which is the interactive process of information and opinion exchange among individuals, groups, and institutions. Risk communication includes the transfer of risk information from expert to nonexpert audiences. In order to be effective, risk communication must provide a forum for balanced discussions of the nature of the risk, lending a perspective that allows the benefits of reducing the risk to be weighed against the costs.

In the United States, the passage of federal and state laws to protect public health and the environment has expanded the application of risk assessment. Major federal agencies that routinely use risk analysis include the Food and Drug Administration (FDA), the Environmental Protection Agency (EPA), and the Occupational Safety and Health Administration (OSHA). Together with state agencies, these regulatory agencies use risk assessment in a variety of situations:

- Setting standards for concentrations of toxic chemicals or pathogenic microorganisms in water or food.
- Conducting baseline analyses of contaminated sites or facilities to determine the need for remedial action and the extent of cleanup required.
- Performing cost/benefit analyses of contaminated-site cleanup or treatment options (including

Table 22-2 Factors affecting risk perception and risk analysis. Source: Covello *et al.*, 1988.

Factor	Conditions Associated with Increased Public Concern	Conditions Associated with Decreased Public Concern
Catastrophic potential	Fatalities and injuries grouped in time and space	Fatalities and injuries scattered and random
Familiarity	Unfamiliar	Familiar
Understanding	Mechanisms or process not understood	Mechanisms or process understood
Controllability (personal)	Uncontrollable	Controllable
Voluntariness of exposure	Involuntary	Voluntary
Effects on children	Children specifically at risk	Children not specifically at risk
Effects manifestation	Delayed effects	Immediate effects
Effects on future generations	Risk to future generations	No risk to future generations
Victim identity	Identifiable victims	Statistical victims
Dread	Effects dreaded	Effects not dreaded
Trust in institutions	Lack of trust in responsible institutions	Trust in responsible institutions
Media attention	Much media attention	Little media attention
Accident history	Major and sometimes minor accidents	No major or minor accidents
Equity	Inequitable distribution of risks and benefits	Equitable distribution of risks and benefits
Benefits	Unclear benefits	Clear benefits
Reversibility	Effects irreversible	Effects reversible
Origin	Caused by human actions or failures	Caused by acts of nature

treatment processes to reduce exposure to pathogens).

- Developing cleanup goals for contaminants for which no federal or state authorities have promulgated numerical standards; evaluating acceptable variance from promulgated standards and guidelines (*e.g.*, approving alternative concentration limits).
- Constructing "what-if" scenarios to compare the potential impact of remedial or treatment alternatives and to set priorities for corrective action.
- Evaluating existing and new technologies for effective prevention, control, or mitigation of hazards and risks.
- Articulating community public health concerns and developing consistent public health expectations among different localities.

Risk assessment provides an effective framework for determining the relative urgency of problems and the allocation of resources to reduce risks. Using the results of risk analyses, we can target prevention, remediation, or control efforts toward areas, sources, or situations in which the greatest risk reductions can be achieved with the resources available. However, risk assessment is not an absolute procedure carried out in a vacuum; rather, it is an evaluative, multifaceted, comparative process. Thus, to evaluate risk, we must inevitably compare one risk to a host of others. In fact, the comparison of potential risks associated with several problems or issues has developed into a subset of risk assessment called **comparative risk assessment**. Some commonplace risks are shown in Table 22-1. Here we see, for example, that risks from chemical exposure are fairly small relative to those associated with driving a car or smoking cigarettes.

Comparing different risks allows us to comprehend the uncommon magnitudes involved and to understand the level, or magnitude, of risk associated with a particular hazard. But comparison with other risks cannot itself establish the *acceptability* of a risk. Thus, the fact that the chance of death from a previously unknown risk is about the same as that from a known risk does not necessarily imply that the two risks are equally acceptable. Generally, comparing risks along a single dimension is not helpful when the risks are widely perceived as qualitatively different. Rather, we must take account of certain qualitative factors that affect risk perception

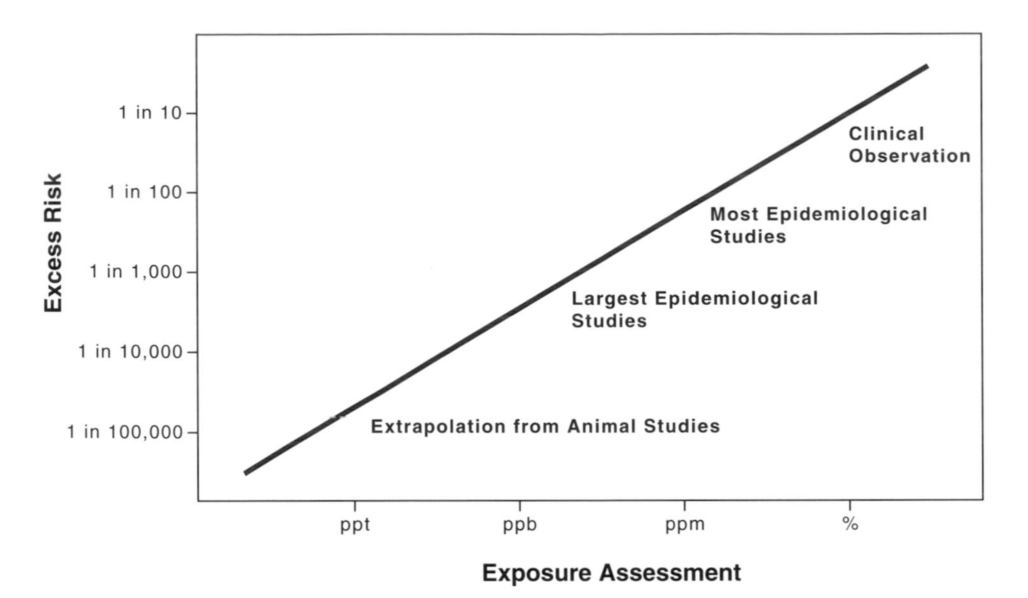

Figure 22-1 Sensitivity of epidemiology in detecting risks of regulatory concern. The generalized units of ppt (parts per trillion), ppb (parts per billion), ppm (parts per million), and percent are used here for comparative purposes—see also Footnote 1 on page 324. Adapted from National Research Council, 1993.

and evaluation when selecting risks to be compared. Some of these qualifying factors are listed in Table 22-2 on page 347. We must also understand the underlying premise that *voluntary risk is always more acceptable than involuntary risk.* For example, the same people who cheerfully drive their cars every day—thus incurring a 2:100 lifetime risk of death by automobile—are quite capable of refusing to accept the 6:10,000 involuntary risk of eating peanut butter contaminated with aflatoxin.

In considering risk, then, we must also understand another principle—the *de minimis* principle, which means that there are some levels of risk so trivial that they are not worth the bother. However attractive, this concept is hard to define, especially if we are trying to find a *de minimis* level acceptable to an entire society. Understandably, regulatory authorities are reluctant to be explicit about an "acceptable" risk. (How much aflatoxin would you consider acceptable in *your* peanut butter and jelly sandwich? How many dead insect parts?) But it is generally agreed that a lifetime risk on the order of one in a million (or in the range of 10^{-5} to 10^{-6}) is trivial enough to be acceptable for the general public. Although the origins and precise meaning of a one-in-a-million acceptable risk remain obscure, its impact on product choices, operations, and costs is very real—running, for example, into hundreds of billions of dollars in hazardous waste site cleanup decisions alone. However, the levels of acceptable risk can vary. Higher levels of risk (10^{-4} rather than 10^{-6}) may be acceptable to just a few people voluntarily exposed as opposed to an entire populace being involuntarily or even covertly exposed to the same hazard. For example, workers dealing with paint solvents often accept higher levels of risk than the public at large. These higher levels are acceptable, because workers tend to be a relatively homogeneous, healthy group and because employment is voluntary; however, the sum level of risks would not be acceptable for those same paint solvents to the general public.

22.2 The Process of Risk Assessment

22.2.1 Hazard Identification

The first step in risk assessment is to determine the nature of the hazard. For pollution-related problems, the hazard in question is usually a specific chemical, a physical agent (such as irradiation), or a microorganism identified with a specific illness or disease (Figure 22-1). Thus the hazard identification component of a pollution risk assessment consists of a review of all relevant biological and chemical information bearing on whether or not an agent poses a specific threat. For example, in the *Guidelines for Carcinogen Risk Assessment* (U.S. EPA, 1986), the following information is evaluated for a potential carcinogen:

- physical/chemical properties, routes, and patterns of exposure
- structure/activity relationships of the substance
- absorption, distribution, metabolism, and excretion characteristics of the substance in the body
- the influence of other toxicologic effects
- data from short-term tests in living organisms
- data from long-term animal studies
- data from human studies

Once these data are reviewed, the animal and human data are both separated into groups characterized by degree of evidence:

- sufficient evidence of carcinogenicity
- limited evidence of carcinogenicity
- inadequate evidence
- no data available
- no evidence of carcinogenicity

The available information on animal and human studies is then combined into a weight-of-evidence classification scheme to assess the likelihood of carcinogenicity. This scheme—which is like that developed by the EPA—gives more weight to human than to animal evidence (when it is available) and includes the following groups:

- Group A—human carcinogen
- Group B—probable human carcinogen
- Group C—possible human carcinogen

Figure 22-2 Risk assessment requires the determination of pollutant levels in environmental samples. Here, marine waters off the coast of Honolulu, Hawaii, USA, are sampled for enteroviruses. Photograph: C.P. Gerba.

- Group D—not classifiable as to human carcinogenicity
- Group E—evidence of noncarcinogenicity toward humans.

As Figure 22-1 shows, clinical studies of disease can be used to identify very large risks (between 1/10 and 1/100), most epidemiological studies can detect risks down to 1/1,000, and very large epidemiological studies can examine risks in the 1/10,000 range. However, risks lower than 1/10,000 cannot be studied with much certainty using epidemiological approaches. Since regulatory policy objectives generally strive to limit risks below 1/100,000 for life-threatening diseases like cancer, these lower risks are often estimated by extrapolating from the effects of high doses given to animals.

22.2.2 Exposure Assessment

Exposure assessment is the process of measuring or estimating the intensity, frequency, and duration of human exposures to an environmental agent. Exposure to contaminants can occur via inhalation, ingestion of water or food, or absorption through the skin upon dermal contact. Contaminant sources, release mechanisms, transport, and transformation characteristics are all important aspects of exposure assessment, as are the nature, location, and activity patterns of the exposed population (Figure 22-2). (This explains why it is critical to understand the factors

Table 22-3 EPA standard default exposure factors. Modified from Kolluru, 1993.

Land use	Exposure Pathway	Daily Intake	Exposure Frequency (days year^{-1})	Exposure Duration (years)
Residential	Ingestion of potable water	2 L day^{-1}	350	30
	Ingestion of soil and dust	200 mg (child)	350	6
		100 mg (adult)		24
	Inhalation of contaminants	20 m^3 (total)	350	30
		15 m^3 (indoor)		
Industrial and commercial	Ingestion of potable water	1 liter	250	25
	Ingestion of soil and dust	50 mg	250	25
	Inhalation of contaminants	20 m^3 (workday)	250	25
Agricultural	Consumption of homegrown produce	42 g (fruit)	350	30
		80 g (vegetable)		
Recreational	Consumption of locally caught fish	54 g	350	30

and processes influencing the transport and fate of a contaminant, as discussed in Chapters 2–7, beginning on page 9.)

An **exposure pathway** is the course that a hazardous agent takes from a source to a receptor (*e.g.*, human or animal) via environmental carriers or media—generally, air (volatile compounds, particulates) or water (soluble compounds). An exception is electromagnetic radiation, which needs no medium. The **exposure route**, or intake pathway, is the mechanism by which the transfer occurs—usually by inhalation, ingestion, and/or dermal contact. Direct contact can result in a local effect at the point of entry and/or in a systemic effect.

The quantitation of exposure, intake, or potential dose can involve equations with three sets of variables:

- concentrations of chemicals or microbes in the media
- exposure rates (magnitude, frequency, duration)
- quantified biological characteristics of receptors (*e.g.*, body weight, absorption capacity for chemicals; level of immunity to microbial pathogens).

Exposure concentrations are derived from measured, monitored, and/or modeled data. Ideally, exposure concentrations should be measured at the points of contact between the environmental media and current or potential receptors. It is usually possible to identify potential receptors and exposure points from field observations and other information. However, it is seldom possible to anticipate all potential exposure points or to measure all environmental concentrations under all conditions. In practice, a combination of monitoring and modeling data, together with a great deal of professional judgment, is required to estimate exposure concentrations.

In order to assess exposure rates via different exposure pathways, we have to consider and weigh many factors. For example, in estimating exposure to a substance via drinking water, we first have to determine the average daily consumption of that water. But this isn't as easy as it sounds. Studies have shown that daily fluid intake varies greatly from individual to individual. Moreover, tapwater intake depends on how much fluid is consumed as tapwater and how much is ingested in the form of soft drinks and other non-tapwater sources. Tapwater intake also changes significantly with age, body weight, diet, and climate. Because these factors are so variable, the EPA has suggested a number of very conservative "default" exposure values that can be used when assessing contaminants in tapwater, vegetables, soil, and the like (see Table 22-3).

One important route of exposure is the food supply. Toxic substances are often bioaccumulated, or concentrated, in plant and animal tissues, thereby

Table 22-4 Bioconcentration factors (BCFs) for various organic and inorganic compounds. Source: U.S. EPA, 1990.

Chemical	BCF (L kg^{-1})
Aldrin	28
Benzene	44
Cadmium	81
Chlordane	14,000
Chloroform	3.75
Copper	200
DDT	54,000
Formaldehyde	0
Nickel	47
PCBs	100,000
Trichloroethylene	10.6
Vinyl chloride	1.17

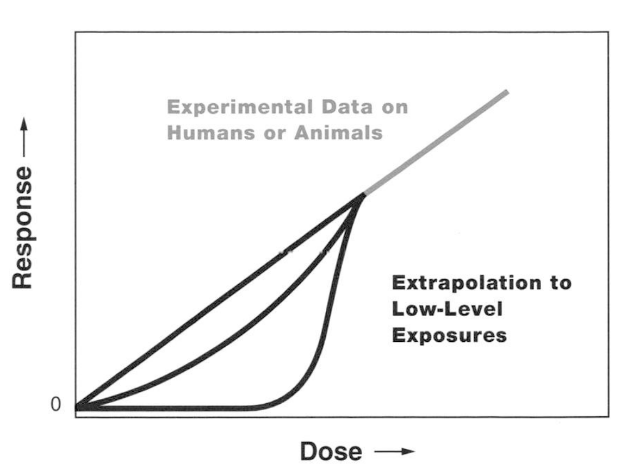

Figure 22-3 Extrapolation of dose-response curves. Adapted from U.S. EPA, 1990.

exposing humans who ingest those tissues as food. Moreover, many toxic substances tend to be biomagnified in the food chain, so that animal tissues contain relatively high concentrations of toxins. Take fish, for example. It is relatively straightforward to estimate concentrations of contaminants in water. Thus, we can use a **bioconcentration factor (BCF)** to estimate the tendency for a substance in water to accumulate in fish tissue. The concentration of a chemical in fish can be estimated by multiplying its concentration in water by the bioconcentration factor. The greater the value of the BCF, the more the chemical accumulates in the fish, and the higher the risk of exposure to humans.

The units of BCF—liters per kilogram (L kg^{-1}) are chosen to allow the concentration of a chemical to be expressed as milligrams per liter (mg L^{-1}) of water and the concentration in fish to be in milligrams per kilogram (mg kg^{-1}) of fish body weight. In Table 22-4, we see the BCFs of several common organic and inorganic chemicals. Note the high values of BCF for the chlorinated hydrocarbon pesticides such as dichlorodiphenyltrichloroethane (DDT) and polychlorinated biphenyls (PCBs).

22.2.3 Dose-Response Assessment

Chemical or other contaminants are not equal in their capacity to cause adverse effects. To determine the capacity of agents to cause harm, we need quantitative toxicity data. Some toxicity data are derived from occupational, clinical, and epidemiological studies (see Section 21.4, beginning on page 326). Most toxicity data, however, come from animal experiments in which researchers expose laboratory animals, mostly mice and rats, to increasingly higher concentrations or doses and observe their corresponding effects. The result of these experiments is the **dose-response relationship**—a quantitative relationship that indicates the agent's degree of toxicity to exposed species. Dose is normalized as milligrams of substance, inhaled, or absorbed (in the case of chemicals) through the skin per kilogram of body weight per day (mg kg^{-1} day^{-1}). Responses or effects can vary widely—from no observable effect, to temporary and reversible effects (*e.g.*, enzyme depression caused by some pesticides or diarrhea caused by viruses), to permanent organ injury (*e.g.*, liver and kidney damage caused by chlorinated solvents, heavy metals, or viruses), to chronic functional impairment (*e.g.*, bronchitis or emphysema arising from smoke damage), to death.

The goal of a dose-response assessment is to obtain a mathematical relationship between the amount (concentration) of a toxicant or microorganism to which a human is exposed and the risk of an adverse outcome from that dose. The data resulting from experimental studies is presented as a dose-response curve, as shown in Figure 22-3. The abscissa describes the dose, while the ordinate measures the risk that some adverse health effect will occur. In the case of a pathogen, for instance, the ordinate may represent the risk of infection but not necessarily illness.

Table 22-5 Primary models used for assessment of nonthreshold effects. Modified from Cockerham and Shane, 1994.

Model[a]	Comments
One-hit	Assumes (1) a single stage for cancer and (2) malignant change induced by one molecular or radiation interaction *Very conservative*
Linear multistage	Assumes multiple stages for cancer *Fits curve to the experimental data*
Multihit	Assumes several interactions needed before cell becomes transformed *Least conservative model*
Probit	Assumes probit (log-normal) distribution for tolerances of exposed population *Appropriate for acute toxicity; questionable for cancer*

[a] All these models assume that exposure to the pollutant will always produce an effect, regardless of dose.

Table 22-6 Lifetime risks of cancer derived from different extrapolation models. Source: U.S. EPA, 1990.

Model Applied	Lifetime Risk (1.0 mg kg^{-1} day^{-1}) of Toxic Chemical[a]	
One-hit	6.0×10^{-5}	(1 in 17,000)
Multistage	6.0×10^{-6}	(1 in 167,000)
Multihit	4.4×10^{-7}	(1 in 2.3 million)
Probit	1.9×10^{-10}	(1 in 5.3 billion)

[a] All risks for a full lifetime of daily exposure. The lifetime is used as the unit of risk measurement because the experimental data reflect the risk experienced by animals over their full lifetimes. The values shown are upper confidence limits on risk.

However, dose-response curves derived from animal studies must be interpreted with care. The data for these curves are necessarily obtained by examining the effects of large doses on test animals. Because of the costs involved, researchers are limited in the numbers of test animals they can use—it is both impractical and cost-prohibitive to use thousands (even millions) of animals to observe just a few individuals that show adverse effects at low doses (*e.g.*, risks of 1:1000 or 1:10,000). Researchers must therefore extrapolate low-dose responses from their high-dose data. Consequently, dose-response curves are subject to controversy because their results change depending on the method chosen to extrapolate from the high doses actually administered to laboratory test subjects to the low doses humans are likely to receive in the course of everyday living.

This controversy revolves around the choice of several mathematical models that have been proposed for extrapolation to low doses. Unfortunately, no model can be proved or disproved from the data, so there is no way to know which model is the most accurate. The choice of models is therefore strictly a policy decision, which is usually based on understandably conservative assumptions. Thus, for noncarcinogenic chemical responses, the assumption is that some *threshold* exists below which there is no toxic response; that is, no adverse effects will occur below some very low dose (say, one in a million). Carcinogens, however, are considered *nonthreshold*—that is, the conservative assumption is that exposure to any amount of carcinogen creates some likelihood of cancer. This means that the only "safe" amount of carcinogen is zero, so the dose-response plot is required to go through the origin (0), as shown in Figure 22-3 on page 351.

There are many mathematical models to choose from, including the one-hit model, the multistage model, the multihit model, and the probit model. The characteristics of these models for nonthreshold effects are listed in Table 22-5.

The **one-hit model**, the simplest mechanistic model of carcinogenesis assumes (1) that a single chemical "hit," or exposure, is capable of inducing malignant change, *i.e.*, a single hit causes irreversible damage of DNA, leading to tumor development (once the biological target is hit, the process leading to tumor formation continues independently of dose); (2) that this change occurs in a single stage. In this commonly used model, the relationship between dose d and lifetime risk (probability) of cancer, $P(d)$, is given as

$$P(d) = 1 - \exp[-(q_0 - q_1 d)] \quad (22\text{-}1)$$

where q_0 and q_1 are parameters picked to fit the experimental data. At low doses the function is linear.

The **multistage model** assumes that tumors are the result of a sequence of biological events, or stages. In simplistic terms, the biological rationale for the multistage model is that there are a series of biological stages which a chemical must pass through (*e.g.*, metabolism, covalent bonding, DNA repair, *etc.*) without being deactivated, before the expression of a tumor is possible. In this model the relationship between risk and dose is expressed as

$$P(d) = 1 - \exp\left[-\sum_{i=0}^{n} q_i d^i\right] \qquad (22\text{-}2)$$

where the individual parameters q_i are positive constants picked for "best fit" with the dose-response data. The rate at which the cell passes through one or more of these stages is a function of the dose rate. The multistage model also has the desirable feature of producing a linear relationship between risk and dose.

The **multihit model** assumes that a number of dose-related hits are needed before a cell becomes malignant. The most important difference between the multistage and multihit model is that in the multihit model all hits must result from the dose, whereas in the multistage model, passage through some of the stages can occur spontaneously. The practical implication of this is that the multihit models are generally much flatter at low doses, and consequently predict a lower risk than the multistage model.

The **probit model** is not derived from mechanistic assumptions about the cancer process. It may be thought of as representing distributions of tolerances to carcinogens in a large population. The model assumes that the probability of the response (cancer) is a linear function of the log of the dose (log normal—see Section 8.2.4, beginning on page 103). While these models may be appropriate for acute toxicity they are considered questionable for carcinogens. These models would predict the lowest level of risk of all the models.

The effect of models on estimating risk for a given chemical is shown in Table 22-6. As we can see, the choice of models results in order-of-magnitude differences in estimating the risk at low levels of exposure.

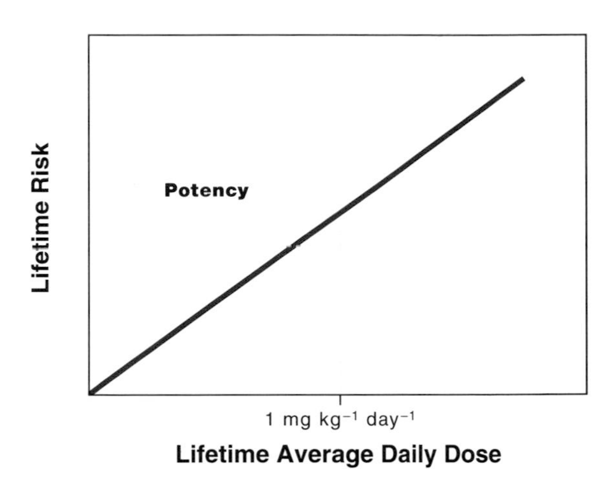

Figure 22-4 Potency factor is the slope of the dose-response curve at low doses. At low doses, the slope of the dose-response curve produced by the multistage model is called the potency factor. It is the risk produced by a lifetime average dose of 1 mg kg^{-1} day^{-1}. Adapted from U.S. EPA, 1990.

The **linear multistage model,** a modified version of the multistage model, is the EPA's model of choice because this agency chooses to err on the side of safety and overemphasize risk. This model assumes that there are multiple stages for cancer (*i.e.*, a series of mutations or biotransformations) involving many carcinogens, cocarcinogens, and promoters (see Section 21.6, beginning on page 335) that can best be modeled by a series of mathematical functions. At low doses, the slope of the dose-response curve produced by the linear multistage model (Figure 22-4) is called the **potency factor** (PF), or **slope factor**, which is the reciprocal of the concentration of chemical measured in milligrams per kilogram of animal body weight per day, that is, 1/(mg kg^{-1} day^{-1}) = mg^{-1} kg day, or the risk produced by a lifetime average daily dose (AD) of 1 mg kg^{-1} day^{-1}. Thus the dose-response equation for a carcinogen is

$$\text{Lifetime Risk} = \text{AD} \times \text{PF} \qquad (22\text{-}3)$$

The probability of *getting* cancer (not the probability of *dying* of cancer) and the associated dose comprise an average taken over an assumed 70-year human lifetime. This dose is called the lifetime average daily dose or **chronic daily intake**.

The dose-response effects for noncarcinogens allow for the existence of thresholds; that is, a certain

Table 22-7 Chemical RfDs for chronic noncarcinogenic effects of selected chemicals. Source: U.S. EPA, 1990.

Chemical	RfD (mg kg^{-1} day^{-1})
Acetone	0.1
Cadmium	0.0005
Chloroform	0.01
Methylene chloride	0.06
Phenol	0.04
Polychlorinated biphenyl	0.0001
Toluene	0.3
Xylene	2.0

quantity of a substance or dose below which there is no observed adverse effect (NOAEL; see Section 21.4.1, beginning on page 327) by virtue of the body's natural repair and detoxifying capacity. Examples of toxic substances that have thresholds are heavy metals and polychlorinated biphenyls (PCBs). These thresholds are represented by the **reference dose**, or **RfD**, of a substance, which is the intake or dose of the substance per unit body weight per day (mg kg^{-1} day^{-1}) that is likely to pose no appreciable risk to human populations, including such sensitive groups as children. A dose-response plot for carcinogens therefore goes through this reference point.

In general, substances with relatively high slope factors and low reference doses tend to be associated with higher toxicities. The RfD is obtained by dividing the NOAEL (see Section 21.4.1, beginning on page 327) by an appropriate uncertainty factor, sometimes called a **safety factor**. A 10-fold uncertainty factor is used to account for differences in sensitivity between the most sensitive individuals in an exposed human population. These include pregnant women, young children, and the elderly, who are more sensitive than "average" people. Another factor of 10 is added when the NOAEL is based on animal data that are extrapolated to humans. In addition, another factor of 10 is sometimes applied when questionable or limited human and animal data are available. The general formula for deriving an RfD is

$$\text{RfD} = \frac{\text{NOAEL}}{VF_1 \times VF_2 \times \ldots \times VF_n} \qquad (22\text{-}4)$$

where VF_i are the uncertainty factors. As the data become more uncertain, higher safety factors are applied. For example, if data are available from a high-quality epidemiological study, a simple uncertainty factor of 10 may be used by simply dividing the original value for RfD by 10 to arrive at a new value of RfD, which reflects the concern for safety. The RfDs of several noncarcinogenic chemicals are shown in Table 22-7.

The RfD can be used in quantitative risk assessments by using the following relationship:

$$\text{Risk} = \text{PF}\,(\text{CDI} - \text{RfD}) \qquad (22\text{-}5)$$

where CDI is the chronic daily intake and the potency factor is the slope of the dose-response curve. But this type of risk calculation is rarely performed. In most cases, the RfD is used as a simple indicator of potential risk in practice. That is, the chronic daily intake is simply compared with the RfD; then, if the CDI is below the RfD, it is assumed that the risk is negligible for almost all members of an exposed population.

22.2.4 Risk Characterization

22.2.4.1 Uncertainty Analysis

Uncertainty is inherent in every step of the risk assessment process. Thus, before we can begin to characterize any risk, we need some idea of the nature and magnitude of uncertainty in the risk estimate. Sources of uncertainty include:

- extrapolation from high to low doses
- extrapolation from animal to human responses
- extrapolation from one route of exposure to another
- limitations of analytical methods
- estimates of exposure.

Although the uncertainties are generally much larger in estimates of exposure and the relationships between dose and response (*e.g.*, the percent mortality), it is important to include the uncertainties originating from all steps in a risk assessment in risk characterization.

Two approaches commonly used to characterize uncertainty are sensitivity analyses and Monte Carlo simulations. In **sensitivity analyses**, we simply vary the uncertain quantities of each parameter (*e.g.*, average values, high and low estimates), usually one at a time, to find out how changes in these quantities affect the final risk estimate. This procedure gives us a range of possible values for the overall risk and tells us which parameters are most crucial in determining the size of the risk. In a **Monte Carlo simulation**, however, we assume that all parameters are random or uncertain. Thus, instead of varying one parameter at a time, we use a computer program to select distributions randomly every time the model equations are solved, the procedure being repeated many times. The resulting output can be used to identify values of exposure or risk corresponding to a specified probability, say the 50th percentile or 95th percentile.

22.2.4.2 Risk Projections and Management

The final phase of the risk assessment process is risk characterization. In this phase, exposure and dose-response assessments are integrated to yield probabilities of effects occurring in humans under specific exposure conditions. Quantitative risks are calculated for appropriate media and pathways. For example, the risks of lead in water are estimated over a lifetime assuming (1) that the exposure is two liters of water ingested per day over a 70-year lifetime and (2) that different concentrations of lead occur in the drinking water. This information can then be used by risk managers to develop standards or guidelines for specific toxic chemicals or infectious microorganisms in different media, such as the drinking water or food supply.

22.3 Ecological Risk Assessment

Ecological risk assessment is a process that evaluates the probability that adverse ecological effects will occur as the result of exposure to one or more stressors. A **stressor** is a substance, circumstance, or energy field that has the inherent ability to impose adverse effects upon a biological system. The environment is subject to many different stressors, including chemicals, genetically engineered microorganisms, ionizing radiation, and rapid changes in temperatures. Ecological risk assessment may evaluate one or more stressors and ecological components (*e.g.*, specific organisms, populations, communities, or ecosystems). Ecological risks may be expressed as true probabilistic estimates of adverse effects (as is done with carcinogens in human health risk assessment), or they may be expressed in a more qualitative manner.

In the United States, the Comprehensive Environmental Response Compensation and Liability Act (CERCLA) (otherwise known as the "Superfund"), the Resource Conservation and Recovery Act (RCRA), and other regulations require an ecological assessment as part of all remedial investigation and feasibility studies (see also Section 11.1, beginning on page 151). Pesticide registration, which is required under the Federal Insecticide, Fungicide, and Rodenticide Act (FIFRA), must also include an ecological assessment (see Section 23.5 on page 368). In the CERCLA/RCRA context, a typical objective is to determine and document actual or potential effects of contaminants on ecological receptors and habitats as a basis for evaluating remedial alternatives in a scientifically defensible manner.

The four major phases or steps in ecological assessment (Figure 22-5 on page 356) are as follows:

- Problem formulation and hazard identification
- Exposure assessment
- Ecological effects/toxicity assessment
- Risk characterization

An ecological risk assessment may be initiated under many circumstances—the manufacture of a new chemical, evaluation of cleanup options for a contaminated site, or the planned filling of a marsh, among others. The problem-formulation process begins with an evaluation of the stressor characteristics, the ecosystem at risk, and the likely ecological effects. An endpoint is then selected. An **endpoint** is a characteristic of an ecological component (*e.g.*, the mortality of fish) that may be affected by a stressor. Two types of endpoints are generally used: assessment endpoints and measurement endpoints. **Assessment endpoints** are particular environmental values to be protected. Such endpoints, which are recognized and valued by the public, drive the decisions made by official risk managers. **Measurement end-**

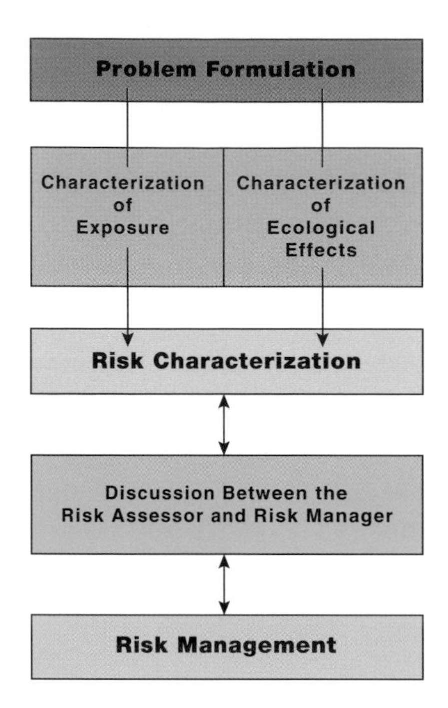

Figure 22-5 Framework for ecological risk assessment. Adapted from U.S. EPA, 1992b.

points are qualitatively or quantitatively measurable factors. Suppose, for example, a community that values the quality of sports fishing in the area is worried about the effluent from a nearby papermill. In this case, a decline in the trout population might serve as the assessment endpoint, while the increased mortality of minnows, as evaluated by laboratory studies, might be the measurement endpoint. Thus, risk managers would use the quantitative data gathered on the surrogate minnow population to develop management strategies designed to protect the trout population.

Exposure assessment is a determination of the environmental concentration range of a particular stressor and the actual dose received by the **biota** (all the plants and animals) in a given area. The most common approach to exposure analysis is to measure actual concentrations of a stressor and combine these measurements with assumptions about contact and uptake by the biota. For example, the exposure of simple aquatic organisms to chemicals can often be measured simply as the concentration of that chemical in the water because the systems of these organisms are assumed to be in equilibrium with the surrounding water. Stressor measurements can also be combined with quantitative parameters describing the frequency and magnitude of contact. For example, concentrations of chemicals or microorganisms in food items can be combined with ingestion rates to estimate dietary exposure. Exposure assignment is, however, rarely straightforward. Biotransformations may occur, especially for heavy metals such as mercury (see Section 7.5.2, beginning on page 90). Such transformations may result in the formation of even more toxic forms of the stressor. Researchers must therefore use mathematical models to predict the fate and resultant exposure to a stressor and to determine the outcome of a variety of scenarios.

The purpose of evaluating ecological effects is to identify and quantify the adverse effects elicited by a stressor and, to the extent possible, to determine cause-and-effect relationships. During this phase, toxicity data are usually compiled and compared. Generally, there are acute and chronic data for the stressor on one or several species. Field observations can provide additional data, and so can controlled-microcosm and large-scale tests. The combination of the exposure-analysis data with the ecological-effects data results in a **stressor-response profile**. This profile represents an attempt to match ecosystem impacts to the levels of stressor concentration under study.

But the process of developing a stressor-response profile is complex because it inevitably requires models, assumptions, and extrapolations. For example, the relationship between measurement and assessment endpoint is an assumption: it is often expressly stated in the model used; but when it is not specifically stated, it is left to professional judgment. In addition, the stressor-response profile is analogous to a dose-response curve in the sense that it involves extrapolations; in this case, though, a single-species toxicity test is extrapolated to the community and ecosystem level. One of the difficulties in the quantitation of the stressor-response profile is that many of the quantitative extrapolations are drawn from information that is qualitative in nature. For example, when we use **phylogenic extrapolation** to transfer toxicity data from one species to another species—or even to a whole class of organisms—we are assuming a degree of similarity based on qualitative characteristics. Thus, when we use green algal toxicity test data to represent all photosynthetic eukaryotes (which we often do), we must remember that all photosynthetic eukaryotes are not, in fact, green algae. Because many of the responses are extrapolations based on models ranging from the molecular to the ecosystem level, it is critically important that uncertainties and assumptions be clearly delineated.

Risk assessment consists of comparing the exposure and stressor-response profiles to estimate the probability of effects, given the distribution of the stressor within the system. And, as you might expect, this process is extraordinarily difficult to accomplish. In fact, our efforts at predicting adverse effects have been likened to the weather forecaster's prediction of rain (Landis and Ho-Yu, 1995). Thus, the predictive process in ecological risk assessment is still very much an art form, largely dependent on professional judgment (see *Case Study: Ecological Risk Assessment of a New Chemical Solvent* on page 359).

Figure 22-6 Risk assessment requires sophisticated techniques that enable trace amounts of pollutants to be analyzed. Above, environmental samples are analyzed for *Cryptosporidium*, an intestinal parasite. Photograph: D.C. Johnson.

22.4 Microbial Risk Assessment

Outbreaks of waterborne disease caused by microorganisms usually occur when the water supply has been obviously and significantly contaminated. In such cases where high-level contamination occurs, the exposure is manifest, and cause and effect are relatively easy to determine. However, exposure to low-level microbial contamination is difficult to determine epidemiologically. We know, for example, that long-term exposure to microbes may have a significant impact on the health of individuals within a community, but we need a way to measure that impact.

For some time, methods have been available to detect the presence of low levels (one organism per 1000 liters) of pathogenic organisms in water, including enteric viruses, bacteria, and protozoan parasites (see Figure 22-6 and Chapter 19, beginning on page 279). The trouble is that the risks posed to the community by these low levels of pathogens in a water supply over time are not like those posed by low levels of chemical toxins or carcinogens. For example, it takes just one amoeba in the wrong place at the wrong time to infect one individual, whereas that same individual would have to consume some quantity of a toxic chemical to be comparably harmed. Microbial risk assessment is therefore a process that allows us to estimate responses in terms of the risk of infection in a quantitative fashion. Although no formal framework for microbial risk assessment exists, it generally follows the steps used in other health-based risk assessments—hazard identification, dose-response assessment, exposure assessment, and risk characterization. The differences are in the specific assumptions, models, and extrapolation methods used.

Hazard identification in the case of pathogens is complicated because several outcomes—from asymptomatic infection to death (see Figure 22-7 on page 358) are possible, and these outcomes depend upon the complex interaction between the pathogen-

Table 22-8 Ratio of clinical to subclinical infections with enteric viruses. Source: Gerba and Rose, 1993.

Virus	Frequency of Clinical Illness[a] (%)
Poliovirus 1	0.1–1
Coxsackie	
A16	50
B2	11–50
B3	29–96
B4	30–70
B5	5–40
Echovirus	
overall	50
9	15–60
18	Rare–20
20	33
25	30
30	50
Hepatitis A (adults)	75
Rotavirus	28–60
Astrovirus (adults)	12.5

[a] The percentage of those individuals infected who develop clinical illness.

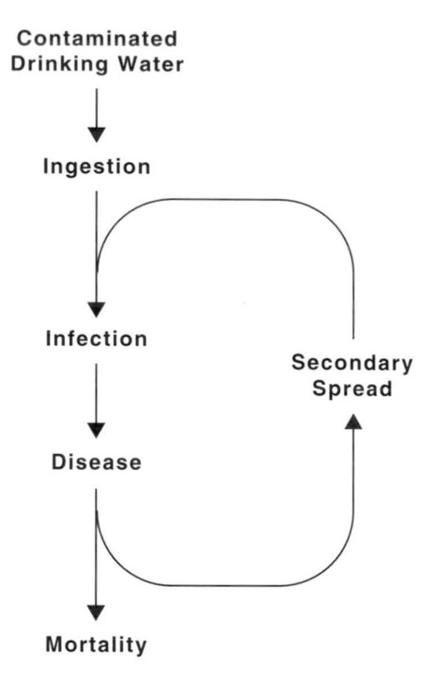

Figure 22-7 Outcomes of enteric viral exposure.

ic agent (the "infector") and the host (the "infectee"). This interaction, in turn, depends on the characteristics of the host as well as the nature of the pathogen. Host factors, for example, include preexisting immunity, age, nutrition, ability to mount an immune response, and other nonspecific host factors. Agent factors include type and strain of the organism as well as its capacity to elicit an immune response.

Among the various outcomes of infection is the possibility of **subclinical illness**. Subclinical (**asymptomatic**) infections are those in which the infection (growth of the microorganism within the human body) results in no obvious illness such as fever, headache, or diarrhea. That is, individuals can host a pathogenic microorganism—and transmit it to others—without ever getting sick themselves. The ratio of clinical to subclinical infection varies from pathogen to pathogen, especially in viruses, as shown in Table 22-8. Poliovirus infections, for instance, seldom result in obvious clinical symptoms; in fact, the proportion of individuals developing clinical illness may be less than 1%. However, other enteroviruses, such as the coxsackie viruses, may exhibit a greater proportion. In many cases, as in that of rotaviruses, the probability of developing clinical illness appears to be completely unrelated to the dose an individual receives via ingestion. Rather, the likelihood of developing clinical illness depends upon the type and strain of the virus as well as host age, nonspecific host factors, and possibly preexisting immunity. The incidence of clinical infection can also vary from year to year for the same virus, depending on the emergence of new strains.

Another outcome of infection is the development of clinical illness. Several host factors may play a major role in this outcome. The age of the host is often a determining factor. In the case of hepatitis A, for example, clinical illness can vary from about 5% in children less than 5 years of age to 75% in adults. Similarly, children are more likely to develop rotaviral gastroenteritis than are adults. Immunity is also an important factor, albeit a variable one. That is, immunity may or may not provide long-term protection from reinfection, depending on the enteric pathogen. It does not, for example, provide long-term protection against the development of clinical illness in the case of the Norwalk virus or *Giardia*. However, for most enteroviruses and for the hepatitis A virus, immunity from reinfection is believed to be lifelong. Other undefined host factors may also

Case Study: Ecological Risk Assessment of a New Chemical Solvent

Coat O'Many Colors, a large paint manufacturer, has just developed a new solvent for use in the production of its new, environmentally friendly line of house paints. Most of Coat O'Many's processing plants are located on or near rivers, into which residual amounts of the solvent will be discharged. Thus, both the company and the communities concerned are anxious about the risk of environmental pollution—especially the risk of damaging the quality of sports fishing in the region. Equally, however, all the parties involved are eager to spur economic growth.

Under the Toxic Substances Control Act (TSCA), the new solvent must be evaluated to determine if it poses any unreasonable risks to the environment. In deference to the needs of the affected communities, the assessment endpoint selected is the protection of all riverine aquatic life from unreasonable risks. To achieve this assessment endpoint, it is decided to measure the mortality of one species of fish, one type of aquatic invertebrate, and an alga (measurement endpoints). That is, these species are selected to serve as surrogate species for the variety of aquatic life in the rivers. Finally, it is agreed that these species will be tested for 1 week.

Disappointingly, at the end of the test period, the new solvent is found to be highly toxic for all the selected species (producing adverse effects in concentrations of 1 mg L^{-1} or less). Coat O'Many must report this finding, including in its characterization of risk a discussion of the toxicity of the chemical to the various surrogate species. Such a finding is likely to support the conclusion that the compound would be toxic to a wide array of aquatic life, including primary producers (algae), primary consumers (aquatic invertebrates), and carnivores (fish). However, this finding may not be definitive: uncertainty exists because the tests do not include a wide range of concentrations of the compound. Moreover, the test data are short-term test data whose endpoint is mortality; thus, they do not reveal sublethal effects on growth and reproduction that could occur at concentrations lower than those tested.

Such results suggest that Coat O'Many's new solvent presents a distinct risk to aquatic life in certain rivers (assessment endpoint). However, this conclusion is based on a number of assumptions: (1) that the test species are as sensitive as other organisms present in the rivers; (2) that the solvent would be present long enough and at sufficient concentrations to cause the measured adverse effects; and (3) that the aquatic organisms would come in contact with the solvent. Nor is that the end of the matter. Even though the solvent may pose a definite risk to some forms of aquatic life, a risk manager would have to weigh the economic and/or social benefits of using the chemical against the risk. Risk management decisions might therefore involve setting requirements for disposal through a waste-treatment process, prohibiting the discharge of the chemical, or even requesting additional testing.

Table 22-9 Case: fatality rates for enteric viruses and bacteria. Source: Gerba and Rose, 1993; Gerba *et al.*, 1995.

Organism	Case: Fatality Rate (%)
Virus	
Poliovirus 1	0.90
Coxsackie	
A2	0.50
A4	0.50
A9	0.26
A16	0.12
Coxsackie B	0.59–0.94
Echovirus	
6	0.29
9	0.27
Hepatitis A	0.60
Rotavirus	
(total)	0.01
(hospitalized)	0.12
Norwalk	0.0001
Adenovirus	0.01
Bacteria	
Shigella	0.2
Salmonella	0.1
Escherichia coli 0157:H7	0.2
Campylobacter jejuni	0.1

control the odds of developing illness. For example, in experiments with the Norwalk virus, human volunteers who did not become infected upon an initial exposure to the virus also did not respond to a second exposure. In contrast, those volunteers who developed gastroenteritis upon the first exposure also developed illness after the second exposure.

The ultimate outcome of infection—mortality—can be caused by nearly all enteric organisms. The factors that control the prospect of mortality are largely the same factors that control the development of clinical illness. Host age, for example, is significant. Thus, mortality for hepatitis A and poliovirus is greater in adults than in children. In general, however, we can say that the very young, the elderly, and the immunocompromised are at the greatest risk of a fatal outcome of most illnesses. For example, the case fatality rate (%) for *Salmonella* in the general population is 0.1%, while it has been observed to be as high as 3.8% in nursing homes. In North America and Europe, the reported case:fatalities (*i.e.*, the ratio of cases to fatalities reported as a percentage of persons who die) for enterovirus infections range from less than 0.1% to 0.94%, as shown in Table 22-9. The case: fatality rate for common enteric bacteria ranges from 0.1 to 0.2% in the general population. Enteric bacterial diseases can be treated with antibiotics while no treatment is available for enteric viruses.

Recognizing that microbial risk involves myriad pathogenic organisms capable of producing a variety of outcomes that depend on a number of factors—many of which are undefined—we must now face the problem of exposure assessment, which has complications of its own. Unlike chemical-contaminated water, microorganism-contaminated water does not have to be consumed to cause harm. That is, individuals who do not actually drink, or even touch, contaminated water also risk infection because pathogens—particularly viruses—may be spread by person-to-person contact or subsequent contact with contaminated inanimate objects (such as toys). This phenomenon is described as the **secondary attack rate**, which is reported as a percentage. For example, one person infected with poliovirus virus can transmit it to 90% of the other people with whom they associate. This secondary spread of viruses has been well documented for waterborne outbreaks of several diseases, including that caused by Norwalk virus, whose secondary attack rate is about 30%.

The question of dose is another problem in exposure assessment. How do we define "dose" in this context? To answer this question, researchers have conducted a number of studies to determine the infectious dose of enteric microorganisms in human volunteers. Such human experimentation is necessary because determination of the infectious dose in animals and extrapolation to humans is often impossible. In some cases, for example, humans are the primary or only known host. In other cases, such as that of *Shigella* and or Norwalk virus, infection can be induced in laboratory-held primates, but it is not known if the infectious-dose data can be extrapolated to humans. Much of the existing data on infectious dose of viruses have been obtained with attenuated-vaccine viruses or with avirulent laboratory-grown strains, so that the likelihood of serious illness is minimized.

Next, we must choose a dose-response model, whose abscissa is the dose and whose ordinate is the

Table 22-10 Best-fit dose response parameters for enteric pathogen ingestion studies. Modified from Regli *et al.*, 1991.

Microorganism	Best Model	Model Parameters
Echovirus 12	Beta-poisson	$\alpha = 0.374$ $\beta = 186.69$
Rotavirus	Beta-poisson	$\alpha = 0.26$ $\beta = 0.42$
Poliovirus 1	Exponential	$r = 0.009102$
Poliovirus 1	Beta-poisson	$\alpha = 0.1097$ $\beta = 1524$
Poliovirus 3	Beta-poisson	$\alpha = 0.409$ $\beta = 0.788$
Cryptosporidium	Exponential	$r = 0.004191$
Giardia lamblia	Exponential	$r = 0.02$
Salmonella	Exponential	$r = 0.00752$
Escherichia coli	Beta-poisson	$\alpha = 0.1705$ $\beta = 1.61 \times 10^6$

risk of infection (see Figure 22-3 on page 351). The choice of model is critical so that risks are neither greatly over- or underestimated. A modified exponential (beta-poisson distribution) or a log-probit (simple log-normal, or exponential distribution) model may be used to describe this relationship for many enteric microorganisms (Haas, 1983). For the beta model the probability of infection from a single exposure, P, can be described as follows:

$$P = 1 - \left(1 + \frac{N}{\beta}\right)^{-\alpha} \qquad \text{(22-6)}$$

where N is the number of organisms ingested per exposure, and α and β represent parameters characterizing the host-virus interaction (Haas, 1983). Some values for α and β for several enteric waterborne pathogens are shown in Table 22-10; these values were determined from human studies. For some microorganisms, an exponential model may better represent the probability of infection.

$$P = 1 - \exp(-rN) \qquad \text{(22-7)}$$

In this equation r is the fraction of the ingested microorganisms that survive to initiate infections (host-microorganism interaction probability).

When we use these models, we are estimating the probability of becoming infected after ingestion of various concentrations of microorganisms. The risk of acquiring a viral infection from consumption of contaminated drinking water containing various concentrations of enteric viruses is first determined by using Eq. (22-6) or Eq. (22-7). Annual and lifetime risks can also be determined, again assuming a Poisson distribution of the virus in the water consumed (assuming daily exposure to a constant con centration of viral contamination), as follows:

$$P_A = 1 - (1 - P)^{365} \qquad \text{(22-8)}$$

where P_A is the annual risk (365 days) of contracting one or more infections, and

$$P_L = 1 - (1 - P)^{25,220} \qquad \text{(22-9)}$$

where P_L is the lifetime risk (assuming a lifetime of 70 years = 25,550 days) of contracting one or more infections.

Risks of clinical illness and mortality can be determined by incorporating terms for the percentage of clinical illness and mortality associated with each particular virus:

$$\text{Risk of clinical illness} = PI \qquad \text{(22-10)}$$

$$\text{Risk of mortality} = PIM \qquad \text{(22-11)}$$

where I is the percentage of infections that result in clinical illness and M is the percentage of clinical cases that result in mortality.

Application of these models allows us to estimate the risks of becoming infected, development of clinical illness, and mortality for different levels of exposure. As shown in Table 22-11 on page 362, for example, the estimated risk of infection from 1 rotavirus in 100 liters of drinking water (assuming ingestion of 2 liters per day) is 1.2×10^{-3} using the beta-poisson model, or almost one in a thousand for a single-day exposure. This risk would increase to 3.6×10^{-1}, or approximately one in three, on an annual basis. Risks of the development of clinical illness and mortality also appear to be significant for exposure to low levels of rotavirus in drinking water.

The EPA has recently recommended that any drinking water treatment process should be designed to ensure that human populations are not subjected

Table 22-11 Risk of infection, disease, and mortality for rotavirus. Modified from Gerba and Rose, 1993.

Virus Concentration per 100 L	Risk	
	Daily	Annual
	Infection	
100	9.6×10^{-2}	1.0
1	1.2×10^{-3}	3.6×10^{-1}
0.1	1.2×10^{-4}	4.4×10^{-2}
	Disease	
100	5.3×10^{-2}	5.3×10^{-1}
1	6.6×10^{-4}	2.0×10^{-1}
0.1	6.6×10^{-5}	2.5×10^{-2}
	Mortality	
100	5.3×10^{-6}	5.3×10^{-5}
1	6.6×10^{-8}	5.3×10^{-5}
0.1	6.6×10^{-9}	2.5×10^{-6}

to risk of infection greater than 1:10,000 for a yearly exposure. To achieve this goal, it would appear from the data shown in Table 22-11 that the virus concentration in drinking water would have to be less than one per 1000 liters. Thus, if the average concentration of enteric viruses in untreated water is 1400 in 1000 liters, then treatment plants should be designed to remove at least 99.99% of the virus present in the raw water (see also Section 20.5, beginning on page 315).

In summary we can see that risk assessment is a major tool for decision making in the regulatory arena. This approach is used to explain chemical and microbial risks as well as ecosystem impacts. The results of such assessments can be used to inform risk managers of the probability and extent of environmental impacts resulting from exposure to different levels of stress (contaminants). Moreover, this process, which allows for the quantitation and comparison of diverse risks, lets risk managers utilize the maximum amount of complex information in the decision-making process. This information can also be used to weigh the cost and benefits of control options and to develop standards or treatment options.

References and Recommended Reading

Cockerham L.G. and Shane B.S. (1994) *Basic Environmental Toxicology.* CRC Press, Boca Raton, Florida.

Covello V., von Winterfieldt D., and Slovic P. (1988) Risk Communication: A Review of the Literature. *Risk Analyses.* **3**, 171–182.

Gerba C.P., and Rose J.B. (1993) Estimating viral disease risk from drinking water. In *Comparative Environmental Risks* (C.R. Cothern, Editor), pp. 117–135. Lewis Publishers, Boca Raton, Florida.

Gerba C.P., Rose J.B, and Haas C.N. (1995) Waterborne disease—Who is at risk? In *Water Quality Technology Proceedings*, pp. 231–254. American Water Works Association, Denver, Colorado.

Haas C.N. (1983) Estimation of risk due to low levels of microorganisms: A comparison of alternative methodologies. *American Journal of Epidemiology.* **118**, 573–582.

Kolluru R.V. (1993) *Environmental Strategies Handbook.* McGraw-Hill, New York.

Landis W.G. and Yu M.H. (1995) *Introduction to Environmental Toxicology.* Lewis Publishers, Boca Raton, Florida.

National Research Council (1983) *Risk Assessment in the Federal Government: Managing the Process.* National Academy Press. Washington, D.C.

National Research Council. (1989). *Improving Risk Communication.* National Academy Press, Washington, D.C.

National Research Council (1991) *Frontiers in Assessing Human Exposure.* National Academy Press, Washington, D.C.

National Research Council (1993) *Managing Wastewater in Coastal Urban Areas.* National Academy Press, Washington, D.C.

Regli S., Rose J.B., Haas C.N., and Gerba C.P. (1991) *Modeling the risk from* Giardia *and viruses in drinking water.* Journal of the American Water Works Association. **83**, 76–84.

Ricci P.F. and Molton L.S. (1985) Regulating cancer risks. *Environmental Science and Technology.* **19**, 473–479.

Rodricks J.V. (1992) *Calculated Risks: Understanding the Toxicity and Human Health Risks of Chemicals in Our Environment.* Cambridge University Press, Cambridge.

U.S. EPA (1990) *Risk Assessment, Management and Communication of Drinking Water Contamination.* EPA/625/4-89/024. U.S. Environmental Protection Agency, Washington, D.C.

U.S. EPA (1992a) *Dermal Exposure Assessment: Principles and Applications.* EPA 600/8-91/011B. U.S. Environmental Protection Agency, Washington, D.C.

U.S. EPA (1992b) *Framework for Ecological Risk Assessment.* EPA 1630/R-92/001. U.S. Environmental Protection Agency, Washington, D.C.

U.S. EPA (1986) *Guidelines for Carcinogen Risk Assessment.* Federal Register, September 24. 1986.

Wilson R., and Crouch E.A.C. (1987) Risk assessment and comparisons: An introduction. *Science.* **236**, 267–270.

Problems and Questions

1. What are some differences between the risks posed by chemicals and those posed by microorganisms?
2. What are some of the potential applications of risk assessment?
3. What is the difference between risk assessment and risk management?
4. Why is the selection of the dose-response curve so important in risk assessment?
5. What is the most conservative dose-response curve? What does this mean?
6. What is meant by a threshold dose-response curve? What types of toxicants have a threshold dose-response curve?
7. Why do we use safety factors in risk assessment?
8. What is the difference between a voluntary and an involuntary risk? Give some examples of both.
9. List the fours steps in a formal health-risk assessment.
10. What are some of the most common exposure pathways? When do we need to consider bioaccumulation in exposure?

11. Why are some chemicals more likely to bioaccumulate than others? Give some examples.
12. What are the different kinds of stressors to which an ecosystem might be exposed?
13. Does infection always lead to illness with enteric pathogens? What are the factors that determine morbidity and mortality outcomes with microbial infections?
14. What types of dose-response curve best reflect pathogen exposure?
15. What are some potential applications of microbial risk assessment?

Regulations are routinely imposed at the state or federal level. The U.S. Congress relies on responsible research provided by environmental scientists to design such regulation. Photograph: J.W. Brendecke.

23.1 A Regulatory Overview

In the United States, most environmental legislation is federal legislation that has been enacted since the mid 1960s. In addition, there are many state environmental laws and regulations that have been patterned after—and work together with—the federal programs. Federal legislation is, however, the basis for the development of regulations designed to protect our air, water, and food supply and to control pollutant discharge. Enactment of new legislation empowers the executive branch to develop environment-specific regulations and to implement their enforcement. Among other things, regulations may involve the development of standards for waste discharge, requirements for the cleanup of polluted sites, or guidelines for waste disposal. The Environmental Protection Agency (EPA) is the federal agency responsible for development of environmental regulations and enforcement.

Federal authority to establish standards for drinking water systems originated with the Interstate Quarantine Act of 1893. This provision resulted in the first water-related regulation in 1912: the use of the common drinking cup on interstate carriers was prohibited. Shortly thereafter, it was recognized that, however sanitary the drinking cup, its cleanliness would be valueless if the water placed in it was unsafe. Thus, in 1914, the first official drinking water standard—a bacteriological standard—was adopted. From 1914 to 1975, federal, state, and local health authorities, in conjunction with water works officials, used this standard to improve the nation's community water systems and to protect the public against waterborne disease.

A brief sketch of subsequent federal environmental legislation is presented here and summarized in Table 23-1 on page 367.

23.2 The Safe Drinking Water Act

In 1974, the passage of the Safe Drinking Water Act (SDWA) gave the federal government overall authority for the protection of drinking water. Prior to that time, the individual states had primary authority

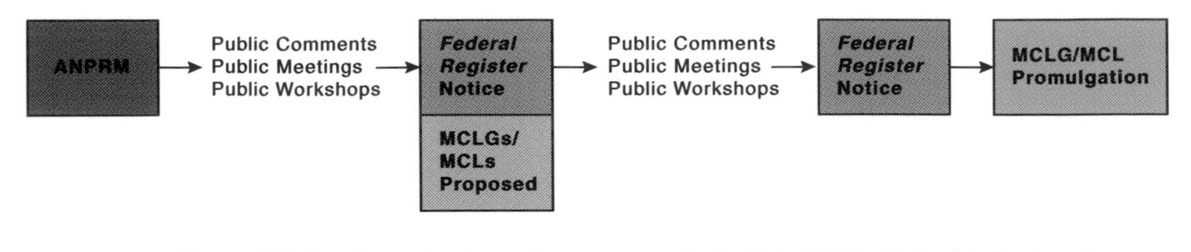

Figure 23-1 Regulatory development processes under the Safe Drinking Water Act. A series of steps is followed in the development of new regulations for contaminants in drinking water or processes for their control. These steps usually involve a series of public notices in the publication, the *Federal Register*, and meetings to allow for comment from the public, environmental groups, and the regulated Industry, before the final regulation is developed.

for development and enforcement of standards. Under this authority, specific standards were promulgated for contaminant concentrations and minimum water treatment. Maximum contaminant levels (MCL) or maximum contaminant level goals (MCLG) were set for specific contaminants in drinking water. Whereas an MCL is an achievable, required level, an MCLG is a desired goal, which may or may not be achievable. For example, the MCLG for enteric viruses in drinking water is zero since only one ingested virus may cause illness. We can't always reach zero, but we are obliged to try. Within the provisions of the SDWA, there are a number of specific rules such as the Surface Treatment Rule, which requires all water utilities in the United States to provide filtration and disinfection to control waterborne disease from *Giardia* and enteric viruses. These rules are developed through a process that allows input from the regulated community, special interest groups, the general public, and the scientific community. This process is outlined in Figure 23-1. To aid utilities, the EPA has published guidance manuals on the types of treatment that have been developed. Other key sections of the SDWA provide for the establishment of state programs to enforce the regulations. Individual states are responsible for developing their own regulatory programs, which are then submitted to the EPA for approval. There, state programs must set drinking water standards equal to, or more stringent than, the federal standards. Such programs must also issue permits to facilities that treat drinking water supplies and develop wellhead protection areas for groundwater drinking supplies.

23.3 The Clean Water Act

The Clean Water Act began with the Water Pollution Control Act of 1948 and was the first law to deal with comprehensive water pollution control. Designed to control the discharge of effluents into surface waters, this legislation focuses largely on point sources of pollution.

Under the terms of this act, effluent standards are set and permits are issued on the basis of these standards for existing and new sources of water pollution. These are source-specific limitations. Also, the act lists categories of such point sources as sewage treatment plants, for which the EPA must issue standards of performance. The EPA may provide a list of toxic pollutants and set effluent limitations based on the best available technology economically achievable for designated point sources. In addition, the EPA may issue pretreatment standards (*i.e.*, treatment preceding discharge of wastes into sewers) for toxic pollutants.

The chief enforcement tool of the Clean Water Act is the permit. Anyone—public or private—engaged in construction or operation of a facility that may discharge anything—animal, vegetable, or mineral (or even energetic, as in the case of heat)—into navigable fresh waters must first obtain a permit (see Section 16.2, beginning on page 238). Permit applications must include a certification that the discharge meets applicable provisions of the act under the National Pollution Discharge Elimination System (NPDES). (Permits for a discharge into ocean

Table 23-1 Scope of federal regulations governing environmental pollution in the United States.

Federal Regulation	Purpose/Scope
Policy	
National Environmental Policy Act (NEPA). Enacted 1970.	This act declares a national policy to promote efforts to prevent or eliminate damage to the environment. It requires federal agencies to assess environmental impacts of implementing major programs and actions early in the planning stage.
Pollution Prevention Act of 1990.	The basic objective is to prevent or reduce pollution at the source instead of an end-of-pipe control approach.
Water	
Clean Water Act. Enacted 1948.	Eliminates discharge of pollutants into navigable waters. It is the prime authority for water pollution control programs.
1977 Amendment	Covers regulation of sludge application by federal and state government.
Safe Drinking Water Act (SDWA). Enacted 1974.	Protects sources of drinking water and regulates proper water-treatment techniques using drinking water standards based on maximum contaminant levels (MCLs).
Clean Air	
Clean Air Act. Enacted 1970.	This act, which amended the Air Quality Act of 1967, is intended to protect and enhance the quality of air sources. Sets a goal for compliance with ambient air quality standards.
Amendments of 1977.	To define issues to prevent industries from benefiting economically from non-compliance.
Amendments of 1990	Basic objectives address acid precipitation and power plant emissions.
Hazardous Waste	
Comprehensive Environmental Response, Compensation, and Liability (CERCLA). Enacted 1980. Amended in 1986 to include Superfund.	The act, known as Superfund or CERCLA, provides an enforcement agency the authority to respond to releases of hazardous wastes. The act amends the Solid Waste Disposal Act.
Superfund Amendments and Reauthorizaton Act (SARA). Enacted 1986.	This act revises and extends CERCLA by the addition of new authorities known as the Emergency Planning and Community Right-to-Know Act of 1986. Involves toxic chemical recall reporting.
Toxic Substances Control Act. Enacted in 1976.	This act sets up the toxic substances program administered by the EPA. The act also regulates labeling and disposal of PCBs.
Amendment of 1986	Addresses issues of inspection and removal of asbestos.
Resource Conservation and Recovery Act (RCRA). Enacted in 1976. Amended in 1984.	An amendment that completely revised the Solid Waste Disposal Act. As it exists now, it is a culmination of legislation dating back to the passage of the Solid Waste Disposal Act of 1965. Defines hazardous wastes. Requires tracking of hazardous waste. Regulates facilities which burn wastes and oils in boilers and industrial furnaces. Requires inventory of hazardous waste sites.
Federal Insecticide, Fungicide, and Rodenticide Act (FIFRA). Enacted 1947.	Regulates the use and safety of pesticide products.
Amendments of 1972	Intended to ensure that environmental harm does not outweigh the benefits.

Figure 23-2 Superfund does not address the issue of financial compensation for suffering caused by illegally dumped hazardous wastes. Such matters, and possibly even the constitutionality of Superfund and other environmental laws, are left for the courts, such as the U.S. Supreme Court (*pictured above*), to decide. Photograph: J.W. Brendecke.

water are issued under separate guidelines from the EPA.)

As with the SDWA, states are responsible for the enforcement of the Clean Water Act; that is, they must develop and submit to the EPA a procedure for applying and enforcing these standards. Finally, the Clean Water Act makes provision for direct grants to states to help them in administering pollution-control programs. It also provides grants to assist in the development and implementation of waste-treatment management programs, including the construction of waste-treatment facilities.

23.4 Comprehensive Environmental Response, Compensation and Liability Act

In the late 1970s, when the now-infamous Love Canal landfill in upstate New York was revealed to be a major environmental catastrophe, the attendant publicity spurred Congress to pass the Comprehensive Environmental Response, Compensation and Liability Act (CERCLA) of 1980. This act makes owners and operators of hazardous waste disposal sites liable for cleanup costs and property damage. Transporters and producers must also bear some of the financial burden. This legislation also established a $1.6 billion cleanup fund (known as Superfund) to be raised over a five-year period. Of that sum, 90% was to come from taxes levied on the production of oil and chemicals by U.S. industries and the rest from taxpayers. This level was later increased to $8.5 billion.

CERCLA establishes *strict liability* (liability without proof of fault) for the cleanup of facilities on "responsible parties." The courts have found under CERCLA that these parties are "jointly and severally" liable; that is, each and every ascertainable party is liable for the full cost of removal or remediation, regardless of the level of "guilt" a party may have in creating a particular polluted site.

The Superfund is exclusively dedicated to clean up sites that pose substantial threats to human health and habitation, that is, *imminent* hazards. Moreover, the Superfund provides only for the cleanup of contaminated areas and compensation for damage to property. It cannot be used to reimburse or compensate victims of illegal dumping of hazardous wastes for personal injury or death. Victims must take their complaints to the courts (Figure 23-2).

23.5 Federal Insecticide and Rodenticide Act

Because of the potentially harmful effects of pesticides on wildlife and humans, Congress enacted the Federal Insecticide and Rodenticide Act (FIFRA) in 1947. This act, which has been periodically amended, requires that all commercial pesticides (including disinfectants against microorganisms) be approved and registered by the EPA.

FIFRA include several key provisions: (1) studies by the manufacturer of the risks posed by pesticides (requiring registration); (2) classification and certification of pesticides by specific use (as a way to control exposure); (3) restriction (or suspension) of the use of pesticides that are harmful to the environment; and (4) enforcement of the above requirements through inspections, labeling, notices, and state regulation. For example, every pesticide container must have a label indicating that its contents have been registered with the EPA.

Figure 23-3 Clean air legislation has led to substantial improvements in air quality worldwide. Westminster Abbey in London, England, long blackened by centuries of coal burning, will retain its actual color since cleaning in the early 1980s (*pictured above*) due to laws restricting the burning of coal in London and many other parts of England. Photograph printed by permission; copyright © 1995 by Jeffrey W. Brendecke. All rights reserved.

23.6 Clean Air Act

The Clean Air Act, originally passed in 1970, required the EPA to establish National Ambient Air Quality Standards (NAAQS) for several outdoor pollutants, such as suspended particulate matter, sulfur dioxide, ozone, carbon monoxide, nitrogen oxides, hydrocarbons, and lead (of these, ozone is the most pervasive and least likely to meet the standard; see also Section 12.2.2, beginning on page 175). Each standard specifies the maximum allowable level, averaged over a specific period of time. These primary ambient air quality standards are designed to protect human health. Secondary ambient air quality standards are designed to maintain visibility and to protect crops, buildings, and water supplies (Figure 23-3).

This act, which sets air pollution control requirements for various geographic areas of the United States, also deals with the control of tailpipe emissions for motor vehicles. Requirements compel automobile manufacturers to improve design standards to limit carbon monoxide, hydrocarbon, and nitrogen oxide emissions. For cities or areas where the ozone and carbon monoxide concentrations are high, reformulated and oxygenated gasolines are required. This act also addresses power-plant emissions of sulfur dioxide and nitrogen oxide, which can generate acid rain (see Section 12.2, beginning on page 172).

23.7 The Pollution Prevention Act

The Pollution Prevention Act of 1990 established as national policy the following waste-management hierarchy designed to prevent pollution and encourage recycling:

1. Prevention—to eliminate or reduce pollution at the source whenever feasible
2. Recycling—to recycle unpreventable wastes in an environmentally safe manner whenever feasible
3. Treatment—to treat unpreventable, unrecyclable wastes to applicable standards prior to release or transfer
4. Disposal—to safely dispose of wastes that cannot be prevented, recycled, or treated. The EPA, which is integrating pollution prevention into all its programs and activities, has developed unique voluntary reduction programs with public and private sectors.

References and Recommended Reading

Calabrese E.J., Gilbert C.E., and Pastides H. (1989) *Safe Drinking Water Act*. Lewis Publishers, Ann Arbor, Michigan.

Jain R.K. *et al.* (1994) Regulatory framework: United States, Canada, and Mexico. In *Environmental Strategies Handbook* (R.V. Kolluru, Editor), pp. 67–117. McGraw-Hill, New York.

Outwater A.B. (1994) *Reuse of Sludge and Minor Wastewater Residuals*. Lewis Publisher, Ann Arbor, Michigan.

The Hyperion Treatment Plant in Los Angeles, California, USA, processing 1.4 billion liters of wastewater per day, must implement wastewater treatment technology at its most advanced level to handle increasing volumes of wastewater under tighter regulations. Photograph: A.B. Brendecke.

24.1 Measurement of Environmental Quality: How Clean Is Clean?

We know that environmental problems—some of them quite severe—exist at all levels, local, state, national, and global. But, as we have seen, new technologies and approaches to mitigate the effects of pollution are also available in increasing abundance. In general, we have the knowledge and the technology to solve many pollution problems. What is it, then, that prevents us from making the earth a pollution-free zone? The answer is complex, of course, and depends on a number of different factors, including financial, political, and societal parameters. Ultimately, however, we can say that the level of pollution around us is a function of how serious we perceive the problem to be, and how much we are willing to pay for clean air, water, and land.

As our technologies improve, they are continually redefining our understanding of the term "clean." Using new instruments and innovative techniques, we are increasingly capable of measuring environmental parameters with greater sensitivity and accuracy. For example, researchers used to determine metal concentrations in water by atomic absorption (AA) spectroscopy, which gave measurements at the level of milligrams per liter. Then, new flameless AA techniques (using a graphite furnace) came along, improving detection limits to the level of micrograms per liter. Now, researchers can use **inductively coupled plasma (ICP) spectroscopy**, which can measure the concentrations of certain elements in nanograms per liter.

Another instrument that has undergone significant refinements is the gas chromatograph (GC), which is primarily used to analyze organic compounds. The first GC detector was the thermal conductivity detector, whose detection limit was 10–100 mg L^{-1}. Newer detectors include the flame ionization detector, which has a detection limit of ~100 μg L^{-1}, and the electron capture detector, whose limit is in the nanogram range.

We also have promising new instruments, such as the **atomic force microscope (AFM)** and its kin.

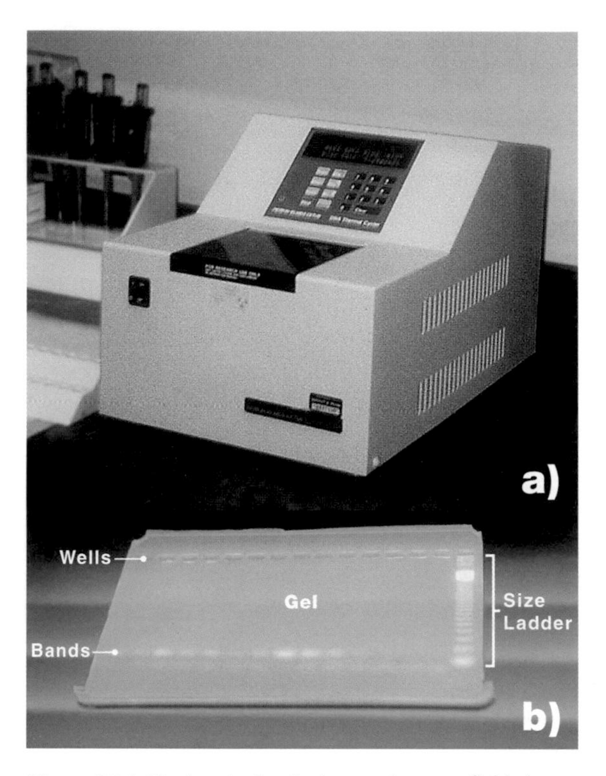

Figure 24-1 Modern technologies are in many fields becoming widely accepted by regulatory agencies and even supplanting older forms of analysis. (a) The thermocycler, pictured above, can aid in the detection of pathogens in water samples by amplifying attogram-quantities (10^{-18} g) of microbial DNA to detectable levels in a process known as the polymerase chain reaction (PCR). (b) Gel electrophoresis, which separates DNA into distinct, identifiable bands visible after staining with a fluorescent dye, such as ethidium bromide, makes PCR quicker and more sensitive than conventional, cultural techniques. Photographs: R.F. Walker.

With the AFM, it is now possible to obtain images of single molecules. For example, AFM and scanning tunneling microscopes are currently used in materials science to detect defects on surfaces of materials used for electronics. Today, such instruments are beginning to be used to characterize the molecular-scale properties of environmental surfaces, such as clay minerals. In the future, we may be able to use AFM for identifying and measuring contaminants at the molecular level.

With the changes and improvements in technology—which allow us to measure parameters unheard of a mere ten years ago—have come a host of exciting new techniques from all fields in science. For example, we can now "see" microorganisms without microscopes. How do we do it? We look for specific nucleic acids or DNA sequences associated with specific organisms. Using a revolutionary technique known as the **polymerase chain reaction (PCR)** method, we can make multiple copies of DNA sequences. First we place a sample into a relatively simple instrument called a PCR thermocycler (see Figure 24-1a). During each cycle of PCR (which takes about 4 minutes), we can enzymatically make a copy, or double the amount, of DNA in a sample. Thus, after repeating the process 25 times, wc can generate 2^{25} copies. This technique allows us to detect an organism's DNA at the attogram level—that is, we can detect 10^{-18} grams, or a millionth of a millionth of a millionth of a gram. After PCR amplification, the DNA is analyzed by gel electrophoresis. The DNA, which is stained with ethidium bromide, shows up as pink bands when viewed under ultraviolet light (Figure 24-1b).

The advent of these supersensitive technologies and exquisitely precise techniques signals the need for a reexamination not only of our definition of "clean," but also of the framework of our environmental questions. Take heavy metals in soil for example. In 1985, we would have declared "pristine" any soil that showed no detectable levels of, say, lead. Today, however, we can detect trace levels of lead in the very same soil. Is that soil still pristine? Was is it ever? Now, we must ask a new question: Is it pristine *enough*? And this question changes the debate. Thus, current solutions must be formulated that help us decide what level of contaminant we can live with, and what limits we should place on cleanup of particular contaminated sites.

24.2 The Role of Government: Who Is Responsible?

While pollution science can provide an answer to the question of how clean is clean and can offer us the technological ability to mitigate the effects of environmental pollution, it cannot resolve the major questions. That is: What clientele are we trying to satisfy? And who should do the satisfying? The first of these questions currently revolves around the needs of the general public, corporations, and environmental groups. While none of these groups is well-defined, it is generally fair to say that the general public is often caught up in conflict between the other two. On the one hand, corporate entities tend to focus on the economic prosperity of the general public. On the other hand, environmental groups, such as Greenpeace and the Sierra Club, tend to focus on ecological and health-related concerns. Both of these foci are equally legitimate, and each is fundamentally informed by the other.

In addition, the news media—the so called "fourth estate"—play a critical role in defining both the needs and desires of the general public. With the help of the news media, many disparate environmental groups have succeeded in making the public more environmentally conscious. Moreover, the media have been instrumental in promoting environmentally responsible behavior on the part of corporations and governments. But the news media, which tend to focus on attention-getting headlines, have also exacerbated the conflict between environmental groups and corporate (and sometimes state and national) entities. Arguably, this inevitable tension among environmental activists and corporate groups, filtered through the broad lens of the media, will shape the agenda of politicians who must deal with environmental problems.

Generally, it is government—the ultimate manifestation of public opinion—that must control the level of pollution in individual countries. This control, however, can take several forms, from outright compulsion and taxes to economic incentives. In the United States, for example, the federal government is currently making tough choices between enhanced or relaxed environmental standards. Among the decision-making tools being used is the concept of cost:benefit ratios, which have been used for many years to frame corporate decisions. As applied to environmental remediation, for example, decision makers attempt to place a dollar value on the cost of a particular cleanup action, and likewise a dollar value on its associated benefits. In theory, this method should result in actions whose implementation would yield a positive net benefit. As attractive as this concept is in theory, it is limited in practice because it requires the quantitation of qualitative factors. How, for instance, do you put a dollar value on breathing clean air, drinking clean water, and eating clean food? On the other hand, can we justify paying so much for these benefits that nobody can afford to enjoy them?

At present, we have very few answers to this conundrum, except to say that it must be unraveled and government must be the primary unraveler. It seems clear that we all want a clean environment. Politicians may therefore choose to pay for environmental remediation and pollution control directly from general and corporate tax revenues or indirectly through corporate regulations and compulsions; but in so doing they are only deciding *how* we must pay, not *if.* It remains to be decided *how much.*

24.3 Trends in Environmental Regulations/Activities: What's Being Done?

24.3.1 Clean Air

A recently published EPA guide to the Clean Air Act states that "the 1990 Act includes a list of 189 hazardous air pollutants selected by Congress on the basis of potential health and/or environmental hazard." It goes on to say that emission levels of these pollutants will be "strictly curtailed" over the next 5 to 10 years. In 1994, industry retaliated to this statement in the popular press: "Unfortunately, the same legislation provides no technological or scientific resolution for how these requirements are to be accomplished." In other words, the EPA is setting these goals, but it's up to industry to figure out how to meet them (Glanz, 1994).

This little by-play illustrates the environmental regulators' dilemma: the need to maintain flexibility while still ensuring compliance with the rules and protecting the public interest. Large industrial complexes operate multiple plants at various sites, and each of these operations has many potential sources of pollution, including smokestacks, production processes, and pipes that carry liquid and gaseous products. At one time, environmental rules required that each of these pollution sources conform to specific, mandated standards. This requirement made enforcement extremely difficult and compliance a virtual impossibility. In order to resolve this dilemma, regulators have introduced aggregated emissions plans, such as the bubble concept.

Under the **bubble concept**, regulators treat a single facility as if it were surrounded by an invisible bubble, and they measure only the pollution arising from the whole bubble. This approach means that even though the emission from one or more of the smokestacks or discharge pipes may actually exceed the standards of the law, it is allowable because the emissions from the other discharge points are low enough to keep the total emissions below the overall standard. The same principle applies to the notion of pollution "credits." An overall standard is set for a specific area. Industries within that area must meet this standard collectively, thus permitting some companies to buy or sell their "shares" as long as the sum of emissions does not exceed the mandated level.

As might be expected, environmental groups have expressed concern that bubbles and other aggregated emissions plans give businesses too much flexibility and discretion. They argue that governments should not hand over such decision-making authority to the very entities it is trying to regulate. Generally, however, this argument has failed in the face of cost: benefit calculations, which demonstrate the monetary advantages to the firm and to the public concerned. For example, this flexible approach to regulation gives firms the opportunity to phase in pollution-control measures, such as smokestack cleanups, on a planned basis over time (Post, 1994). With time to plan, firms can implement and pay for necessary changes without undue economic hardship to themselves or to the community.

24.3.2 Safe Drinking Water

Currently, the Safe Drinking Water Act (SDWA) sets a specific number of chemical contaminants to regulate, and the EPA must come up with regulations to meet that quota. According to the water industry and many public officials, this method is little more than a giant "numbers game." Priorities for regulating contaminants, they say, should instead be based on how harmful a substance is and how often it occurs in the environment. They should also take into account the costs of obtaining occurrence data as well as the costs of removal. Although most water providers in the United States meet all current federal standards for water safety, they face several problems. For example, the costs of modern-day detection are not inconsiderable. Moreover, long delays (five or more years) may occur between discovering a contaminating substance in the water and building or installing new systems to treat that contaminant.

The American Water Works Association (AWWA) believes that the recent outbreaks of *Giardia* and *Cryptosporidium* in Milwaukee and Washington are indicative of serious problems. Having previously issued a call for its water utility members (4,000 utilities in the United States and Canada) to voluntarily test for *Cryptosporidium,* the AWWA is now urging its members to take extra precautions in treatment procedures and to pay special attention to signs of increasing turbidity—a condition that may signal the presence of pathogens and hence require treatment adjustments. Unfortunately, information is lacking on the occurrence of *Cryptosporidium, Giardia,* and enteric viruses in the raw and untreated water supplies in the United States. Such information is needed to determine the adequacy of current and proposed requirements for drinking water treatment.

Yet another problem has been raised by concerns over the potential health risks posed by the chemicals used to disinfect drinking water. Thus, while offering proven protection from microbial infection, some chemicals may also be producing carcinogenic or toxic by-products (see Section 20.5, beginning on page 315). The need to balance the two threats—microbes versus chemicals—has led to a call for better understanding, and quantitation, of the risks involved. In 1996, the water industry begins a major effort called the **Information Collection Rule**, whereby they will collect data on the occurrence of major waterborne disease pathogens and disinfection by-products in the United States. The results of this effort may have a significant impact on future U.S. regulations for water supply and treatment in the next century.

24.3.3 Earth, Air, and Water: A Multimedia Approach

When examining the contamination caused by a particular activity or at a specific site, we tend to focus on a single medium. For example, we often concentrate on groundwater contamination when evaluating hazardous waste sites. However contrived, this separation of land, water, and air media is especially pervasive in the regulatory arena. Separate environmental laws exist for controlling pollution in each medium (see Table 23-1 on page 367). In some cases, this compartmentalization of systems, which has facilitated the growth of the existing regulatory structure, has resulted in sets of fractionated, and sometimes overlapping and/or contradictory, rules (see Section 13.2 on page 190 for an example).

An alternative approach—one that can integrate the several media—is finding increasing acceptance in the scientific and economic community. Termed **multimedia**, this holistic approach considers each medium as part of a whole and allows us to evaluate the synergistic and antagonistic interactions that can and do occur in real systems. Such an approach, however, is fundamentally multi- and interdisciplinary in nature, requiring the active collaboration of many individuals trained in such disparate disciplines as science, engineering, and economics, and public policy. Although difficult to achieve, this approach will be important for future pollution management and regulation.

24.3.4 Pollution Prevention

In this book, we have seen that many human activities in the environment can create pollution problems that are exceedingly hard to solve and even harder to pay for. It should therefore be obvious that the most beneficial and cost-effective approach to pollution control is to stop it before it starts. Thus pollution prevention has become a widely accepted goal—a goal that is being implemented to an increasing degree. Pollution prevention can take many forms. For example, best management practices (see Section 14.2.5, beginning on page 216) have been developed to optimize the amount of fertilizers and pesticides for crop production—that is, to use only the amounts necessary to support growth. Industry, too, is developing optimization procedures to use chemicals efficiently and minimize waste generation. As the world's population continues to rise and as natural resources become scarcer, prevention strategies will become increasingly important in maintaining a healthy economy, environment, and society.

24.4 Planning for the Twenty-First Century: What's Next?

Although the government of a nation can regulate and control production, it is rarely a primary producer of goods. Instead, industry initiates, develops, and manufactures the products upon which modern society depends. It also controls, to a large extent, the production of agricultural products, from cereal grains and cattle to fruit and flowers. Therefore, industry will play a pivotal role in environmental protection in the next century. What, then, can industry do to ensure that this role will be a beneficial one? Let's look at a few key possibilities.

- *Assume responsibility.* Set up detailed programs to improve the environmental literacy of the entire workforce; build a corporate culture that includes a strong environmental ethic; shift environmental responsibilities from special departments to all personnel in all departments.
- *Demand accountability.* Recognize that pollution prevention requires sustained support from upper management; devise precise, accurate systems that measure performance and provide credible data for stockholders and consumers; consider changes in production systems that enhance environmental performance.
- *Innovate.* Improve the life cycle environmental performance of existing products and processes; assess new products on the basis of their capacity to satisfy both market needs and public demand for effective environmental solutions; encourage research and seek to commercialize true techno-

logical innovations with environmentally competitive advantages.

- *Communicate.* Become active in public policy discussions; eschew adversarial confrontation with government agencies and environmental groups in favor of proactive participation and cooperation; work to ensure that governmental policies and program adequately reflect business needs and opportunities.

- *Think globally.* Seek out opportunities to become involved with public- and private-sector programs focusing on pollution prevention and clean technologies in developing countries; be sensitive to different kinds of environmental problems and priorities in other countries.

- *Demonstrate positive leadership.* Promote the attitude that what's good for the environment is good for business, and be prepared to make the case that what's good for business is also good for the environment (Hirschlorn, 1994).

Finally, it is important to remember that governments, industries, and environmental groups—as well as the scientific community and the popular media—are individually and collectively parts of the "general public." Moreover, it is very much in the general public interest to preserve natural resources, remediate contaminated areas, and guard against environmental health hazards. Thus, while each of these societal components has a crucial role to play in achieving these goals, each is also dependent on the others for success. Industry, for example, may preempt the government's regulatory responsibility by offering workable solutions to environmental problems based on sound scientific principles. Environmental groups may work with industry to allocate scarce resources on the basis of health and safety risks. The government may cooperate with both environmental and industrial groups in order to find flexible, affordable solutions. And the news media may work with the scientific community to provide a forum for balanced discourse on proven risks, rather than imaginary or remote threats.

Epilogue

In 1996, the EPA announced that 40% of the lakes and rivers in the United States were unfit for swimming or drinking. Forty percent is almost half. Will we reach the halfway mark? Or exceed it? And if so, when?

The answers to these questions are now up to those of us who work in the field of Pollution Science, and to those who will do so in the future. The enormous problems that face this planet at times seem overwhelming, but with access to correct knowledge, along with level-headed thinking, the proper decisions can be made that will benefit the natural world as well as the people who have to live in it and prosper from it.

Scientific, political, social, and monetary constraints all affect the fate and mitigation of pollutants, but are very difficult to successfully blend with each constraint competing for prominence. But the integration of these ideologies and needs is not much different from the complex ecosystems that we are finally beginning to learn so much about; each piece of the puzzle is crucial to the success of the whole.

References and Recommended Reading

Glanz J. (1994) *Researchers Scramble to Meet Clean Air Act Goals.* R & D Magazine. **44**, 121–122.

Hirschlorn J.S. (1994) Business and the Environment. In *Environmental Strategies Handbook* (R.V. Kolluru, Editor), pp. 67–117. McGraw-Hill, New York.

Post J.E. (1994) Business, environment, and stewardship. In *Environmental Strategies Handbook* (R.V. Kolluru, Editor), pp. 11–30. McGraw-Hill, New York.

Index

A

B

C

D

E

F

G

H

I

J

K

L

M

N

O

P

T

U

V

W

X

Y

Z

Periodic Table of the Elements*

Key: Atomic Number (3), Symbol (Li), Atomic Mass** (6.941)

IA	IIA	IIIB	IVB	VB	VIB	VIIB	VIIIB	VIIIB	VIIIB	IB	IIB	IIIA	IVA	VA	VIA	VIIA	VIIIA
1 H 1.0079																	2 He 4.0026
3 Li 6.941	4 Be 9.0122											5 B 10.811	6 C 12.011	7 N 14.007	8 O 15.999	9 F 18.998	10 Ne 20.180
11 Na 22.990	12 Mg 24.305											13 Al 26.982	14 Si 28.086	15 P 30.974	16 S 32.066	17 Cl 35.453	18 Ar 39.948
19 K 39.098	20 Ca 40.078	21 Sc 44.956	22 Ti 47.867	23 V 50.942	24 Cr 51.996	25 Mn 54.938	26 Fe 55.845	27 Co 58.933	28 Ni 58.693	29 Cu 63.546	30 Zn 65.39	31 Ga 69.723	32 Ge 72.61	33 As 74.922	34 Se 78.96	35 Br 79.904	36 Kr 83.80
37 Rb 85.468	38 Sr 87.62	39 Y 88.906	40 Zr 91.224	41 Nb 92.906	42 Mo 95.94	43 Tc (98)	44 Ru 101.07	45 Rh 102.906	46 Pd 106.42	47 Ag 107.87	48 Cd 112.41	49 In 114.82	50 Sn 118.71	51 Sb 121.76	52 Te 127.62	53 I 126.90	54 Xe 131.29
55 Cs 132.90	56 Ba 137.33	57 †La 138.90	72 Hf 178.49	73 Ta 180.95	74 W 183.84	75 Re 186.21	76 Os 190.23	77 Ir 192.22	78 Pt 195.08	79 Au 196.97	80 Hg 200.59	81 Tl 204.38	82 Pb 207.2	83 Bi 208.98	84 Po (209)	85 At (210)	86 Rn (222)
87 Fr (223)	88 Ra (226)	89 ‡Ac (227)	104 Rf (261)	105 Ha (262)													

† Lanthanides	58 Ce 140.12	59 Pr 140.91	60 Nd 144.24	61 Pm (145)	62 Sm 150.36	63 Eu 151.96	64 Gd 157.25	65 Tb 158.92	66 Dy 162.50	67 Ho 164.93	68 Er 167.26	69 Tm 168.93	70 Yb 173.04	71 Lu 174.97
‡ Actinides	90 Th 232.04	91 Pa 231.04	92 U 238.03	93 Np (237)	94 Pu (244)	95 Am (243)	96 Cm (247)	97 Bk (247)	98 Cf (251)	99 Es (252)	100 Fm (257)	101 Md (258)	102 No (259)	103 Lw (262)

* Data adapted from Lide, D.R., Editor (1995) *CRC Handbook of Chemistry and Physics*, 76th Edition. CRC Press, Boca Raton, Florida.

** Atomic mass values in parentheses denote the atomic mass of the most stable isotope of those elements too unstable for the determination of a standard atomic mass.

Symbol	Chemical Name	Atomic Number
Ac	Actinium	89
Ag	Silver	47
Al	Aluminum	13
Am	Americium	95
Ar	Argon	18
As	Arsenic	33
At	Astatine	85
Au	Gold	79
B	Boron	5
Ba	Barium	56
Be	Beryllium	4
Bi	Bismuth	83
Bk	Berkelium	97
Br	Bromine	35
C	Carbon	6
Ca	Calcium	20
Cd	Cadmium	48
Ce	Cerium	58
Cf	Californium	98
Cl	Chlorine	17
Cm	Curium	96
Co	Cobalt	27
Cr	Chromium	24
Cs	Cesium	55
Cu	Copper	29
Dy	Dysprosium	66
Er	Erbium	68
Es	Einsteinium	99
Eu	Europium	63
F	Fluorine	9
Fe	Iron	26
Fm	Fermium	100
Fr	Francium	87
Ga	Gallium	31
Gd	Gadolinium	64
Ge	Germanium	32
H	Hydrogen	1
Ha	Hahnium	105
He	Helium	2
Hf	Hafnium	72
Hg	Mercury	80
Ho	Holmium	67
I	Iodine	53
In	Indium	49
Ir	Iridium	77
K	Potassium	19
Kr	Krypton	36
La	Lanthanum	57
Li	Lithium	3
Lr	Lawrencium	103
Lu	Lutetium	71
Md	Mendelevium	101
Mg	Magnesium	12
Mn	Manganese	25
Mo	Molybdenum	42
N	Nitrogen	7
Na	Sodium	11
Nb	Niobium	41
Nd	Neodymium	60
Ne	Neon	10
Ni	Nickel	28
No	Nobelium	102
Np	Neptunium	93
O	Oxygen	8
Os	Osmium	76
P	Phosphorus	15
Pa	Protactinium	91
Pb	Lead	82
Pd	Palladium	46
Pm	Promethium	61
Po	Polonium	84
Pr	Praseodymium	59
Pt	Platinum	78
Pu	Plutonium	94
Ra	Radium	88
Rb	Rubidium	37
Re	Rhenium	75
Rf	Rutherfordium	104
Rh	Rhodium	45
Rn	Radon	86
Ru	Ruthenium	44
S	Sulfur	16
Sb	Antimony	51
Sc	Scandium	21
Se	Selenium	34
Si	Silicon	14
Sm	Samarium	62
Sn	Tin	50
Sr	Strontium	38
Ta	Tantalum	73
Tb	Terbium	65
Tc	Technetium	43
Te	Tellurium	52
Th	Thorium	90
Ti	Titanium	22
Tl	Thallium	81
Tm	Thulium	69
U	Uranium	92
V	Vanadium	23
W	Tungsten	74
Xe	Xenon	54
Y	Yttrium	39
Yb	Ytterbium	70
Zn	Zinc	30
Zr	Zirconium	40